Ultradian Rhythms from Molecules to Mind

David Lloyd • Ernest L. Rossi
Editors

Ultradian Rhythms from Molecules to Mind

A New Vision of Life

Springer

Editors
Professor David Lloyd
Cardiff University
Microbiology (BIOSI 1)
Main Building
Cardiff CF10 3TL
Wales, UK
Email: lloydd@cf.ac.uk

Dr Ernest L. Rossi
125 Howard Ave.
Los Osos
CA 93402, USA

ISBN 978-1-4020-8351-8 e-ISBN 978-1-4020-8352-5

Library of Congress Control Number: 2008930863

Printed on acid-free paper

9 8 7 6 5 4 3 2 1

springer.com

David Lloyd would like to dedicate this book to the memory of Gregorio Weber, Geoffrey Callely and David E. Hughes, and to two other exceptional experimental scientists J. Woody Hastings and Britton Chance on their 80th and 95th birthdays respectively.

Finally to Dr R.R. Klevecz, who made enormous contributions to our fundamental insights into the dynamic complexity of biological systems. Sadly, Bob passed away on 13.05.2008 whilst this volume was in the course of preparation.

Beauty is the proper conformity of the parts to one another and to the whole

HEISENBERG: The Meaning of Beauty in the Exact Sciences

The almost frightening simplicity and wholeness of the relationships which nature suddenly spreads out before us and for which none of us was in the least prepared

HEISENBERG in discussion with EINSTEIN

Acknowledgements

DL would like to record his indebtedness to all those who helped this volume through its production, especially Mrs. Drusilla Lewis for interpretation of my script, Dr. Douglas Murray for many of the figures and members of my family (Margaret, Alun, Siôn and Anvita) for their assistance.

Finally we thank those at Springer, Fabio de Castro, Marlies Vlot, Meran Owen, Tanja van Gaans, André Tournois, S. Padmasani, and V. Ramakrishnan for their help at all stages of production of this book.

December 27, 2007

David Lloyd
Ernest L. Rossi

Contents

Introduction: The Temporal Organization of Living Systems from Molecule to Mind

D. Lloyd[1], E.L. Rossi[2] and M.R. Roussel[3]

It is the pattern maintained by this homeostasis, which is the touchstone of our personal identity. Our tissues change as we live: the food we eat and the air we breathe become flesh of our flesh, and bone of our bone, and the momentary elements of our flesh and bone pass out of our body every day with our excreta. We are but whirlpools in a river of ever-flowing water. We are not the stuff that abides, but patterns that perpetuate themselves.

Wiener, 1954

What are called structures are slow processes of long duration, functions are quick processes of short duration.

Von Bertalanffy, 1952

Images of Earth taken from an orbiting satellite emphasize the sharp transition in light intensity, temperature and relative humidity that almost every point on our planet experiences twice daily. The crescent of dawn sweeping the earth, as it has done about a million million times since life began, has imprinted itself on our evolutionary trajectory and the physiological characteristics of our ancestral lineage. We are born into regularly periodic circumstances; the capabilities of living organisms to use the separation of environmental light from dark have yielded an extraordinary repertoire of means for its exploitation. The properties of the organism are enmeshed with the planet's geophysical parameters by the circadian (τ about 24 h) clock, an internal temperature-compensated biological mechanism which is both re-settable and entrainable. Evolution of the processes whereby matching of the endogenous biochemical network to the daily cycles of night and day occurs is but one aspect of the self-optimization of the living system. The Circadian clock, ubiquitous at least from cyanobacteria to humans is underpinned by an enormously complex time structure, so that if we consider typical

[1] Microbiology (BIOSI 1), Main Building, Cardiff School of Biosciences, PO Box 915, Cardiff CF10 3TL, Wales, UK

[2] 125 Howard Avc. Los Osos, CA 93402, USA

[3] Department of Chemistry and Biochemistry, University of Lethbridge, Lethbridge, Alberta, TIK 3M4, Canada

D. Lloyd, E. Rossi (eds.), *Ultradian Rhythms from Molecules to Mind*,

relaxation times of component reactions, these range over many logarithmic decades of time (Fig. 1a). We must thus be prepared to study the processes which are relevant to life on timescales ranging from attoseconds to gigaseconds and beyond, from those events completed on sub-molecular scales to rates of evolutionary change. Seen in this way, circadian control represents a very small window in an immense panoply, one that is still only dimly comprehended, because it is not studied as part of an enormously more complicated whole.

Key to the understanding of that whole is a more central hub, a more primeval time-base, one that encompasses both biophysics and biochemistry, both cellular biology and cell reproduction (Lloyd et al., 1982; Lloyd and Gilbert, 1998; Lloyd and Murray, 2005, 2006, 2007). This is the ultradian clock; more precisely designated the circahoralian clock (Brodsky, 1975), as its period is about 1 h. This time-base has been recognized only since its discovery in the soil amoeba *Acanthamoeba castellanii* in 1979, again as a temperature-compensated mechanism, which is therefore more than an oscillation and even more than a rhythm. Whereas in *A. castellanii* its period is about 69 min (Edwards and Lloyd, 1978, 1980), in other organisms this value is different. In one species of yeast, *Candida utilis*, it is 30 min (Kader and Lloyd, 1979), in another, *Schizosaccharomyces pombe*, 40 min (Poole et al., 1973). The ciliate protozoon, *Tetrahymena pyriformis* has a 50 min clock (Lloyd et al., 1978); in the flagellate *Crithidia fasciculata* it is 66 min (Edwards et al., 1975, Fig. 1b). These values have no special significance *per se*, other than as an expression of a self-optimized selection, especially appropriate to the somewhat differing networks characteristic of these different species. In humans, the 90 min basic rest activity rhythm (BRAC) first observed in 1953, by Aserinsky and Kleitman probably represents an analogous phenomenon.

Except in organisms in their cryptobiotic states, where biochemical network activities are minimized to very low levels of activity, it seems likely that this ultradian clock operates in every unicellular organism, in every metazoan and plant cell, and thus in every organism at almost all times. Its pervasive nature underlines its central importance at the core of coherence, the insistent drumbeat to which all metabolic and biosynthetic pathways, events and processes are tied. The intracellular network, involving as it does not merely sequences of enzyme-catalyzed steps, but also binding of controlling ligands both stimulatory and inhibitory, as well as protein-DNA interactions and the less well-understood protein-protein interactions is orchestrated by reference to this timekeeper.

Just as clocks can be built which indicate the year, date, hour, minute and second, living organisms also need to coordinate activities over a range of time scales. Away from the equator, the year is a biologically important time scale, and it is well known that plants and animals are very adept at preparing for the changes of the seasons. Arguably the best-studied biological clock is the circadian, which allows a variety of organisms, from bacteria to humans, to cope with the daily cycles of illumination and temperature. The circahoralian clock provides organisms with a biological hour, with consequences that are explored in detail in this book. What of biological minutes and seconds? Do our internal clocks have fast rhythms that allow for the measurement of intervals of time shorter than the biological hour of the circahoralian clock?

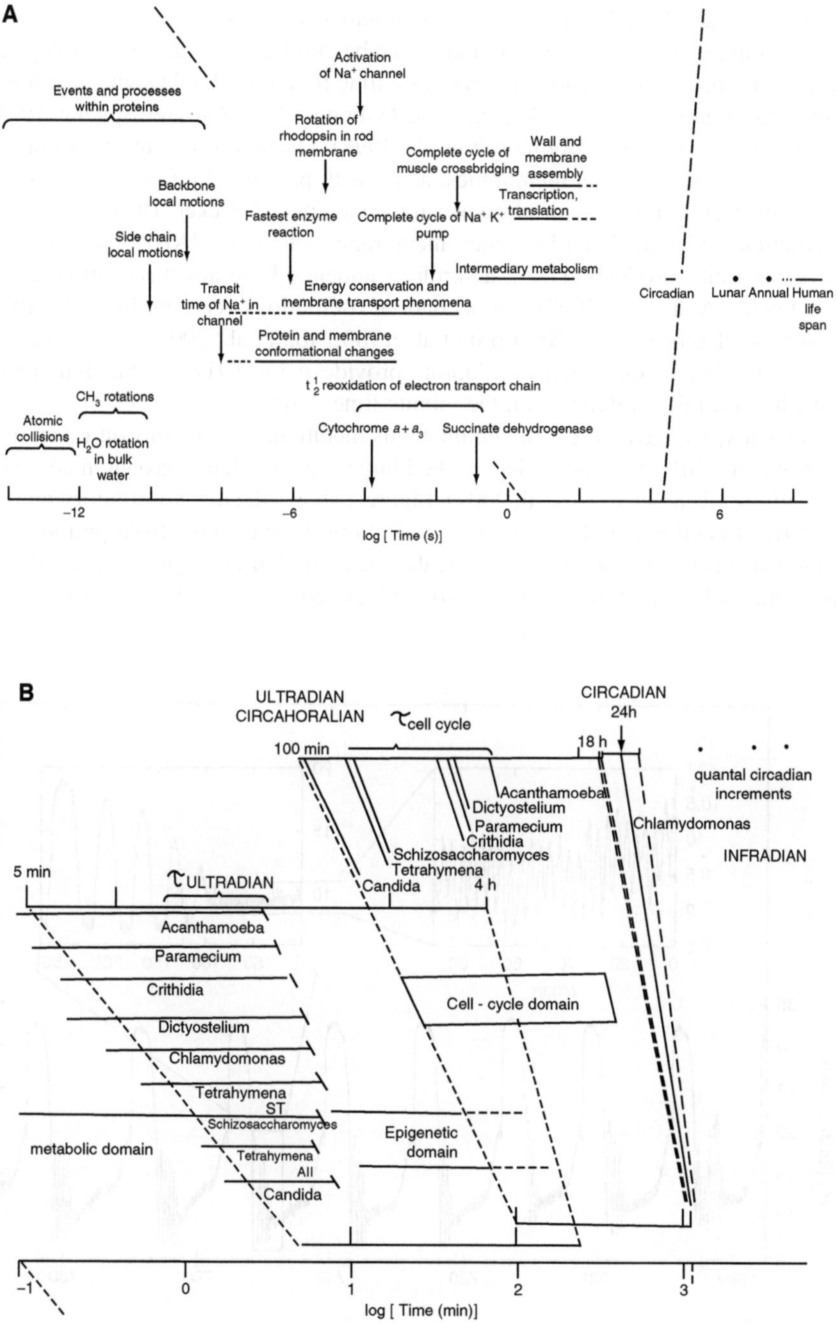

Fig. 1 The time domains of living systems: (**a**) Approximate relaxation times (to a return of 1/e of a small perturbative effect (Goodwin, 1963) are indicated. (**b**) Ultradian clock and cell division cycle times in lower eukaryotes

The biological clock may have several minute hands. Calcium oscillations have periods ranging from seconds to minutes, can be initiated by a variety of biological stimuli (hormones, etc.) and may serve as a time base for reliable signal transmission and targeting *in vivo* (Berridge and Galione, 1988; Gu and Spitzer, 1995; Hajnóczky et al., 1995; Dolmetsch et al., 1998). Oscillations in the transport of ions across mitochondrial membranes, again with periods of a few minutes, have been observed in mitochondrial suspensions (Gooch and Packer, 1974). Sustained oscillations of mitochondrial inner membrane potential, NADH and reactive oxygen species production with a similar frequency have also been observed in intact cells (Aon et al., 2003). Mitochondrial oscillations occuring in undisturbed cells have also been seen (Bashford et al., 1980; Aon et al., 2007). However, it is not inconceivable that such oscillations provide a local (i.e. within individual mitochondria) time reference on the minute time scale.

Recently, we have observed simultaneous metabolic oscillations on three, and perhaps four different time scales in a budding yeast population grown in continuous culture (Roussel and Lloyd, 2007). Figure 2 shows the three oscillatory modes we have studied in detail. The longest period observed resulted from partial cell-cycle synchrony. The well-known circahoralian rhythm also appeared clearly in our data, although it was periodically obliterated by the cell-cycle-associated

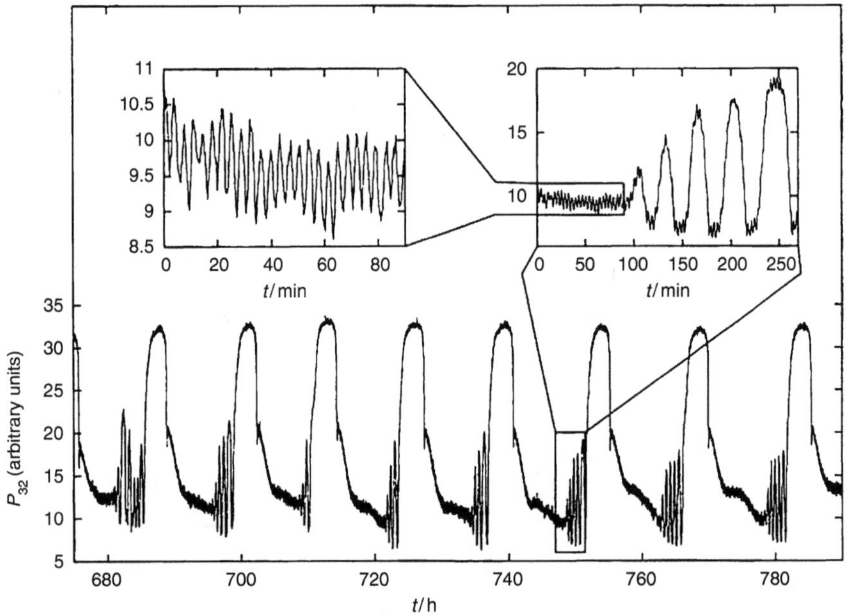

Fig. 2 The *m/z* = 32 signal from the experiment reported by Roussel and Lloyd (2007). This signal corresponds to dissolved oxygen in the fermentor medium. We see a hierarchy of oscillations associated, from slowest to fastest, with the cell cycle, the cirahoralian clock, and an as-yet unidentified biochemical oscillator with a period of approximately 4 min

oscillations. We also observed oscillations of small amplitude with a period of 4 min. These oscillations appear to have a cellular origin (i.e. they are not strictly a collective phenomenon of the culture, nor are they a physico-chemical artifact) and may be the manifestation at the population level of the mitochondrial oscillations described above.

In addition to the three periodic modes described above, we observed highly irregular fluctuations in the CO_2 and H_2S signals with a characteristic time on the order of 1 min. Unfortunately, the time scale of these fluctuations was similar to our sampling time (12 s) so that it is difficult to make any definitive statements about the nature of this oscillatory (chaotic?) mode. Interestingly, fast, irregular fluctuations (with a characteristic time of a few seconds) have also been observed in plant leaves after transfer to a low-CO_2 environment (Roussel et al., 2007). In both cases, these rapid fluctuations arise in healthy organisms. In addition to clocks with regular outputs, it thus appears that cells can display irregular oscillatory modes. Whether these modes are chaotic or stochastic is unknown at this time, as are their possible mechanisms and physiological roles, if any.

What of faster clocks, with periods on the second time scale? Due to the inherent difficulties in studying such fast rhythms, very little is known of cellular timekeeping on this time scale, and even less on subsecond time scales. Calcium oscillations operate in different tissues and situations over a range of time scales, right down to the second range (Berridge and Galione, 1988). This versatile timekeeper may thus function as the cell's second hand in some situations.

In man-made timepieces, typically there is a single fast oscillator which is geared down several times to generate second, minute, hour and day chronometers. What of biological clocks? It is known theoretically that fast oscillators can be coupled to generate slower rhythms whose period is determined in part by coupling topology (Barrio et al., 1997). There are also a few lines of evidence which suggest that the circahoralian and circadian oscillators are related in some way (Dowse and Ringo, 1992; Fuentes-Pardo and Hernández-Falcón, 1993; Lloyd and Murray, 2005). It seems unlikely however that all of the cell's timekeeping functions derive from a single primitive clock. Even circadian timekeeping may involve more than one clock in some organisms (Morse et al., 1994; Johnson, 2001; Lillo et al., 2001). The metaphor we should perhaps have in mind for a living organism's timekeeping functions is thus that of a scientific lab in which multiple timepieces exist to satisfy various needs: our data acquisition systems usually have built-in clocks. We typically have some simple bench-top chronometers for timing routine tasks. And of course, the experimenter usually possesses a wristwatch to remind him or herself of teatime.

The "time domains of living systems" presented in Fig. 1a and b are obviously a "bottom-up" perspective from molecules to cells and organisms illustrating how we typically construct a reductionist view of life. While this bottom-up perspective is impeccably scientific it is not always meaningful to the philosophical mind searching for the meaning and significance of it all for understanding human life and experience (Rossi, 2007). For meaning and understanding, our consciousness often seems to prefer a "top-down" perspective wherein we can utilize the reductionist "facts" to construct an ever more useful comprehension of ourselves in the world. This is what we hope to achieve in section four of this volume: Ultradian and Circadian Rhythms in Human Experience.

Charles Darwin, in a little noticed statement on natural selection in chapter four of *The Origin of Species*, comments on the significance of what we now designate as ultradian and circadian rhythms in human experience.

> It may be said that natural selection is daily and hourly scrutinising, throughout the world, every variation, even the slightest; rejecting that which is bad, preserving and adding up all that is good; silently and insensibly working, whenever and wherever opportunity offers, at the improvement of each organic being in relation to its organic and inorganic conditions of life. We see nothing of these slow changes in progress, until the hand of time has marked the long lapses of ages, and then so imperfect is our view into long past geological ages, that we only see that the forms of life are now different from what they formerly were.

We take this statement by Darwin very seriously, indeed, to help us unify the bottom-up and top-down perspectives of this volume. In our final chapter we generalize the implications of Darwin's natural selection to explore a new view of the significance of art, beauty, and truth on the molecular-genomic level. We utilize current research on the evolutionary origins of consciousness that distinguishes between the experience and behavior of humans and other primates at the levels of brain anatomy, neuronal activity, and gene expression. We take a few tentative steps toward creating a new science of psychosocial genomics to resolve the now hoary nature-nurture controversy with recent epigenetics research on the molecular pathways of communication between mind, brain, body, and gene.

In all of this we must beg the reader's indulgence for we truly identify with Isaac Newton when he describes the scientific ethos.

> I don't know what I may seem to the world, but as to myself, I seem to have been only like a boy playing on the sea-shore and diverting myself in now and then finding a smoother pebble or a prettier shell than ordinary, whilst the great ocean of truth lay undiscovered all before me.

We can only hope that the creative uncertainty we all naturally experience with our bumbling efforts in the laboratory, the clinic, and public outreach will be assuaged at least in part by the earnest efforts and imagination of the excellent contributors to this volume.

References

Aon MA, Cortassa S, Marbán E, O'Rourke B (2003) Synchronized whole cell oscillations in mitochondrial metabolism triggered by a local release of reactive oxygen species in cardiac myocytes. J Biol Chem 278:44735–44744

Aon MA, Cortassa S, Lemar KM, Hayes AJ, Lloyd D (2007) Single and cell population respiratory oscillations in yeast: A 2-photon scanning laser microscopy study. FEBS Lett 581:8–14

Aserinsky E, Kleitman N (1953) Regularly occurring periods of eye mobility, and concomitant phenomena, during sleep. Science 118: 273–274

Barrio RA, Zhang L, Maini PK (1997) Hierarchically coupled ultradian oscillators generating robust circadian rhythms. Bull Math Biol 59:517–532

Bashford CL, Chance B, Lloyd D, Poole RK (1980) Oscillations of redox states in synchronously dividing cultures of *Acanthasmoeba castellanii* and *Schizosaccharomyces pombe*. Biophys J 29: 1–12

Berridge MJ, Galione A (1988) Cytosolic calcium oscillators. FASEB J 2:3074–3082

Brodsky VY (1975) Protein synthesis rhythm. J Theor Biol 55: 167–200

Dolmetsch RE, Xu K, Lewis RS (1998) Calcium oscillations increase the efficiency and specificity of gene expression. Nature 392:933–936

Dowse HB, Ringo JM (1992) Do ultradian oscillators underlie the circadian clock in Drosophila? In: Lloyd D, Rossi EL (eds) Ultradian Rhythms in Life Processes. Springer, London, pp 105–117

Edwards C, Statham M, Lloyd D (1975) The preparation of large-scale synchronous cultures of the trypanosomatid *Crithidia fasciculata* by cell-size selection: changes in respiration and adenylate charge through the cell cycle. J Gen Microbiol 88:141–152

Edwards SW, Lloyd D (1978) Oscillations of respiration and adenine nucleotides in synchronous cultures of *Acanthamoeba castellanii*: mitochondrial respiratory control in vivo. J Gen Microbiol 108:197–204

Edwards SW, Lloyd D (1980) Oscillations in protein and RNA during synchronous growth of *Acanthamoeba castellanii*: evidence for periodic turnover of macromolecules during the cell cycle. FEBS Lett 109:21–26

Fuentes-Pardo B, Hernández-Falcón J (1993) Neurobiology of the circadian clock of crayfish. Trends Comp Biochem Physiol 1:635–673

Gooch VD, Packer L (1974) Oscillatory systems in mitochondria. Biochim Biophys Acta 346:245–260

Gu X, Spitzer NC (1995) Distinct aspects of neuronal differentiation encoded by frequency of spontaneous Ca^{2+} transients. Nature 375:784–787

Hajnóczky G, Robb-Gaspers LD, Seitz MB, Thomas AP (1995) Decoding of cytosolic calcium oscillations in the mitochondria. Cell 82:415–424

Johnson CH (2001) Endogenous timekeepers in photosynthetic organisms. Annu Rev Physiol 63:695–728

Kader J, Lloyd D (1979) Respiratory oscillations and heat evolution in synchronous cultures of *Candida utilis*. J Gen Microbiol 114:455–461

Lillo C, Meyer C, Ruoff P (2001) The nitrate reductase circadian system. The central clock dogma contra multiple oscillatory feedback loops. Plant Physiol 125:1554–1557

Lloyd D, Gilbert D (1998) Temporal organization of the cell division cycle in eukaryotic microbes. Symp Soc Gen Micro 56: 251–278

Lloyd D, Murray DB (2005) Ultradian metronome: timekeeper for orchestration of cellular coherence. Trends Biochem Sci 30:373–377

Lloyd D, Murray DB (2006) The temporal architecture of eukaryotic growth. FEBS Lett 580:2830–5

Lloyd D, Murray DB (2007) Redox rhythmicity: clocks at the core of temporal coherence. Bioessays 29:1–7

Lloyd D, Phillips CA, Statham M (1978) Oscillations of respiration adenine nucleotides and heat evolution in synchrous cultures of *Tetrahymena pyriformis* ST prepared by continuous flow size-selection. J Gen Microbiol 106:19–26

Lloyd D, Edwards SW, Fry JC (1982) Temperature-compensated oscillations in respiration and cellular protein content in synchronous cultures of *Acanthamoeba castellanii*. Proc Natl Acad Sci USA 79:3785–3788

Morse D, Hastings JW, Roenneberg T (1994) Different phase responses of the two circadian oscillators in Gonyaulax. J Biol Rhythms 9:263–274

Poole RK, Lloyd D, Kemp RB (1973) Respiratory oscillations and heat in synchronously dividing cultures of the fission yeast, *Schizosaccharomyces pombe* 972h^{-}. J Gen Microbiol 77:209–220

Rossi EL (2007) The Breakout Heuristic: The New Neuroscience of Mirror Neurons, Concionsness and Creativity in Human Relationships. Phoenix, AZ: Milton H Erickson Foundation Press
Roussel MR, Lloyd D (2007) Observation of a chaotic multioscillatory metabolic attractor by real-time monitoring of a yeast continuous culture. FEBS J 274:1011–1018
Roussel MR, Ivlev AA, Igamberdiev AU (2007) Oscillations of the internal CO_2 concentration in tobacco leaves transferred to low CO_2. J Plant Physiol 164:1188–1196
Von Bertalanffy L (1952) Problems of Life. New York: Harper & Brothers, p 216
Wiener N (1954) The Human Use of Human Beings. New York: Doubleday, p 96

Part I
The Molecular-Genetic-Cellular Level

Chapter 1
The Ultradian Clock (~40 min) in Yeast

(*Saccharomyces cerevisiae*)

D. Lloyd[1], D.B. Murray[2], R.R. Klevecz[3*], J. Wolf[4], and H. Kuriyama[5]

Abstract A precisely controlled, continuously grown, aerobic yeast culture system shows an autonomous sustained respiratory oscillation (i.e. a high amplitude change in dissolved O_2 levels: the residual O_2 that remains after the organisms use what they require). This spontaneously organized synchronous state can be maintained for extended periods (months) and continuously monitored for intracellular redox state (by direct fluorimetric output for nicotinamide nucleotides: excitation 366 nm, emission 450 nm), and dissolved gases (O_2 electrode and/or direct membrane inlet mass spectrometry for O_2, CO_2 and H_2S, sampled on a 15 s cycle). The whole culture population (~5 × 10^8 organisms/ml) is behaving metabolically as if it were a single cell, so that analysis for metabolic intermediates (keto and amino-acids, carboxylic acids), redox components (glutathione, cysteine, NAD(P)H and reactive O_2 species) and products (acetaldehyde, acetic acid and ethanol), with sampling at frequent time intervals gives reliable information on phase relationships on the approximately 40 min ultradian cycle. Lipid peroxidation levels indicate the changing levels of oxidative stress. Microarray analysis shows a genome-wide oscillation in transcription, with expression maxima at three nearly equally spaced intervals on the 40 min time-base. The first temporal cluster (4,679 of 5,329 genes expressed) occurred maximally during the reductive phase, whereas the remaining 650 transcripts were detected maximally in the oxidative phase. Furthermore, when fixed samples of yeasts were analysed for DNA using flow cytometry, synchronous bursts in the initiation of DNA replication (occurring in about 8% of the total population) were shown to coincide with decreasing respiration rates. A precisely defined appearance of mitochondrial energy generation-dependent conformational

*Dr Bob Klevecz passed away on 13.05.2008

[1] Microbiology (BIOSI 1) Main Building, Cardiff School of Biosciences, PO Box 915, Cardiff CF10 3TL, Wales, UK

[2] Institute for Advanced Biosciences, Keio University, Tsuruoka, Yamagata, 997–0017, Japan

[3] Dynamics Group, Department of Biology, Beckman Research Institute of the City of Hope Medical Center, Duarte, CA 91010, USA

[4] Humboldt-University, Berlin Invalidenstr 42/43, D-10115, Germany

[5] Biochemical Engineering Lab., National Institute of Advanced Industrial Science and Technology, AIST, Tsukuba, Ibaraki, Japan

D. Lloyd, E. Rossi (eds.), *Ultradian Rhythms from Molecules to Mind*,

changes, the machinery for protein synthesis and degradation and for mitochondrial ribosome assembly are also all locked on the ultradian clock cycle phase. This indicates the pervasive and all-embracing nature of temporal co-ordination of cellular growth and division. The major players in this yeast model system are identified as some of the most highly conserved components common to all eukaryotes. Thus a primeval, fundamental and insistent drum-beat reverberates in all eukaryotic systems, in microbes, animals and plants. The ultradian clock that co-evolved with increasing atmospheric O_2 predates and underpins the circadian clock at the central core of endogenous intracellular timekeeping. It represents the prototype of all biological rhythmicity: circadian clocks evolved much more recently, almost as the icing on the cake.

Keywords Yeast, self-synchrony, respiration, redox state, metabolic switch, transcriptional regulation, cell division cycle ultradian clock

1.1 Introduction: Continuous Culture of Yeast – An Ideal System for Study

Detailed understanding of the time structure of the intricately ordered processes of energy generation, metabolic transformations, synthesis and assembly of membranes and organelles, as well as the organization of chromosomal dynamics and the cell division cycle, requires studies either in a single cell or in synchronous populations of cells or organisms. Whilst the former is even now restricted (despite enormous strides in optical microscopy, image processing, microfluidics and atomic force microscopy), and the latter is usually problematic because of the difficulties associated with the preparation of material without perturbatively disturbed "normality". However, in a seminal contribution published in 1992, the group of Kuriyama discovered a method which gave spontaneous self-synchrony of dense (~5 × 10^8 organisms/ml) cultures of *Saccharomyces cerevisiae* (Satroutdinov et al., 1992). The strain employed (IFO-0233, IFO, Institute of Fermentation, Osaka, Japan), is a distiller's yeast. During batch-growth on glucose-containing complex growth medium at pH 3.4 with a constant supply of air (calculated for each reactor according to its specific oxygen transfer coefficient), glucose is consumed to produce biomass, CO_2, H_2O and ethanol as well as many other fermentation products. Following ethanol utilization, depletion of trehalose and glycogen in the second stage of this diauxic growth results in the initiation of oscillatory respiration, as indicated by 40 min cycles of dissolved O_2 and CO_2 Commencement of continuous culture (Fig. 1.1) at this stage, gives a continuous supply of material for long-term monitoring of the organisms through many thousands of generations over a period of many months. The rapidly-responding probes immersed in the culture give highly sampled intervals, e.g. for NAD(P)H fluorescence 50 k samples/s, for dis-a

b

Fig. 1.1 (**a**) Dr. Hiroshi Kuriyama working with a continuous culture of *Saccharomyces cerevisiae*: (**b**) The continuous culture system. The impellors were driven by a 24 V motor (K; green line) and controlled by a power supply(**A**). Temperature was measured using a Pt 100 probe (yellow line) and controlled (**B**) to a user set-point by a 333 W heater (pink line). A water jacket (at 10°C) consisting of rubber tubing wrapped around the glass vessel (**G**) provides cooling (not shown). pH was measured using a steam sterilizable gel electrode connected (cyan line) to a

(continued)

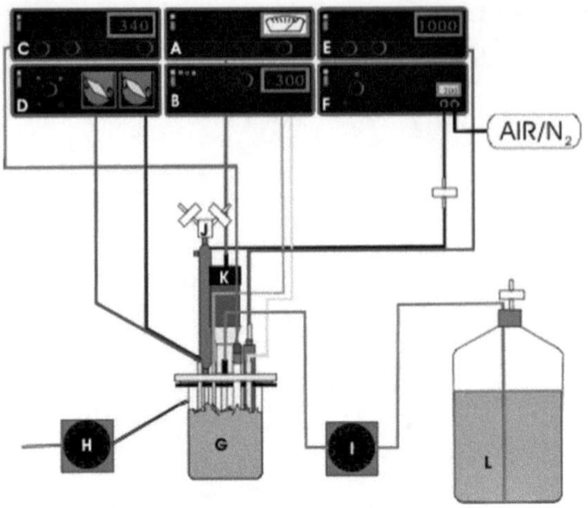

Fig. 1.1 (continued) pH meter (**C**). pH was automatically controlled (**D**) by the addition of 2.5 M NaOH (dark blue line) or 2 N phosphoric acid (red line). An oxygen meter (**E**) was connected to a polarographic oxygen electrode (magenta line). Sterile dry air was fed (black line) to the fermentor via a sparge arm, its flow rate regulated by an Aalborg mas flow controller (**F**). Waste air was passed through a condenser (10°C) and exited via two sterile filters (**J**).Sterile medium was stored in a resevoir (10 l; **L**) and pumped into the fermentor by low flow peristaltic pump (I; Microperpex, LKB) To avoid back contamination the medium was pumped through an anti-growback device prior to entry to the vessel(**G**). Waste was collected from the fermentor by a glass arm weir. A peristaltic pump (**H**; 101U Watson- Marlow) then pumped the waste into a collection vessel (white line). In addition there was a double-valved sample assembly and three auxillary ports with rubber septa that were used for extra apparatus or pulse addition of purturbating agents. The data was acquired by an in-house software that controlled the data acquistion board (DAS 16, Computer Boards) and was programmed by the user to collect and store data (up to 1 s intervals). In some experiments (not shown), a membrane-inlet mass spectrometric probe was used (for measuring gases and low molecular mass volatiles directly), either by immersion in the culture, or exposed to the headspace gas, and a rotating disc dual-channel fluorimeter (for measuring redox levels of NAD(P)H and/or flavins; Johnson Foundation, University of Pennsylvania) was linked to the vessel using light guides Standard conditions employed were: working vol. 800 ml; dilution rate, 0.085 h^{-1}; stirrer rate, 800 rpm; airflow rate, 180 ml min^{-1}; temperature 30°C (±0.2) and pH controlled at 3.4. Pulse injections were carried out through a sterile filter (0.2 µm porosity)

solved O_2 10 samples/s (Murray et al., 2007). The most convenient readout, dissolved O_2, can also be monitored by membrane inlet mass spectrometry, as can CO_2, H_2S and ethanol. Near-infra-red spectroscopy can be used on-line for ethanol, glucose, NH_3, glutamine and biomass (unpublished data). The oscillatory state of this autonomously self-sustained system is largely independent of glycolysis, and gates the cell cycle (Klevecz et al., 2004); its approximately 40 min period is temperature compensated (Fig. 1.2; Murray et al., 2001). Thus its timekeeping characteristics place it in the ultradian time domain (as defined by its characteristic of cycling many times during a day). That similar clock control can be demonstrated when ethanol is used as the dominant carbon source in the growth medium shows that these oscillations are not the well-known glycolytic or cell-cycle associated oscillations (Keulers et al., 1996).

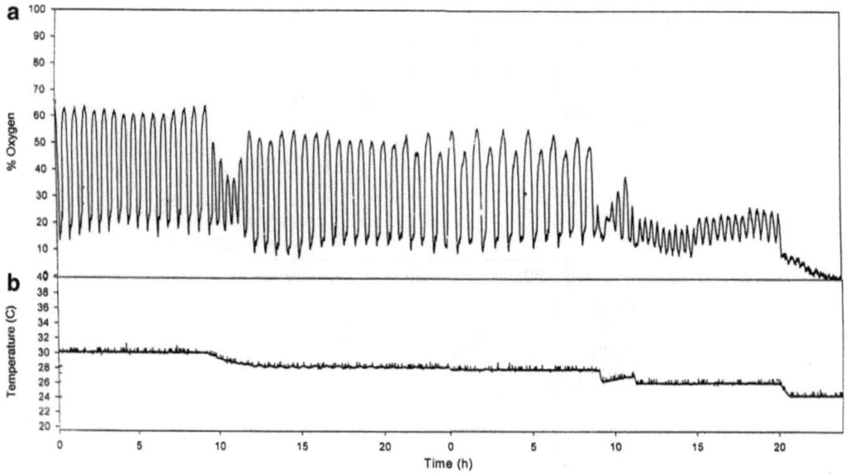

Fig. 1.2 Effects of step temperature changes on the respiratory oscillations in a continuous culture of *S. cerevisiae* grown in ethanol-containing medium. (a) Dissolved O_2, and (b) temperature were measured continuously on-line (Murray et al., 2001)

1.2 Discovery of the Ultradian Clock

It is their sustained nature that confers upon the 40 min periodicity the status of a rhythm, and temperature independence further elevates it to that of a biological clock. It thus shares important properties with the more intensively-studied Circadian clock as well as physical clocks (Lloyd et al., 2002a): The latter was first recognized as an endogenous timekeeper in 1729 (de Mairan). The Ultradian Clock, initially called The "Epigenetic Clock" was finally characterized in 1982 as a temperature-compensated ultradian timekeeper in the soil amoeba, *Acanthamoeba castellanii* (Edwards and Lloyd, 1978, 1980; Lloyd et al., 1982). Respiratory oscillations during the cell division cycle of synchronous cultures of this organism that had been produced by a minimally-perturbing size selection procedure were shown to be phase-coupled to high amplitude changes in adenine nucleotide pool sizes (ATP, ADP and AMP) and to cycles of accumulation of total cellular RNA and protein. The Q_{10} values for the periods of the O_2 consumption and protein rhythms were respectively 1.03 and 1.05 over the temperature range from 20°C to 30°C whereas the Q_{10} for the cell division cycle (16 and 7.8 h respectively) was just more than 2 over the same temperature range (Fig. 1.3a; Marques et al., 1987). These observations clearly established a timekeeping function for this Ultradian Clock, so that a quantal relationship between this clock and the cell cycle time, as earlier proposed for mammalian cells (Klevecz, 1976) and subsequently shown in lower eukaryotes (Chisholm and Costello, 1980; Lloyd et al., 1982; Homma and Hastings, 1988) is evident. Similar results previously obtained for synchronous cultures of other lower eukaryotes (the yeasts, *Candida utilis* (Kader and Lloyd,

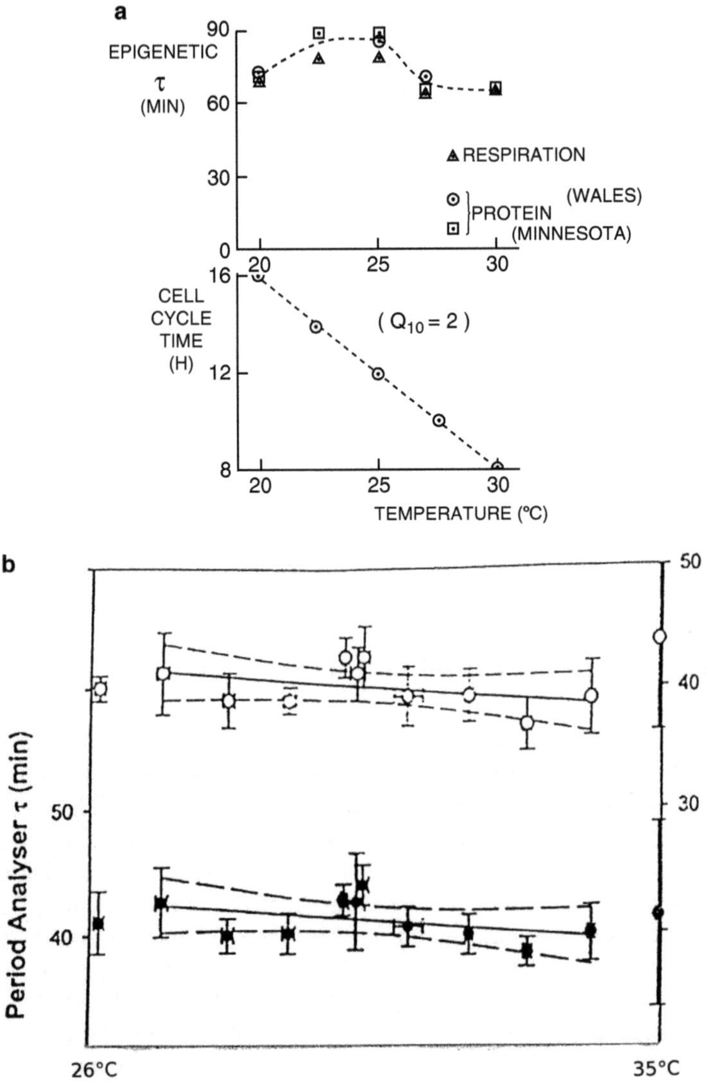

Fig. 1.3 Temperature dependence of the respiratory oscillation in (**a**) Periods of the ultradian clock-driven outputs in *Acanthamoeba castellanii*, contrasted with cell cycle times (Marques et al., 1987); analysis of protein data calculated by independent methods in Wales (**a**) and Minnesota (**b**). Periods of the respiratory oscillation in yeast calculated by Period Analyser and fast Fourier transform analysis (FFT) (Murray et al., 2001)

1979) and *Schizosaccharomyces pombe* (Poole et al., 1973), for a flagellate protist (*Crithidia fasciculata*, Edwards et al., 1975) and a ciliate (*Tetrahymena pyriformis*, Lloyd et al., 1978) indicated that the Ultradian Clock is a timer in many systems. Analogies with the 90-min BRAC (basic rest-activity cycle) observed during sleep studies on human subjects in the Chicago laboratory of Nathanial Kleitman (Aserinsky and Kleitman, 1953) are clear, and suggest that the Ultradian Clock is a universal timekeeper from amoeba to humans. The period of the Ultradian Clock is different in each of the organisms studied, presumably representing a self-optimized value for the intracellular network that is distinctive for each species.

Of all the species studied, it seemed that the asymmetrical cellular proliferation of baker's or brewer's yeasts by budding would make studies of ultradian rhythms more complicated. However, the serendipitous observations of Satroutdinov et al. (1992) altered this view, and it soon became clear that the spontaneously self-organizing continuous culture system offered huge possibilities for continuous monitoring of the Ultradian Clock over extended time spans and by non-invasive methods. The temperature compensated nature of the yeast 40 min oscillator was intensively investigated (Figs. 1.2, 1.3b; Murray et al., 2001). Thus glucose-fed cultures showed a Q_{10} of 1.06 between 27°C and 34°C whereas ethanol-grown cultures gave a value of 0.85 (i.e. were somewhat overcompensated) Homeodynamic control as a function of dilution rate of the continuous cultures was also noted.

1.3 Metabolic Studies of the Yeast Ultradian Clock

1.3.1 Carbon Metabolism

The aerobic continuous culture of yeast can be supplied with glucose (Satroutdinov et al., 1992; Keulers et al., 1996), ethanol (Keulers et al., 1996), or acetaldehyde (Keulers and Kuriyama, 1998) as major carbon sources (in addition to those present in the yeast extract common to all the growth media employed. Whereas for growth on glucose or ethanol the period of the "metabolic" oscillation was about 40 min, with 290 mM acetaldehyde as primary carbon source it was approximately 80 min. Acetaldehyde, acetate, ethanol, dissolved O_2 and CO_2 production were all oscillatory variables (Fig. 1.4), with the acetaldehyde and acetate approximately in phase with O_2 uptake rate (i.e. these products were at a maximum during low dissolved oxygen concentration). This enhanced respiration indicated that acetaldehyde and acetate never exceeded their potentially inhibitory levels, and that this was the same irrespective of which of the three principle carbon sources were employed. Flux balance calculations indicated that ethanol production from acetaldehyde occurred periodically during the oscillation, and was accompanied by decreased acetaldehyde conversion to acetate and acetyl-CoA. It was suggested that the diminished rates of acetaldehyde flux to the TCA cycle were related to the decreased respiratory activity due to some inhibitory effect on the mitochondrial respiratory chain. The importance of acetaldehyde (but not O_2 or ethanol) as a specific synchronizing agent had been previously suspected (Keulers et al., 1996), when it was observed that high aeration rates leading to loss of CO_2 and the volatile aldehyde led to loss of the oscillatory state. The confirmation that the synchronizer

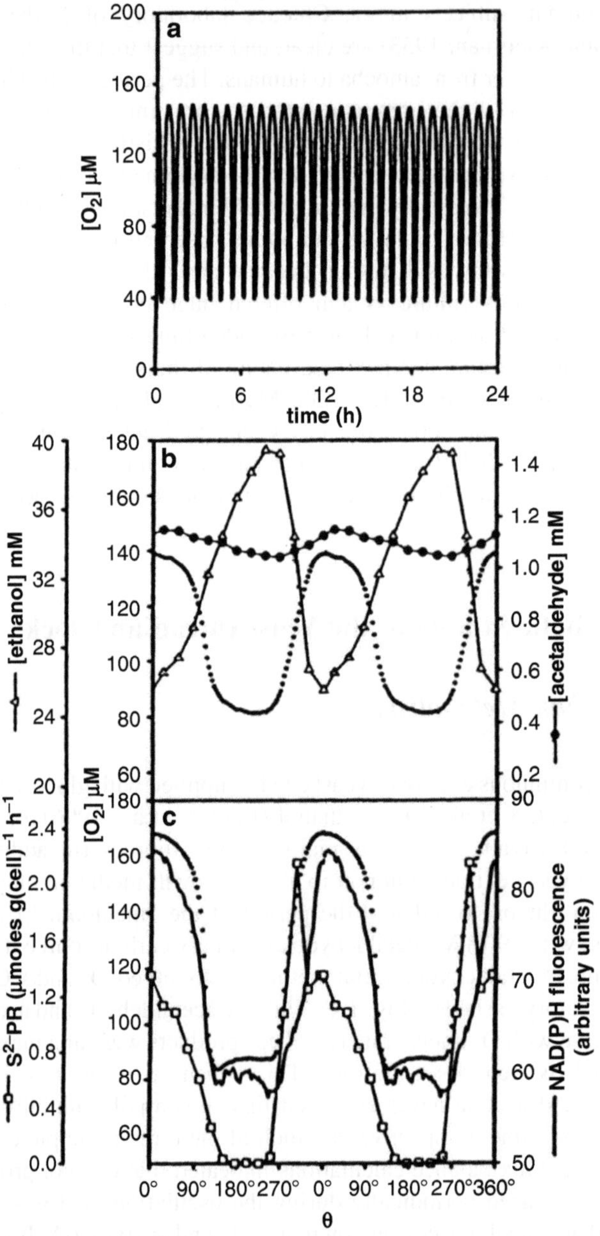

Fig. 1.4 Continuously monitored oscillatory variables (dissolved O_2 and NAD(P)H and metabolites sampled from the aerobic continuous cultures of yeast. (**a**) The respiratory oscillation. (**b**, **c**) Phase angle plots with respect to dissolved O_2 (···), of acetaldehyde (▲); (**b**), ethanol (•; **b**) S^{2-}PR (sulphide production rate, ■; **c**), and NAD(P)H fluorescence (-; **c**). The data are double-plotted (side-by-side) in order to clarify phase-related events (Murray et al., 2003)

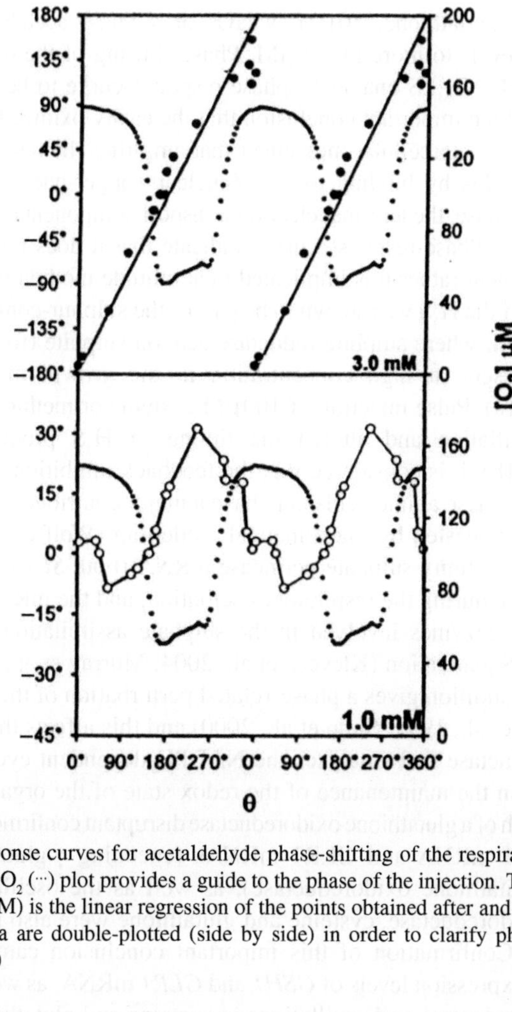

Fig. 1.5 Phase response curves for acetaldehyde phase-shifting of the respiratory oscillation in yeast. The dissolved O_2 (···) plot provides a guide to the phase of the injection. The straight line on the upper panel (3 mM) is the linear regression of the points obtained after and before the "break-point" (0°). The data are double-plotted (side by side) in order to clarify phase-related events (Murray et al., 2003)

was acetaldehyde rather than CO_2 awaited the definitive experiments of Murray et al. (2003) who demonstrated the phase-shifting capacity of acetaldehyde (Fig. 1.5, see also Chapter 3).

1.3.2 *Sulphur Metabolism*

A breakthrough in understanding of the mechanism of the oscillation came with the discovery that H_2S is produced at the onset of lowered O_2 uptake rates. This gas, a potent respiratory inhibitor, reached a maximum (1.5 μM) just prior to minimum respiration rates (Fig. 1.4; Sohn et al., 2000). After this, the concentration of H_2S decreased to 0.2 μM before the restoration of high respiration again. Perturbants such as 50 μM glutathione, 50 μM $NaNO_2$ or 4.5 mM acetaldehyde transiently increased H_2S levels to more than 6 μM. Phase-shifting of the oscillation by additions of 0.77 μM $(NH_4)_2S$ enabled a phase response curve to be obtained (Murray et al., 2003) and the important conclusion that the easily oxidized, rapidly diffusing gas H_2S acts as an intercellular messenger that amplifies the respiratory oscillation. It accomplishes this by binding to the binuclear copper-haem reaction centre of cytochrome *c* oxidase, the terminal electron transport component of the mitochondrial respiratory chain. Phase-response curves indicate that it does not act by itself as a synchronizing agent rather it is implicated in amplitude modulation of the cycles.

The source of the H_2S was shown to be part of the sulphur-containing amino acid synthesis network where sulphite reductase acts on sulphite (formed by reduction of sulphate present at high concentration in the growth medium (Sohn and Kuriyama, 2001a). Pulse injection of 100 μM cysteine or methionine perturbed the respiratory oscillation and altered the timing of H_2S production (Sohn and Kuriyama, 2001b): it is considered that the feedback inhibition of sulphate uptake into the yeast by these amino acids may have a major contribution to the respiratory oscillations as suggested by mathematical modelling (Wolf et al., 2001; Henson, 2004). The high-affinity sulphate permease mRNA (both *SUL1* and *SUL2*) shows cyclic expression during the respiratory oscillation, and the mRNA expression patterns of all the enzymes involved in the sulphate assimilation pathway precede endogenous H_2S generation (Klevecz et al., 2004; Murray et al., 2007).

Glutathione addition gives a phase-related perturbation of the respiratory oscillation (Murray et al., 1999; Sohn et al., 2000) and this affects the essential role of glutathione reductase that catalyses the NADPH-dependent cycling of the GSH-GSSG system in the maintenance of the redox state of the organism (Sohn et al., 2005 a,b) Growth of a glutathione oxidoreductase disruptant confirmed this. Expression levels of GSH1 mRNA and GLR1 mRNAs (encoding γ-glutamylcysteine synthetase and glutathione oxidoreductase), as well as the oscillating activities of glutathione oxidoreductase, cysteine and glutathione were also in agreement with this proposal. Confirmation of this important conclusion came from extensive studies on the expression levels of *GSH1* and *GLR1* mRNA, as well as the activities of glutathione reductase and oscillations in cysteine and glutathione. The effects of pulse injection of thiol redox modifying agents (diethylmaleate, *N*-ethylmaleimide), of inhibitors of glutathione reductase (DL-butathionine [*S, R*]-sulphoxamine) or of glutathione synthesis (5-nitro-2-furaldehyde) further strengthened these data. The network of sulphate uptake and sulphur amino acid interactions is shown (Fig. 1.6). That this network is intricately interwoven with the regulation of redox state was further demonstrated (Kwak et al., 2003). Cellular per-oxidative adducts, as measured by the levels of lipid peroxidation products, oscillates out-of-phase with levels of dissolved O_2. Pulse addition at minima of dissolved O_2 of 100 μM *N*-acetylcysteine (which scavenges H_2O_2 and hydroxyl radicals) perturbed the respiratory oscillation and attenuated H_2S production to 63% of its normal amplitude in the next 40 min cycle. Then the respiratory oscillation damped out, only to be regained 20 h later. The non-toxic free radical scavenger, ascorbic acid as well as the inhibitor of catalase (3-aminotriazole) or superoxide dismutase (*N-N*[1] diethyldithio carbamate) gave further evidence for this. H_2O_2 (0.5 mM) added at a minimum of dissolved O_2

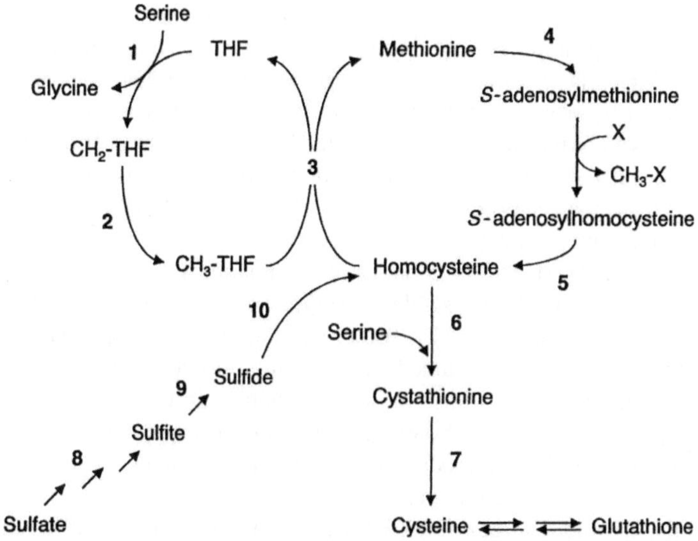

Fig. 1.6 Sulphate uptake and metabolism in yeast genes and enzymes catalyzing individual reactions are: 1. SHM1 and SHM2: serine hydroxymethyltransferase (SHMT; EC 2.1.2.1); 2. MET12 and MET13: methylenetetrahydrofolate reductase (MTHFR; EC 1.5.1.20); 3. MET6: methionine synthase (EC 2.1.1.14); 4. SAM1 and SAM2: S-adenosylmethionine synthetase (EC 2.5.1.6); 5. SAH1: S-adenosylhomocysteine hydrolase (EC 3.3.1.1); 6. CYS4: cystathionine β-synthase (CBS; EC 4.2.1.22); 7. CYS3: cystathionine γ-lyase (EC 4.4.1.1); 8. MET14: adenylsulfate kinase (EC 2.7.1.25); 9. MET10: sulfite reductase (EC 1.8.1.2); and 10. MET17: O-acetylhomoserine (thiol)-lyase (EC 2.5.1.49). "X" represents any methyl group acceptor (Chan and Appling 2003)

both perturbed the respiratory oscillation and elevated H_2S production in the subsequent cycle; 50 μM menadione, a superoxide generating agent, mimicked these effects but at two orders of magnitude less concentration, indicating that cellular reactive oxygen species defences were active throughout the cycle. However an increase ROS localised in the mitochondria caused by menadione could perturb the system at low concentrations.

Further work on the effects of glutathione perturbation on branched-chain and sulphur-containing amino acids tends to suggest that the observed effects are more closely related to amino acid metabolism and H_2S generation than with cellular redox state *per se* (Sohn et al., 2005 a,b). Confirmation of this suggestion awaited more in-depth analyses of the transcriptome and metabolome (Murray et al., 2007).

1.4 A Precise Transcriptional Oscillator Defines the Dynamic Architecture of Phenotype

It is becoming clear that regulation of levels of individual genes (and ultimately their proteins) is primarily a transcriptional process. With increasing genomic complexity of metazoans there is a dramatic increase of the fraction of the genome that is transcribed but not translated (Taft et al., 2007). In *S. cerevisiae* the role of non-coding RNAs appears to be less dominant than in some other model eukaryotes studied, and this allows us to uncover the architecture and dynamics of phenotype unencumbered by what is estimated, in humans, to be more than a million small regulatory RNAs. Analysis of genome-wide oscillations in transcription (Fig. 1.7) reveals that the entire unicellular organism or animal cell is an oscillator, and because its trajectory can progress as series of period doublings during the course of its changing pattern of expression, it is most parsimoniously seen as an attractor (Klevecz and Li, 2007). The simulations of the dynamics that would permit period doublings yet still allow both a stable phenotype and relatively precise timekeeping have been published (Bolen et al., 1993). A chaotic attractor from which much of the unpredictability could be tuned involves appropriate arrangements of phase among the thousands of its transcriptional oscillators. In essence the maintenance of a stable phenotype requires that expression maxima in expression in clusters of transcripts must be poised at antipodal phases around the steady-state – this is the dynamic architecture of phenotype.

Recently, Li and Klevecz (2006) further characterized the first example of period doubling behaviour in the genome-wide transcriptional oscillation (Salgado et al., 2002) seen in yeast cultures growing at high cell densities and exhibiting continuous gated synchrony, and suggested that this behaviour was closely modelled by a simple modification of the Rossler attractor (Rossler, 1976). From the dynamic systems perspective we can take this rigorous model for describing the global behaviour of the timekeeping oscillator and begin to build a gene-by-gene or regulon-by-regulon dynamic systems network. The model also immediately offers an accessible graphical representation of the genome scale changes that can lead from a high frequency timekeeper to the cell-cycle and circadian rhythms. Cell-to-cell signalling in continuous cultures of *S. cerevisiae* leads to mutual entrainment or synchronization that is manifested as an oscillation in redox state (Lloyd et al., 2003) and a genome-wide oscillation in transcription and metabolism (Klevecz et al., 2004; Murray et al., 2007). In turn, this transcriptional redox attractor cycle times or gates, DNA replication and other cell cycle events. In this regard the cell cycle is a developmental process timed by the attractor. DNA replication is restricted to the reductive phase of the cycle and is initiated as levels of hydrogen sulphide rise. This is seen as an evolutionarily important mechanism for preventing oxidative damage to DNA during replication (Klevecz et al., 2004). Both exposure to drugs (e.g. Li^+ or Type A monoamine oxidase inhibitors) known to alter circadian rhythms (Fig. 1.8; Salgado et al., 2002) and deletions of known clock genes in *S. cerevisiae* yield an increase in the period of the attractor that follows a bifurcation path from 40 min to 3–4 h. This finding resonates with an earlier observation of a 3–4 h oscillation that gated cell cycle events in higher organisms (Klevecz, 1976). We suggest that the genome-wide oscillation discovered in yeast is a primordial oscillator, and that the differentiation pathways to multiple phenotypes (e.g. formation of spores or filaments) maybe through processes characterized by period doubling and period three bifurcations.

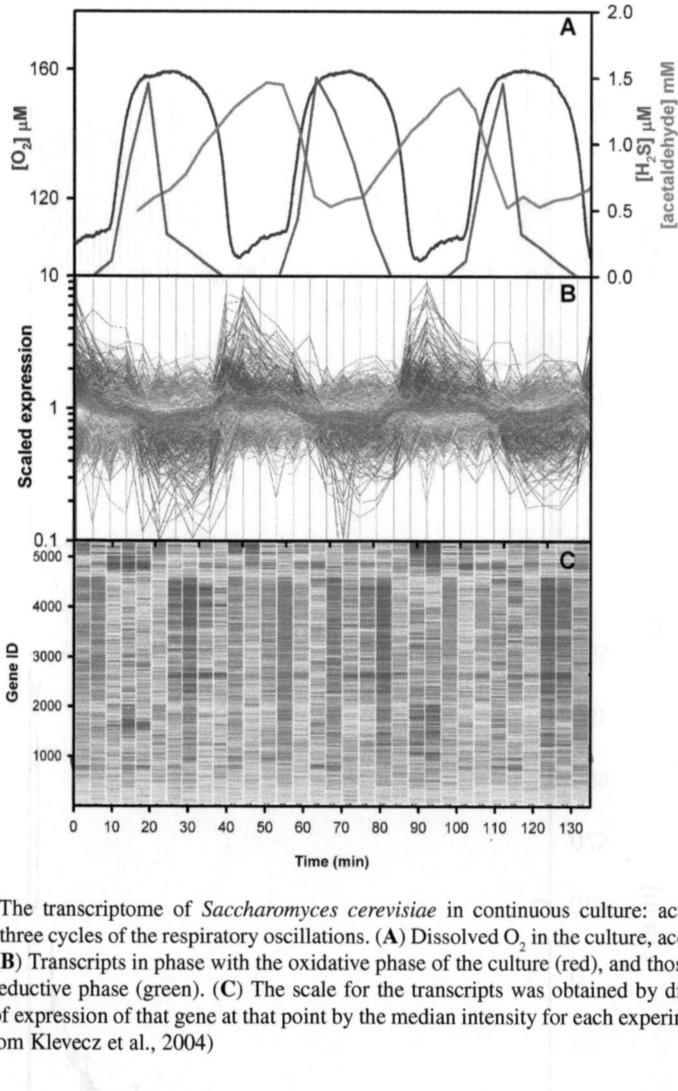

Fig. 1.7 The transcriptome of *Saccharomyces cerevisiae* in continuous culture: activity with respect to three cycles of the respiratory oscillations. (**A**) Dissolved O_2 in the culture, acetaldehyde and H_2S. (**B**) Transcripts in phase with the oxidative phase of the culture (red), and those in phase with the reductive phase (green). (**C**) The scale for the transcripts was obtained by dividing the intensity of expression of that gene at that point by the median intensity for each experiment (Data derived from Klevecz et al., 2004)

Prior to the development of methods for examining genome-wide expression, clues to the dynamic structure of the cell could only be assessed by measurement of small subsets of cellular constituents from synchronous cultures (Lloyd et al., 1982; Klevecz and Ruddle, 1968) or by intentional perturbation and measurement of the phase response curves of events such as DNA replication and cell division to the perturbation (Mitchison, 1971; Klevecz et al., 1984). These early studies provided a sketch of the dynamic architecture of phenotype and offered a rigorous alternative to simple branched sequential models based on mutational analysis of the cell cycle (Hartwell et al., 1974). Quantized generation times, together with perturbation analyses, formed the experimental foundation of efforts to synthesize a model of the cell cycle in which such disparate concepts as check points, and limit

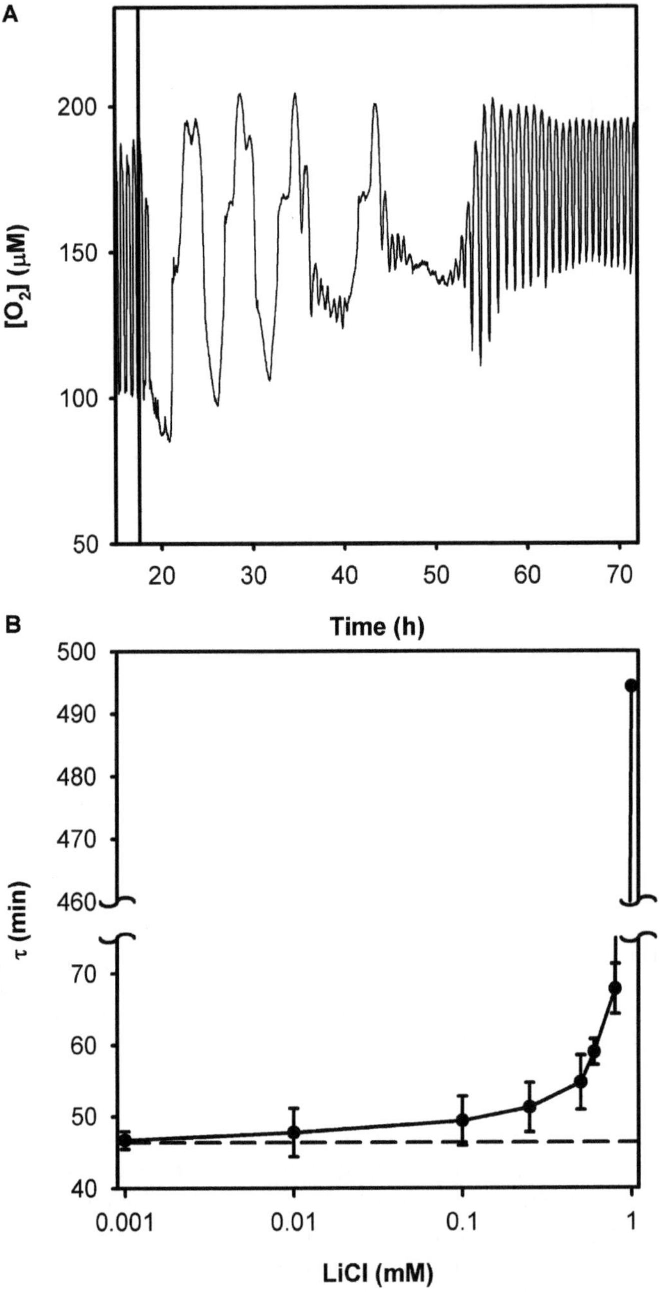

Fig. 1.8 Effects of Li^+ on the respiratory oscillation in yeast. (**a**) The perturbation of the ultradian oscillation produced during continuous culture of *S. cerevisiae* by 1.0 mM LiCl. The vertical bar indicates the points of injection. (**b**) The influence of LiCl concentration on oscillation period. The horizontal dashed line represents periodicity with no LiCl injection (Salgado et al., 2002)

cycles or complex attractors were fused (Shymko et al., 1984). The basic idea was that checkpoints represent sub-threshold oscillations in an attractor that underlies the cell cycle (Klevecz et al., 1984). The oscillator that gave rise to gated cell divisions in mammalian cells was shown to be phase responsive and temperature compensated (Klevecz et al., 1978; Klevecz and King, 1982). The quantized generation time model was extended to other unicellular eukaryotes and to gating of circadian rhythm-based cell division in plants, dinoflagellates and a variety of mammalian cells in culture. One prediction of the attractor models was that all cell cycle events would be gated by the attractor, and this period would be an integral sub-multiple of the cell cycle or circadian rhythm it timed.

These high frequency oscillations in mammalian cells (Klevecz and Stubblefield, 1967) were later codified as quantized generation times and the more recent finding indicates that the continuous culture system in yeast appears to be timed by a similar oscillator that can be tuned or driven to "fold" (i.e., undergo a series of period two or period three bifurcations). Thus that cell cycle events in *S. cerevisiae* are gated by this transcriptional cycle suggests that a similar phenomenon is operating in all systems from yeast to mammalian cells. This realization has opened a new and experimentally more accessible path to investigations of synchronous gating and the role of oscillations in generating and maintaining a stable phenotype. There is a strong evolutionary argument too for expecting that the circadian clock started out as a high frequency oscillator.

1.5 The Spatio-Temporal Self-Organisation of the Reactome

The prevailing view during the last century has been that once a biological system has been broken down to its "important" components and these components have each been characterised, then a systems properties will somehow become evident. This approach has lead to a static and linear view of biology, where for the most part, research has become fragmented, focusing on individual proteins or rather small modules (e.g. isolated membranes or organelles). With the advent of high-throughput analyses, e.g. microarray-based multiple assays and mass spectrometric metabolomics, this viewpoint has become progressively less tenable. Conservative estimates during its lifecycle and in its response to environmental challenges (e.g. on exposure to heat) indicates that between 20–50% of the yeast system alters. This has been confirmed by genome-wide and metabolome-wide elucidation of properties of the respiratory oscillations in yeast (Klevecz et al., 2004). In conjunction with the advent of high-throughput analyses, computational approaches are being developed to curate and better understand the implications of vast array of new data obtained. Construction of cellular networks and elucidation of the logic that underlies the cellular processes shows that these networks are now known to be of diverse architecture (Murray et al., 2007). Data obtained from high-throughput chromatin immunoprecipitates hybridised to DNA microrray containing the probes for the upstream regulatory sequences or DNA tiling microarrays (ChIP-chip) reveal that the underlying

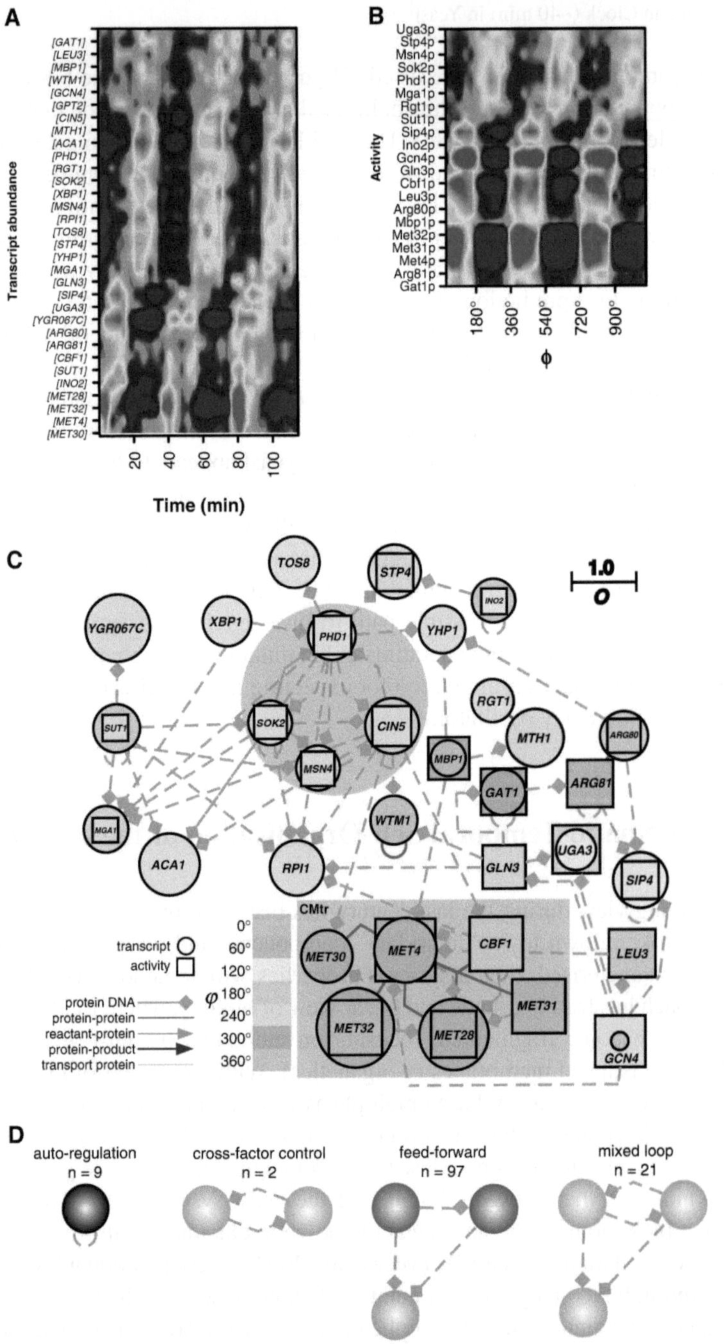

Fig. 1.9 Transcriptional regulation of the yeast respiratory oscillation. (**a**) The transcriptional regulatory core network involved in the regulation of the yeast respiratory oscillation was derived from the top oscillating transcript concentrations (oscillation strength ≥ 0.75; Murray et al., 2007). (**b**) Activities of the transcriptional regulators were derived from the statistical analysis of their target transcripts (oscillation strength ≥ 0.75). Both heat maps were ordered according to the phase angle of the measurements' peak production. A ball-and-stick network representation of the transcriptional regulators shown in (**a**) and (**b**). (**c**) The figure provides a guide to the network.

system structures involved in global transcription can be profoundly altered in response to environmental stimuli (MacIsaac et al., 2006). The topology of the protein-protein interaction network has been approached by series of immuno-precipitation, 2-hybrid and MS analyses (Ito et al., 2001; Gavin et al., 2006; Krogan et al., 2006). In conjunction with this has been a series of advances in informatics that have revolutionised the manipulation and correlation of data obtained from prior biochemical information using appropriated selected on-line resources (e.g. KEGG, SGD or SwissPROT).

Although these advances have led to a deeper understanding of the structure of the cell network, they have done little to advance our understanding of cellular dynamics. Moreover, little has been done to combine the derived networks together, and even less has been done to model large scale datasets to produce a coherent view of the formation of cellular phenotypes.

With this in mind, we recently combined computational and statistical network approaches, with high quality transcriptional and metabolomic data, to analyse the global landscape of the yeast oscillatory phenotype (Murray et al., 2007) (Fig. 1.9). This revealed that the entire biochemical network or reactome self-organises into two distinctive regions; oxidative and reductive (Fig. 1.10). Transcription during the oxidative phase was almost exclusive focussed on biosynthesis, and a clear temporal program was initiated starting with nucleotide biosynthesis and ending with acetate metabolism (Fig. 1.10). This program spanned 10–12 min, and metabolic, physiological and morphological processes (e.g. sulphate uptake, amino acid biosynthesis, S-phase, etc.) occurred 10–14 min after the peak in their transcriptional activity. Statistical analysis of transcription factor binding targets and the reconstruction of a yeast protein-protein interaction map implicated the temporal construction and destruction of a transcription factor complex comprising Cbf1 (centromere binding factor), Met4, Met28, Met31 and Met32 (methionine regulation factors), and Gen4 (general control protein involved in nitrogen catabolite repression); these processes orchestrated the majority of transcriptional changes. Network analysis of the oscillatory transcription factor network (comprising some 33 transcriptional regulators) indicated that this works by spacing gene transcription *via* the formation of multiple input feed-forward genetic circuits (Fig. 1.10).

Out-of-phase with oxidative phase transcription, a much larger group of transcripts (~80% of the most oscillatory transcripts) showed a peak production in the reductive phase. Many transcripts encode for proteins that are involved in sulphur

Fig. 1.9 (continued) Circles represent transcript abundance and squares represent the transcription factor activity. If the transcription factor activity of a node was greater than its transcript concentration, the square was placed behind the circle, otherwise the square was placed in front of the circle. The nodes are shaded according to the phase angle (Ψ) and the oscillation strength (O) is indicated by the size of the node. The shaded box indicates the CMtr complex, and the shaded circle shows an area enriched with cross-factor control loops. (**d**) The gene network motifs ($p_{val} < 0.01$) were scored for quality ($z_{score} > 7$) where the number of nodes was 3. The colour of the circle represents what phase in which the motif is enriched; where autoregulatory units are distributed throughout all phases they are black

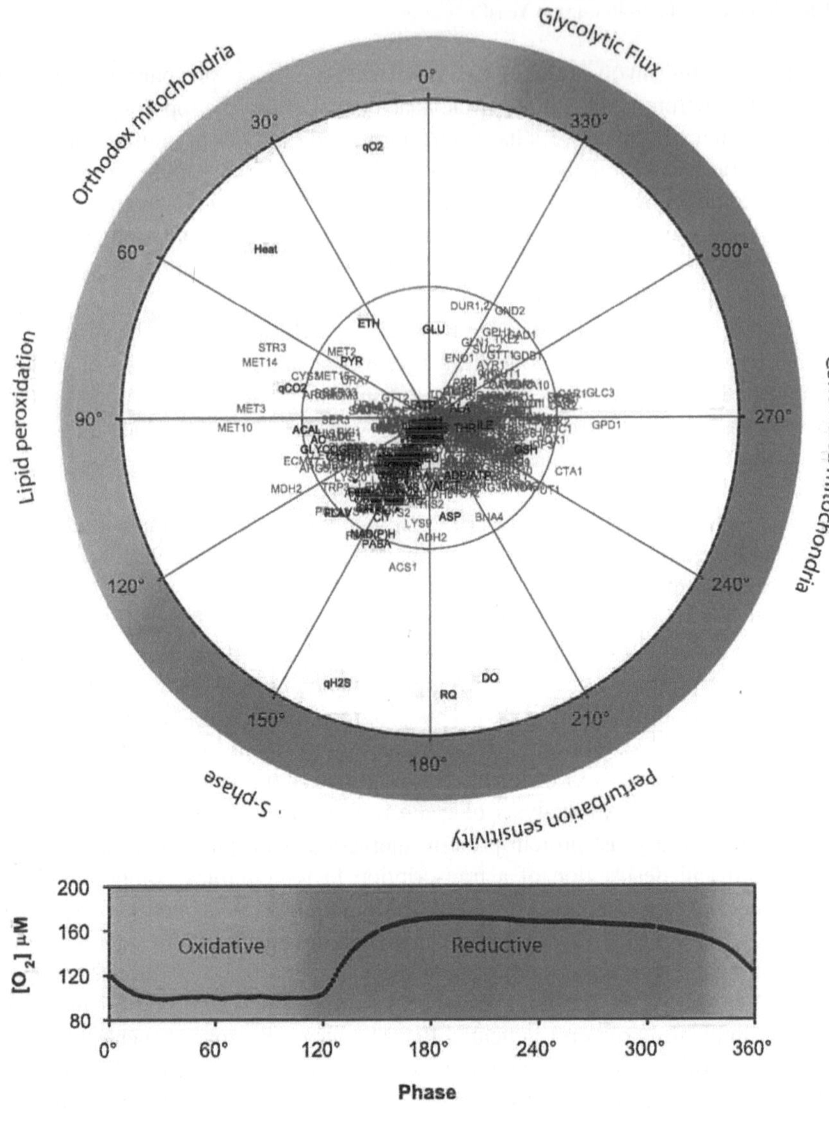

Fig. 1.10 A map of the Ultradian Clock ($\tau \simeq 40$ min) in *Saccharomyces cerevisiae*. Phase reconstruction of the signal-to-noise ratio (S/N) was measured for variables during the respiratory oscillation in continuously growing yeast. The polar plot was constructed from the fast Fourier transform analyses on numerous datasets. The red text represents the online parameters where the S/N had to be divided by 10. The blue text represents metabolites measured in low throughput enzymatic or HPLC methods. The black text shows data from high throughput GC-MS measurements. The grey text represents transcripts involved in metabolism measured using microarray studies. The timing of transcription constructed from phase angles and oscillation strength values. The blue region indicates the oxidative phase and the red region represents reductive phase. Physiological markers are indicated in text centred on the peak of the phenotype. For example, perturbation sensitivity refers to the sensitivity of this region to redox altering compounds such as ROS and glutathione, as well as other chemical agents. The lower plot represents a phase normalised cycle of dissolved oxygen and is meant to guide the reader. The transcript names are common names found in the yeast genome database and the metabolite names are from a model of yeast metabolism (Data from Murray et al., 2007)

assimilation (Fig. 1.11) mitochondrial assembly, respiration, carbohydrate catabolism and the stress response (Fig. 1.11) and these processes are shown to occur 14–18 min later. It was clear from the annotation and statistical analysis of network structure that this group of transcriptional regulatory proteins controlling the expression of these processes are the most highly interconnected nodes in the yeast network. These have complex regulatory patterns and are key components of the differentiation responses in yeast (for pseudohyphae and spore formation). However, the functions of most of the target transcripts, which show a strong oscillatory pattern and peak in the reductive phase, remain largely an enigma. This

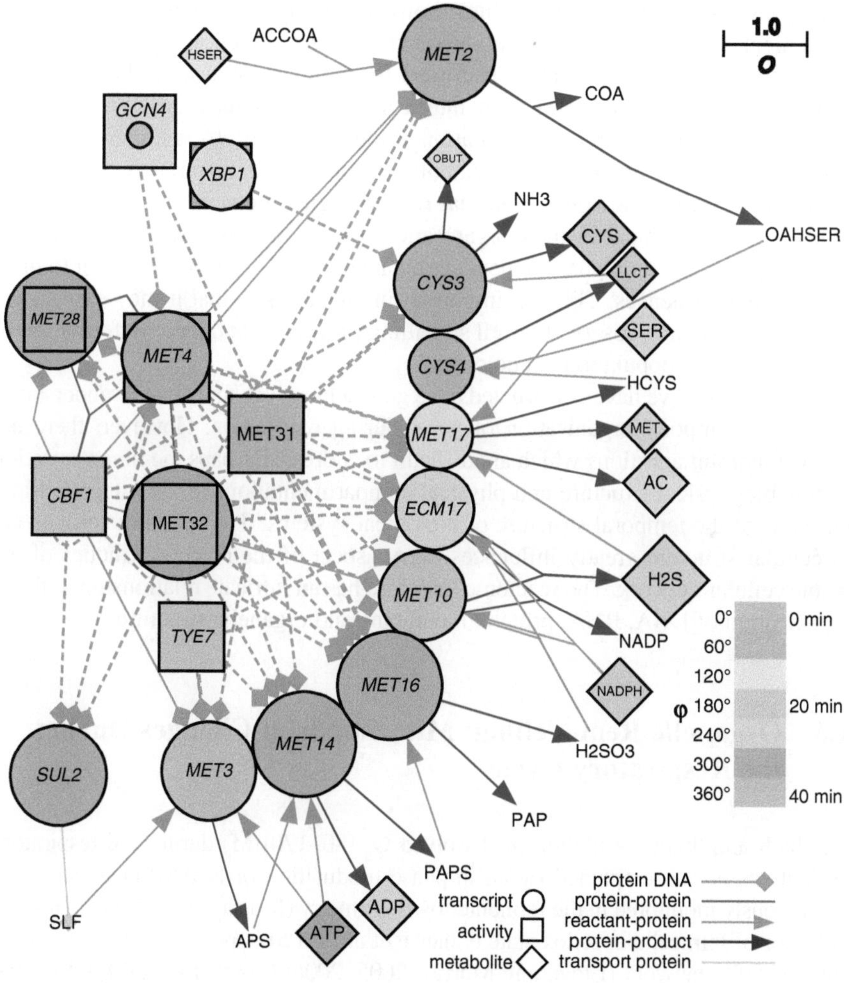

Fig. 1.11 Network derived for sulphur assimilation from top oscillating transcripts and metabolites. The figure key provides a guide to the network SLF, sulphate; LLCT, cystathione; HSER, homoserine; OASER, o-acetylhomoserine; OBUT, 2 oxobutanoate; AC, acetate; ACCOA, acetyl-coenzyme A; COA, coenzyme A (For further details see Murray et al., 2007)

makes it difficult to analyse the transcriptional regulatory network for temporal programs. The lack of annotation for many of the reductive phase transcripts may be due in part to the focus of research activity the yeast community on glucose repressed growth, where differentiation programs and respiration are repressed (Peña-Castillo and Hughes, 2007).

The respiratory oscillation percolates through almost every facet of the biochemistry and cell biology of cellular processes, and although transcriptional feed-forward is important for the regulation and long term stability of the oscillation, it does not act in isolation. Current dogma dictates that there is a distinctive hierarchy in cellular processes where DNA forms RNA to form proteins that biochemically alter other proteins or produce metabolites; however biological networks cannot be so conveniently modularised. As mentioned in the previous section, metabolic regulation of the respiratory oscillation is critical for the fast response of the oscillation to perturbation, and for intra- and inter-cellular communication. We have also shown that post-transcriptional regulation of the GCN4 mRNA is important for the correct timing of amino acid metabolism during the respiratory oscillation. Therefore we drew the conclusion that real biological networks are heterarchical rather than hierarchical in their organisation. Thus a highly interconnected reactome responds in redox phase dependent manner, which can robustly adapt to most perturbative influences. This has important implications in that it may explain the context-dependency observed in cell signalling processes, and thereby formation of independent sub-populations in colonies.

The networks we have constructed have gone a long way towards the understanding of the temporal organisation of yeast during respiration. However, there are many important questions which are difficult to address. Perhaps the major question is how biophysical structure and physical compartmentation can be integrated into our view of the temporal structure of biological systems. The spatial heterogeneity of cellular structure greatly influences every aspect of the reaction kinetics of the whole cellular network. The reactome therefore has an intricate relationship with the organization of DNA, RNA, protein, membrane and organelle structure.

1.6 Organelle Remodelling: Mitochondrial Changes During the Respiratory Cycle

The high amplitude oscillation of dissolved O_2 (80–170 μM) during the respiratory oscillations are accompanied by an in-phase modulation of NAD(P)H directly and continuously monitored in the fermenter by fluorimetry (Murray et al., 1998). This key indicator of intracellular redox state (Chance et al., 2005) is pivotal at the core of the entire cellular network (Lloyd and Murray, 2005, 2006; Lloyd et al., 2003). Electron micrographs of thin sections of yeasts rapidly fixed at different stages of the 40 min cycle showed marked ultra-structural changes in the mitochondria (Lloyd et al., 2002a, b). The extremes of conformational state correspond to those originally described for liver mitochondria (Hackenbrock, 1968) as "orthodox" or "condensed". In the former, a relatively large matrix volume with the inner membrane closely apposed to the outer membrane closely opposed to the outer membrane: this state was identified at high levels of dissolved O_2 (when respiration rate was low). In the condensed form, the cristae became more clearly defined as the inter-membrane compartment was larger this corresponds to the energized state (Chance and Williams, 1956). It has been understood for many years that massive and rapid changes in ion concentrations between the two mitochondrial compartments and the cytosol drive these changes, and that they are perturbed by protonophores and ionophores that uncouple mitochondrial energy conservation from electron transport by collapse of inner membrane electrochemical membrane potential (Hackenbrock, 1968; Mitchell and Moyle, 1969). Further evidence for changing mitochondrial function was provided by a study of the cytochromes of the respiratory chain during the 40 min ultradian clock cycle. Determination of total mitochondrial content of cytochromes b, c_1, c and aa_3 as measured in difference spectra at 77 K showed that any changes were below the level of detectability. However, the physiological redox states of cytochromes c and aa_3 *in vivo* indicated that high respiration was associated with elevated reduction of these two redox components. Effects of two protonophores (*m*-chlorocarbonylcyanide phenylhydrazone, CCCP, and 5-chloro-*t*-butyl-2^1-chloro-4^1-nitrosalicylanilide, S-13) (Fig. 1.12) were dramatic and similar. The well-established acceleration of respiratory rates on addition of uncouplers initially and rapidly drives down the dissolved O_2 in the fermenter. At sub-optimal concentrations, the subsequent cycle is prolonged and more than five cycles were required for recovery to the normal cycle time. At higher concentrations (Fig. 1.12) both uncouplers produced very interesting effects that provide new insights (Lloyd, 2003). For instance, at 10 μM FCCP the uncoupling effect is more evident and the dissolved O_2 remains low for more than 5 h, during which no oscillation was observed. Recovery to normal amplitudes required more than 20 h, although the respiratory oscillations were restored before this, albeit with greatly diminished amplitude. Another very significant observation was that uncoupler treatment gives a lasting complex waveform with the cell cycle-dependent oscillation modulating the amplitude of the 40 min clock so as to produce an 8 h envelope. Thus, interference with mitochondrial energy generation can induce an alignment of cell division cycle controls with ultradian clock control. Further addition of uncoupler apparently stops the clock (or desynchronizes the organisms) and prolongs the cell division cycle time: which of these two alternative explanations explains those data requires further research.

In terms of the kinetics of mitochondrial processes, it is quite remarkable that these organelles characteristically associated with an extremely dynamic potentiality (as determined by their rapid metabolic responses *in vivo* after extraction from organisms should show such slowly changing characteristics *in vitro*). *In vitro* they behave quite differently and their changing properties and functions during the 40 min clock cycle are evidently constrained (Lloyd, 2003). We have suggested that inside cells or organisms the observation that mitochondria do not display rapid kinetics suggest that "they dance to a tune that is played by a piper performing

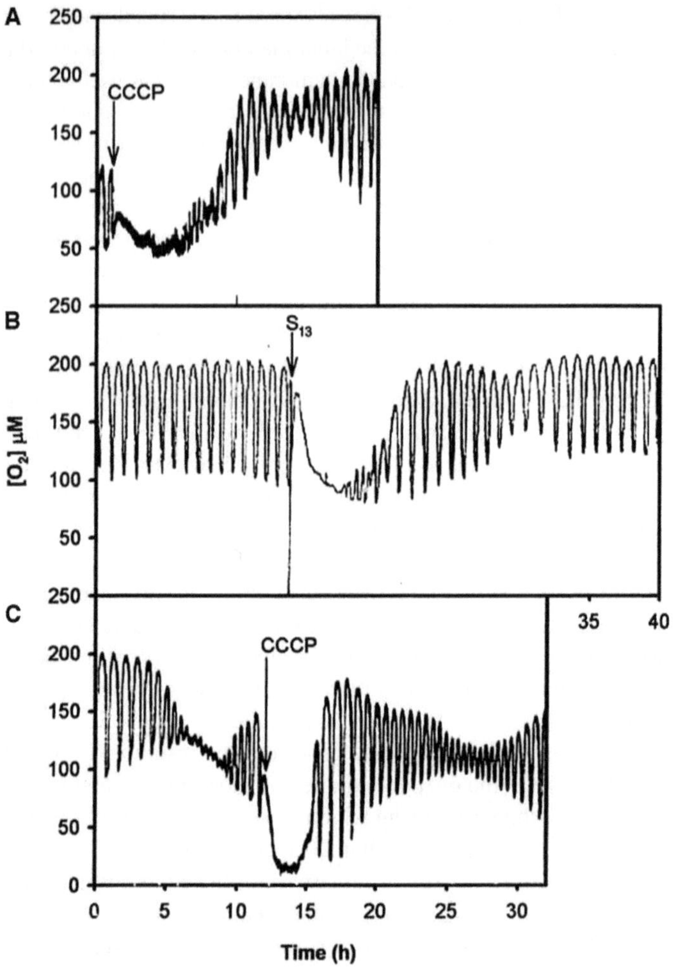

Fig. 1.12 Effects of protonophore additions to continuous cultures of *Saccharomyces cerevisiae* (Lloyd, 2003)

elsewhere in the cell". *In situ* the mitochondrion is enslaved to the slower beat of the nucleo-cytosolic system. Further testimony to this is the temporal compartmentation between the mitochondrial and nuclear-cytostolic systems. It has been suggested (Searcy et al., 1981; Searcy, 2003; Lloyd, 2006) that the earliest eukaryote consisted of an Archeal host capable of producing H_2S from environmental sulphate and a proteobacterial H_2S oxidizing endosymbiont that had become engulfed by a process of phagocytosis (Fig. 1.13). An echo of those primeval times exists even today, and has been revealed not only in the organized orchestration of phase-separated H_2S cycling in the yeast organism but also in ciliated protozoa (Searcy and Peterson, 2004).

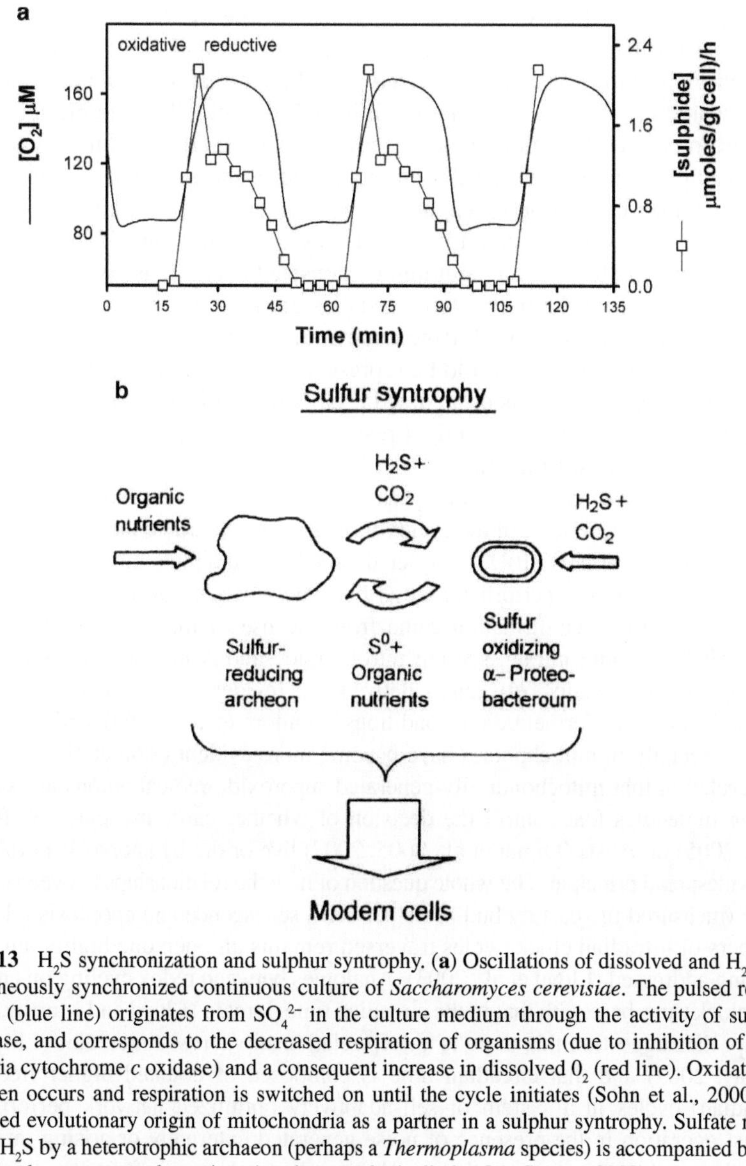

Fig. 1.13 H_2S synchronization and sulphur syntrophy. (**a**) Oscillations of dissolved and H_2S in a spontaneously synchronized continuous culture of *Saccharomyces cerevisiae*. The pulsed release of H_2S (blue line) originates from SO_4^{2-} in the culture medium through the activity of sulphite reductase, and corresponds to the decreased respiration of organisms (due to inhibition of mitochondria cytochrome *c* oxidase) and a consequent increase in dissolved 0_2 (red line). Oxidation of H_2S then occurs and respiration is switched on until the cycle initiates (Sohn et al., 2000). (**b**) Proposed evolutionary origin of mitochondria as a partner in a sulphur syntrophy. Sulfate reduction to H_2S by a heterotrophic archaeon (perhaps a *Thermoplasma* species) is accompanied by H_2S oxidation by an *a*-proteobacterium (or photosynthetically by O_2), Searcy (2003)

1.7 The Redox Core of the Oscillator

That redox cycling underlies the progression of growth was originally proposed by Rapkine (1931), who noticed that in synchronously-dividing sea urchin eggs, total acid-soluble thiol levels as measured by a colorimetric nitroprusside-based reaction varied cyclically with the interdivision time. Mano (1970) went on to show that fertilization initiates a series of cycles of rapid cell division in these eggs, virtually in the absence of net cell growth, and that a soluble cytosolic factor is characterized by a changing thiol/disulphide ratio. The soluble factor was thought to be the glutathione redox couple (GSH/GSSG) which interacts with a thiol-disulphide interchanging enzyme localized in the ribosomes (Mano, 1977) and which may control protein synthesis activity by periodic activation of membrane-bound ribosomes. In retrospect it was especially interesting that these "mitotic" cycles showed temperature-compensated characteristics over the range 15–25°C (Mano, 1975). In theoretical studies Sel'kov (1970) and Gilbert (1974) proposed redox mechanisms for the control of the cell division cycle such that any stage of the cell cycle would be represented by a unique combination of the oxidized and reduced forms of a cofactor. Gilbert (1974) went on to suggest that perturbation of the limit cycle might result in non-cycle arrest in the G_0 state to be followed by transformation.

In the continuous culture system, yeast too shows redox state changes (Fig. 1.14). As well as the direct effects of the cycles or respiration, central to the oscillatory process is the cycling of NAD(P)H (Lloyd et al., 2002a, b) and glutathione (Murray et al., 1999). As well as the perturbants described above when considering the sulphur metabolism, further confirmation came from the use of the NO^+ donors (Murray et al., 1998), sodium nitrate, sodium nitroprusside and *N*-nitrosoglutathione. More recently the relationships between oxidative stress (evident to some extent during our standard operating fermentation conditions, Ohmori et al., 1999) and nitrosative stress especially in mitochondria have become more evident (Aon et al., 2006), and the revelation that mitochondrially-generated superoxide radical anions are key signalling molecules that control the decision of whether cardiomyocytes (O'Rourke et al., 2005) or yeasts (Lemar et al., 2005, 2007) live or die by apoptosis is validated as a widespread principle. The whole question of how the relationships between cellular aging (measured in yeast by bud-scars per cell), senescence and apoptosis related to numbers of ultradian clock cycles traversed remains an open one highly amenable to being addressed (Lloyd et al., 2003). Analogies between redox cycling on an ultradian clock time-base with that in the circadian time domain (Lloyd and Murray, 2005) prompt the suggestion that there is a deep relationship across timescales (Lloyd and Murray, 2007) and that circadian time is composed of counted higher frequency (ultradian) cycles in a system of self-adaptively optimized network performance. Robust operation in the presence of noise generated externally or arising internally from the stochastic processes that involve small numbers of effector molecules is more easily modelled in control circuits that are not implicitly 24h in duration (Paetkau et al., 2006). These ideas are especially attractive with the growing realization that simple auto-dynamics of canonical circadian protein-gene product feedback circuits cannot explain the growing evidence for a whole-systems involvement in the generation of circadian rhythmicity in systems from cyanobacteria to humans (Lakin-Thomas, 2006).

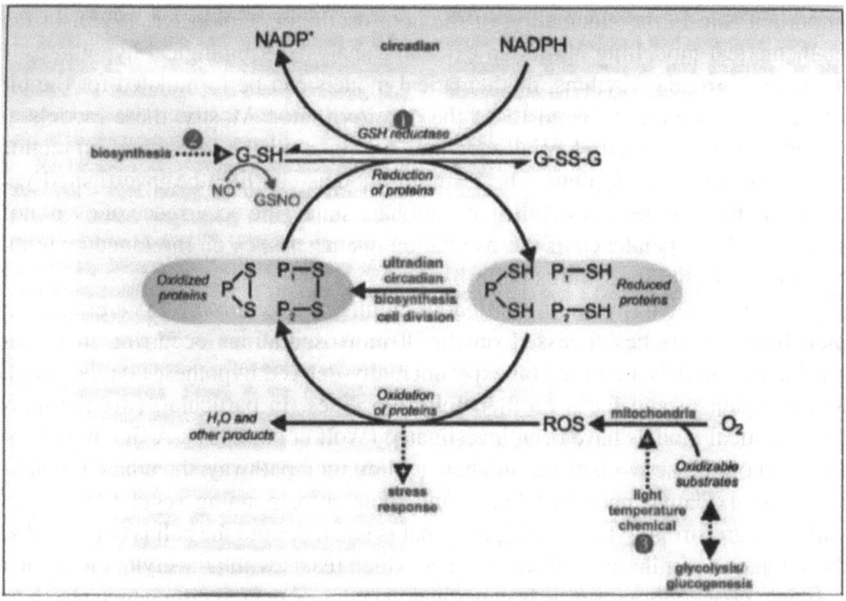

Fig. 1.14 The redox core of the 40 min ultradian clock. Redox cycling of intracellular thiols at the core of rhythmicity. Generation of rhythms entails the cycling of proteins between their oxidized and reduced states. Oxidation is by reactive oxygen species (ROS), produced by respiratory processes especially electron transport in mitochondria. Stress responses can ensue. These processes can be enhanced by illumination, temperature changes or chemical perturbation. Oxidation of some small proteins (e.g. thioredoxin or glutaredoxin) occurs during biosynthetic reactions, especially DNA synthesis. Ultimately the major species involved in reducing these proteins is reduced glutathione (GSH), which itself becomes oxidised. G-S-S-G is in turn reduced by NAD(P)H, the other major redox buffer in the organism. Numerous processes (magenta) have proven to be oscillatory and we propose that ensembles of oscillators are coupled via this primordial mechanism. Perturbation analysis of the yeast ultradian system utilising NO^+ donors, 5-nitro-2-furaldehyde (1), D,L-butathionine (S,R)-sulphoximine (2), or electron transport chain uncoupling agents (e.g. carbonyl cyanide m-chlorophenylhydrazone) (3), confirms the central role of this redox system; the number on the figure represents the site of perturbation (Lloyd and Murray, 2007)

1.8 Mathematical Modelling of Ultradian Rhythmicity in Continuous Cell Culture of *Saccharomyces cerevisiae*

A multitude of processes is involved in ultradian rhythmicity. A crucial, but still open question is that of the molecular mechanism generating the rhythm. Which processes and regulations are critical for the oscillator? Which of them determine the oscillatory characteristics such as the period and phase shifts? Related questions have been addressed for other cellular rhythms, e.g. circadian rhythms, Calcium oscillations or the cell cycle. In all these cases, mathematical modelling played an important role in the elucidation of the core oscillator (Goldbeter, 1996). In fact, mathematical modelling allows an extensive analysis of biological hypotheses. In the case of cellular rhythms, mathematical models can be formulated for the biochemical network considered to be the core oscillator. Mostly, these models are systems of ordinary differential equations and take all the important interactions and regulations into account. The resulting models can then be investigated with respect to the number and stability of stationary states, the occurrence of oscillatory solutions, the dependence of the oscillation characteristics on the kinetic parameters, or the sensitivity towards perturbations.

For the ultradian rhythm in continuous culture of *Saccharomyces cerevisiae* first modelling approaches focussed on the 40 min oscillations occurring in ethanol media. The models are based on experimentally-derived hypotheses on the oscillatory mechanism and focus on metabolic processes. In particular, two alternative mathematical models have been investigated (Wolf et al., 2001). A first hypothesis focussed on the network of the sulphate assimilation pathway, the ethanol metabolism, the oxidative phosphorylation and their corresponding regulations. These pathways are strongly linked. Sulphate that is taken up into the cell is converted via the sulphate assimilation pathway to H_2S, which reacts with *O*-acetylhomoserine to cysteine. *O*-acetylhomoserine is generated via ethanol degradation which also leads to the production of acetyl-CoA, a substrate of the citrate acid cycle and therefore respiration. In the context of oscillations two regulations are critical within this network, these are: (i) the inhibition of the sulphate uptake by cysteine, and (ii) the inhibitory feedback from hydrogen sulphide to oxidative phosphorylation. The model based on these processes and regulations indeed shows sustained oscillatory dynamics. Moreover, it explains most phase relations observed in experiments, e.g. those of ethanol and NAD(P)H to DOT. However, the out-of-phase oscillation of ATP and DOT is in contrast to the experimental data.

An alternative model focused on central regulatory loops in the carbon metabolism as an underlying mechanism for the rhythmic dynamics. As an example, the indirect negative feedback of ATP on the citrate circle was analysed. This feedback comes about because the citrate acid cycle and several biosynthetic pathways share the substrate acetyl-CoA. Whereas biosynthetic processes consume ATP, the citric acid cycle together with the respiratory chain leads to ATP-production. Simulations show that the resulting indirect regulatory loop can give rise to limit cycle oscillations. But many of the experimentally observed phase relations, e.g. those of ATP or NADPH with DOT, respectively, are not reproduced by this mechanism.

For both models it was shown that the phase relations between oscillatory components is specific for the structure of the underlying network. Variations in the kinetic parameters are not sufficient to shift these phase relations significantly. Taken together, mathematical modelling indicated that the set of pathways and regulations of the core oscillator was not completely determined and has to be further investigated. In fact, this conclusion was later confirmed by experiments showing that a larger network is involved in the generation of the oscillatory dynamics. These experiments have shown that the underlying mechanism is far more complex and includes a large-scale yeast interaction network with different layers of regulation, e.g. translational, proteomic and metabolic regulatory mechanisms. Therefore, a revision of the processes and regulations underlying the core oscillator is necessary. This might be hampered by the fact that genetic and metabolic processes are often considered as separated layers and cross-regulations are less understood then regulations within the individual layers. However, the investigation of this complex dynamical behaviour will promote our understanding of the architecture of cellular networks.

Yet, another layer of complexity arises from the fact that all observations of the ultradian rhythm in continuous culture of *Saccharomyces cerevisiae* are made on the population level in a suspension of millions of cells. In order to observe oscillatory behaviour under these conditions the individual cells must be synchronised. Acetaldehyde and hydrogen sulphide are discussed as diffusing agents that mediate cell-cell communication. Population modelling for cells coupled via H_2S provided insights for conditions under which yeast cells can synchronise their dynamics (Henson, 2004). The example of glycolytic oscillations, where the oscillatory mechanism was discovered much earlier, shows that a quantitative understanding of synchronisation is not always straightforward and can reveal additional aspects of regulation (Wolf and Heinrich, 2000; Hynne et al., 2001; Dano et al., 2007).

1.9 Non-linear Dynamics of the Ultradian Clock

Chaotic operation has been experimentally demonstrated in continuous yeast cultures under several conditions:

1. Under "permittistatical" control (i.e. using output from an AC impedance measuring device to control the rate of growth medium supply (Davey et al., 1996).
2. Stepwise decreases in culture pH reveals a progressive increase in complexity of trajectories and the eventual uncovering of a strange attractor (Murray and Lloyd, 2007) with increasing cellular stress.
3. A long term (3 months) experiment yielding almost 40,000 data points using a direct immersion mass spectrometric probe (dissolved O_2, CO_2, H_2S; Fig. 1.15) indicated a low-dimensional chaotic attractor (Roussel and Lloyd, 2007).
4. After addition of the type A monoamine oxidase inhibitor, phenelzine, to cultures, period 2 doubling occurs (Salgado et al., 2002; Li and Klevecz, 2006).

These results support that the proposal that a controlled chaotic attractor enables the coordination of systems operation so as to produce tuneable multioscillatory outputs on many time scales (AL Lloyd and D Lloyd, 1993) with all the biological advantages that this may confer (AL Lloyd and D Lloyd, 1995). Furthermore the interdependence of operation across multiple time scales as a heterarchical whole (Yates, 1992) is evidenced by the recent discovery that the dynamics of both yeast and cardiomyocyte systems show self-similar (fractal) organisation in time (MA Aon, unpublished).

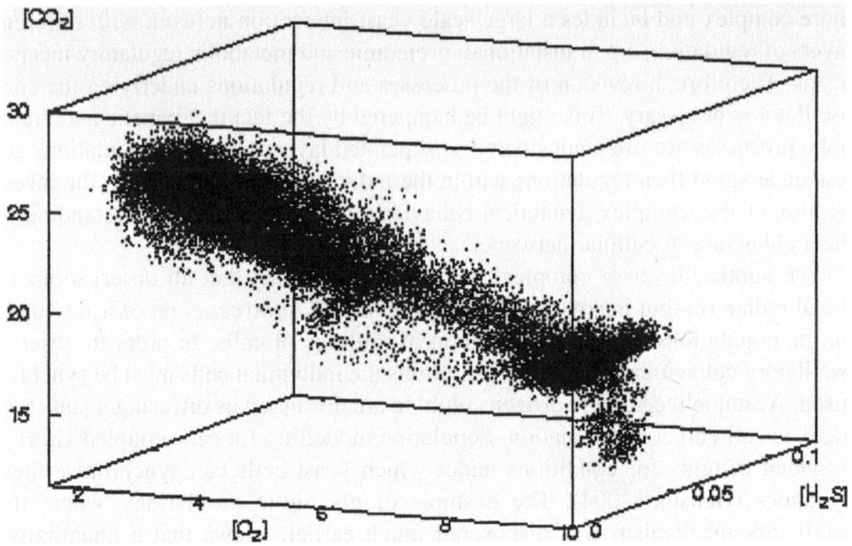

Fig. 1.15 A chaotic attractor operates in yeast continuous cultures. The plot was obtained by mass spectrometric monitoring of dissolved O_2, CO_2 and H_2S over a period of 3 months (more than 36,000 points) (Data from Roussel and Lloyd, 2007)

References

Aldrich K, & Bernstein D (1987) The effect of time of day on hypnotizability. *International Journal of Clinical & Experimental Hypnosis,* 35: 141–145

Aon MA, Cortassa S, O'Rourke B (2006) The fundamental organization of cardiac mitochondria as a network of coupled oscillators. Biophys J 91: 4317–4327

Aserinsky E, Kleitman N (1953) Regularly occurring periods of eye motility, and concomitant phenomena during sleep. Science 118: 273–274

Balch W, Morimoto R, Dillin A, Kelly J (2008) Adapting proteostasis for disease intervention. *Science,* 319: 916–919.

Bolen JL, Duran O, Klevecz RR (1993) Amplification and damping of deterministic noise in coupled cellular arrays. Physica D 67: 245–248

Chan SY, Appling DR (2003) Regulation of S-adenosyl methionine levels in *Saccharomyces cerevisiae*. J Biol Chem 278: 43051–43059

Chance B, Williams RG (1956) The respiratory chain and oxidative phosphorylation. Adv Enzymol 17: 65–134

Chance B, Nioka S, Warren W, Yurtsever G (2005) Mitochondrial NADH as the bell-wether of tissue O_2 delivery. Adv Exp Med Biol 566: 231–242

Chisholm SW, Costello JC (1980) Influence of environmental factors and population composition on the finding of cell division in *Thalassiosira fluviatilis* (Bacillariophyceae) grown on light/dark cycles. J Phycol 16: 375–383

Dano S, Madsen MF, Sorensen PG (2007) Quantitative characterization of cell synchronization in yeast. Proc Natl Acad Sci USA 104: 12732–12736

Davey HM, Davey CL, Woodward AM, Edmonds AN, Lee AW, Kell DB (1996) Oscillatory, stochastic and chaotic growth rate fluctuations in permittistatically controlled yeast cultures. Biosystems 39: 43–61

De Mairan J (1729) On diurnal leaf movements. Acad Roy Soc Paris p. 35
Edwards C, Statham M, Lloyd D (1975) The preparation of large-scale synchronous cultures of the trypanosomatid *Crithidia fasciculata* by cell-size selection: changes in respiration and adenylate charge through the cell cycle. J Gen Microbiol 88: 141–152
Edwards R, Gibson R, Illner R, Paetkau V (2007) A stochastic model for circadian rhythms from coupled ultradian oscillators. Theor Biol Med Model 4: 1–12
Edwards SW, Lloyd D (1978) Oscillations of respiration and adenine nucleotides in synchronous cultures of *Acanthamoeba castellanii*: mitochondrial respiratory control in vivo. J Gen Microbiol 108: 197–204
Edwards SW, Lloyd D (1980) Oscillations in protein and RNA during synchronous growth of *Acanthamoeba castellanii*: evidence for periodic turnover of marcromolecules during the cell cycle. FEBS Lett 109: 21–26
Gavin AC, Aloy P, Grandi P, Krause R, Boesche M, Marzioch M, Rau C, Jensen LJ, Bastuck S, Dümpelfeld B, Edelmann A, Heurtier MA, Hoffman V, Hoefert C, Klein K, Hudak M, Michon AM, Schelder M, Schirle M, Remor M, Rudi T, Hooper S, Bauer A, Bouwmeester T, Casari G, Drewes G, Neubauer G, Rick JM, Kuster B, Bork P, Russell RB, Superti-Furga G (2006) Proteome survey reveals modularity of the yeast cell machinery. Nature 440: 631–636
Gilbert DA (1974) The nature of the cell cycle and the control of cell proliferation. Biosystems 5: 197–204
Goldbeter A (1996) Biochemical Oscillations and Cellular Rhythms: The Molecular Bases of Periodic and Chaotic Behaviour. Cambridge University Press, Cambridge
Hackenbrock CR (1968) Ultrastructural bases for metabolically-linked mechanical activities in mitochondria. II Electron transport-linked ultrastructure transformations in mitochondria. J Cell Biol 37: 345–369
Hartwell LH, Culotti J, Pringle JR, Reid BJ (1974) Genetic control of the cell division cycle in yeast. Science 183: 46–51
Henson MA (2004) Modelling the synchronization of yeast respiratory oscillation. J Theor Biol 231: 443–458
Homma K, Hastings JW (1988) Cell cycle synchronization of *Gonyaulax polyedra* by filtration: quantized generation times. J Biol Rhythms 3: 49–81
Hull C (1933/1968) *Hypnosis and suggestibility*. New York: Appleton-Century-Crofts.
Hynne F, Dano S, Sorensen PG (2001) Full-scale model of glycolysis in *Saccharomyces cerevisiae*. Biophys Chem 94: 121–163
Ito K, Chiba T, Ozawa R, Yoshida M, Hattori M, Sakaki Y (2001) A comprehensive two-hybrid analysis to explore the yeast protein interactome. Proc Natl Acad Sci USA 98: 4569–4574
Kader J, Lloyd D (1979) Respiratory oscillations and heat evolution in synchronous cultures of *Candida utilis*. J Gen Microbiol 114: 455–461
Keulers M, Kuriyama H (1998) Acetaldehyde influence on metabolic oscillation of *Saccharomyces cerevisiae*. In: Information Processing in Cells and Tissues. Plenum, New York
Keulers M, Satroutdinov AD, Suzuki T, Kuriyama H (1996) Synchronization affector of autonomous short-period sustained oscillation of *Saccharomyces cerevisiae*. Yeast 12: 673–682
Klevecz RR (1976) Quantized generation times in mammalian cells as an expression of the cellular clock. Proc Natl Acad Sci USA 73: 4012–4016
Klevecz RR, King GA (1982) Temperature compensation in the mammalian cell cycle. Exp Cell Res 140: 307–313
Klevecz RR, Li CA (2007) Evolution of the clock from yeast to man along a period doubling path. Cold Spring Harbour Symposium
Klevecz RR, Ruddle FH (1968) Cyclic changes in enzyme activity in synchronized mammalian cell cultures. Sciences 159: 634–636
Klevecz RR, Stubblefield E (1967) RNA synthesis in relation to DNA replication in synchronized Chinese hamster cell cultures. J Exp Zool 165: 259–268
Klevecz RR, Kros J, Gross SD (1978) Phase response versus positive and negative division delay in animal cells. Exp Cell Res 116: 285–290

Klevecz RR, Kauffman SA, Shymko RM (1984) Cellular clocks and oscillators. Int Rev Cytol 86: 97–128

Klevecz RR, Bolen J, Forrest G, Murray DB (2004) A genomewide oscillation in transcription gates DNA replication and cell cycle. Proc Natl Acad Sci USA 101: 1200–1205

Krogan NJ, Cagney G, Yu H, Zhong G, Guo X, Ignatchenko A, Li J, Pu S, Datta N, Tikuisis AP, Punna T, Peregrín-Alvarez JM, Shales M, Zhang X, Davey M, Robinson MD, Paccanaro A, Bray JE, Sheung A, Beattie B, Richards DP, Canadien V, Lalev A, Mena F, Wong P, Starostine A, Canete MM, Vlasblom J, Wu S, Orsi C, Collins SR, Chandran S, Haw R, Rilstone JJ, Gandi K, Thompson NJ, Musso G, St Onge P, Ghanny S, Lam MH, Butland G, Altaf-Ul AM, Kanaya S, Shilatifard A, O'Shea E, Weissman JS, Ingles CJ, Hughes TR, Parkinson J, Gerstein M, Wodak SJ, Emili A, Greenblatt JF (2006) Global landscape of protein complexes in the yeast *Saccharomyces cerevisiae*. Nature 440: 637–643

Kwak W-J, Kwon G-S, Jin I, Kuriyama H, Sohn H-Y (2003) Involvement of oxidative stress in the regulation of H_2S production during oscillation of *Saccharomyces cerevisiae*. FEMS Microbiol Lett 219: 99–104

Lakin-Thomas PL (2006) Circadian clock genes frequency and white collar-1 are not essential for entrainment to temperature cycles in *Neurospora crassa*. Proc Natl Acad Sci USA 103: 4469–4474

Lamb J, Crawford E, Peck D, Modell J, Blat I, Wrobel M, Lerner J, Brunet J, Subramanian A, Ross K, Reich M, Hieronymus H, Wei G, Armstrong S, Haggarty S, Clemons P, Wei R, Carr S, Lander E, Golub T (2006) The Connectivity Map: Using gene-expression signatures to connect small molecules, genes, and disease. *Science*, 313: 1929–1935

Lemar KM, Passa O, Aon MA, Cortassa S, Müller CT, Plummer S, O'Rourke B, Lloyd D (2005) Allyl alcohol and garlic (*Alium sativum*) extract produce oxidative stress in *Candida albicans*. Microbiology 151: 3257–3265

Lemar KM, Aon MA, Cortassa S, O'Rourke B, Plummer S, Müller CT, Lloyd D (2007) Diallyl disulphide depletes glutathione in *Candida albicans*: oxidative stress mediated cell death studied by two-photon microscopy. Yeast 24: 695–706

Li CM, Klevecz RR (2006) A rapid genome-scale response of the transcriptional oscillator to perturbation reveals a period-doubling path to phenotypic change. Proc Natl Acad Sci USA 103: 16254–16259

Lin L, Chen H, Kuang D, Wang D, Tsien J (2007) Neural encoding of the concept of nest in the mouse brain. *Proceedings of the National Academy of Sciences USA,* 104: 6066–6071.

Lisman J, Morris G (2001) Why is the cortex a slow learner? *Nature*, 411: 248–249.

Lloyd D, Rossi E (Eds.) (1992). *Ultradian Rhythms in Life Processes: An Inquiry into Fundamental Principles of Chronobiology and Psychobiology.* New York: Springer-Verlag

Lloyd AL, Lloyd D (1993) Hypothesis: the central oscillator of the circadian clock is a controlled chaotic attractor. Biosystems 29: 77–85

Lloyd AL, Lloyd D (1995) Chaos: its detection and significance in biology. Biol Rhythm Res 26: 233–252

Lloyd D (2003) Effects of uncoupling mitochondrial energy conservation on the ultradian clock-driven oscillations in *Saccharomyces cerevisiae* continuous culture. Mitochondrion 3: 139–146

Lloyd D (2006) Hydrogen sulphide: clandestine microbial messenger? Trends Microbiol 14: 456–462

Lloyd D, Murray DB (2005) Ultradian metronome: timekeeper for orchestration of cellular coherence. Trends Biochem Sci 30: 373–377

Lloyd D, Murray DB (2006) The temporal architecture of eukaryotic growth. FEBS Lett 580: 2830–2835

Lloyd D, Murray DB (2007) Redox rhythmicity: clocks at the core of temporal coherences. BioEssays 29: 465–474

Lloyd D, Phillips CA, Statham M (1978) Oscillations of respiration adenine nucleotides and heat evolution in synchrous cultures of *Tetrahymena pyriformis* ST prepared by continuous flow size-selection. J Gen Microbiol 106: 19–26

Lloyd D, Edwards SW, Fry JC (1982) Temperature-compensated oscillations in respiration and cellular protein content in synchronous cultures of *Acanthamoeba castellanii*. Proc Natl Acad Sci USA 79: 3785–3788

Lloyd D, Poole RK, Edwards SW (1984) The Cell Division Cycle: Temporal Organisation of Growth and Reproduction. Academic, London

Lloyd D, Salgado LE, Turner MP, Suller MT, Murray D (2002a) Cycles of mitochondrial energization driven by the ultradian clock in a continuous culture of *Saccharomyces cerevisiae*. Microbiology 148: 3715–3724

Lloyd D, Salgado LE, Turner MP, Suller MT, Murray DB (2002b) Respiratory oscillations in yeast: clock driven mitochondrial cycles of energization. FEBS Lett 519: 41–44

Lloyd D, Lemar KM, Salgado LEJ, Gould TM, Murray DB (2003) Respiratory oscillations in yeast: mitochondrial reactive oxygen species, apoptosis and time: a hypothesis. FEMS Yeast Res 3: 333–339

MacIsaac KD, Wang T, Gordon DB, Gifford DK, Stormo G, Fraenkel E (2006) An Improved Map of Conserved Regulatory Sites for *Saccharomyces cerevisiae*. BMC Bioniformatics 7: 113

Mano Y (1970) Cytoplasmic regulation and cyclic variation in protein synthesis is in the early cleavage stage of the sea urchin embryo. Dev Biol 22: 433–460

Mano Y (1975) Systems constituting the metabolic sequence in the cell cycle. Biosystems 7: 51–65

Mano Y (1977) Interaction between glutathione and the endoplasmic reticulum in cyclic protein synthesis in sea urchin eggs. Dev Biol 61: 273–286

Marques N, Edwards SW, Fry JC, Halberg F, Lloyd D (1987) Temperature-Compensated Variation in Cellular Protein Content of *Acanthamoeba castellanii* Revisited. Adv Chronobiol Part A (J Pauly, LE Scheving, eds.), pp. 131–151, Alan R Liss, New York

Mitchell P, Moyle J (1969) Estimation of membrane potential and pH difference across the cristal membrane of rat liver mitochondria. Eur J Biochem 7: 471–478

Mitchison JM (1971) The Biology of the Cell Cycle. Cambridge University Press, Cambridge

Murray DB, Lloyd D (2007) A tuneable attractor underlies yeast respiratory dynamics. BioSystems 90: 287–294

Murray DB, Engelen FA, Keulers M, Kuriyama H, Lloyd D (1998) NO^+, but not NO, inhibits respiratory oscillations in ethanol-grown chemostat cultures of *Saccharomyces cerevisiae*. FEBS Lett 431: 297–299

Murray DB, Engelen F, Lloyd D, Kuriyama H (1999) Involvement of gluthathione in the regulation of respiratory oscillation during a continuous culture of *Saccharomyces cerevisiae*. Microbiology 145: 2739–2745

Murray DB, Roller S, Kuriyama H, Lloyd D (2001) Clock control of ultradian respiratory oscillations found during yeast continuous culture. J Bacteriol 183: 7253–7259

Murray DB, Klevecz RR, Lloyd D (2003) Generation and maintenance of synchrony in *Saccharomyces cerevisiae* continuous culture. Exp Cell Res 286: 10–15

Murray DB, Beckmann M, Kitano H (2007) Regulation of yeast oscillatory dynamics. Proc Natl Acad Sci USA 104: 2241–2246

Ohmori S, Nawata Y, Kiyono K, Murata H, Tsuboi S, Ikeda M, Agaki R, Morohashi K, Ono B (1999) Cultured under aerobic and anaerobic conditions: air-level oxygen stress. Biochem Biophys Acta 1472: 587–594

O'Rourke B, Cortassa S, Aon MA (2005) Mitochondrial ion channels: gatekeepers of life and death. Physiology (Bethesda) 20: 303–315

Paetkau V, Edwards R, Illner R (2006) A model for generating circadian rhythms by coupling ultradian oscillators. Theor Biol Med Model 3: 12–23

Pavlov I (1927) *Conditioned reflexes: an investigation of the physiological activity of the cerebral cortex*. Anrep, G. (Trans.). Oxford: Oxford University Press.

Peña-Castillo L, Hughes TR (2007) Why Are There Still over 1000 Uncharacterized Yeast Genes? Genetics 176: 7–14

Poole RK, Lloyd D, Kemp RB (1973) Respiratory oscillations and heat in synchronously dividing cultures of the fission yeast *Schizosaccharomyces pombe* 972h. J Gen Microbiol 77: 209–220

Rapkine L (1931) Sur les processes chimiques au course de la division cellulaire. Ann Physiol Physiocochem 7: 382–418

Ribeiro S, Simões C, Nicolelis M (2008) Genes, Sleep and Dreams. In Lloyd, D. and Rossi, E. (2008). *Ultradian Rhythms from Molecules to Mind: a New Vision of Life.* New York: Springer.

Rosbash M, Takahashi J (2002) Circadian rhythms: the cancer connection. *Nature,* 420: 373–374.

Rossi E (2008) The neuroscience of therapeutic hypnosis, psychotherapy, and rehabilitation. In Rossi E, Erickson-Klein R, & Rossi K, (Eds.), *The Complete Works of Milton H. Erickson, M.D. Volume 1: The Nature of Hypnosis.* Phoenix: The MHE Foundation Press.

Rossler OE (1976) An equation for continuous chaos. Phys Lett 35A: 397–39

Roussel MR, Lloyd D (2007) Observation of a chaotic multioscillatory metabolic attractor by real-time monitoring of a yeast continuous culture. FEBS J 274: 1011–1018

Salgado LEJ, Murray DB, Lloyd D (2002) Some anti-depressant agents (Li^+, monoamine oxidase type A inhibitors) perturb the ultradian clock in *Saccharomyces cerevisiae*. Biol Rhythm Res 33: 351–361

Satroutdinov DV, Kuriyama H, Kobayashi H (1992) Oscillatory metabolism of *Saccharomyces cerevisiae* in continuous culture. FEMS Microbiol Lett 98: 261–268

Searcy DG (2003) Metabolic integration during the evolutionary origin of mitochondria. Cell Res 13: 229–238

Searcy DG, Patterson MA (2004) Hydrogen sulphide consumption measured at low steady state concentrations using a sulfidostat. Anal Biochem 324: 269–275

Searcy DG, Stein DB, Searcy KB (1981) A mycoplasma-like archaebacterium possibly related to the nucleus and cytoplasms of eukaryotic cells. Ann NY Acad Sci 361: 312–324

Sel'kov EE (1970) Two alternative self-oscillating stationary states in thiol metabolism-two alternative types of cell division normal and malignant ones. Biophysika 15: 1065–1073

Shymko RM, Klevecz RR, Kauffman SA (1984) The cell cycle as an oscillatory system. In Cell Cycle Clocks (LN Edmunds, Jr, ed.), p. 273, Marcel Dekker, New York

Sohn HY, Murray DB, Kuriyama H (2000) Ultradian oscillation of *Saccharomyces cerevisiae* during aerobic continuous culture: hydrogen sulphide mediates population synchrony. Yeast 16: 1185–1190

Sohn H-Y, Kuriyama H (2001a) Ultradian metabolic oscillation of *Saccharomyces cerevisiae* during aerobic continuous culture: hydrogen sulphide, a population synchronizer, is produced by sulphite reductase. Yeast 18: 125–135

Sohn H-Y, Kuriyama H (2001b) The role of amino acids in the regulation of hydrogen sulfide production during ultradian oscillation of respirationin Saccharomyces cerevisiae. Arch Microbiol 176: 69–78

Sohn HY, Kum EJ, Kwon GS, Jin I, Kuriyama H (2005a) Regulation of branched-chain and sulphur-containing amino acid metabolism by glutathione during ultradian metabolic oscillation of *Saccharomyces cerevisiae*. J Microbiol 43: 375–380

Sohn H-Y, Kum EJ, Kwon GS, Jin I, Adams CA, Kuriyama H (2005b) GLR1 plays an essential role in the homeo dynamics of glutathione and the regulation of the H_2S production during respiratory oscillation of *Saccharomyces cerevisiae*. Biosci Biotechnol Biochem 69: 2450–2454

Taft RJ, Pheasant M, Mattick JS (2007) The relationship between non-protein-coding DNA and eukaryotic complexity. Bioessays 29: 288–299

Wolf J, Heinrich R (2000) The effect of cellular interaction on glycolytic oscillations in yeast. A theoretical investigation. Biochem J 345: 321–333

Wolf J, Sohn HY, Heinrich R, Kuriyama H (2001) Mathematical analysis of a mechanism for autonomous metabolic oscillations in continuous cultures of *Saccharomyces cerevisiae*. FEBS Lett 499: 230–234

Yates FE (1992) Fractal applications in biology: scaling time in biochemical networks. Meth Enzymol 210: 630–675

Chapter 2
ENOX Proteins, Copper Hexahydrate-Based Ultradian Oscillators of the Cells' Biological Clock

D.J. Morré and D.M. Morré

Abstract A homodimeric, growth-related and time-keeping hydroquinone [NAD(P)H] oxidase of the mammalian (and plant) cell surface with a binuclear copper center and protein disulfide-thiol interchange activity has characteristics of an ultradian driver of the biological 24 h circadian clock. A member of the ECTO-NOX or ENOX protein family, the constitutive ENOX, ENOX1 (EntrezGene ID 55068 NCBI Gen Bank accession Number EF432052), exhibits a recurring complex 2 + 3 set of oscillations in secondary structure, enzymatic activity and redox potential with a period length of 24 min (repeats 60 times over 24 h). The period length is temperature independent and entrained by light, melatonin and low frequency EMF. COS cells transformed with ENOX2 (EntrezGene ID 10495) where specific cysteine codons were replaced by alanine codons, ENOX2 oscillations with period lengths of 22, 36 or 42 min yielded circadian periods of 22, 36 or 42 h respectively, based on activity of glyceraldehyde-3-phosphate dehydrogenase. The oscillations require bound copper and are recapitulated in solution by copper salts. The period length of both the ENOX and copper II oscillations in D_2O is increased to 30 min (Organisms grown in deuterium oxide exhibit a 30 h circadian day.). The oscillatory pattern appears to be determined by periodic variations in the ratios of ortho and para nuclear spins of the paired hydrogen or deuterium atoms of the elongated octahedral structure of the protein bound copper II hexahydrate as determined by spectroscopic analyses. That the oscillations result from physical rather than chemical events account, for the first time, for the temperature independence of the period length of clock-related pheonomena.

Keywords Biological clock; Circadian/Ultradian rhythms; Copper II; ECTO-NOX (ENOX); Hydroquinone (NADH) oxidase (CNOX; tNOX); Growth; Low frequency electromagnetic fields; protein disulfide-thiol interchange; Time keeping

---

Department of Medicinal Chemistry and Molecular Pharmacology, Purdue University, 201 S. University Street, West Lafayette, IN 47907, United States

D. Lloyd, E. Rossi (eds.), *Ultradian Rhythms from Molecules to Mind*,

2.1 NOX Discovery

NOX proteins were discovered as the result of efforts in our laboratories to discover one or more cell surface proteins involved in cell enlargement and cancer. The search was for plasma membrane-associated redox enzymes (pyridine nucleotide oxidoreductases) that were inhibited by adriamycin in both plants and animals and stimulated by auxins in plants (Morré et al., 1986; Brightman et al., 1988; Morré, 1998). The results of that search yielded an adriamycin- inhibited, auxin-stimulated NADH oxidase of soybean plasma membranes and an unregulated adriamycin-inhibited NADH oxidase of tumor plasma membranes (ENOX2) designated tNOX (for tumor-associated NADH oxidase). An adriamycin resistant (of mammalian and plant cells) and 2,4-D-insensitive (of plants) counter-part to tNOX (ENOX1) was subsequently identified and designated as the constitutive ECTO-NOX (CNOX) now designated ENOX1.

While the first described and characterized external oxidases of the cell surface (ENOX proteins) were assayed based on an ability to oxidize NADH, a second activity, that of protein disulfide exchange, was subsequently associated with the auxin-stimulated activity of plants (Morré et al., 1995), the adriamycin-inhibited activities of cancer cells and tissues (Morré et al., 1998) and the constitutive CNOX of both plants and animals (Morré et al., 1999a). The protein disulfide thiol interchange activity was the first observed to be periodic (Morré et al., 1995). Extension of these observations revealed that both activities, NADH oxidation and protein disulfide interchange, were catalyzed by a single protein but that two activities alternated to create a complex ultradian rhythm with maxima separated by 24 min (ENOX1) or 22 min (ENOX2) (Morré, 1998).

2.2 ECTO-NOX Proteins Defined

By definition, ECTO-NOX proteins are NAD(P)H oxidases with protein disulfide-thiol interchange activity located at the external cell surface (Fig. 2.1). They lack transmembrane binding motifs (Morré et al., 2001b) and are released from the cell surface of HeLa cells, for example, by treatment with 0.1 M sodium acetate, pH 5 (del Castillo et al., 1998). In vertebrate cells, they are released from cells as is a characteristic of ectoproteins in general, where they are carried by the blood throughout the body. The periodic oscillations resulting from the alternation of the two activities they catalyze further distinguishes ECTO-NOX proteins. They lack flavin-binding motifs and require neither flavin nor ancillary proteins for activity. These latter characteristics serve to clearly differentiate the ECTO-NOX proteins from other NADH or NAD(P)H oxidases including the PHOX-NOX proteins of host defense located at the cytosolic plasma membrane surface (Lambeth et al., 2000).

The ECTO-NOX proteins are included in the HUGO Human Gene Nomenclature Data Base (http://www.gene.ucl.ac.uk/nomenclature/) and are designated as ENOX proteins. CNOX is ENOX1 or ECTO-NOX disulfide-thiol exchange 1 (EntrezGene ID: 55068). The tumor-associated tNOX is designated ENOX2 or ecto-NOX disulfide-thiol exchanger 2 (EntrezGene ID: 10495).

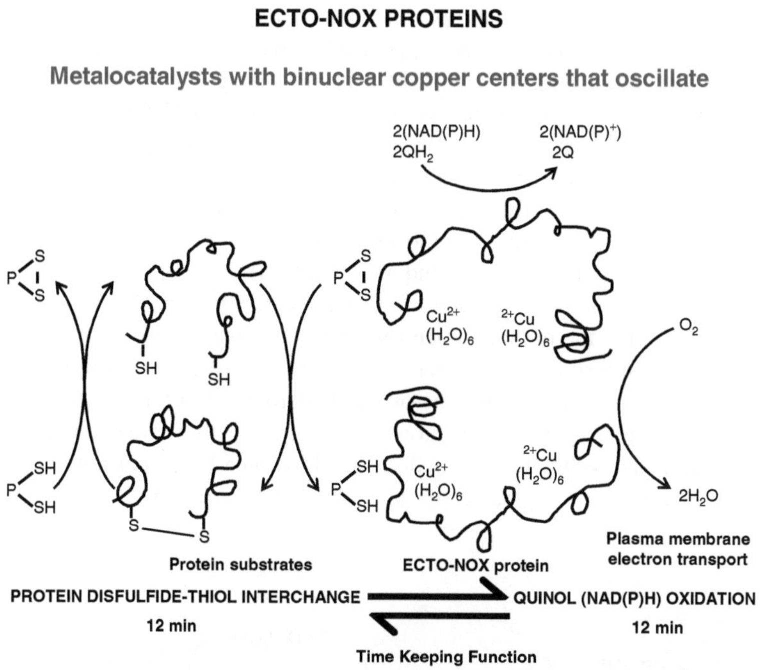

Fig. 2.1 ECTO-NOX (ENOX) proteins are dimeric oxidases with binuclear copper centers that lack flavin. They catalyze four electron transfers from hydroquinones (QH_2) in the membrane or from external NADH to oxygen to form 2 H_2O during one part of their catalytic cycle and carry out protein disulfide interchange (a protein disulfide isomerase-like activity) in the second part of the catalytic cycle. The latter is despite the fact that ENOX proteins lack the CXXC motif common to most, if not all, of the classic protein disulfide isomerase family members (Chueh et al., 2002a)

While NADH as the substrate offers ease of assay, hydroquinones (coenzyme Q for animals, phylloquinone for plants) of the plasma membrane, reduced through the action of quinone reductases at the cytosol-plasma membrane interface, emerge as the physiological substrates for the oxidative portion of the ECTO-NOX cycle (Kishi et al., 1999; Bridge et al., 2000).

2.3 Mechanisms

The ECTO-NOX proteins appear to carry out four electron transfers from 2 $NADH^+$ $-2H^+$ or hydroquinone (2 QH_2) to oxygen to form 2 H_2O. NADH would be encountered infrequently, if at all, at the cell's exterior where the ECTO-NOX proteins were located. In the absence of flavin, a metal involvement in the catalytic mechanism has been postulated (Fig. 2.1). Both the cloned ENOX1 NCBI GenBank accession number EF432052 and ENOX2 NCBI GenBank accession number AF207881 contain a H546-V-H (in ENOX2) motif conserved in periplastic copper oxidases together with Y556 or H562 to form copper binding ligands that are required for enzymatic activity as determined by site-directed mutagenesis and are potentially important to oxygen binding. The H546-V-

H-P-F-G motif is conserved in the copper binding site of both human and chicken superoxide dismutase. A candidate second copper binding sequence is H582-T-F. In ENOX1, one copper site required for activity is the conserved H579-V-H-P-F-G motif while a second potential copper site required for activity is provided by H260-Y-S.

Subsequent study has led to the hypothesis that ECTO-NOX proteins function as core oscillators of the cells' biological clock (Morré et al., 2002a). This follows from the observed alternation of the two enzymatic activities they catalyze, hydroquinone (NADH) oxidation and disulfide-thiol interchange, giving rise to a regular period length of 24 min for CNOX (Morré and Morré, 2003a) (Fig. 2.2). CNOX carries out hydroquinone oxidation or oxidation of external NAD(P)H for 12 min and then that activity rests. While the hydroquinone (NADH) oxidative activity rests, the CNOX proteins engage in disulfide-thiol interchange activity for 12 min. That activity then rests as the 24 min cycle repeats. The overall pattern, however, is much more complex. Within the hydroquinone oxidation part of the cycle, there are two maxima separated by 6 min and within the protein disulfide-thiol interchange portion of the cycle, there are three maxima separated by intervals of 4.5 min (Fig. 2.3). The periodic

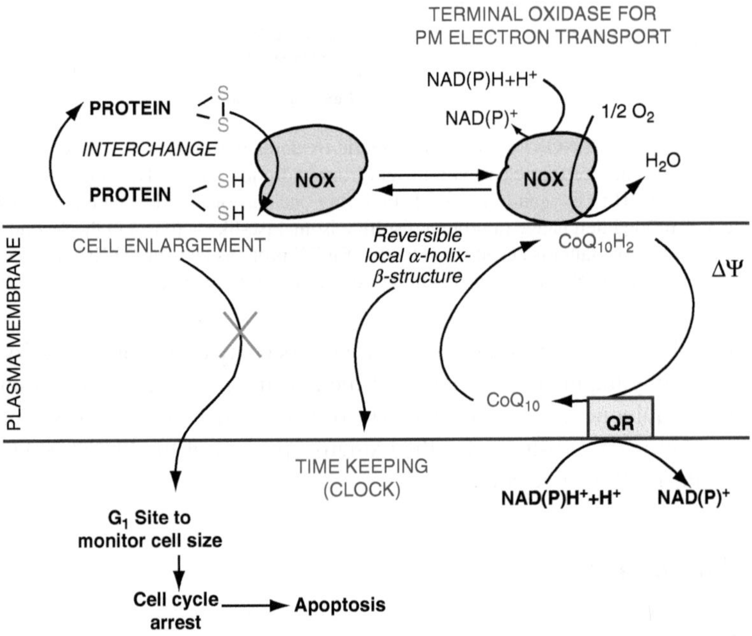

Fig. 2.2 Relationships among ENOX proteins of the cell surface, the inside NAD(P)H quinone reductase and the membrane pool of coenzyme Q (Q_{10}) (Vitamin K_1 in plants) to shuttle electrons and protons across the plasma membrane. In this manner, the ENOX proteins function as terminal oxidases of plasma membrane electron transport donating electrons from cytosolic NAD(P)H either to molecular oxygen in a four electron transfer or to reduce protein disulfides. The left hand portion of the diagram summarizes reactions catalyzed by the protein disulfide-thiol interchange activities which occur in the absence of external reductants and have a role in growth. Growth factors appear to serve as switches to couple this part of the mechanism to cell enlargement. The alternation of these two activities within a 24 min (ENOX1) or a 21–22 min (ENOX2) cycle gives rise to the periodic clock function of these unusual proteins

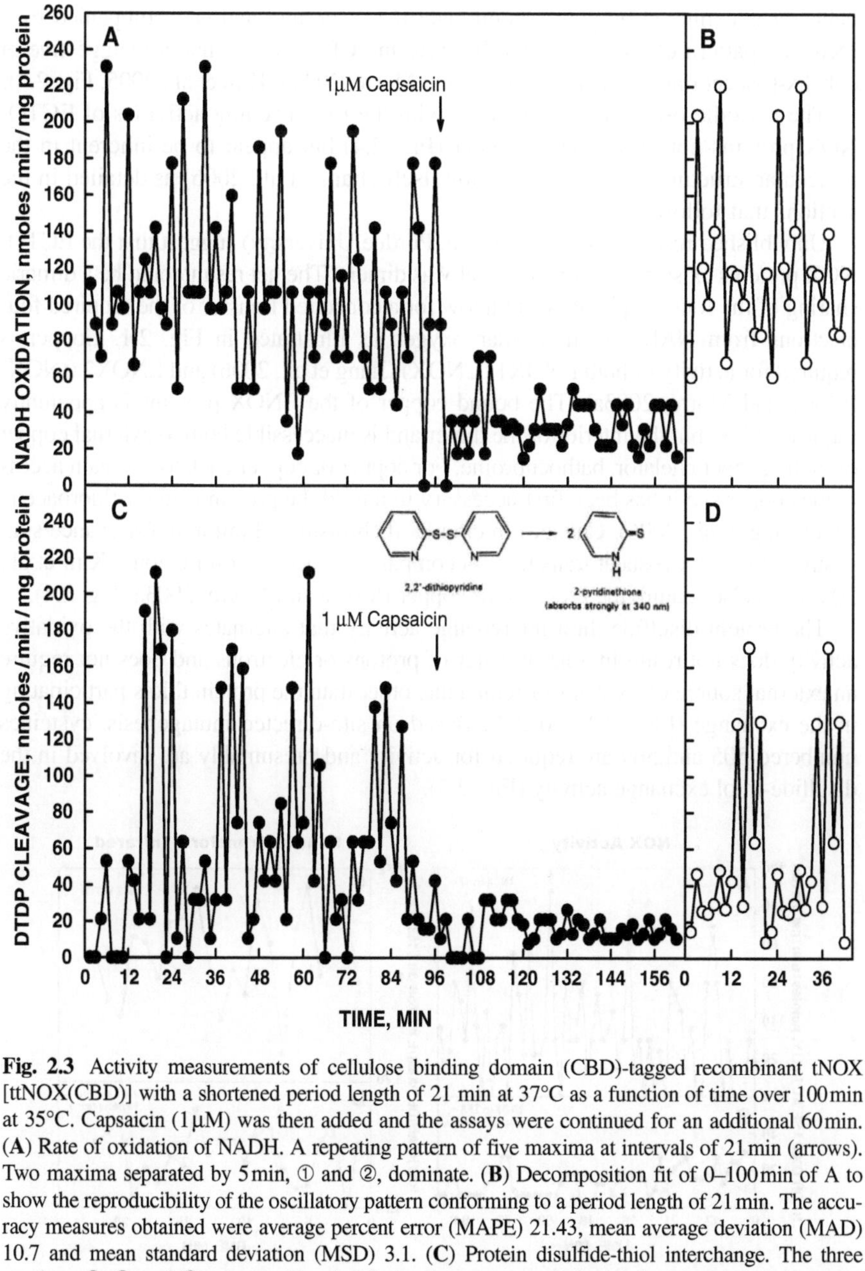

Fig. 2.3 Activity measurements of cellulose binding domain (CBD)-tagged recombinant tNOX [ttNOX(CBD)] with a shortened period length of 21 min at 37°C as a function of time over 100 min at 35°C. Capsaicin (1 μM) was then added and the assays were continued for an additional 60 min. (**A**) Rate of oxidation of NADH. A repeating pattern of five maxima at intervals of 21 min (arrows). Two maxima separated by 5 min, ① and ②, dominate. (**B**) Decomposition fit of 0–100 min of A to show the reproducibility of the oscillatory pattern conforming to a period length of 21 min. The accuracy measures obtained were average percent error (MAPE) 21.43, mean average deviation (MAD) 10.7 and mean standard deviation (MSD) 3.1. (**C**) Protein disulfide-thiol interchange. The three maxima, ③, ④, and ⑤, separated by 4–4.5 min alternate as a triad with the two maxima in rate of NADH oxidation. (**D**) Decomposition fit of 0–100 min of C to show the reproducibility of the oscillatory pattern over two full cycles of 21 min each. The accuracy measures obtained MAPE 50.0, MAD 19.0 and MSD 2.1. The estimation of NADH oxidation is based on the decrease in A_{340} as described (Morré et al., 1995). Protein disulfide-thiol interchange is determined from the cleavage of an artificial dithiodipyridine (DTDP) substrate indicated in C (Morré et al., 1999a). Both activities were inhibited completely by 1 μM capsaicin added after 100 min. Decomposition fits for two full periods show the reproducibility of the oscillations and a period length of 21 min (Kim et al., 2005)

behavior is exhibited by pure recombinant tNOX protein and is accompanied by a recurring pattern of amide I/amide II FTIR and CD spectral changes suggestive of α-helix-β-sheet transformations (Morré and Morré, 2003a; Kim et al., 2005; Fig. 2.4).

The periodic oscillations associated with the time keeping activities of ECTO-NOX proteins not only require copper (Fig. 2.4) but appear to be inherent in the molecular structure of the Cu^{II} aqua ion itself (Jiang et al., 2006) as detailed in the sections that follow.

Unpublished results (Nina Gorenstein, Purdue University) suggest that the ECTO-NOX proteins exist in solution principally as dimers. The net result could be a dimeric sharing of the four copper atoms to allow for a concerted transfer of the required four electrons from NADH to molecular oxygen as illustrated in Fig. 2.1. Copper is required for activity of both ENOX1 (CNOX) (Jiang et al., 2006) and ENOX2 (tNOX) (Morré and Morré, 2003a). The bound copper of the ENOX proteins is apparently sequestered within the interior of the dimer and is inaccessible both to external copper or to the copper chelator, bathocuproine. For copper or copper chelators to gain access to the copper site it has been first necessary to unfold the protein using trifluoroacetic acid (Jiang et al., 2006). Changes in circular dichroism and infrared absorbance suggestive of α-helix-β-sheet transitions accompany the activity oscillations (Kim et al., 2005) and also require the presence of copper (Morré and Morré, 2003a; Fig. 2.4).

The protein disulfide thiol interchange activity that alternates with the oxidative activity does not result in a net transfer of protons or electrons and does not require an external source of oxidants or reductants other than the protein thiols participating in the exchange (Figs. 2.1 and 2.2). Based on site-directed mutagenesis, cysteines numbered 505 and 569 are required for activity and presumably are involved in the disulfide-thiol exchange activity (Fig. 2.5).

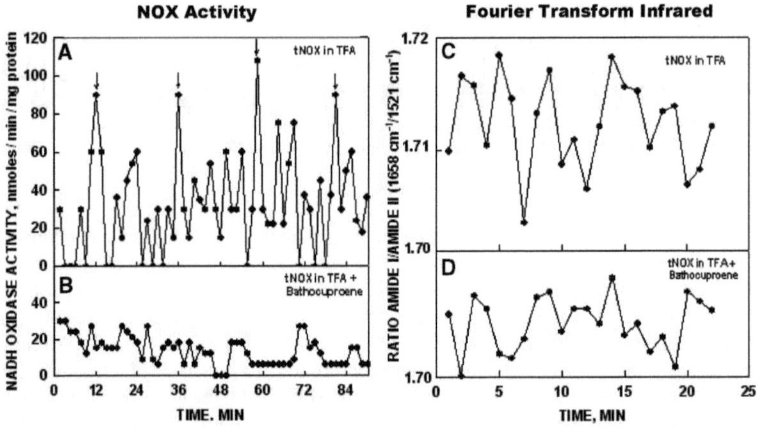

Fig. 2.4 Activity of recombinant ENOX2 protein unfolded in the presence of 0.1% trifluoroacetic acid (TFA) alone (**A, C**) or in the presence of 25 μM bathocuproine to remove bound copper (**B, D**). Upon removal of the TFA by dialysis, activity in the absence of bathocuproine was retained (**A, C**). Upon removal of TFA from the bathocuproine-treated sample no activity was observed (**B, D**) but activity was restored by addition of 100 μM $Cu^{II}Cl_2$ (not shown). Similar results were obtained for recombinant human CNOX (Jiang et al., in preparation) and for a plant CNOX isolated from plasma membranes of soybean (Jiang et al., 2006)

Effect of site-directed mutagenesis of ttNOX on NADH oxidase enzyme activity, period length and inhibition by capsaicin

Mutation	Enzymatic activity	Period length	Capsaicin inhibition
None	+	22 min	+
C505A*	–		
C510A	+	36 min	+
C558A	+	42 min	+
C569A*	–		
C575A	+	36 min	+
C602A	+	36 min	+
M396A	+	22 min	–
H546A	–		
H562A	–		
G592V	–		

*Absolute specific activities determined from detailed kinetic analyses of both NADH oxidation and protein disulfide-thiol interchange for each mutant (i.e,. with C505A and C569A replacements both activities were absent).

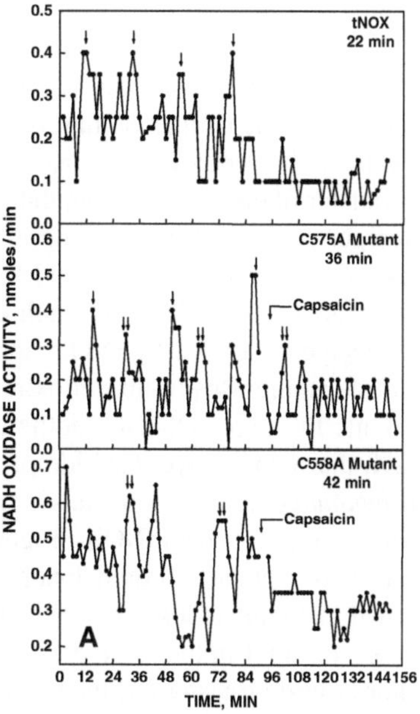

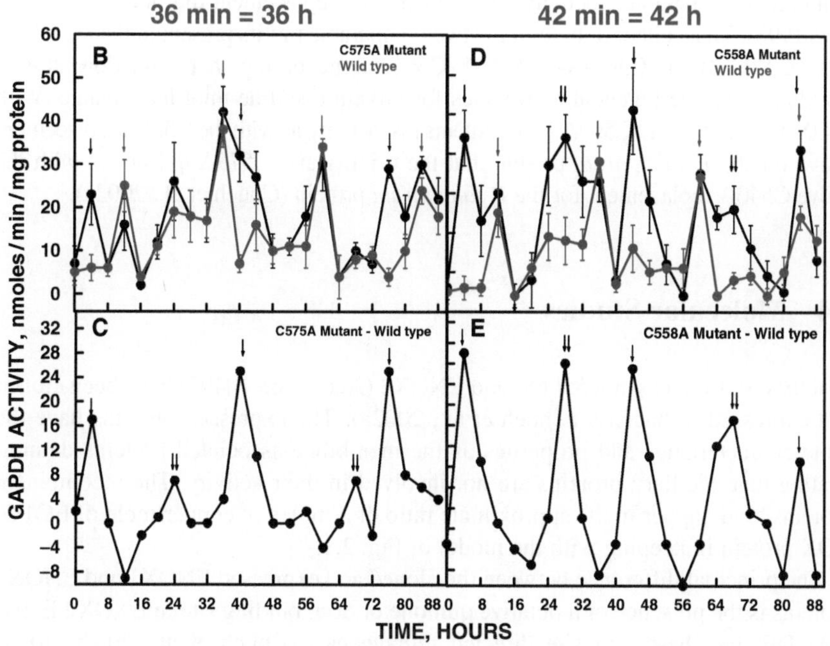

Fig. 2.5 Effect of cysteine replacements with alanine on the period length of ENOX2 (**A**) and the corresponding effect on the length of the circadian period of COS cells in response to transfection (**B–E**). The C575A replacement with an ENOX period length of 36 min, exhibited GAPDH (**B**) maxima (solid arrows) absent from wild type (open symbols, dashed lines and broken arrows). Maxima and minima (asterisks) were significantly different ($p < 0.005$). (**B**) GAPDH activities of wild type COS cells subtracted from those carrying the C575A replacement both showed a major (single arrows) and a minor (double arrows) circadian period length of 36 h (ECTO-NOX period X 60) (**C**). As in B and D but for the C558A replacement having a period length of 42 min and a circadian period length of 42 h (**D, E**) ($p < 0.005$) (Morré et al., 2002a).

The characteristic C-X-X-C motif common to most, if not all, members of the protein disulfide isomerase family of proteins (Ohnishi et al., 1995) and is present, as well, in thioredoxin reductase and related proteins where it appears to catalyze the transfer of electrons in conjunction with bound flavin (Russel and Model, 1988; Gilberger et al., 1997) is lacking in tNOX. In addition to lacking C-X-X-C, tNOX does not appear to contain bound flavin nor is its activity dependent upon addition of flavins (FAD or FMN) (Chueh et al., 2002a). Thus, the protein disulfide-thiol interchange catalyzed by tNOX appears to be unique from that of classic protein disulfide isomerases or thioredoxin reductases (Gilberger et al., 1998).

While the two C-X-X-C motifs characteristic of flavoproteins are missing from tNOX, the redox-active disulfide of thioredoxin reductase from the malaria parasite *Plasmodium falciparum* contains a motif, C88-X-X-X-X-C93 (Gilberger et al., 1998), similar to one found in tNOX. Together with a downstream His 509, this motif was shown to be a putative proton donor/acceptor. Either a C88A or the C93A replacement resulted in complete loss of enzymatic activity (Gilberger et al., 1998). A C535-X-X-X-X-C540 motif in the same protein was crucially involved in substrate coordination and/or electron transfer (Kliman and Hey, 1993). A C535A replacement did result in diminution of enzymatic activity, but the C5440A replacement did not (Kliman and Hey, 1993). Thus, either or both of the two comparable motifs present in tNOX, C505-X-X-X-X-C510 or C569-X-X-X-X-X-C575, alone or together with downstream histidines, provide potential active sites for protein disulfide-thiol interchange. With tNOX, the C505A and C569A replacements lost activity as with the C535A replacement above for the *P. falciparum* protein, but the C510A and C575A did not as with the above C540A replacement for the *P. falciparum* protein (Chueh et al., 2002a).

2.4 Molecular Studies

Both ENOX1 (Accession 55068) and ENOX2 (Accession 10495) have been cloned and expressed in bacteria (Chueh et al., 2002a). The expressed proteins have the same characteristics and properties as the membrane-associated proteins demonstrating that ancillary proteins are not involved in their activity. The recombinant proteins bind copper in the approximate ratio of 2 moles of copper/mole of ECTO-NOX protein in keeping with the model of Fig. 2.1.

The principal difference between the cloned and expressed ENOX1 and ENOX2 proteins is the presence of a putative quinone or drug binding site in ENOX2 E394-E-M-T-E that, based on site directed mutagenesis (Chueh et al., 2002a, b), is responsible for the drug-inhibition characteristic of ENOX2. Also confirmed by site-directed mutagenesis is the requirement of the H546-V-H copper site for activity and that of the T589-G-V-G-A-S-L and an E605 adenine nucleotide (NADH) binding site near the protein's C terminus (Chueh et al., 2002a). A second potential copper site at H582-T-F has also been confirmed by site-directed mutagenesis.

To determine if ECTO-NOX proteins might represent the ultradian time keepers (pacemakers) of the biological clock, COS cells were transfected with cDNAs encoding tNOX proteins having a period length of 22 min or with C575A or C558A cysteine to alanine replacements having period lengths of 36 or 42 min (Morré et al., 2002a). Such transfectants exhibited 36, or 40–42 h circadian patterns in the activity of glyceraldehyde-3-phosphate dehydrogenase, a common clock-regulated protein, in addition to the endogenous 24 h circadian period length (Fig. 2.5). The fact that the expression of a single oscillatory ECTO-NOX protein determined the period length of a circadian biochemical marker (60 X the ECTO-NOX period length) provides compelling evidence that ECTO-NOX proteins represent the biochemical ultradian drivers of the cellular biological clock.

When assayed for GAPDH activity at 4 h intervals over 76 or 88 h, experiments in triplicate reproduced a major 24 h circadian rhythm of alternating maxima and minima with nontransfected (wild type) COS cells. The GAPDH activity of COS cells transfected with tNOX retained the wild-type pattern plus a second set of maxima with a period length of 22 h reflective of the 22 min period length of tNOX (ENOX2) (Morré et al., 2002a).

The circadian pattern of GAPDH activity for transfectants carrying the C575A replacement with a 36 min period showed a pattern of circadian activity oscillations distinct from that of wild type. When the wild-type activity for that particular experimental series was subtracted from that of the C575A replacement activity, the differences due to the C575A replacement were seen as a new set of major oscillations spaced at intervals of 36 h (ECTO-NOX period X 60) (Fig. 2.5 B,C). The C558A replacement yielded a period length of 42 min and a circadian period length of 42 h (Fig. 2.5 D,E).

The period lengths of the activity oscillations of the NOX proteins are independent of temperature (temperature compensated) (Morré and Morré, 1998; Pogue et al., 2000; Wang et al., 2001; Morré et al., 2002b; Table 2.1), and their phases are entrainable (Morré et al., 1999b, 2002b, c; Morré and Morré, 2003b). These two

Table 2.1 ECTO-NOX period length is independent of temperature (temperature compensated

Temperature	Soybean plasma Membrane[a]	Milk-fat globule Membrane[b]	HeLa plasma Membrane[c]
		Period length (min)	
17°	24.4 ± 1.3	23.8	24.5 ± 0.0
27°	24.1 ± 1.9	23.8	24.7 ± 1.1
37°	23.8 ± 2.4	23.5	24.0 ± 0.9
Average	24.1 ± 0.3	23.7 ± 0.2	24.4 ± 0.4

[a] Morre and Morré (1998)
[b] Morre et al. (2002b)
[c] Wang et al. (2001)

Period length also temperature-independent for:

- ECTO-NOX activity and enlargement of CHO cells and ECTO-NOX activity of CHO cell plasma membranes (Pogue et al., 2000b)
- tNOX activity and enlargement of HeLa cells (Wang et al., 2001)
- Enlargement of soybean cells and elongation of soybean stem sections (Morré et al., 2001a)

characteristics, temperature compensation and entrainment (coupling the intrinsic clock to environmental cues), are two defining hallmarks of the biological clock (Dunlap, 1996; Edmunds, 1988). ECTO-NOX synchrony is achieved through autosynchrony in solution (Morré et al., 2002b), by coupling to red (plants) (Morré et al., 1999b) and blue (plants and animals) (Morré et al., 2002c; Morré and Morré, 2003b) light photoreceptors and in direct response to melatonin (Morré and Morré, 2003b) or to electromagnetic fields (Morré et al., 2008b).

2.5 Studies with Deuterium Oxide

Deuterium oxide (^2H_2O, D_2O, heavy water) is the most impressive example of a small number of substances that, of themselves, modify significantly, reproducibly and in a wide range of organisms, the length of the circadian day. Thus, it is significant that deuterium oxide lengthens the ECTO-NOX period in proportion to its effects on the circadian day (Fig. 2.6). That deuterium oxide reversibly lengthens both the period and the phase of the period of the circadian day has been established for a variety of organisms (Bruce and Pillendrigh, 1960; Suter and Rawson, 1968). In E*uglena gracilis* adapted for long periods of several months to D_2O, the period of the running circadian rhythm of phototaxis was lengthened from its normal value (close to 23 h) to 28 or 29 h (Bruce and Pillendrigh, 1960). Many subsequent studies including those of Suter and Rawson (1968) and Enright (1997) indicated that this

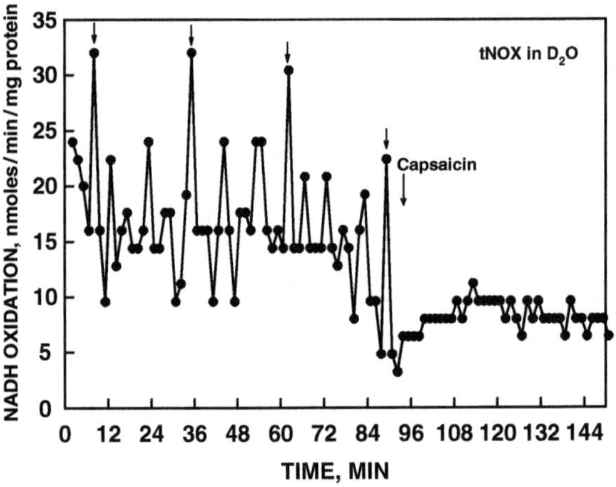

Fig. 2.6 NADH oxidation of nus-tagged recombinant tNOX at 37°C measured in the presence of a reaction mixture in which water was replaced by D_2O. The final NADH concentration was 150 μM and the entire system was buffered at pH 6.5 with 50 mM Tris-HC1. The period length was proportionately increased by 25% from 22 to 25 min in D_2O compared to water. With the constitutive ENOX1 or CNOX, the period length in D_2O was increased from 24 to 30 min in keeping with the response of organisms to D_2O where the length of the circadian day is increased from 24 h to about 30 h (see text)

effect of D_2O was of widespread occurrence. D_2O lengthened the period in green plants (Bunning and Baltes, 1963), in isopods (Enright, 1997), birds (Palmer and Dowse, 1969), mice (Suter and Rawson, 1968; Palmer and Dowse, 1969; Dowse and Palmer, 1972; Richter, 1977), insects (Pittendrigh et al., 1973) and unicellular organisms (Bruce and Pittendrigh, 1960; McDaniel et al., 1974). The effect may be general since no exceptions were found in 12 different species examined (Pittendrigh et al., 1973).

2.6 The Role of Copper

As detailed in the preceding sections, protein bound copper is required for ECTO-NOX activity and the characteristic 2 + 3 oscillatory pattern. In the copper depleting experiment of section 3, the material removed by chelation with bathocuproine was checked and exhibited a 2 + 3 pattern of NADH oxidation similar to that of the ECTO-NOX proteins except that NADH was reduced to nearly equal levels by all five oscillatory maxima instead of the ①, ② pattern of the oxidative portion of the ECTO-NOX cycle (Jiang et al., 2006). The copper solutions of themselves exhibited the characteristic 2 + 3 oscillatory pattern based on rates of NADH oxidation (Fig. 2.7). The major difference is that in the absence of protein, the oscillations of the 2 + 3 pattern are approximately of equal magnitude. Assymetry is still maintained with maxima ① and ② being separated by six min whereas the maxima ③, ④ and ⑤ are separated from each other and from ① and ② by 4.5 min (2 X 6 + 4 X 4.5 = 24). Apparently, the principal difference with the protein present is that redox potential of oscillations ③, ④ and ⑤ are diverted into protein disulfide-thiol interchange. The period length of the copper oscillations is not only independent of temperature but also independent of pH, copper concentration and concentration of NADH (Jiang et al., 2006). As will be developed in the sections that follow, a possible underlying pattern of distortion in the four equatorial oxygens of the Cu^{II} aqua ion at a close distance relative to the two axial oxygen atoms at a longer distance indicated from extended X-ray absorption fine structure (EXAFS) studies did correlate with changes in redox potential sufficient to catalyze NADH oxidation in solution. The findings were consistent with changes in the coordinated water molecules of the Cu^{II} aqua ion possibly related to a metastable equilibrium condition in the ratio of ortho to para nuclear spin orientation of the water-associated hydrogen or deuterium atoms (Morré et al. 2008a).

2.7 The Copper Clock

Based on the above findings that solvated Cu^{II} as the chloride or other salts alone exhibited the property of catalysis of NAD(P)H oxidation, a copper "clock" was devised as illustrated in Fig. 2.8. By consulting the clock, one could accurately

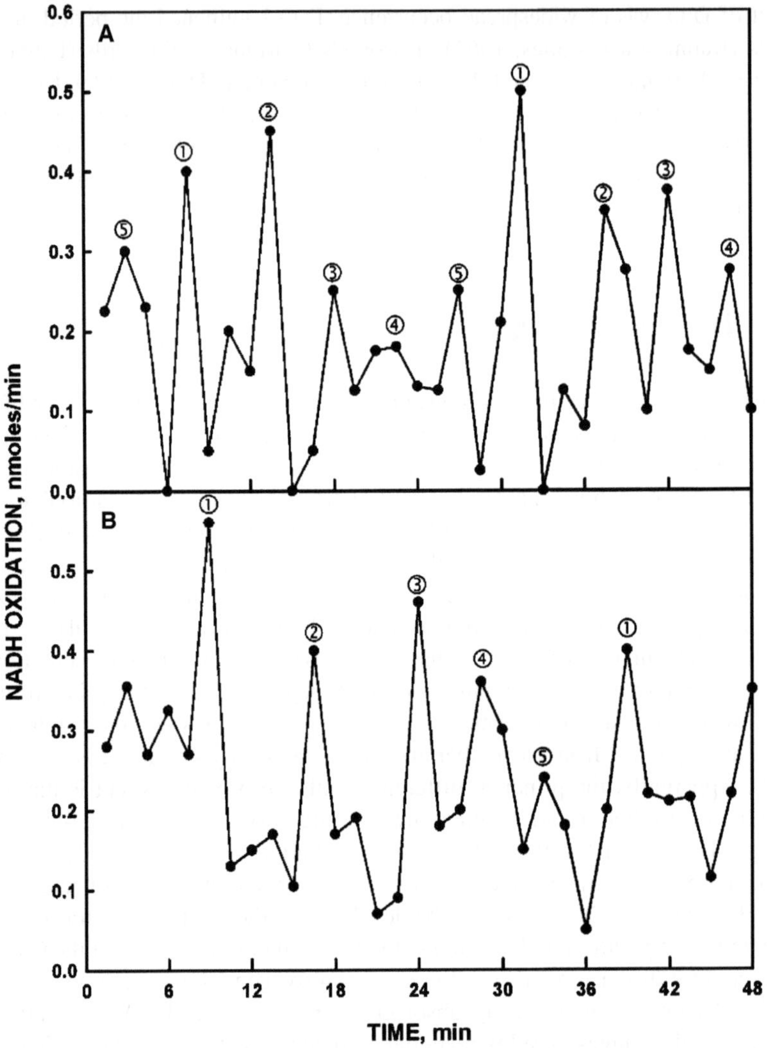

Fig. 2.7 Periodic catalysis of the oxidation of NADH by 25 μM aqueous $Cu^{II}Cl_2$ (**A**) with a period length of 24 min and of 30 min with $Cu^{II}Cl_2$ solvated in D_2O (**B**). The final NADH concentration was 150 μM. The solution was buffered at pH 6.0 using 50 mM Tris-HC1. Within each period of 24 min two maxima are separated by 6 min (7.5 min for D_2O) and the remaining maxima are separated by 4.5 min (5.5 min for D_2O). The first maximum of the doublet separated by 6 min (7.5 min for D_2O) is designated as ① and the pattern repeats after the fifth maximum to generate the 24 min period (30 min for D_2O) (Morré et al., 2007)

predict the precise times of initiation of new 24 min cycles. The basis for copper catalyzed oxidation of NAD(P)H appears to be the periodic oscillations in redox potential exhibited by aqueous copper solutions (Fig. 2.9). The oscillatory 2 + 3

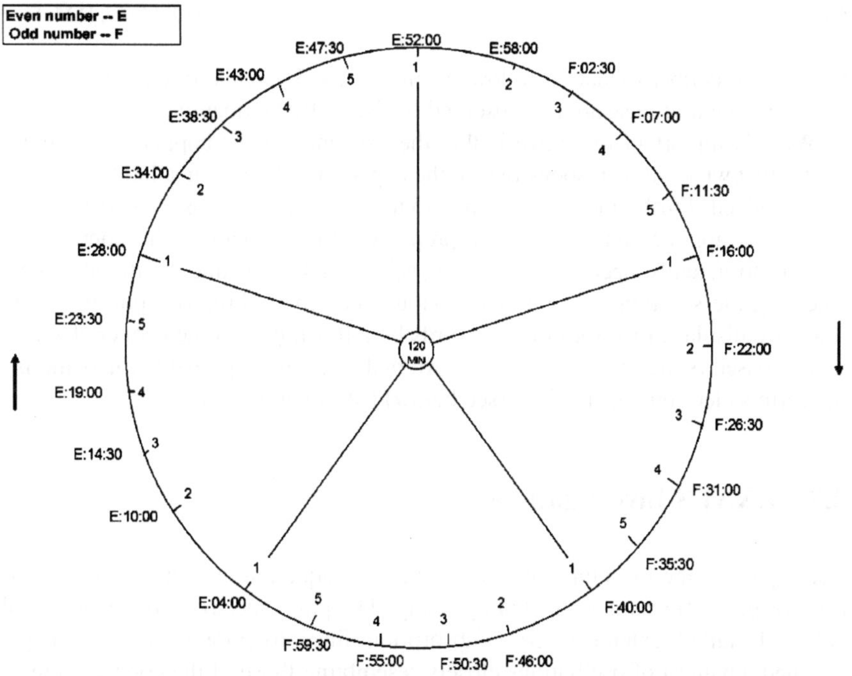

Fig. 2.8 The copper clock (Morré et al., 2008b). The phase of the oscillations in NADH oxidation were determined as described by Jiang et al. (2006). The clock time in minutes is shown for each of the five maxima labeled 1 to 5 over 120 min after which the cycle repeats. Those times marked E are for even numbered clock time hours. Those times marked F are for odd number clock time hours

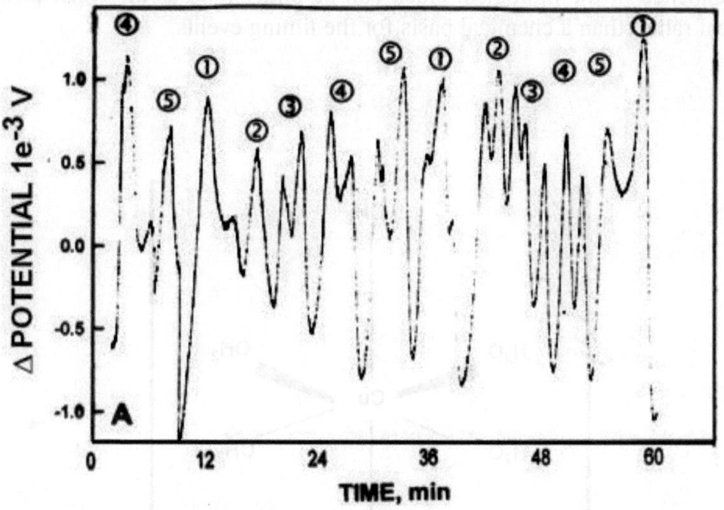

Fig. 2.9 Oscillations in redox potential of an unbuffered 10 mM solution of $Cu^{II}Cl_2$ as determined using a platinum electrode and a CH Instruments (Austin, TX), Model 600 B Series Electrochemical Analyzer/Workstation system as a function of time under oxygen-depleted conditions. The five maximum pattern is reproduced except that several of the maxima appear to bifurcate and are represented as doublets which may be reflective of the enhanced resolution of the method (Morré et al., 2007)

catalysis of copper oxidation follows a similar 2 + 3 oscillatory pattern of change in redox potential measured as observed for NAD(P)H oxidation.

What is important to realize is that the existence of the copper clock demonstrates that whatever is responsible for the copper oscillations in solution are highly synchronized. Furthermore, all copper solutions including those stored for weeks or months are the same as freshly prepared. When shielded, the copper clock appears to actually become less well synchronized with time. In the absence of shielding the solutions remain synchronous but depart abruptly from an existing sequence displaced forward or backwards by a few min as if the copper clock was being phased (synchronized) by some external force, most probably fluctuations in the earth's electromagnetic field (see below) (Morré et al., 2008b).

2.8 EXAFS Investigations

Consequently, the local environment of the Cu^{2+} aqua ion in copper chloride solutions (Fig. 2.10) was investigated by X-ray absorption spectroscopy (Morré et al., 2007). Detailed extended X-ray absorption fine structure (EXAFS) analyses revealed a pattern of oscillations closely resembling those of the copper-catalyzed oxidation of NADH. With $CuCl_2$ in D_2O, a pattern with a period length of 30 min was observed. The findings suggest a regular pattern of distortion in the axial and/or equatorial oxygen atoms of the coordinated water molecules which correlate with redox potential changes sufficient to oxidize NADH. Thus, the temperature independence of the biological clock can be understood as the consequence of a physical rather than a chemical basis for the timing events.

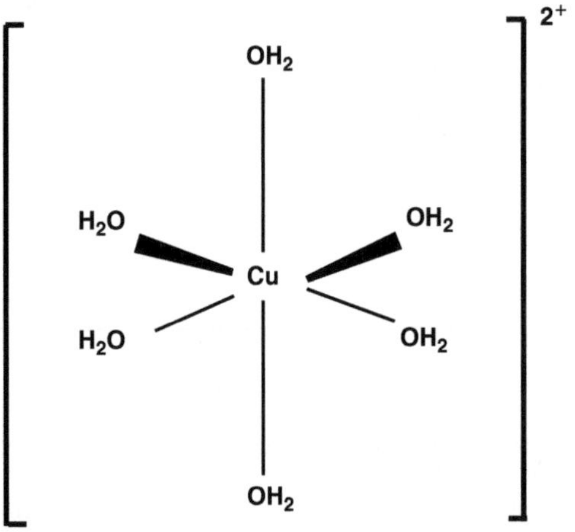

Fig. 2.10 The Cu^{II} hexaaqua ion

When individual values for aqueous $Cu^{II}Cl_2$ were plotted as a function of time for a k (Å^{-1}) value of 9, an oscillatory pattern was observed (Fig. 2.11) reminiscent of the rates of NADH oxidation. An alternating pattern yielded maxima and minima with two of the maxima separated by an interval of 6 min and the remaining maxima separated by intervals of 4.5 min. Fourier analysis confirmed a 24 min periodic spacing (Fig. 2.12A). Statistical analysis of the changes in amplitude were evaluated by decomposition fits to create the overall pattern of alternating maxima and minima at k (Å^{-1}) of 9 (Fig. 2.12B) based on a total time period of data collection of 92 min. When several such decomposition fits were superpositioned and averaged for 8.9–9.1 k (Å^{-1}) and standard errors were calculated, statistically different maxima and minima were observed that recapitulated the oscillations observed with Cu^{II}-catalyzed oxidation of NADH (Fig. 2.7) and in redox potential of aqueous $Cu^{II}Cl_2$ (Fig. 2.9).

For a similar experiment carried out in D_2O, the Fourier analysis now revealed a period length of 30 min instead of 24 min (Fig. 2.14A). The decomposition fits yielded patterns of oscillations similar to those observed for $Cu^{II}Cl_2$ in water except that the intervals between maxima and minima were increased by about 25% to generate an overall period shift of 6 min from 24 to 30 min (Fig. 2.14B). Averages of several such decomposition fits when superpositioned revealed significant differences between maxima and minima within the repeating five-peak pattern of oscillations (Fig. 2.15).

Previous EXAFS studies of $Cu^{II}Cl_2$ solutions indicated sixfold oxygen coordination with four short equatorial bonds and two longer axial bonds (Filipponi et al.,

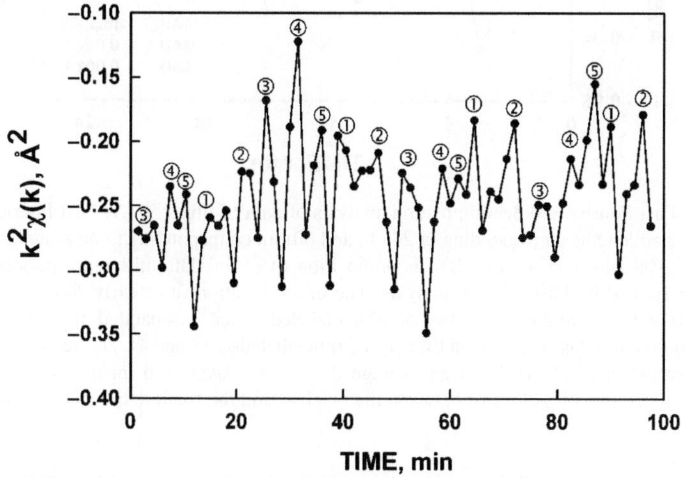

Fig. 2.11 EXAFS absorbance data of K^2 weighted $\chi^{(k)}$ spectra for a 0.1 M aqueous solution of $Cu^{II}Cl_2$ at 25°C and 170 bar plotted as a function of time plotted at a k (Å^{-1}) of 8.9. Measurements were collected at ca. 1.5 min intervals (± 0.1 min) over 90 min. Plotted values are real times with the plotted times corresponding to the times at the end of the measurements. Maxima ① and ② are separated by 6 min whereas maxima ③, ④ and ⑤ are separated from each other and from ① and ② by 4.5 min (6 + 4 X 4.5 min = 24 min) (Morré et al., 2007)

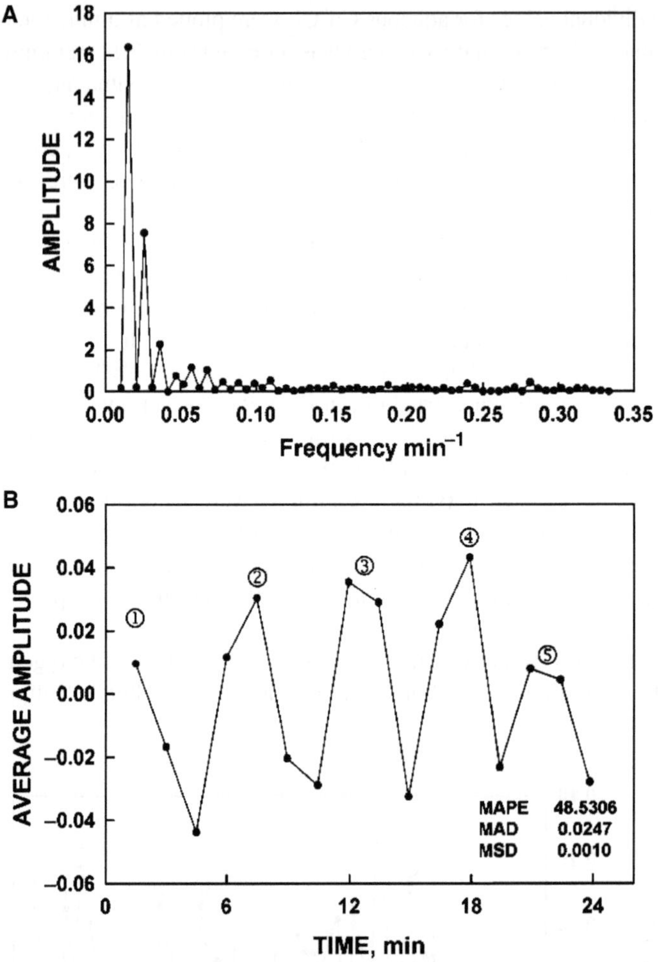

Fig. 2.12 Fast Fourier and decomposition analysis of data of Fig. 2.9. (**A**) Fast Fourier analysis revealed period lengths corresponding of 24, 36 and 60 min corresponding to measured frequencies of 0.017, 0.108 and 0.042 min^{-1}. (**B**) Decomposition fit of the data utilizing a period length of 24 min determined by Fast Fourier analysis. The decomposition fits clearly show a patterns a k (Å^{-1}) of 9.0 with recurrent maxima, two of which labeled ① and ② separated in time by 6 min and three additional maxima separated in time by 4.5 min labeled ③, ④ and ⑤. The accuracy measures, average percent error (MAPE), mean average deviation (MAD) and mean standard deviation (MSD) are indicative of a close fit between the original and the fitted data (Morré et al., 2007)

1994; Fulton et al., 2000; Korshin et al., 1998). The longer axial bonds have a large vibrational amplitude and only contribute to the EXAFS at low k. Because the calculated distance changes between the copper and the water oxygens of the copper hexahydrate were relatively small, and since the changes were located at relatively high *k*, some involvement involving the 4 short bonds of the copper aqua ion was indicated.

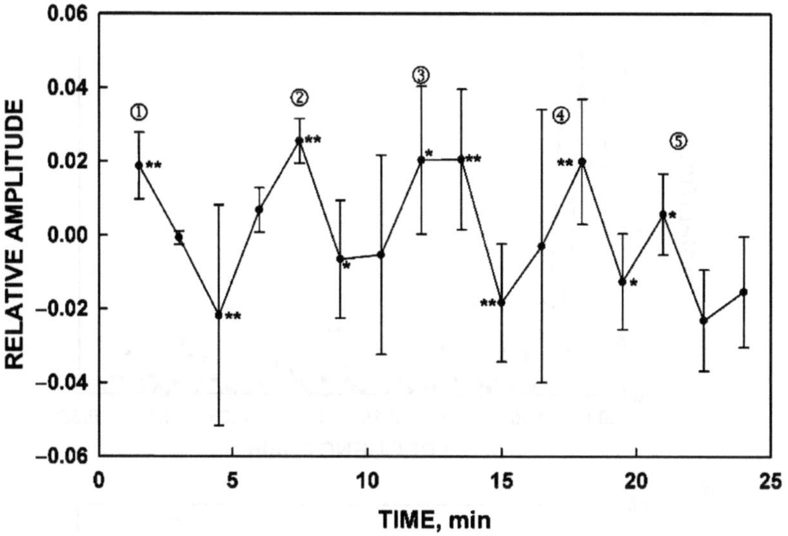

Fig. 2.13 Decomposition fits averaged from data collected at intervals of 0.1 k (Å^{-1}) in the range 8.9–9.1 ± standard deviations. Maxima and minima are significantly different (* $p < 0.006$; ** $p < 0.003$) as determined by two-tailed t test comparisons. Values averaged over the range 8.8–9.4 k (Å^{-1}) were similar except for variations around the maximum labeled ④ corresponding to the latter (Morré et al., 2007)

With copper solutions or with ECTO-NOX proteins, solvation in D_2O increases the period length from 24 to 30 min and for pure water from 20 to 24 min. Heavy water is the only perturbation known to alter the period length of the circadian day and it is significant that both the NOX and copper clocks are similarly altered.

2.9 Oscillations Inherent in the Structure of Water

The oscillations with maxima separated by intervals of 4.5 or 6 min are slow by comparison even to Jahn-Teller equilibria (Bersuker, 1984). The only atomic phenomenon that occurs on a similar time scale is the equilibrium of the two alternative orientations of the nuclear spins (ortho and para) of the hydrogen atoms of water (Tikhonov and Volkov, 2002). The relaxation time to convert spin-modified ice to the 3:1 ortho:para equilibrium ratio has been estimated to require several months. Liquid lifetimes were measured as 26 ± 5 min for para water and 55 ± 5 min for ortho water. Metals and other impurities can result in ortho:para imbalance and conceivably an oscillatory non-equilibrium condition that would translate into a regular periodic pattern of distortion in the four equatorial oxygens of the copper aqua ion and concomitant redox changes ultimately affecting rates of NADH oxidation (Morré et al., 2007). A similar contribution from the two equatorial oxygens might account for the 2 + 3 asymmetry (Morré et al., 2007).

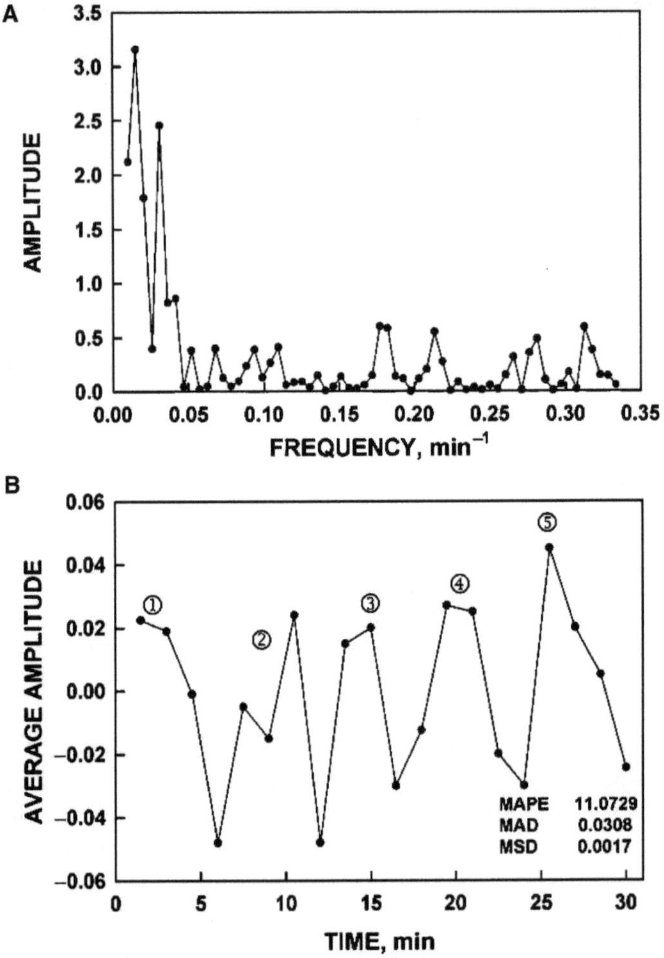

Fig. 2.14 As in Fig. 2.12 except that the $Cu^{II}Cl_2$ was solvated in D_2O and the determinations were for 60 min rather than 90 min. (**A**) Fast Fourier analysis at k ($Å^{-1}$) in the range of $10.0 < k < 10.4$ yielded period lengths of 30 and 60 min corresponding to frequencies of 0.017 and 0.033 min^{-1}. (**B**) Decomposition fits at k ($Å^{-1}$) of 10 utilizing a period length of 30 min revealed a five peak pattern within the 30 min period with maxima labeled ① and ② separated by 7.5 min and maxima labeled ③, ④ and ⑤ separated by intervals of 5.5 min. Maxima ② was resolved into two maxima not observed in averaged data (e.g., Fig. 2.9). The accuracy measures, mean average percent error (MAPE), mean average deviation (MAD) and mean standard deviation (MSD) are indicative of a close fit between the original and the fitted data (Morré et al., 2007)

The hydrogen atom nuclear spins may either be parallel (total spin is 1) in ortho molecules or antiparallel (total spin is 0) in para molecules (Tikhonov and Volkov, 2002). Each spin isomer has its own system of rotational levels. Optical transitions between levels of different spin isomers are not permitted. The number of ortho isomer molecules is three times greater than the number of para isomer molecules in the thermal equilibrium state at room temperature.

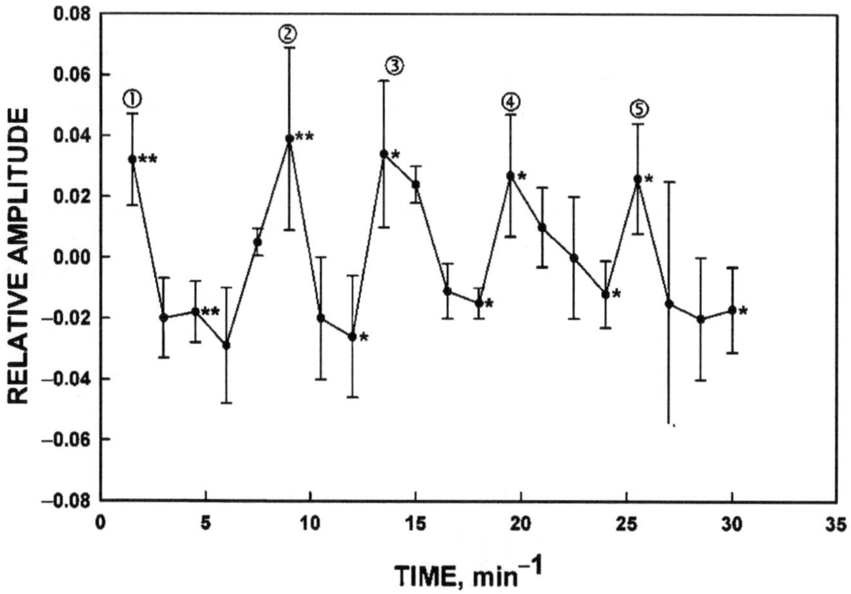

Fig. 2.15 As in Fig. 2.14 for data for $Cu^{II}Cl_2$ solvated in D_2O at intervals of 0.1 k (Å^{-1}) over the range of 9.8–10.1. Maxima and minima were significantly different (*$p < 0.03$; **$p < 0.009$) as determined by two-tailed t test comparisons. Values averaged over the range of 8.8–9.2 as in Fig. 2.10 were similar except for a less pronounced resolution in the regions of maxima labeled ④ and ⑤ (Morré et al., 2007)

The middle infrared spectral region has been reported as suitable as an indirect method to study liquid water states (Binhi and Stepanov, 2000; Mumma et al., 1987). Infrared spectroscopic measurement of the para-H_2O/ortho-H_2O ratio above the water sample surface is based on the fact that the vibrational infrared spectra of many molecules depend on their nuclear spin state. For a narrow spectral range including two close lines belonging to the different water molecule modifications within the same quantum transition, the relative heights of lines will, on average, equal the natural ortho to para composition ratio of water of 3:1 (Binhi and Stepanov, 2000).

Oscillatory patterns similar to those observed with aqueous copper solutions or with the ECTO-NOX proteins were seen to be inherent in the properties of pure water or of D_2O (Fig. 2.16; Fig. 2.17). Because of spectral differences comparing water and D_2O, potential oscillations in the ortho-para hydrogen equilibrium of water and D_2O were recorded at 3,801 and 3,779 cm^{-1} for water and in the range of 2,600–2,650 cm^{-1} or 2,425 to 2,475 cm^{-1}, for D_2O where the oscillatory changes above baseline differed approximately threefold (Fig. 2.18a,b). Repeating patterns of five maxima at approximately 18 min intervals was obtained which, when analyzed by decomposition fits (Figs. 2.18c–e), indicated 2 + 3 patterns of oscillations similar to those observed previously (Jiang et al., 2006; Morré et al., 2007). Results with D_2O were similar except that the period length was extended from 18 to 24 min (Morré et al. 2008a). Why the ortho/para hydrogen compositions should exhibit a regular pattern of oscillation from their equilibrium ratio of 3:1 remains unexplained but such variation would account for the oscillatory behavior of the redox potential and catalytic activity of aqueous solutions in relation to our time keeping model (Morré et al., 2002a).

The explanation as to why the period length of the orotho-para transition would increase from 18 to 24 min in the presence of Cu^{II} appears to reside in the previously observed response to the ionic environment of the water (Tikonov and Volkov,

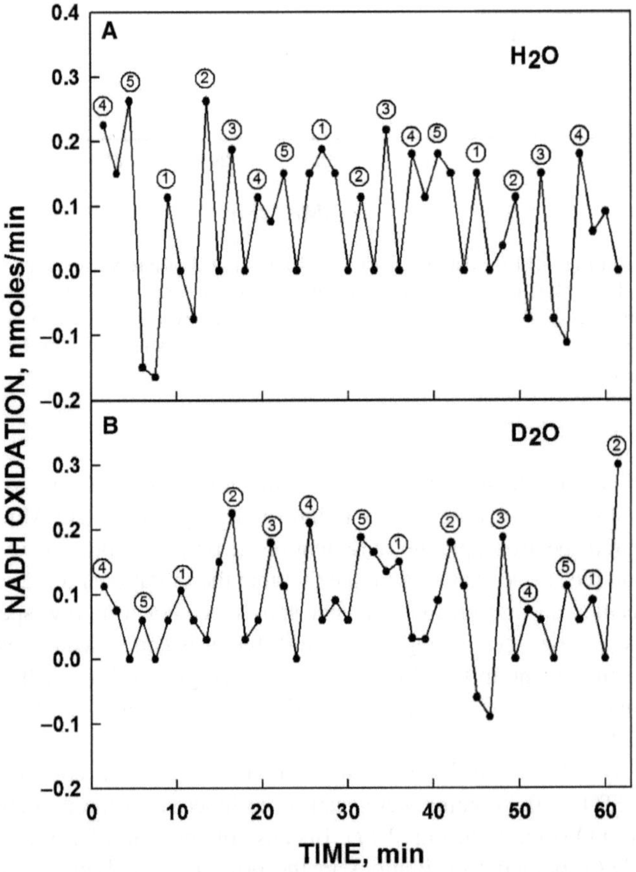

Fig. 2.16 Periodic catalysis of the oxidation of NADH by deionized, distilled water (**A**) with a period length of 18 min and D_2O (**B**) with a period length of 30 min. The final NADH concentration was 150 µM. The solution was unbuffered. Within each period of 18 min maxima labeled ① and ② are separated by 4.5 min for H_2O and by 5.8 min for D_2O. The remaining maxima designated ③, ④ and ⑤ are separated by 3.4 min for H_2O and by 4.4 min for D_2O. The pattern repeats after the fifth maximum to generate the 18 min period for H_2O or 24 min period for D_2O (Morré et al. 2008a)

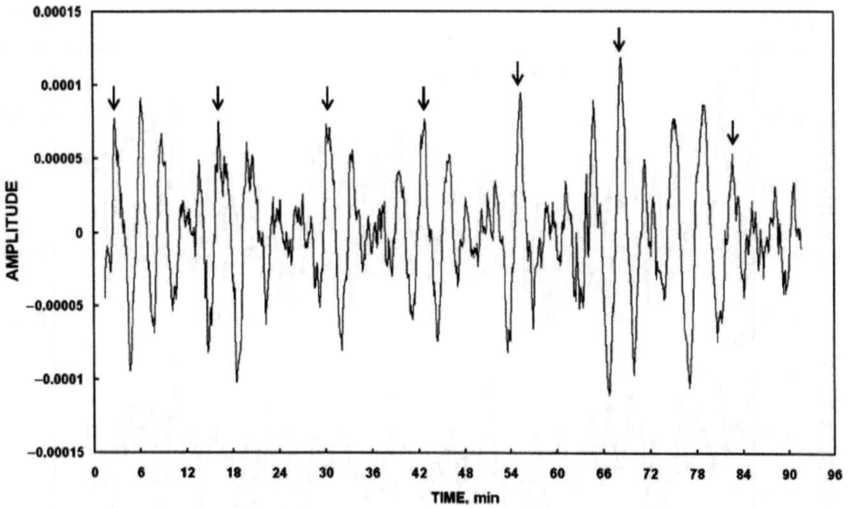

Fig. 2.17 Moving average of EXAFS measurements recorded at intervals of 10 s at a constant k (Å^{-1}) of 9.3 for deionized, distilled water. An oscillatory pattern was observed with a recurring pattern of five maxima (arrows). Fast Fourier analyses and decomposition fits revealed a period length of 18 min. Similar measurements for D_2O yielded a period length of about 24 min (Morré et al. in press)

2002). We have measured period lengths of the oscillatory patterns for a series of ionic solutions and find proportionality of the period length of the associated water to the ionic radius with Cu^{II} uniquely giving rise to a 24 min period (Fig. 2.18). This observation provides a basis for the potential importance of protein bound copper hexahydrate to biological time keeping at the molecular level (Morré et al., 2002a).

Several mechanisms for the ortho/para-conversion have been discussed in the literature. A paramagnetic mechanism for dihydrogen has been developed where the spin conversion is caused by the magnetic interaction of the hydrogen spins with paramagnetic centers such as the unpaired electron spins themselves (Wigner, 1933). Normally the ortho/para-conversion is a slow process but paramagnetic impurities, i.e., unpaired electron spins or electrons with orbital momentum such as provided by metallic cations including copper would serve as catalytic centers for the interconversion process (Kummer, 1962). In a recent model (Buntkowsky et al., 2006), ortho/para-H_2 molecules in one particular spin state were suggested to react with a suitable catalyst to create an X-H_2 complex. Within the complex a postulated evolution of the initial density matrix of the dihydrogen would initiate a partial conversion of one rotational state to the other. Upon decay of the bound state, the final density matrix would then be transferred to the free dihydrogen state.

Water molecules take on the two different spin states in a manner analogous to hydrogen molecules and the net interconversion of the two forms might occur by a similar mechanism even within the Cu^{II} hexahydrate. However, the oscillations we observe with water are apparently not due to the net final equilibrium position

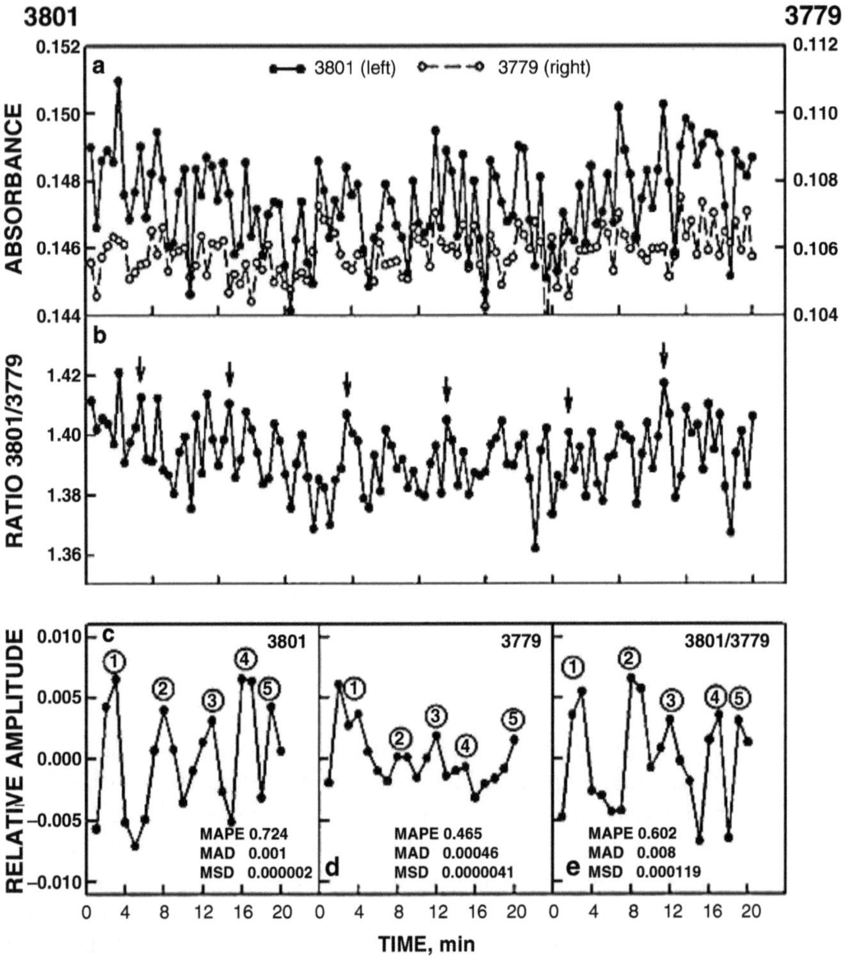

Fig. 2.18 FTIR spectroscopic measurement of the ratio of para-H_2O/ortho-H_2O above a water sample surface determined at 3,801 and 3,779 cm^{-1} respectively (**a**). The ratio of the two wavelengths exhibited a repeating pattern of oscillations of five maxima at intervals of about 18 min (arrows) (**b**). Decomposition fits using an imposed period length of 18 min of data collected at 3,801 cm^{-1} (**c**), and at 3,779 cm^{-1} (**d**) as well as the ratio of the two (e) revealed the oscillatory pattern typical of water with five recurrent maxima, two of which, labeled ① and ②, were separated in time by about 4.5 min and three additional maxima separated in time by 3.4 min labeled ③, ④ and ⑤. The accuracy measures, mean average percent error (MAPE), mean average deviation (MAD) and mean standard deviation (MSD) are indicative of a close fit between the original and the fitted data (Morré et al. 2008a)

determining ortho/para conversions. Net ortho/para conversions are temperature dependent (Milenkl et al., 1997) and reduced in ice (Tikhonov and Volkov, 2002). While amplitudes of the oscillatory phenomena reported here are temperature dependent, the period length by contrast is temperature independent (Jiang et al., 2006) and unchanged even at liquid nitrogen temperatures (Fig. 2.19).

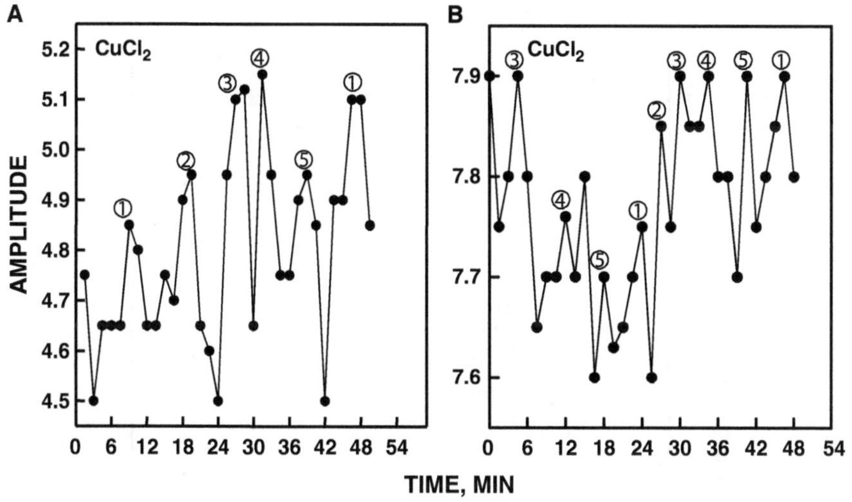

Fig. 2.19 EPR amplitude of unbuffered 10 mM solutions of $Cu^{II}Cl_2$ determined at liquid nitrogen temperature (110 k) oscillated with a period length of 22.8 min. A and B are from two different experiments

The differences in energies of the different spin isomers of water are small (less than 10^{-24} erg) (Emsley et al., 1965) and are much lower than the energy of thermal motion. Therefore, a spin-only interaction would not be expected to affect intermolecular interactions (Emsley et al., 1965). On the other hand, the absorption rates of ortho and para water from water vapor to various organic and inorganic sorbents have been observed to differ markedly. The binding of the para isomer with such preparations is distinctly faster than the binding of the ortho isomer. Binding avidity is most likely determined from differences in quantum statistics for the two different spin isomers as noted by Potekhin and Khusainova (2005). Estimates of the energy barriers that determine rates of absorption suggested by these authors that the difference in free energy barriers may exceed the energy of spin-spin and spin-orbit interaction by many orders (Potekhin and Khusainova, 2005). This raises the possibility that the spin state of water may substantially influence physical, chemical and biological phenomena including redox potential.

2.10 Period Length Determined by Ionic Radius of Liganded Cation

Chlorides in solution of other members of Group II, the so-called coinage metals such as silver and gold, also exhibited periodic oscillations but the period lengths were longer than those of copper. Some property of these different metals was then sought that might correlate with the period length of the oscillations. Ionic radius was found to do so (Fig. 2.20). Period length of the oscillations of an aqueous

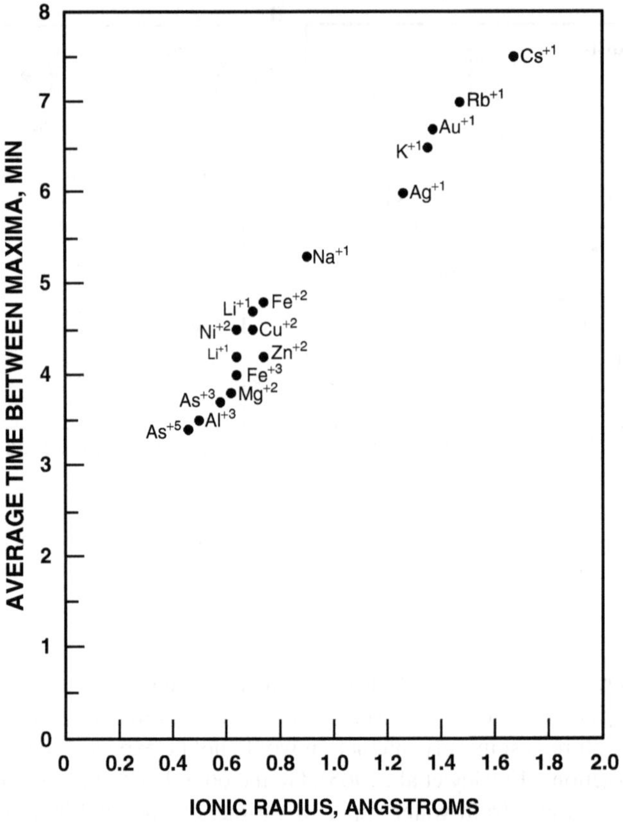

Fig. 2.20 Period length of the oscillations of an aqueous solution is directly proportional to the ionic radius of the cation present and independent of cation concentration. All solutions were tested as the chloride at a final concentration of 10 μM. Only with Cu^{+2} (replaceable by Ni^{+2}) was the asymmetric period length of 5.8 min X 5 = 24 min observed

solution is directly proportional to the ionic radius of the cation present and independent of cation concentration. All solutions were tested as the chloride at a final concentration of 10 μM. Only with Cu^{II} (replaceable by Ni^{II}) was the asymmetric period length of 5.8 min X 5 = 24 min observed.

2.11 Relationship to Cell Enlargement and Growth

Cell enlargement of both animal (Pogue et al., 2000; Wang et al., 2001; Fig. 2.21) and plant cells (Morré et al., 2001a) and tissues (Millet and Badot, 1996; Morré et al., 2002d) is periodic exhibiting a 24 min period. The linear phase of cell enlargement correlates with the protein-disulfide thiol interchange portion of the ENOX cycle (Morré and Morré, 2003a; Fig. 2.2). The period length is independent of temperature (Table 2.1) and in cancer cells expressing ENOX2 (tNOX) the

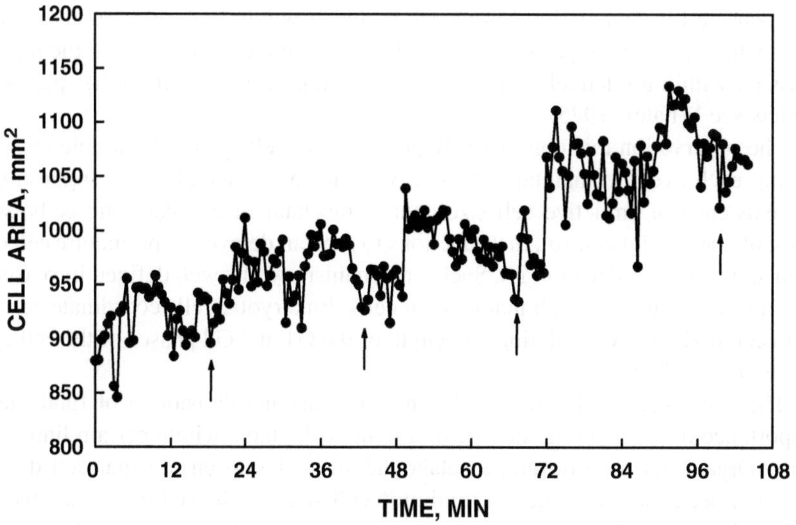

Fig. 2.21 Rate of cell enlargement of animal and plant cells oscillates with a period length of 24 min. Illustrated is the enlargement of a single cyto cell measured by automated image enhanced light microscopy. Cell enlargement proceeds in bursts over 12 min separated by rest periods of about 12 min where the cells actually shrink (arrows). The maxima contained within the elongation phase correspond to the protein disulfide-thiol interchange portion of the NOX cycle when determined in parallel (Morré and Morré, 2003a)

oscillations as well as cell enlargement are blocked by ENOX2 inhibitors, EGCg and phenoxodiol (Yagiz et al., 2007).

A model has been proposed involving a transplasmamembrane located AAA-ATPase whereby the interchange activity is responsible for linking the AAA-ATPase to membrane proteins in a process involving force-extension and membrane protein displacement (Morré et al., 2006). Oscillatory membrane vesicle enlargement with a period length equal to that of the ECTO-NOX protein present has been reconstituted in a completely cell-free system containing purified lipids, albumin as a source of protein thiols, recombinant AAA-ATPase and recombinant ECTO-NOX protein (Morré et al., 2006).

Cell growth (cell enlargement) is as fundamental for growth of organisms as is cell division. Implicit in cell growth is an increase in cell size, i.e., cell enlargement. Without cell enlargement, no organism can continue to grow. Cells unable to enlarge eventually are unable to divide such that growth will cease. Yet, compared to cell division, the enlargement phase of cell growth has been largely neglected by cell biologists. Moreover, it is this part of the growth process where the ECTO-NOX proteins appear to be most closely involved.

Proliferating cells in culture tend to double both their size and their mass before each division (Mitchison 1971). A dividing cell, if spherical, in order to double in volume, must increase the surface covered by the plasma membrane by a factor of 1.6 (Graham et al., 1973). In addition, they must produce membrane material to compensate for degradation (turnover). Included within this category are dividing cells which form large parts of their surface membrane de novo along the plane of division within a relatively short time such as during mitosis in higher plants (for review see Whaley, 1975).

The observation that continuously proliferating cells precisely double their size during each cycle to maintain constant volumes has resulted in the suggestion of the existence of an active cell size control mechanism in eukaryotic cells, some form of checkpoint control, that functions to prevent delayed or premature cell division at inadequate size or mass. Such a mechanism seems well defined in yeasts but remains an open issue with mammalian cells. Eukaryotic cells coordinate cell size with cell division by regulating the length of the G1 and G2 phases of the cell cycle (Sweiczer et al., 1996).

The coordination of cell growth (enlargement) and division is of fundamental importance in maintaining the size of a living cell stable within certain limits over time. Organisms have developed elaborate strategies to ensure that cell division does not occur until a particular minimum cell size is achieved that guarantees the production of viable progeny cells after cell division. The pathways that regulate growth are often deranged in human diseases such as cancer.

For rat hepatocytes, the absolute rate of plasma membrane protein synthesis has been estimated to be about 7 fg/cell per min (Franke and Kartenbeck, 1976).

A function as an integral part of the driving mechanism for cell enlargement emerges as one primary function of the ultradian ENOX cycle. The alternation of activities likely is essential to its participation in cell enlargement. Cycle maxima ③, ④ and ⑤ result in increased cell surface area and, correspondingly, cell volume. During cycle maxima ① and ② cell enlargement rests and the cells may actually shrink. It would seem reasonable that this oxidative portion of the ENOX cycle would serve some essential function in the cell enlargement process as well involving plasma membrane electron transport per se. As protons and electrons from cytosolic NADH are moved across the plasma membrane sufficient energy is generated to create a substantial membrane potential or even to generate ATP or ATP equivalents required in the force extension part of the cycle linked to maxima ③, ④ and ⑤. The metabolic advantage of the 2 + 3 pattern of oscillations must have a fundamental basis in evolution and may rest with the fundamental cellular process of increase in cell size prior to cell division. As such, a clock function would be a secondary, albeit important, function of synchronizing the activities of living organisms to the 24 h day/night earth's rotational cycle.

2.12 Phasing of the Rhythm

Since the operation of the "copper clock" requires a high degree of synchrony in solution, there must be some means of simultaneously phasing the oscillatory pattern. In cells and tissues red (Morré et al., 1999b) and blue (Morré et al., 2002c; Morré and Morré, 2003b) light set the rhythm so that at day break all of

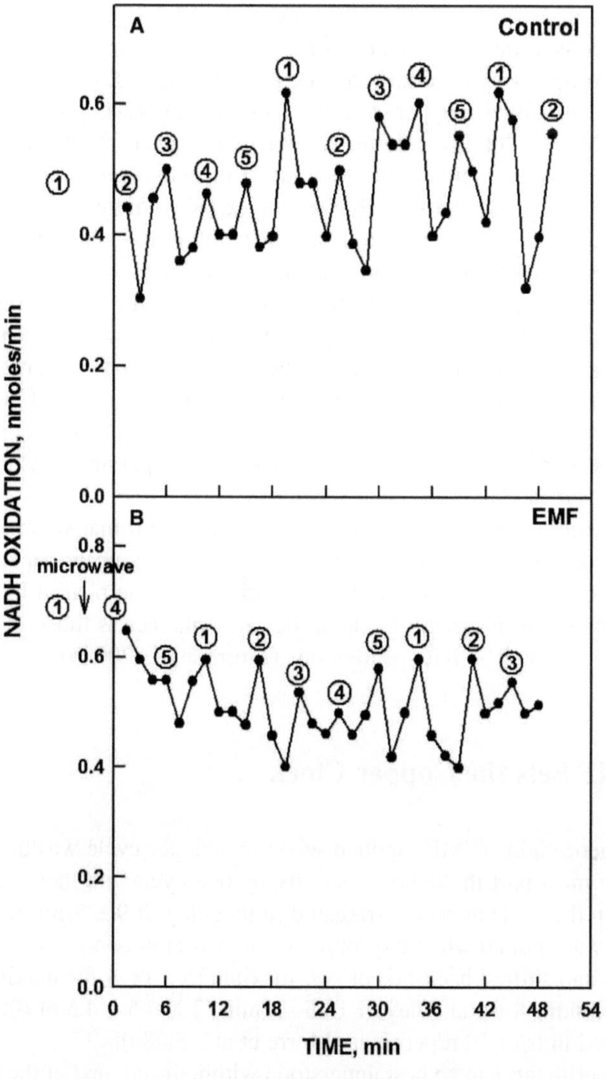

Fig. 2.22 The copper clock is phased (set) by EMF. Oscillatory pattern of NADH (120 μM) oxidation catalyzed by buffered (50 mM Tris-MES) 100 μM copper at pH 6.0 in the absence of protein altered by EMF. Activity maxima labeled ① and ② are separated by 6 min and are followed by activity maxima labeled ③, ④ and ⑤ which are separated by 4.5 min. The two activity maxima separated by 6 min recur every 24 min to establish the 24 min periodic activity. The two assays were carried out in parallel with the same copper solution using two paired spectrophotometers with the only difference being the EMF exposure as indicated for b which displaced activity peak ① ahead by about 15 min or back by about 9 min to coincide with peak ④ in the control preparation. The actual shift depended where in the cycle the EMF was initiated. Similar results were observed with both ENOX1 and ENOX2 preparations both as soluble proteins and associated with cells and membranes (Morré et al., 2008b)

the cells in a sun-exposed plant are in synchrony. Blue light also synchronizes in mammalian cells as one potential basis for the successful use of blue light therapy to treat pathological skin conditions such as psoriasis. Light, however, has no direct effect on either copper solutions or ECTO-NOX proteins. Melatonin (Morré and Morré, 2003b), lithium (Kromkowski et al., 2008) and caffeine (J. Kromkowsky, Purdue University, 2007 unpublished results) do directly affect CNOX by phasing the rhythm. Presumably CNOX proteins specifically bind these substances in some manner to result in synchronization of function. ECTO-NOX protein in solution also autoentrain (Morré et al., 2002b). If two out of phase preparations are mixed and incubated together for several hours to overnight, the combined preparations will adjust their oscillatory cycles to generate a new period with a maximum located midway between the two original periods. However, autoentrainment, melatonin, lithium or caffeine, all of which entrain CNOX (ENOX2) for example, have no effect on entrainment of copper in solution. Some other means of achieving synchrony must operate to ensure synchrony of the copper clock.

Findings from several laboratories suggest that electromagnetic fields affect a number of processes that appear to be clock-related. In a recent paper, we provide results from $Cu^{II}Cl_2$ solutions that the oscillations may be phased by exposure to low frequency electromagnetic fields in the same manner as those responsible for ECTO-NOX-mediated activity oscillations (Morré et al., 2008b).

2.13 EMF Sets the Copper Clock

Electromagnetic fields (EMF) applied when the copper cycle was at maxima②, ③ or ④, for the most part the delayed activity by one cycle and then resumed at the same point in the cycle as when irradiated [total delay of 9 ± 5 min (2 X 4.5 min)]. When EMF was applied when the copper cycle was at maxima ① or ⑤, the activity was delayed and shifted back one or two maxima to ④ or ⑤ for maxima ① or to ③ or ④ for maximum ⑤ (total delay of 13.5–16 min (2 X 4.5 + 4.5 or 6). This pattern was confirmed in over 30 repetitions (Morré et al., 2008b).

These experiments can be best understood within the context of the copper clock illustrated in Fig. 2.8. Within a narrow span of several days, the copper clock will keep very accurate time. However, abrupt phase shifts of 5–15 min were observed on occasion in preparations unshielded from environmental microwave sources. Shielded preparations remained synchronous longer. Synchrony was not restricted to a particular copper solution or source.

The oscillatory behavior of ECTO-NOX activity of HeLa cells grown in culture, ECTO-NOX protein released from the surface of HeLa cells by treatment with 0.1 M sodium acetate, pH 4.5 or recombinant ECTO-NOX (tNOX or ENOX2), all showed a phase shift when exposed to 2 min of 50 Hz EMF exposure (Morré et al., 2008b). These experiments were carried out in parallel on two aliquots of the same preparation, one exposed to EMF and one not. Analyses were in side by side

Hitachi U3210 spectrophotometers with EMF exposure being the only variable. Similarly ECTO-NOX-containing plasma membrane vesicles isolated from a plant source (soybean) also exhibited an oscillatory pattern of NADH oxidation phase-shifted by microwave irradiation. The period length was 24 min. As with the human ECTO-NOX, the period length was unaltered by microwave irradiation. However, the phase of the pattern of oscillation was shifted by about 6 min in the experiment and by about 12 min in another experiment.

What we have proposed as the basis for the copper clock is a relatively slow periodic change in the shape of the orbitals occupied by the valence electrons of Cu^{II} that gives rise to a progression of $copper^{II}$ aqua ion species with varying redox potential that reach a maximum every 4.5 min (Morré et al., 2007). Asymmetry to the pattern of periodicity appears to be imparted by interaction with the paired (ortho-para equilibration) spins of the protons of the six bound waters (Morré et al. 2008a).

There has been public concern about the possibility that exposure to extremely low frequency magnetic fields, particularly those generated by electrical transmission lines, distribution lines, and electrical appliances, may influence human carcinogenesis (Fedrowitz et al., 2002; Thun-Battersby et al., 1999; Wolf et al., 2005; Yoshizawa et al., 2002; Simko et al., 2001; Henshaw, 2002). Some but not all epidemiological studies have suggested that electromagnetic fields may be associated with increased incidence of cancer (Simko et al., 2001). Thun-Battersby et al. (1999) showed that 50-Hz magnetic fields of low (100 microTesla = 100 μT) flux density enhanced mammary gland tumor development and growth in the 7,12-dimethylbenz (a) anthracene model of breast cancer in female Sprague-Dawley rats. Elevated magnetic field exposures have been associated with increased risk of childhood leukemia as well (Henshaw, 2002).

HL-60 leukemia cells, Rat-1 fibroblasts and WI-38 diploid fibroblasts when exposed for 24 72 h to 0.5–1.0 mT of a 50-Hz magnetic field exhibited dose-dependent increases in the proliferation rate up to about 30% after 72-h of exposure (Wolf et al., 2005). An increased percentage of cells in the S-phase accompanied the increase in proliferation. A dose-dependent increase in DNA strand breaks in response to EMF was observed along with increased formation of 8-OHdG adducts by pretreatment of cells with the antioxidant alpha-tocopherol as was an increase in dichlorofluorescein-detectable reactive oxygen species in Raf-1 fibroblasts suggesting involvement of free radical species. Cells also exhibited modifications of NFκB-related proteins (p65-p50 and IκBα), suggestive of increased formation of p65-p50 or p65-p65 active forms, a process they attributed to redox reactions possibly involving free radical species as well.

The ability of weak electromagnetic fields to affect human cells and human health has been the subject of numerous cellular and epidemiological studies, with a variety of results. Reactive oxygen species, immunological responses, inflammation, cell proliferation, wound healing, developmental processes, and nerve regeneration may be affected by electromagnetic fields (Byus et al., 1988; Goodman et al., 1993; Lyle et al., 1988; Matanoski, 1995; Savitz, 1995; Sisken et al., 1993).

A sinusoidal 50 Hz magnetic field with a magnetic flux density of 0.5 mT induced variations in the expression of cell adhesion molecules (CAMs) in two human osteosarcoma cell lines (MG-63 and SAO-2) (Santini et al., 2003). Among the CAMs effected were VLA-2, the integrin receptor for collagen, and VLA-5, the integrin receptor for fibronectin. Angiogenesis was analyzed by Ruggiero et al. (2004) using the chick embryo chorioallantoic membrane assay. They reported that exposure to a 0.2 T static magnetic field blocked the angiogenic response induced by treatment with prostaglandin E_1 and fetal calf serum. Chakkalakal et al. (1999) reported that a relatively low dose (0.1 mg/ml) of adriamycin combined with a magnetic field inhibited proliferation of human osteosarcoma cells in culture by 82%, whereas the magnetic field or adriamycin alone caused only 19% and 44% inhibition, respectively. Such observations offer the possibility of lowering the therapeutic dose of drugs by combination with EMF.

To what extent these EMF effects on biological systems may also relate to altered states of liquid water (Binhi and Stepanov, 2000) is not known. The alterations induced by the EMF might be transduced on a biological level as a result of participation of water in various metabolic reactions as we proposed for the ECTO-NOX related oscillations.

Metastable changes in physical properties of liquid water have been discussed earlier in both experimental and theoretical works as the target for EM fields. Low frequency spectra of the electric conductivity of liquid water under EM field exposure were studied by Fesenko and Gluvstein (1995) and ascribed to formation of clusters of water molecules. Metastable changes of water structure were observed (Lobyshev et al., 1999) as changed UV luminescence spectra and assumed to stem from defects of water structure at the microscopic level. Nuclear spins of protons of water were considered as primary target for the external magnetic field by Binhi and Stepanov (2000). It was suggested that proton spins take part in spin-orbit interactions thus modifying proton motion and influencing the rate of formation and breakup of water clusters.

A central problem is how the copper based oscillations become synchronized in nature even in the absence of light and outside a living organism. Synchrony seems to occur at irregular intervals, perhaps every 3–5 days. Since synchrony is achieved by exposure to low frequency electromagnetic fields, it may result in nature by sensing the earth's electromagnetic field. Large changes (ca. 800 nT) in the earth's magnetic field occur globally (http://www.antidiv.gov.au/default.asp?casid = 2140) at irregular intervals similar to those we have observed for the natural resetting of the copper clock.

Minorsky (2007) reported that infradian rhythms in bean seed inhibition correlate with extremely small oscillations in the vertical vector (Bz) of the interplanetary magnetic field. He concluded that the positive correlations between the lunar phase and seed inhibition that were reported previously (Brown and Chow, 1973; Brown, 1977; Spruyt et al., 1987) were purely coincidental. The alternative hypothesis proposed by Minorsky (2007) was that the 7-day and 14-day rhythms in bean seed inhibition were related to subharmonics of the 27.3-day solar rotation cycle and perhaps synchronized by EMF in much the same manner as proposed by us for the copper clock and for ECTO-NOX proteins (Morré et al., 2008b).

2.14 ENOX Clock and Cancer

Several carefully designed studies beginning in the 1960s (see Levi 2002 for literature) document that disruption of circadian rhythms affects tumor development. Constant light exposure (as in people that work predominantly at night) or pinealectomy accelerates breast epithelial stem cell proliferation, induces mammary-gland development, increases the formation of mammary tumors in rodent models and is correlated with increased incidence and development of mammary tumors. The robustness of diurnal rythmicity also emerges as a significant predictor for survival of patients with breast cancer (Sephton et al., 2000). In patients with metastatic colorectal cancer, a 2-year study showed that patients with clearly defined rest and activity rhythms has a fivefold higher survival rate and experienced less fatigue than patients with diminished rest/activity rhythms (Mormont et al., 2000).

The dominant ECTO-NOX form of cancer cells (ENOX2 or tNOX) has a period length of 21–22 min rather than 24 min. A 21 or 22 min NOX period would be expected to generate a 21 or 22 h circadian day as observed in activity patterns of some cancer patients (Levi, 2000). ENOX at the surface of cancer cells is drug responsive and it should be possible to time chemotherapy, for example, to where the cancer cells and tissues are maximally sensitive to the drug compared to normal cells and tissues to increase the margin of safety. This is the principle of chronotherapy in oncology which is to take advantage of the asynchronies in cell proliferation and drug metabolic rhythms between normal and malignant tissues by administering therapy at a specific time of day. This has been accomplished on an empirical basis with patients and could potentially minimize the damage of chemotherapies to host tissue and maximize toxicity in tumors (Levi, 2002).

2.15 Are the NOX Oscillators Linked to the Drivers of the Circadian Clock and How Are They Linked?

The NOX oscillations appear somehow to be linked to the drivers of the circadian 24 h clock. Such indications come from two sources:

1. Transfection experiments with tNOX cysteine replacements
2. Correlative experiments with heavy water, lithium and caffeine, among the few substances known to alter the period length of the circadian day

The transformation experiments were with COS cells transfected with tNOX cDNAs in which codons for specific cysteines were replaced by alanine codons. The result was tNOX proteins having periodic oscillations of 22 (normal tNOX), 36 and 42 min. The circadian period of COS cells was synchronized by exposure to light. The activity of a common glycolytic house keeping protein glyceraldehyde-3-phosphate dehydrogenese (GAPDH), which normally exhibits a 24 h periodicity, was used as the measure of changes in circadian day length. In all three instances, the 24 h circadian period (corresponding to the 24 min period of the constitutive CNOX) was retained by the transfected cells but, in addition, 22, 34 and 42 h circadian daily rhythms were superimposed corresponding, respectively, to the 22, 34 and 42 min period lengths exhibited by the tNOX proteins generated as a result of the transfection.

Equally or perhaps even more indicative are the results from substances known to alter the period length of the circadian day in living organisms. Paramount are the studies with heavy water (D_2O) summarized in Section 5 where both the ENOX period and the copper clock period lengths are increased by 30% from 24 min to about 30 min in keeping with the response of the circadian day to D_2O which is to increase the period length from about 24 h to about 30 h.

Lithium, used in the treatment of bipolar disorder, also has been reported to lengthen the circadian day by about 3 h and alters the ENOX and copper clock cycles accordingly from 24 to 27 min (Kramkowsky et al., 2008). Observations of rapidly-cycling manic-depressive patients have suggested that some may have free-running circadian rhythms which cycle faster than once per 24 h (Abe et al., 2000; Kripke et al., 1978). The therapeutic effect of lithium could be to slow or delay these rhythms (Kripke et al., 1978; Welsh and Moore-Ede, 1990).

Lithium does delay the sleep-wake rhythm both in healthy normal subjects (Kripke et al., 1979) and since lithium may be changing the circadian clock, bipolar disorder could, in fact, be a circadian clock disorder. The relationship of lithium to the treatment of bipolar disorder is based on the hypothesis that individuals with bipolar disorder suffer from a disturbance of the master clock in the brain and that sleep cycles, body temperature and metabolism are not properly synchronized (Abe et al., 2000; Yin et al., 2006). Lithium is thought to restore these daily rhythms (Atkinson et al., 1975; Kripke et al., 1978; Wehr and Goodwin, 1983; Yin et al., 2006). Similarly sleep phase advance has been suggested to sustain the effects of total sleep deprivation both with or without a combined antidepressant drug treatment (Benedetti et al., 2001). A single sleep deprivation followed by 3 days of sleep phase advance (beginning with sleep allowed from 17:00 until 24:00, with daily shift backs of 2 h) in consecutively admitted bipolar depressed inpatients who were taking chronic lithium salts treatments resulted in sustained acute antidepressant effect of the total sleep deprivation and lithium enhancement of the effect of the chronobiological treatment. The improved clinical outcome observed in lithium-treated patients was attributed to the phase delaying effect of lithium on biological rhythms, leading to a better synchronization of biological rhythms within the sleep-wake cycle (Benedetti et al., 2001). Lithium delays circadian rhythms in the isolated *Aplysia* eye (Engelmann, 1973) and slows or delays circadian rhythms in *Kalanchoe* (Engelmann, 1972), a plant, in cockroaches (Engelmann, 1973; Hofmann et al., 1978) and in rodents (Kripke and Wyborney, 1980). With *Kalanchoe*, a concentration of 1 mM LiCl was demonstrated to lengthen the period by more than 1 h from 22.3 h in the control to 23.6 h for the lithium-treated plants (Engelmann, 1972). With rats, lithium increased the free-running circadian wheel-running rhythm in the laboratory from 24 to 24.4 h (Kripke and Wyborney, 1980).

That lithium increases the length of the circadian day by about 4% supports our earlier work that constitutive ECTO-NOX proteins are the drivers of the cells circadian clock (Morré et al., 2002a). A second effect of lithium to induce a new set of oscillations with a period length slightly shorter than 24 min (Kromkowsky et al., 2008) was unexpected but might also have relevance to bipolar disorder and the use of lithium in its treatment.

Elevated anxiety symptoms have been observed among consumers of large amounts of caffeine and cases of caffeine-induced manic depression with improvements related to discontinuation of caffeine in the diet have been reported (Iancu et al., 2007). In unpublished work, J. Kromkowsky et al. (2007 Purdue University) observed a lengthening of the ENOX cycle of about 1 min per period in the presence of 0.4 μM caffeine.

Additional evidence for the ENOX protein being the ultradian driver of the circadian clock comes from light phasing of the ENOX activity and the circadian day by red and blue light in plants and by blue light with mouse skin and clinical patients with blue light responsive skin disorders (J. H. Wilkins, D. J. Morré and D. M. Morre, unpublished results, 2006) as well as from a response to melatonin with both plant and animal CNOX where a new maximum ① appears exactly 24 min after melatonin addition. With blue light, the new maximum appears after 12 + 24 = 36 min after light exposure of the tissues. Neither red nor blue light nor melatonin alters the copper clock in the absence of the ENOX protein. How then are the ultradian ENOX and the circadian clocks linked? Since putative light receptors (phytochrome for red light in plants; crytochrome for blue light) are largely thought to be cytosolic, it would be reasonable to consider a transmembrane receptor system that might link ENOX proteins to the cells interior. Dr. X. Tang in our laboratory generated an anti-idiotypic antibody to a peptide complementary to the selected regions of tNOX which might be expected to recognize a cell surface receptor if one exists. Several candidate proteins have been located using this antibody including a promising candidate that does bind tNOX.

The pathway nevertheless remains obscure as does the relationship, if any, between the ENOX 24 min cycle and the eight or more core circadian genes that have been identified so far along with the attendant control models of transcription-translation feedback loops (See contribution by U. Schibler, this volume). Presumably the mechanism would be somewhat analogous to a traditional mechanical clock that counts pendulum swings (the 24 min ultradian rhythm) to drive a 2 h rhythm or gear, for example, which would engage 12 times in a sequence that would eventually generate the 24 h circadian day (see 2.18).

2.16 ENOX Proteins Have Prion-Like Properties: Prions Are Redox Proteins

No discussion of ENOX proteins would be complete without discussion of their prion-like properties. ENOX proteins exhibit most properties of prions including resistance to proteases, resistance to cyanogen bromide digestion, and an ability to form amyloid filaments closely resembling those of spongiform encephalopathies and all of which are characteristics of PrP^{sc} (PrP^{res}), the presumed infective and proteinase K resistant particle of the scrapie prion (Kelker et al., 2001). The tNOX protein from the HeLa cell surface co-purified with authentic glyceraldehyde-3-phosphate dehydrogenase (muscle form) (GAPDH). Surprisingly, the tNOX-associated muscle GAPDH then became proteinase K resistant. Combination of authentic rabbit muscle GAPDH with tNOX rendered the GAPDH resistant to proteinase K digestion. This property, that of converting the normal form of a protein into a likeness of itself, is one of the defining characteristics of the group of proteins designated as prions.

Prions also exhibit properties characteristic of ENOX proteins (Kim and Morré, 2004). Both recombinant full-length mouse prion protein expressed in *Escherichia coli* and native prion protein (PrP^{sc}) from mouse brain exhibited NADH oxidase and protein disulfide-thiol interchange activities similar to those of ENOX proteins. The two activities exhibited the complex 2 + 3 pattern of oscillations characteristic of ENOX proteins where the two activities alternate to generate a period length of 24 min. The oscillations were augmented by copper and diminished by addition of the copper chelator bathocuproine. That the activity might be attributable to a contaminating protein was ruled out by experiments where the purified recombinant prion-containing extracts were resolved by SDS-PAGE and the activity was restricted to a single band corresponding to the predicted Mr of the recombinant prion as verified by Western blot analyses.

2.17 A Prion Mechanisms Explains How the Herbicide 2,4-D Kills Plants

The hormone-stimulated and growth-related cell surface hydroquinone (NADH) oxidase activity from plants also oscillates with a period of about 24 min or 60 times per 24-h day. Plasma membranes of hypocotyls from dark-grown soybeans contain two such NADH oxidase activities that have been resolved by purification on concanavalin A columns. One in the apparent molecular weight range of 14–17 kDa is stimulated by the auxin herbicide 2,4-dichorophenoxyacetic acid (2,4-D). The other is larger and unaffected by 2,4-D. That the auxin-stimulated cell surface NADH oxidase was distinct from the constitutive oxidase was first indicated by Morré and Grieco (1999) using quassinoid growth inhibitors. The constitutive NADH oxidase activity and the constitutive cell elongation were inhibited by simalikalactone D but not by glaucarubolone, whereas the 2,4-dichlorophenoxyacetic acid- (2,4-D) stimulated NADH oxidase activity and the 2,4-D-stimulated rate of cell enlargement were inhibited by glaucarubolone but not by simalikalactone D (Morré and Grieco, 1999).

The 2,4-D-stimulated activity absolutely requires 2,4-D for activity and exhibits a period length of about 24 min. Also exhibiting 24 min oscillations is the rate of cell enlargement induced by the addition of 2,4-D or the natural auxin growth regulator, indole-3-acetic acid (IAA). Immediately following 2,4-D or IAA addition, a very complex pattern of oscillations is frequently observed (Fig. 2.23). However, after several h a dominant 24 min period emerges at the expense of the constitutive activity. The sum of the two activities remains constant. A recruitment process whereby an altered protein form converts a normal form of a protein into a likeness of itself analogous to that exhibited by prions (Griffith, 1967; Prusiner et al., 1998) has been postulated to explain this unusual behavior to a growth-regulating herbicide (Morré et al., 2003; Fig. 2.24).

2,4-D is a synthetic auxin which was discovered and developed during World War II as a selective herbicide for general weed control in corn, small grains and other grasses. Auxins, both natural and synthetic, as a class, stimulate plant cells to increase in size (enlarge). An auxin-stimulated ENOX activity of plasma membrane fractions of soybean was reported early (Morré et al., 1986). The activity was purified and shown to require auxin addition for activity (Brightman et al., 1988). A variety of mechanisms have been proposed to understand how 2,4-D and the natural auxins such as IAA promote cell enlargement. Our findings implicate surface hydroquinone oxidases (NOX proteins) as drivers of the cell enlargement process, one of which requires auxin for activity. The new 2,4-D-induced period then persists as growth of affected tissues and plant parts becomes unregulated and cancer-like. The ultimate result with the 2,4-D-sensitive species is death of the plant or of the affected plant parts.

2.18 Why Oscillate? A Consequence of Active Sites in Metallo-proteins?

There are no clear explanations as to why ultradian rhythmic activities may offer adaptive advantages over continuous manifestations of the same activities. One of these might be that periodic function provides opportunities for sequential alternations of activities. One such example is the molecular segmentation clock involved in the formation of somites – regions of pre-somitic mesoderm that give rise to the precursors of the vertebral column and segmented muscles (Andrade et al., 2007; Shifley and Cole, 2007). These structures appear with an ultradian period of one every 2 h (120 min), i.e., every five ENOX1 24 min oscillations. If the 2 + 3 pattern of periodic oscillations persists through successive additional rounds of signal transduction, a 2 h ultradian period (5 × 24 min) might be anticipated for one of these (see 2.16).

The involvement of ECTO-NOX proteins in growth may actually be the primary functional outcome of the oscillatory pattern at the cellular level with a time-keeping role being a secondary but none-the-less important outcome. During the protein disulfide-thiol interchange portion of the cycle, cells undergo three cycles of cell enlargement. This phase then rests during the oxidative portion of the cycle where plasma membrane electron transport occurs presumably to fulfill some metabolic requirement to continue the cell enlargement process. The 24 min ultradian period

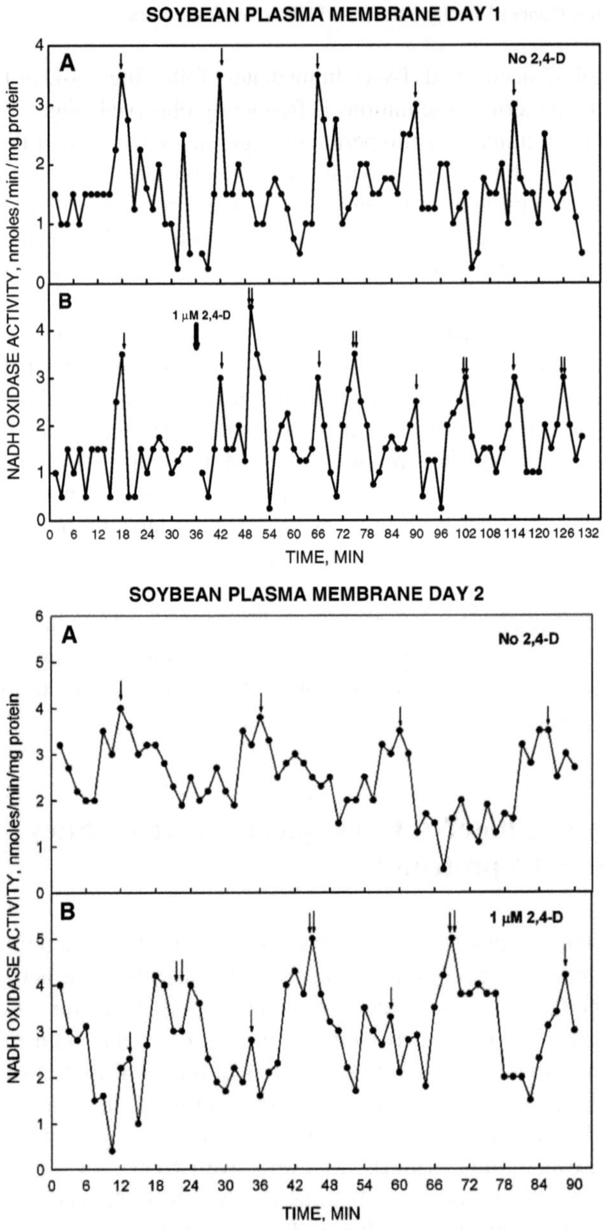

Fig. 2.23 Oscillations in NADH oxidase activity of soybean plasma membranes in the absence (**A**) and presence (**B**) of 1 μM 2,4-D on day 1 (upper panel). The assays were carried out simultaneously with two spectrophotometers in parallel and with the same plasma membrane preparation. The characteristic CNOX period length of 24 min was seen clearly in both preparations (single arrows). In (**B**), a second 2,4-D responsive component, not seen in (**A**) in the absence of 2,4-D appeared with a period length of about 24 min (double arrows). The lower panel shows the activity of the same two preparations measured in parallel on day 2 after 2,4-D addition. The single arrows indicate the original CNOX activity still present in both A and B but very much diminished in B. In contrast, the 2,4-D induced activity illustrated in B is augmented in activity to a value approximating the sum of the two activities originally present (Morré et al., 2003)

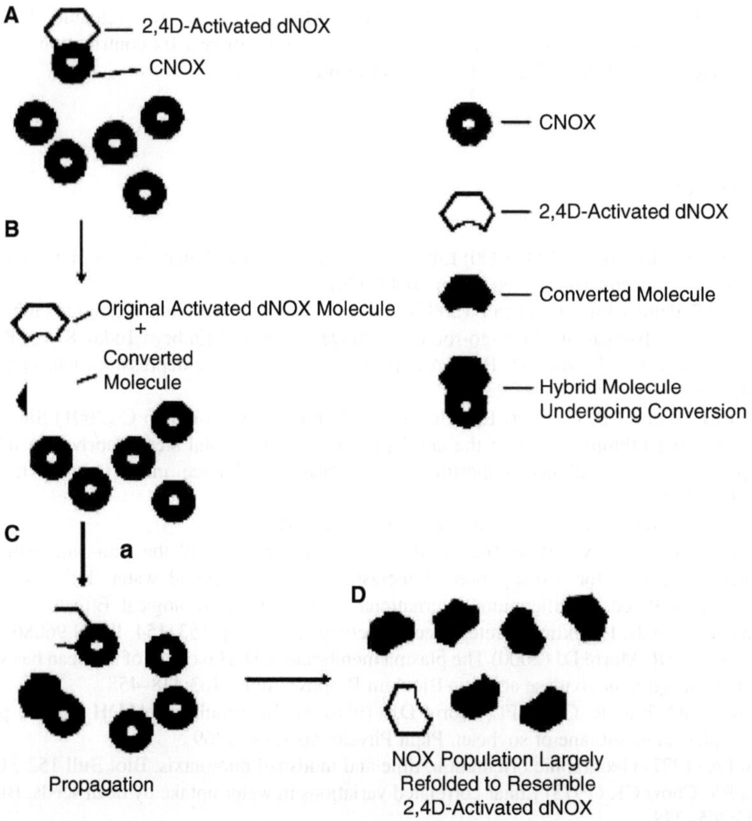

Fig. 2.24 Diagram to explain the observations of Fig. 2.23. An active recruitment of CNOX proteins by dNOX proteins is postulated, followed by a conversion of one (CNOX) into a likeness of the other (dNOX), much in the manner of the creation of infective prion proteins from non-infective prion proteins by an interaction of one with the other. (**A**) A 2,4-D-activated dNOX is postulated to recruit normal CNOX proteins and cause them to undergo a conformational alteration to allow their entrainment with other 2,4-D-stimulated molecules. This process continues long after the 2,4-D has been metabolized and leads to the death of the plant. (**B**) The converted molecules (shaded) are then considered to participate in the conversion of other CNOX proteins. (**C**) The net result is a propagation of the 2,4-D response and a net loss of unconverted CNOX. (**D**) Eventually the bulk of the NOX population becomes refolded through entrainment to produce predominantly a single 2,4-D-stimulated oscillatory NOX activity and growth response resembling that of the original dNOX (Morré et al., 2003)

is apparently unique to copper and possibly nickel. The 22 min tNOX period may be modulated by zinc but not as a redox metal since the valence of zinc does not undergo redox changes. To what extent other metals might drive ultradian rhythms when appropriately linked to catalytic centers of proteins provides an intriguing possibility for future study.

Acknowledgments We thank Natalie Bronstein, Mercy College, for valuable discussions regarding the clock-regulated generation of somites, Elyse Freiberger for contributions to data of Figs 2.9 and 2.19 and Peggy Runck for manuscript preparation.

References

Abe M, Herzog ED, Block GD (2000) Lithium lengthens the circadian period of individual suprachiasmatic nucleus neurons. Neuroreport 11:3261–3264

Andrade RP, Palmeirim I, Bajanca F (2007) Molecular clocks underlying vertebrate embryo segmentation: a 10-year-old hairy-go-round. Birth Defects Res C Embryo Today 81:65–83

Atkinson M, Kripke DF, Wolf SR (1975) Autorhythmometry in manic-depressives. Chronobiologia 2:325–335

Benedetti F, Barbini B, Campori E, Fulgosi MC, Pontiggia A, Colombo C (2001) Sleep phase advance and lithium to sustain the antidepressant effect of total sleep deprivation inbipolar depression: new findings supporting the internal coincidence model? J Psychiatr Res 35:323–329

Bersuker IB (1984) Modern Chemistry. Plenum, New York

Binhi VN, Stepanov EV (2000) Tunable diode-laser spectroscopy of the para- and ortho-water vapour as a tool for investigation of metastable states of liquid water. In: Kostarakis P, Stavrolakis P (eds), Millennium International Workshop on Biological Effects of Electromagnetic Fields, Heraklion, Crete, Greece, October 17–20, pp 153–154, ISBN 9608673305

Bridge A, Barr R, Morré DJ (2000) The plasma membrane NADH oxidase of soybean has vitamin K(1) hydroquinone oxidase activity. Biochim Biophys Acta 1463:448–458

Brightman AO, Barr R, Crane FL, Morré DJ (1988) Auxin-stimulated NADH oxidase purified from plasma membrane of soybean. Plant Physiol 86:1264–1269

Brown FA (1977) Geographic orientation, time and mudsnail phototaxis. Biol Bull 152:311–324

Brown FA, Chow CK (1973) Lunar correlated variations in water uptake by bean seeds. Biol Bull 145:265–278

Bruce VG, Pittendrigh CS (1960) Temperature independence in a unicellular "clock." J Cell Comp Physiol 56:256–231

Bunning E, Baltes J (1963) Zur Wirkung von schwerem Wasser au die endogene Tagersrhythmik. Naturwissenschaften 50:622

Buntkowsky G, Walaszek B, Adamczyk A, Xu Y, Limbach H-H, Chaudret B (2006) Mechanism of nuclear spin initiated para-H2 to ortho-H2 conversion. Phys Chem Chem Phys 8:1929–1935

Byus CV, Kartun K, Pieper S, Adey WR (1988) Increased ornithine decarboxylase activity in cultured cells exposed to low energy modulated microwave fields and phorbol ester tumor promoters. Cancer Res 48: 4222–4226

Chakkalakal DA, Mollner TJ, Bogard MR, Fritz ED, Novak JR, McGuire MH (1999) Magnetic field induced inhibition of human osteosarcoma cells treated with adriamycin. Cancer Biochem Biophys 17:89–98

Chueh P-J, Kim C, Cho N, Morré DM, Morré DJ (2002a) Molecular cloning and characterization of a tumor-associated, growth-related, and time-keeping hydroquinone (NADH) oxidase (tNOX) of the HeLa cell surface. Biochemistry 41:3732–3742

Chueh P-J, Morré DM, Morré DJ (2002b) A site-directed mutagenesis analysis of tNOX functional domains. Biochim Biophys Acta 1594:74–83

Del Castillo-Olivares A, Yantiri F, Chueh P-J, Wang S, Sweeting M, Sedlak D, Morré DJ, Burgess J, Morré DM (1998) A drug-responsive and protease-resistant peripheral NADH oxidase complex from the surface of HeLa S cells. Arch Biochem Biophys 358:125–140

Dowse HB, Palmer JD (1972) The chronomutagenic effect of deuterium oxide on the period and entrainment of a biological rhythm. Biol Bull 143:513–524

Dunlap JC (1996) Genetics and molecular analysis of circadian rhythms. Annu Rev Genetics 30:576–601

Edmunds Jr. LN (1988) Cellular and Molecular Basis of Biological Clocks. Springer, New York/Berlin/Heidelberg, Germany p 497

Emsley JM, Feeney J, Sutcliffe LH (1965). High Resolution Nuclear Magnetic Resonance Spectroscopy, vol 1. Pergamon, Oxford

Engelmann W (1972) Lithium slows down the *Kalanchoe* clock. Z Naturforsch 27B:477

Engelmann W (1973) A slowing down of circadian rhythms by lithium ions. Z Naturforsch 28:733–736

Enright JT (1997) Heavy water slows biological timing processes. Z Vergl Physiol 72:1–16

Fedrowitz M, Westermann J, Loscher W (2002) Magnetic field exposure increases cell proliferation but does not affect melatonin levels in the mammary gland of female Sprague Dawley rats. Cancer Res 62:1356–1363

Fesenko EE, Gluvstein AY (1995) Changes in the state of water, induced by radiofrequency electromagnetic fields. FEBS Lett 367:53–55

Filipponi A, D'Angelo P, Pavel NV, Di Ciecco A (1994) Triplet correlations in the hydration shell of aquaions. Chem Phys Lett 225:150–155

Franke WW, Kartenbeck J (1976) Some principles of membrane differentiation. In: Müller-Bérat (ed), Progress in Differentiation Research. Elsevier/North-Holland, New York, pp 213–243

Fulton JL, Hoffman MM, Darab JG, Palmer BJ, Stein EA (2000) Copper (I) and copper (II) coordination structure under hydrothermal conditions at 325°C: an X-ray absorption fine structure and molecular dynamics study. J Phys Chem A 104:11651–11663

Gilberger T-W, Walter RD, Müller S (1997) Identification and characterization of the functional amino acids at the active site of the large thioredoxin reductase from *Plasmodium falciparum.* J Biol Chem 272:29584–29589

Gilberger T-W, Bergmann B, Walter RD, Müller S (1998) The role of the C-terminus for catalysis of the large thioredoxin reductase from *Plasmodium falciparum.* FEBS Lett 425:407–410

Goodman R, Chizmadzhev Y, Shirley-Henderson A (1993) Electromagnetic fields and cells. J Cell Biochem 51:436–441

Graham JM, Sumner MCB, Curtis DH, Pasternak CA (1973) Sequence of events in plasma membrane assembly during the cell cycle. Nature 246:291–295

Griffith JS (1967) Self replication and scrapie. Nature 215:1043–1044

Henshaw DL (2002) Does our electricity distribution system pose a serious risk to public health? Med Hypotheses 59:39–51

Hofmann K, Gunderoth-Palmowski M, Wiedenmann G, Engelmann W (1978) Further evidence for period lengthening effect of Li^+ on circadian rhythms. Z Naturforsch 33C:231–234

Iancu I, Olmer A, Strous RD (2007) Caffeinism: history, clinical features, diagnosis, and treatment. In: Smith BD, Gupta U, Gupta BS (eds), Cafeine and Activation Theory. CRC, Boca Raton, FL/London/New York, pp 331–347

Jiang Z, Morré DM, Morré DJ (2006) A role for copper in biological time-keeping. J Inorg Biochem 100:2140–2149

Kelker M, Kim C, Chueh P-J, Guimont R, Morré DM and Morré DJ (2001) Cancer isoform of a tumor-associated cell surface NADH oxidase (tNOX) has properties of a prion. Biochemistry 40:7351–7354

Kim C, Morré DJ (2004) Prion proteins and ECTO-NOX proteins exhibit similar oscillating redox activities. J Biochem Biophys Res Comm 315:1140–1146

Kim C, Layman S, Morré DM, Morré DJ (2005) Fourier transform infrared and circular dichroism spectroscopic analysis underlie tNOX periodic oscillations. Nonlinearity Biol, Toxicol Med (now Dose Response) 3: 391–413

Kishi T, Morré DM, Morré DJ (1999) The plasma membrane NADH oxidase of HeLa cells has hydroquinone oxidase activity. Biochim Biophys Acta 1412:66–77

Kliman RN, Hey J (1993) DNA sequence variation at the period locus within and among species of the *Drosophila melanogaster* complex. Genetics 133:375–387

Korshin GV, Frenkal AI, Stern EA (1998) EXAFS study of the inner shell structure in copper (II) complexes with humic substances. Envir Sci Technol 32:2699–2705

Kripke DF, Wyborney VG (1980) Lithium slows rat circadian activity rhythms. Life Sci 26:1319–1321

Kripke DF, Judd LL, Hubbard B, Janowsky DS, Huey LF (1979) The effect of lithium carbonate on the circadian rhythm of sleep in normal human subjects. Biol Psychiatry 14:545–548

Kripke DF, Mullaney DJ, Atkinson ML, Wolf S (1978) Circadian rhythm disorder in manic-depressives. Biol Psychiatr 13:335–351

Kromkowski, J, Hignite H, Morré DM, Morré DJ (2008) Response to lithium of a cell surface ECTO-NOX protein with time-keeping characteristics. Neurosci Lett 438:121–125

Kummer JT (1962) Ortho-para hydrogen conversion by metal surfaces at 21°K. J Phys Chem 66:1715

Lambeth JD, Cheng G; Arnold RS, Edens WA (2000) Novel homologs of gp91phox. TIBS 25:459–461

Levi F, (2000) Marked 24-h rest/activity rhythms are associated with better quality of life, better response, and longer survival in patients with metastatic colorectal cancer and good performance status. Clin. Cancer Res 6:3038–3045

Levi F (2002) From circadian rhythms to cancer chemotherapeutics. Chronobiol Int 25:459–461

Lobyshev VI, Shikhlinskaya RE, Ryzhikov BD (1999) Experimental evidence for intrinsic luminescence of water. J Mol Liquids 82:73–81

Lyle DB, Ayotte RD, Sheppard AR, Adey WR (1988) Suppression of T-lymphocyte cytotoxicity following exposure to 60-Hz sinusoidal electric fields. Bioelectromagnetics 9:303–313

Matanoski GM (1995) Electromagnetic fields: biological interactions and mechanisms. In: Blank M (ed), Advances in Chemistry 250:157–190

McDaniel M, Sulzman FM, Hastings JW (1974) Heavy water slows the *Gonyaulax* clock: a test of the hypothesis that D2O affects circadian oscillations by diminishing the apparent temperature. Proc Natl Acad Sci USA 71:4389–4391

Milenkl YY, Sibileva RN, Strzhemechny MA (1997) Natural ortho-para conversion rate in liquid and gaseous hydrogen. J Low Temp Phys 107:77–82

Millet B, Badot PM (1996) The revolving movement mechanism in *Phaseolus*: new approaches to old questions. In: Greppin H, Agosti RD, Bonzon M (eds), Vistas on Biorhythmicity. University of Geneva Press, Geneva, pp 77–98

Minorsky PV (2007) Solar-terrestrial effects on bean seed inhibition. Poster Abstracts. Am Soc Plant Biologists p186

Mitchison JM (1971) The Biology of the Cell Cycle. Cambridge University Press, Cambridge

Mormont MC, Waterhouse J, Bleuzen P, Giacchetti S, Jami A, Bogdan A, Lellouch J, Misset JL, Touitou Y, Levi F (2000) Marked 24-h rest/activity rhythms are associated with better quality of life, better response, and longer survival in patients with metastatic colorectal cancer and good performance status. Clin Cancer Res 6:3038–3045

Morré DJ (1998) NADH oxidase: a multifunctional ectoprotein of the eukaryotic cell surface. In: Asard H, Bérczi A, Caubergs RJ (eds), Plasma Membrane Redox Systems and Their Role in Biological Stress and Disease. Kluwer, Dordrecht, The Netherlands, pp 121–156

Morré DJ, Morré DM (1998) NADH oxidase activity of soybean plasma membranes oscillates with a temperature compensated period of 24 min. Plant J 16:279–284

Morré DJ, Grieco PA (1999) Glaucarubolone and simalikalactone D, respectively, preferentially inhibit auxin-induced and constitutive components of plant cell enlargement and the plasma membrane NADH oxidase. Int J Plant Sci 160:291–297

Morré DJ, Morré DM (2003a) Cell surface NADH oxidases (ECTO-NOX proteins) with roles in cancer, cellular time-keeping, growth, aging and neurodegenerative diseases. Free Radical Res 37:795–808

Morré DJ, Morré DM (2003b) The plasma membrane-associated NADH oxidase (ECTO-NOX) of mouse skin responds to blue light. J Photochem Photobiol B 70:7–12

Morré DJ, Navas P, Penel C, Castillo FJ (1986) Auxin-stimulated NADH oxidase (semidehydroascorbate reductase) of soybean plasma membrane: role in acidification of cytoplasm? Protoplasma 133:195–197

Morré DJ, de Cabo R, Jacobs E, Morré DM (1995) Auxin-modulated protein disulfide-thiol interchange activity from soybean plasma membrane. Plant Physiol 109:573–578

Morré DJ, Chueh P-J, Lawler J, Morré DM (1998) The sulfonylurea-inhibited NADH oxidase activity of HeLa cell plasma membranes has properties of a protein disulfide-thiol oxidoreductases with protein disulfide-thiol intherchange activity. J Bioenerg Biomembr 30:477–487

Morré DJ, Gomez-Rey ML, Schramke C, Em O, Lawler J, Hobeck J, Morré DM (1999a) Use of dipyridyl-dithio substrates to measure directly the protein disulfide-thiol interchange activity of the auxin stimulated NADH: protein disulfide reductase of soybean plasma membranes. Mol Cell Biochem 200:7–13

Morré DJ, Morré DM, Penel C, Greppin H (1999b) NADH oxidase periodicity of spinach leaves synchronized by light. Int J Plant Sci 160:855–860

Morré DJ, Pogue R, Morré DM (2001a) Soybean cell enlargement oscillates with a temperature-compensated period length of ca 24 min. In Vitro Cell Dev Biol-Plant 37:19–23

Morré DJ, Sedlak D, Tang X, Chueh P-J, Geng T, Morré DM (2001b) Surface NADH oxidase of HeLa cells lacks intrinsic membrane binding motifs. Arch Biochem Biophys 392:251–256

Morré DJ, Chueh P-J, Pletcher J, Tang X, Wu L-Y, Morré DM (2002a) Biochemical basis for the biological clock. Biochemistry 41:11941–11945

Morré DJ, Lawler J, Wang S, Keenan TW, Morré DM (2002b) Entrainment in solution of an oscillating NADH oxidase activity from the bovine milk fat globule membrane with a temperature-compensated period length suggestive of an ultradian time-keeping (clock) function. Biochim Biophys Acta 155:10–20

Morré DJ, Penel C, Greppin H, Morré DM (2002c) The plasma membrane-associated NADH oxidase of spinach leaves responds to blue light. Int J Plant Sci 163:543–547

Morré DJ, Ternes P, Morré DM (2002d) Cell enlargement of plant tissue explants oscillates with a temperature-compensated period length of ca 24 min. In Vitro Cell Dev Biol – Plant 38:18–28

Morré DJ, Morré DM, Ternes P (2003) Auxin-activated NADH oxidase activity of soybean plasma membranes is distinct from the constitutive plasma membrane NADH oxidase and exhibits prion-like properties. In Vitro Cell Dev Biol – Plant 39:368–376

Morré DJ, Kim C, Hicks-Berger C (2006) ATP-dependent and drug-inhibited vesicle enlargement reconstituted using synthetic lipids and recombinant proteins. BioFactors 28:105–117

Morré DJ, Heald S, Coleman J, Orczyk J, Jiang Z, Morré DM (2007) Structural observations of time dependent oscillatory behavior of $Cu^{II}Cl_2$ solutions measured via extended X-ray absorption fine structure. J Inorg Biochem 100:715–726

Morré DJ, Orczyk J, Hignite H, Kim C (2008a) Regular oscillatory behavior of aqueous solutions of Cu^{II} salts related to effects on equilibrium dynamics of ortho/para hydrogen spin isomers of water. J Inorg Biochem 102:260–267

Morré DJ, Jiang Z, Marjanovic M, Orczyk J, Morré DM (2008b) "Response of the regulatory oscillatory behavior of copperII-containing ECTO-NOX proteins and of CuIICl2 in solution to electromagnetic fields" for Journal of Inorganic Biochemistry is now available on line at http://dx/doi.org/10.1016/j.jinorgbio.2008.06.001.

Mumma MJ, Weaver HA, Larson HP (1987) The ortho-para ratio of water vapor in comet P/Halley. Astron Astrophys 187:419–424

Ohnishi K, Niimura Y, Hidaka M, Masaki H, Suzuki H, Uozumi T, Nishino T (1995) Role of cysteine 337 and cysteine 340 in flavoprotein that functions as NADH oxidase from Amphibacillus xylanus studied by site-directed mutagenesis. J Biol Chem 270:5812–5817

Palmer JD, Dowse HB (1969) Preliminary findings on the effect of D_2O on the period of circadian activity rhythms. Biol Bull 137:388 (Abstract)

Pittendrigh CS, Caldarola PC, Cosbey ES (1973) A differential effect of heavy water on temperature-dependent and temperature-compensated aspects of circadian system of *Drosophila pseudoobscura*. Proc Natl Acad Sci USA 70:2037–2041

Pogue R, Morré DM, Morré DJ (2000) CHO cell enlargement oscillates with a temperature-compensated period of 24 min. Biochim Biophys Acta 1498:44–51

Potekhin SA, Khusainova RS (2005) Spin-dependent absorption of water molecules. Biophys Chem 118:84–87

Prusiner SB, Scott MR, DeArmon MR, Cohen FE (1998) Prion protein biology. Cell 93:337–348
Richter CP (1977) Heavy water as a tool for study of the forces that control length of period of the 24-hour clock of the hamster. Proc Natl Acad Sci USA 74:1295–1299
Ruggiero M, Bottaro DP, Liguri G, Gulisano M, Peruzzi B, Pacini S (2004) 0.2 T magnetic field inhibits angiogenesis in chick embryo chorioallantoic membrane. Bioelectromagnetics 25:390–396
Russel M, Model P (1988) Sequence of thioredoxin reductase from Escherichia coli. Relationship to other flavoprotein disulfide oxidoreductases. J Biol Chem 263:9015–9019
Santini MT, Rainaldi G, Ferrante A, Indovina PL, Vacchia P, Donelli G (2003) Effects of a 50 Hz sinusoidal magnetic field on cell adhesion molecule expression in two human osteosarcoma cell lines (MG-63 and Saos-2). Bioelectromagnetics 24:327–338
Savitz DA (1995) Overview of occupational exposure to electric and magnetic fields and cancer: advancements in exposure assessment. Environ Health Perspect 103:69–74
Sephton SE, Sapolsky RM, Kraemer HC, Spegal D (2000) Diurnal cortisol rhythm as a predictor of breast cancer survival. J Natl Cancer Inst 92:994–1000
Shifley ET, Cole SE (2007) The vertebrate segmentation clock and its role in skeletal birth defects. Birth Defects Res C Embryo Today 81:121–133
Simko M, Richard D, Kriehuber R, Weiss GG (2001) Micronucleus induction in Syrian hamster embryo cells following exposure to 50 Hz magnetic fields, benzo(a)pyrene, and TPA in vitro. Mutat Res 22:43–50
Sisken BF, Walker J, Orgel M (1993) Prospects on clinical applications of electrical stimulation for nerve regeneration. J Cell Biochem 52:404–409
Spruyt E, Verbelen JP, DeGruf JA (1987) Expression of circasepan and circammal rhythmicity in the inhibition of dry stored seeds. Plant Physiol 84:707–710
Suter RB, Rawson KS (1968) Circadian activity rhythm of the deer mouse, *Peromyscus*: effect of deuterium oxide. Science 160:1011–1014
Sweiczer A, Novak B, Mitchison JM (1996) The size control of fission yeast revisited. J Cell Sci 109:2947–2957
Thun-Battersby S, Mevissen M, Löscher W (1999) Exposure of Sprague-Dawley rats to a 50-Hertz, 100-microTesla magnetic field for 27 weeks facilitates mammary tumorigenesis in the 7,12-dimethylbenz[a]-anthracene model of breast cancer. Cancer Res 59:3627–3633
Tikhonov VI, Volkov AA (2002) Separation of water into its ortho and para isomers. Science 296:2363
Wang S, Pogue R, Morré DM, Morré DJ (2001) NADH oxidase activity (NOX) and enlargement of HeLa cells oscillate with two different temperature-compensated period lengths of 22 and 24 minutes corresponding to different NOX forms. Biochim Biophys Acta 1539:192–204
Welsh DK, Moore-Ede MC (1990) Lithium lengthens circadian period in a diurnal primate, *Saimiri sciureus*. Biol Psychiatry 28:117–126
Whaley WG (1975) The Golgi Apparatus. Cell Biology Monographs, vol 2. Springer, Wien/New York, pp 190
Wigner EZ (1933) Mechanism of nuclear spin initiated para-H_2 to ortho-H_2 conversion. Phys Chem B 23:28
Wolf FI, Torsello A, Tedesco B, Fasanella S, Boninsegna A, D'Ascenzo M, Grassi C, Azzena GB, Cittadini A (2005) 50-Hz extremely low frequency electromagnetic fields enhance cell proliferation and DNA damage: possible involvement of a redox mechanism. Biochim Biophys Acta 1743:120–129
Yagiz K, Wu L-Y, Kuntz CP, Morré DJ, Morré DM (2007) Mouse embryonic fibroblast cells from transgenic mice overexpressing tNOX express an altered growth and drug response phenotype. J Cell Biochem 101:295–306
Yin L, Wang J, Klein PS, Lazar MA (2006) Nuclear receptor Rev-erbα is a critical lithium-sensitive component of the circadian clock. Science 311:1002–1005
Yoshizawa H, Tsuchiya T, Mizoe H, Ozeki H, Kanao S, Yomori H, Sakane C, Hasebe S, Motomura T, Yamakawa T, Mizuno F, Hirose H, Otaka Y (2002) No effect of extremely low-frequency magnetic field observed on cell growth or initial response of cell proliferation in human cancer cell lines. Bioelectromagnetics 23:355–368

Chapter 3
Self-Organized Intracellular Ultradian Rhythms Provide Direct Cell-Cell Communication

V.Y. Brodsky[1] and D. Lloyd[2]

Abstract The ultradian rhythm of protein synthesis in hepatocytes *in vitro* was used as a marker for direct cell-cell communication. Self-organization of the rhythms leads to cell co-operation and synchronization throughout the population. Gangliosides and/or catecholamines and Ca^{2+} dependent protein kinase mediated protein phosphorylation gives phase modulation for cell-cell interaction during the establishment of synchrony. Thus the pathway involves signalling of gangliosides or other calcium agonists, Ca^{2+} release from intracellular stores, elevation of Ca^{2+} cytosol protein kinase activation steps, protein phosphorylation, synchronisation of protein synthesis rates and results in common rhythm induction throughout the population.

Analogous processes in yeast, as yet not mechanistically investigated, leads to spontaneous self-synchrony in continuously-stirred aerobic cultures.

Messengers responsible for cell-cell interactions in the expression of this 40 min ultradian clock include H_2S and acetaldehyde, as shown by phase-resetting experiments. Cell division cycle synchrony is spontaneously achieved for about 8% of the total population of approximately 5 × 10e8 organisms/ml for each ultradian clock cycle. A faster respiratory oscillation (period 4 min) also occurs concomitantly in these cultures.

This shorter-period oscillation is also evident in monolayers of organisms perfused with glucose and O_2, as revealed by 2-photon excitation. Several mitochondrial activities: (NAD(P)H redox state, inner membrane electrochemical potential and levels of reactive O_2 species) were simultaneously observed. This intracellular rhythm is synchronous amongst the mitochondria within a single yeast as well as across the population of organisms.

We conclude that the development of cellular co-operation leading to the multicellular organization of tissues and the harmonious functioning of the whole organism is

[1]Institute of Developmental Biology, Russian Academy of Sciences, 26 Vavilov Street, 119334 GSP-1, Moscow, Russia
Email: Brodsky.idb@bk.ru

[2]Microbiology (BIOSI 1), Main Building, Cardiff School of Biosciences, PO Box 915, Cardiff CF10 3TL, Wales, UK

based on ancient mechanisms, and that hepatocytes and yeast provide highly convenient systems for study.

Keywords Cell-cell interactions, synchronization, hepatocytes, yeast.

3.1 Spontaneous Self-Synchronization

Synchronization of oscillators is a well-studied and universal natural phenomenon in physical and biological systems, but it is a process that is still incompletely understood. (Pikovsky et al., 2002; Strogatz, 2000; Wiley et al., 2006). Its widespread occurrence in mechanical systems was first studied systematically by Huygens in 1665. … "two clocks hanging next to one another……when in consonance, the pendulums do not oscillate parallel to one another, but instead they approach and separate in opposite directions". He concluded that the clocks might be interacting by way of small vibrations of their common support. Detailed interpretation in terms of nonlinear dynamics is more recent (Bennett et al., 2002), and one important component for self-synchronization in this "weakly coupled" clock system has been shown to be the coupling strength between the oscillators (Klarreich, 2002). Synchronous behaviour in other mechanical structures (and their sometimes catastrophic failure) is a major concern in the structural engineering of boats and bridges. Synchrony of electrons in circuits, molecular flow in gases, and of photons in laser light and other optoelectronic circuits are all processes that indicate the universality of common principles (Strogatz, 2003; Lee et al., 2005). In biology, the most studied phenomenon arising principally on account of its spectacular magnificence, is the synchronous flashing of fireflies, an ensemble of similar self-sustained oscillations all firing at the same time (Buck, 1988). From synchronous beating of protozoal cilia (Gueron and Levit-Gurevich, 1998), pacemaker groups of cells in the suprachiasmatic nucleus (Morin, 2007) or in the sino-atrial node (Michaels et al., 1987), to the coherent oscillatory electrical outputs of the brain, rhythmic function is a centrepiece of biological function. The complexities of biological rhythms and the even greater complexity of population biology (Lloyd and May, 1999) may become fully comprehended and quantitatively described only when analogous behaviour in the phase-coherence of paired electrons in a superconductor (Josephson, 1962, 1974), in electronic networks (Winfree, 2001; Pyragas, 2006), lasers (Lee et al., 2005) and the world-wide web (May and Lloyd, 2001) is elucidated.

The important tendency of interaction between individual micro-organisms to give concerted behaviour finds examples even in the prokaryotes where infection, invasion, and pathogenicity may require collective onslaught and thereby an overwhelmed host immune system (pathogenicity). Quorum sensing, first described as the autoinduction of bioluminescence in various photobacterium species (Hastings and Greenberg, 1999) and recently as a very large scale phenomenon observable from space (Nealson and Hastings, 2006) is pivotal in bacterial cell-cell interactions and organized collective effects. That the developmental aggregation processes in the cellular slime-mold, *Dictyostelium discoideum*, depends on oscillatory release of cAMP is extensively studied, as is the formation of spiral-wave formations of amoebae starved on Petri dishes (Wurster, 1976). In this Chapter, we outline recent discoveries that indicate that communal activities of mammalian cells in culture (Brodsky, 1975, 2006) and of baker's yeast (Murray et al., 2003; Aon et al., 2007) have their origins in intracellular ultradian rhythms.

3.2 Self-Synchronization in Hepatocytes

Direct cell-cell communication occurs by way of tight contacts between adjacent cells (Evans et al., 2006) or by intercellular exchange of ions and small molecules via the aqueous extracellular medium (Berridge and Galione, 1988; Berridge, 1990, 1993; Frame and DeFeijter, 1997; Isakson et al., 2003).

In multicellular organisms such mechanisms act alongside central nervous and hormonal control, whereas in unicells, information exchange is solely dependent on direct mechanisms, either when cell individuals contact one another (even transiently) or where they are communicating at a distance. Although it was shown suggested by Pye (1969) and by Ghosh et al. (1971) that glycolytic oscillations may be involved in yeast intercellular conversations, evidence for a definite role for oscillatory message transmission came years later when Ca^{2+} waves were shown to be involved in mammalian systems. The intervening years have seen the growth of an extensive literature on this topic and even the publication of a specialized journal. All these studies were of "metabolic" oscillators with periods on a scale of minutes. Very recently, new insights into ultradian rhythms (on the circahoralian time scale) (Brodsky, 1975, 1992, 2006; Brodsky et al., 2000, 2003a, b, 2004, 2005, 2007) have led to the establishment of their important newly-recognized functions in intercellular interactions and coherent performance. Here we review the mounting evidence that rhythmicity underlies basic aspects of the evolution of multicellularity.

The possibility for investigations of a role for circahoralian rhythms in direct intercellular communication arose only when a suitable system had been developed. (Brodsky et al., 2000) The protein synthesis rhythm in hepatocyte cultures was compared with that first discovered in ganglion cell growth (Brodsky, 1960) and the early work reviewed (Brodsky, 1975). Hepatocytes are mature and virtually non-proliferating cells. The very existence of ultradian rhythms as an innate cellular characteristic with a fractal structure (Barnsley et al., 1987; Brodsky, 1992, 2006; Gilbert et al., 2000; Aon et al., 2000; Lloyd, 1992, 2005; Lloyd and Lloyd, 1993, 1995) can only come about as a consequence of communication between cells and processes of self-organization. Crucial to this work was the elaboration of a serum-free medium for the culture of primary rat hepatocytes. Furthermore, studies were facilitated by the observation of two types of culture behaviour that depended on their cell densities. In cultures at high cell counts (Fig. 3.1), cells became closely arranged whereas in sparse cultures individuals are widely-separated. The protein synthesis rhythm is displayed in dense cultures soon after medium renewal, and its detection necessarily indicates synchrony (as

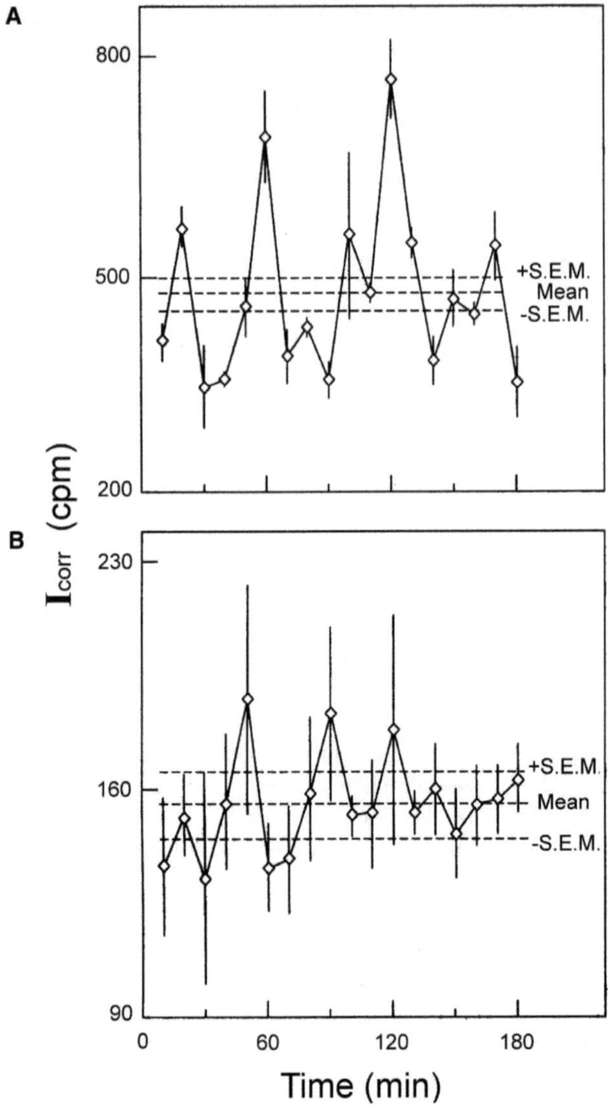

Fig. 3.1 Examples of protein synthesis kinetics in dense hepatocyte cultures with closely arranged cells (**A**), and sparse cultures consisting of widely separated cells (**B**). The cultures were obtained from the same suspension of isolated rat hepatocytes 24 h after inoculation of either 4×10^6 isolated hepatocytes (dense cultures) or 4×10^5 hepatocytes (sparse cultures) on slides coated with collagen. The slides were washed and samples of three cultures each were taken at 10 min intervals simultaneously for (**A**) and (**B**). Icorr is protein synthesis rate computed as ^{3}H-leucine incorporation corrected for the pool size of a given culture. Values are mean ± SEM calculated for three cultures at each time-point. The total mean ± SEM for all values of a given curve is shown by the horizontal dotted lines (Brodsky et al., 2000)

asynchronous cultures would give featureless time-averaged outputs). In sparse cultures (Fig. 3.2), on the other hand, individuals are not synchronous. Thereby two test systems became available: (a) desynchronization of cultures at high cell counts, and (b) synchronization of cells that are far apart and well-separated by medium. Signalling substances that promote synchronization of protein synthesis rhythms were first described (Brodsky et al., 2000) and characterized as gangliosides, although not all gangliosides were effective. Thus, 0.3–0.5 μM of a

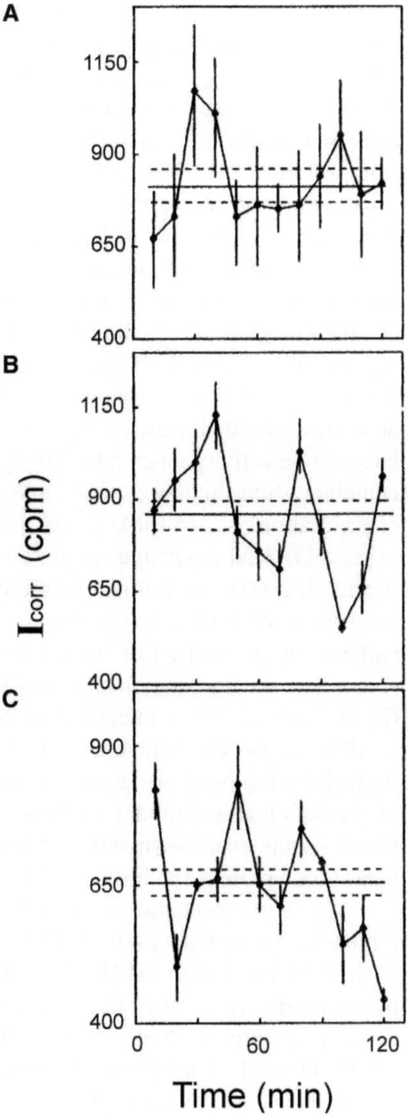

Fig. 3.2 Kinetics of protein synthesis in sparse 1-day primary hepatocyte cultures. (**A**) Control cultures, washed and transferred to fresh normal medium for 2 min, were again washed and transferred to fresh medium of 10 min, following which samples were taken for estimation of protein synthesis from this medium. (**B**) Similar washed cultures transferred to a medium containing 2 μM phenyleprine for 2 min, then washed and transferred to normal medium for 10 min; samples were taken from this medium. (**C**) Same procedures as (**B**) but cultures were treated with 10 μM 2,5-di(tert-butyl)-1,4-benzohydroquinone for 2 min, which stimulates Ca^{2+} elevation in a non-receptor manner. All cultures of a given experiment were obtained from the same suspension of hepatocytes isolated from one liver and assayed at the same time. Icorr is protein synthesis rate computed ^{3}H-leucine incorporation corrected for the pool size of a given culture. Values are mean ± SEM calculated for three cultures at each time-point. The total mean ± SEM for all values of a given curve is shown by the horizontal lines (Brodsky et al., 2003a)

mixture of bovine brain gangliosides to fresh culture medium, along with either 0.06–0.2 μM GM1 or 0.2–0.2 μM GM1 or 0.1–0.2 μM GDIa gave positive results. GTIb and GDIb did not produce oscillations, nor did human liver ganglioside determinants maintained in the conditioned medium with purified polyclonal antibodies to GM1. Cell-cell interaction was coupled with protein synthetic oscillatory activity (metabolic synchronization). Incubation of dense cultures with GM1-antibodies for 24 h decreased the amplitude of the oscillations. Expression of GM1 was low in those sparse cultures maintained in fresh medium and no oscillatory output occurred.

The role of calcium ions in cell-cell cooperation between cultured hepatocytes was established by the enhancing effects on synchronization by 2 min exposure to 2 μM phenylephrine, an α1-adrenoreceptor agonist which elevates $[Ca^{2+}]_{cyt}$ (Woods et al., 1986) by releasing these ions from intracellular stores. Under these conditions stimulation of the protein synthesis rhythm in sparse cultures resulted in initiation of increased cooperative interaction between cells. A similar result was obtained on administration of 10 μM 2,5-di(tertiary-butyl)-1,4-benzohydroquinone for 2 min. This compound increases $[Ca^{2+}]_{cyt}$ by a non-receptor mediated process (Brodsky et al., 2003a). The intracellular calcium chelator,1,2-bis(2-aminophenoxy)ethane-$N_1N_1N_1{}^1N^1$-tetra-aceticacid(acetoxymethy/ester) (BAPTA-AM) at 10–20 μM for 30–60 min resulted in loss of observable protein synthesis rhythm (i.e. loss of cell co-operativity) in dense cultures. Medium pre-conditioned in such cultures initiated rhythms in sparse cultures, whereas similar medium treated with is calcium chelator did not. The GM1 monosialoganglioside content, specifically sequestered into Ca^{2+}-dependent shedded vesicles (but not the remaining culture supernatant) initiated cooperative activity when added to sparse cultures (Brodsky et al., 2003b). GM1-containing liposomes were also able to do this, and they did so at extremely low concentrations (0.3–60 nM). It was pointed out that elevation of $[Ca^{2+}]_{cyt}$ would be expected to lead to changes in protein kinase activity and protein phosphorylation (Carafoli, 2002), and it might be these pathways that modulated the oscillations in protein metabolism. That the composition of the growth medium was most important for the production of these effects was confirmed by experiments with the serum-free medium supplemented with 0.2 mg/ml albumin and 0.5 μg/ml insulin (Brodsky et al., 2004). In cells cultured from 3 month-old rats the amplitude of the oscillations was on average twice that in cells from 2 year old animals. Addition of 0.3 μM bovine brain gangliosides, or 2 μM phenylephrine resulted in restoration of the deficient amplitudes to those of young rat cells. Also addition of 10% rat blood serum to non-synchronous sparse cultures from young rats resulted in a measurable protein synthesis rhythm.

In the most recent studies (Brodsky et al., 2007) treatment of dense cultures with protein kinase inhibitors 40 μM H7 (1-(5-isoquinolinesulphonyl)-5-methylpiperasine dihydrochloride) or 25 μM H8 (N-methylamino-ethyl)-5-isoquinoline-sulphonamide hydrochloride resulted in loss of the protein synthesis rhythm, i.e. suppression of the cell self-organization. Stimulation of protein kinase activity with either 0.5–1.0 μM phorbol 12-myristate 13-acetate (PMA) or with 10 μM forskolin resulted in the appearance of the protein synthesis rhythm in non-synchronous (under normal conditions) sparse cultures. After inhibition of protein kinase activity with H7, signal factors such as gangliosides and phenylephrine did not initiate protein synthesis rhythm in sparse cultures. Activation of protein kinase activity, with PMA shifted the phase pattern of protein synthesis rhythm. Thus protein kinase activity, and consequently protein phosphorylation reactions (Barford, 1999), are the crucial step for the initiation of cascade processes resulting in cell synchronization during self-organization of cells so as to form a coherent rhythm of a cell population. The general pathway (Fig. 3.3) can be presented as follows: signalling of gangliosides or other calcium agonists → efflux of calcium ion from intracellular stores (Parek and Putney, 2005), elevation of calcium concentration in the cytoplasm → activation of protein kinases → protein phosphorylation → synchronization of individual oscillations of the protein synthesis rate → formation of the common rhythm.

In all these studies the ultradian rhythm of protein synthesis was used as the marker for direct cell-cell communication in cultured cells. It was previously established that many biochemical correlates (e.g. respiration and adenylate nucleotide pools) show distinct phase relationships with protein synthesis in synchronized lower eukaryotes and in mammalian cells (Lloyd, 1992; Brodsky 1992, 2006). These intracellular rhythms provide a fine-scale framework for the establishment of integrated outputs on a larger scale in tissues and in organs (cardiodynamics, brain activity) and act as the very basis of temporally-patterned behaviour. Thus the ultradian clock regulates tissue and organ function, and some dysfunctions ("dynamic diseases", e.g. cardiac arrhythmias and epilepsy) result from disrupted rhythmicity (Goldberger and West, 1987; Glass and Mackey, 1988; Skinner et al., 1990; Wheatley, 2000; Wuhl et al., 2005).

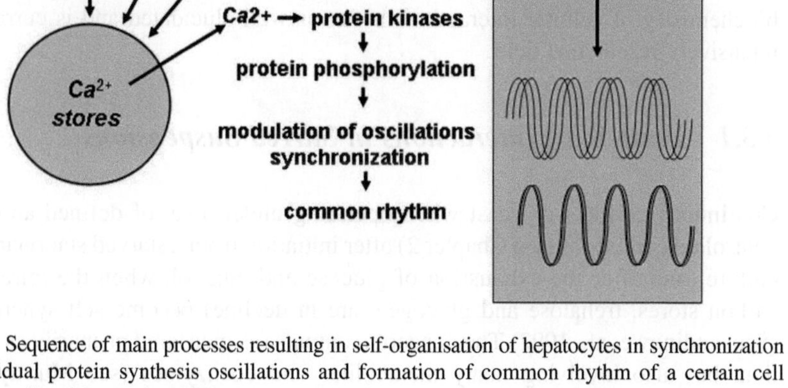

Fig. 3.3 Sequence of main processes resulting in self-organisation of hepatocytes in synchronization of individual protein synthesis oscillations and formation of common rhythm of a certain cell population (Brodsky et al., 2007)

Although the model studies employ cultured hepatocytes, *in situ* studies with denervated liver slices (Brodsky et al., 1995) also indicate ultradian protein synthesis rhythmicity, where native supplies of gangliosides of blood serum (Bergelson, 1995) are supplemented in the experiments described above. Catecholamines (e.g. epinephrine and norepinephrine) have also been determined in blood serum (Prozorovskaya, 1983; Jensen et al., 1993), and hepatocytes are known to possess adrenoreceptors. Among the many self-organizers (autoinducers) of bacterial behaviour neurotransmitters have been shown to be effective. Norepinephrine, serotonin and dopamine provide direct organism-organism and intercolony communication, and also induce bacterial growth (Lyte and Ernst, 1992; Oleskin et al., 1998; Walters and Sperandio, 2006; Freestone et al., 2007). Thus neurotransmitters are seen to arise in evolution long before a nervous system. During animal embryogenesis, neurotransmitters already function at the blastula stage (Buznikov, 1990). In our experiments, norepinephrine and its synthetic analog phenylephrine organized behaviour of hepatocytes *in vitro*; it is very likely that this also occurs *in situ* in the liver (Fig. 3.3). Thus, in the whole organism, cells can receive synchronizing signals. Variations in the degree of synchronization in tissue culture as demonstrated, depend on the separation of the individual cells and on the ages of the rats from which they are cultured. Thus variation in the degree of self-organisation may arise from a variety of influences, but these include the concentration of the synchronizing signals that vary with age, functional stress factors or disease are measured as an integrated "coupling strength". Mathematical modeling of self-organized systems predicts mechanisms (Gelfand and Tsetlin, 1960; Romanovsky and Chernavsky, 1972; Pavlidis, 1973) whereby oscillations of high amplitude propagate and are maintained throughout a population of individuals so as to result in close phase-coherence.

3.3 Self-Synchronization in Yeast

It is quite remarkable that the synchronization phenomena, so important in the establishment and maintenance of the functional organisation of tissues and organs in mammalian systems should have analogies in lower eukaryotes. Here the biochemistry of cellular interactions is not as well elucidated and is currently an intensively-researched field.

3.3.1 Yeast-Yeast Interactions in Stirred Suspensions

Continuous cultures of yeast when operating under a set of defined and highly controlled conditions (see Chapter 2) after initiation from a starved stationary phase culture (and after the exhaustion of glucose and ethanol, when the intracellular carbon stores, trehalose and glycogen, are in decline) become self-synchronized (Satroutdinov et al., 1992) The concerted outputs (as measured primarily as respiration, but also involving many metabolites and coenzymes (see Murray, 2004; Murray et al., 2001; Lloyd and Murray, 2005, 2006 for concise reviews) of the dense population (approx. 5×10^8 organisms/ml) show large-amplitude oscillations (sometimes as much as 40% of air-saturation values). This coherent behaviour indicates that almost the entire population has become highly synchronized to the endogenous circa 40 min ultradian clock period.

Identification of two candidates as synchronizing compounds has been described. Thus proposals that acetaldehyde would be a suitable signalling in anaerobic yeast cultures and suspending substance date back to work in the group of Britton Chance (Chance et al., 1973) soon after glycolytic oscillations had been discovered (Ghosh and Chance, 1964; Pye, 1969). For the respiratory oscillations in the aerobic continuous culture system, conclusive evidence for the same putative synchronizer (Keulers et al., 1996, 1998) came with the observation of phase shifting by pulse additions of acetaldehyde (Murray et al., 2003). A strong response ("Type O", using the nomenclature of Winfree, 2001) was produced at 3 mM acetaldehyde, whereas at 1 mM this compound elicited a weak (Type 1) response.

Another highly volatile, easily diffusible and rapidly disposable metabolic product, H_2S, was identified as a periodically spiked waveform in these cultures (Sohn et al., 2000) as a product of the sulphite reductase reaction of sulphur-containing amino acid synthesis from the inorganic sulphate of the growth medium (Sohn and Kuriyama, 2001). Further details of H_2S production have recently been discovered (Kwak et al., 2003; Sohn et al., 2005a, b).

Using ammonium sulphide as an effector, phase response curves showed that pulse additions only give phase shifting of the respiratory oscillation when made at the times when endogenous sulphide production are maximal (Murray et al., 2003). Only Type 1 responses were observed. Additions of 3 μM sulphite gave maximal phase shift $\Delta\theta$ at 60°, where the value of θ was calculated as 360° $\times$ ($t^* - 3 \times \tau/\tau$), where t^* was the time elapsed between the last dissolved oxygen concentration maximum and the third dissolved oxygen concentration maximum after addition of the effector. It was concluded from these experiments that endogenously generated acetaldehyde and sulphide (which reacts with the aldehyde carbonyl group) tune the oscillation, whereas sulphide is the primary determinant of population synchrony. It was however suggested (Murray et al., 2003) that other small molecules and molecular interactions are involved, and that the complete mechanism is complex with a strong and essential mitochondrial involvement (Lloyd et al., 2002), as H_2S acts as an inhibitor of the respiratory chain at the level of cytochrome *c* oxidase. Reactive oxygen species may also be involved (Lloyd et al., 2003) as is the case in the concerted mitochondrial network that triggers cardiomyocyte synchronous beating (Aon et al., 2003, 2006; Cortassa et al., 2004).

3.3.2 Yeast-Yeast Interactions on a Substratum

More recently, another shorter period oscillation ($\tau \sim 100$ s at 30°C) has been discovered in yeast (Figs. 3.4, 3.5). When aerobically grown *Saccharomyces cerevisiae* is tethered to a poly-lysine coated slide and perfused with phosphate-buffered saline and

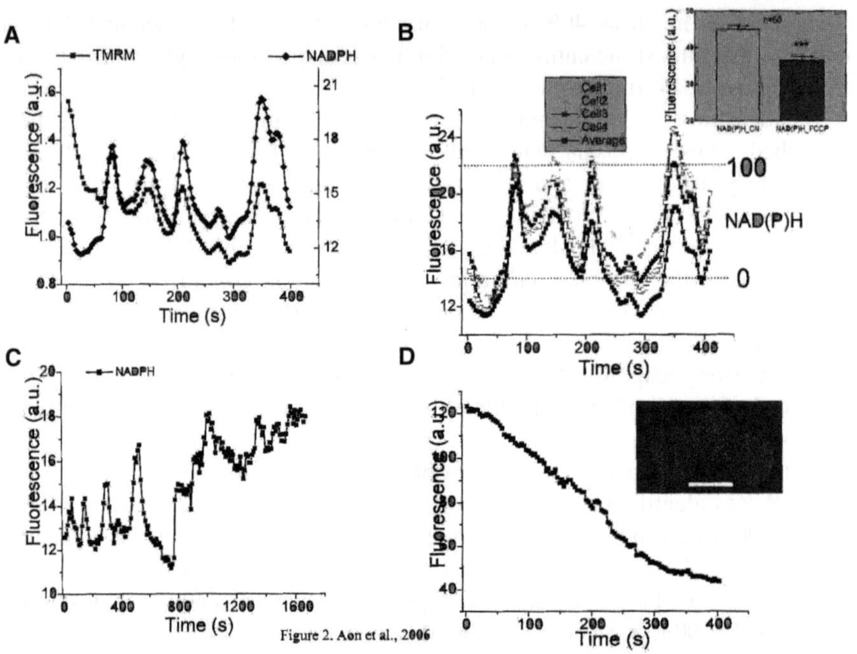

Fig. 3.4 Spontaneous, synchronized oscillations in a contiguous layer of *S. cerevisiae* cells incubated with (**A, B**) or without (**C**) aeration of the perfusion buffer, monitored by 2-photon scanning laser microscopy. (**A**) Aerobic, whole cell Ψm and NAD(P)H oscillations exhibited by a microscopic field of ~30 yeasts perfused with aerated PBS, pH 7.4, in the presence of 5 mM glucose. (**B**) Synchronous NAD(P)H oscillations of individual yeast cells as well as the average fluorescence from the whole microscopic field. The inset (see also Supplementary material) shows a histogram of whole cell NAD(P)H fluorescence intensity in the presence of 5 mM cyanide (CN) or 2 μM FCCP, a Δψm uncoupler, for n = 60 yeasts; see Section 2 for further details. $^{***}P < 0.001$. (**C**) Whole yeast cell NAD(P)H oscillations under perfusion conditions similar as in panel A but in the absence of aeration. (**D**) Fluorescent microspheres (see Section 2 for further details) of approximately the same diameter of a yeast cell were imaged in the same microscopic field of yeast and their fluorescence followed simultaneously during the oscillatory response. A monotonic decrease in the microspheres' fluorescence intensity can be noticed and were essentially due to photobleaching. Three fluorescent microspheres are shown in the inset: the bar corresponds to 5 μm (Aon et al., 2007)

glucose, spontaneous large-scale oscillations of NAD(P)H and inner mitochondrial trans-membrane electrochemical potential (ΔΨm) occur throughout the entire population. (Aon et al., 2006). Synchrony of the blue fluorescence emission from NADH, synchronized in a single layer of yeasts is a spectacular phenomenon (see supplementary material movie on-line in Aon et al., 2007). Both the nicotinamide nucleotide and ΔΨm could be simultaneously monitored by 2-photon excitation at 740 nm using a Ti: Sapphire laser and autofluorescence (<490 nm) and tetramethylrhodamine (ethyl ester) (605 nm) emissions. When perfusion was stopped, the oscillations ceased, and the NAD(P)H pools became completely reduced. Individual mitochondria (Fig. 3.6) oscillate in phase with one another and throughout the cellular population. Within a single

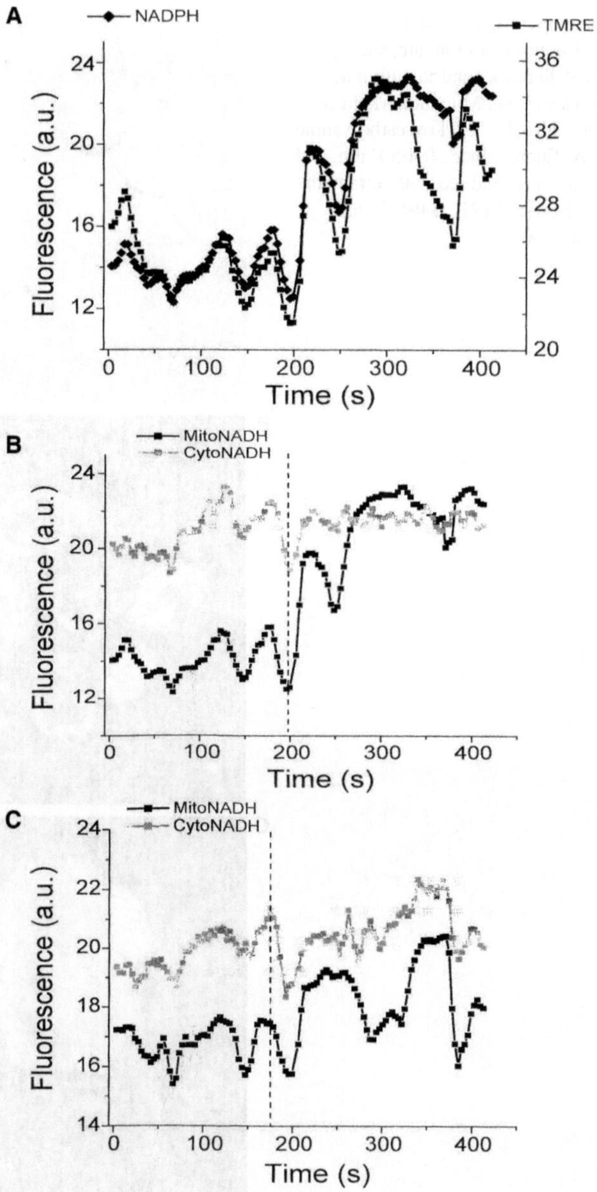

Fig. 3.5 Spontaneous oscillations in TMRE and NAD(P)H fluorescence in a single *S. cerevisiae* cell incubated aerobically with 5 mM glucose. (**A**) Whole cell TMRE, NA(P)H fluorescence, (**B, C**) representative traces of mitochondrial and cytoplasmic NAD(P)H autofluorescence. Cells loaded with TMRE allowed determining the regions of interest corresponding to mitochondria or extra-mitochondrial space in order to quantify autofluorescence in both compartments (Aon et al., 2007)

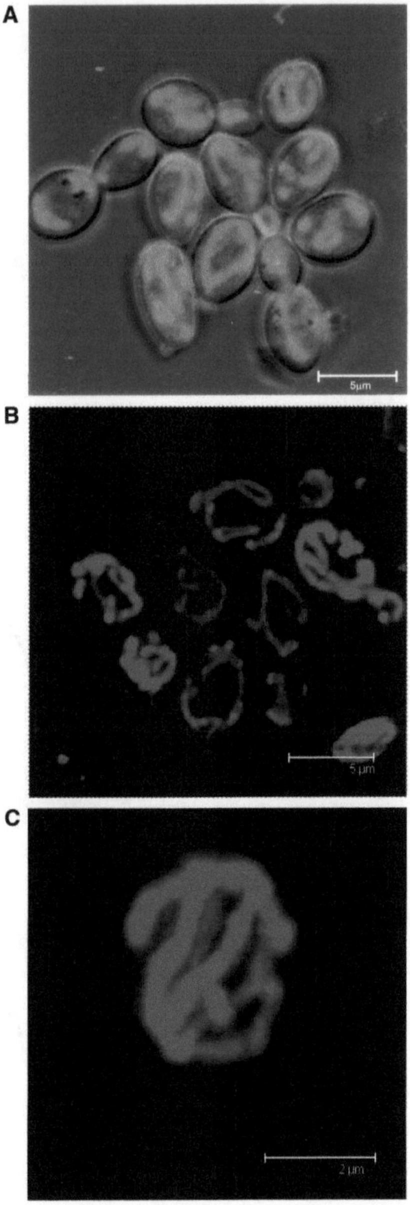

Fig. 3.6 Confocal microscope 'maximum intensity' reconstructions of elongate, sausage-shaped, and reticulate mitochondria within aerobically-grown *S. cerevisiae*. Mitochondria have been stained green with 3,3-dihexyl-oxacarbocyanine iodide, DiOC6 (**A**, fluorescence, 484/501 nm, and transmitted light overlay) and red with tetramethylrhodamine ethyl esterTMRE, 549/573 nm (**B, C**) (Aon et al., 2007)

4 min) at 30°C are produced by the same mechanisms as those recently discovered in continuously-stirred growing cultures of yeast (Roussel and Lloyd, 2007), see p., remains to be investigated. It appears likely that mitochondrial oscillations earlier

yeast (Fig. 3.5) isolated from others, individual mitochondria oscillate in synchrony. Synchronous oscillations of redox state of the fission yeast *Schizosaccharomyces pombe* were earlier reported. (Bashford et al., 1980). Whereas this phenomenon is a startling example of self-synchrony, another kind of experiment indicates that a single localized laser flash, focussed to a spot of <1 μm can trigger mitochondrial oscillations in a single cell, and this suggests that the mitochondrion is behaving as an autonomous oscillator. A similar result has been obtained with cardiomyocytes (Aon et al., 2003), where it has been demonstrated that reactive oxygen species act as signalling molecules for the rapid percolation of synchronised excitation responses (Aon and Cortassa 2006, Aon et al., 2006, 2007). The physiological importance of these processes in ventricular myocytes is of such profound importance, that it is a mitochondrial "critical state" that determines whether the cells remain viable (O'Rourke et al., 2005). In the yeast *Candida utilis* similar techniques have been used to determine oxidative stress induced by allyl alcohol or diallyl disulphide: both reactive oxygen species and glutathione depletion were measured alongside and simultaneously with ρΨm and NAD(P)H redox state (Lemar et al., 2005, 2007). Similar fluorophores (carboxymethyl-dichloro-dihydrofluorescein) and monochlorobimane have been used respectively to monitor oscillatory reactive oxygen species generation and glutathione in *S. cerevisiae* (Fig. 3.7, KM Lemar and MA Aon, 2006 unpublished results). Whether these oscillations (τ about

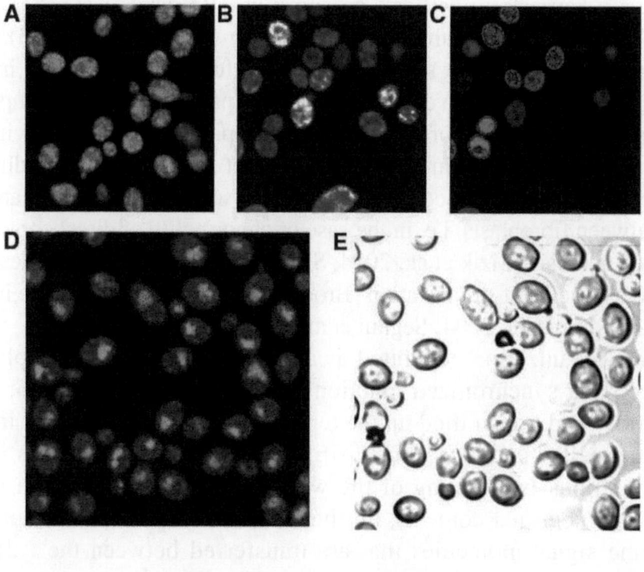

Fig. 3.7 Images of yeasts obtained using 2-photon laser-scanning microscopy. (**a**) Autofluorescence of NADH. Fluorescent probes used were: (**b**) tetramethylrhodamine ethyl ester (inner mitochondrial membrane potential), (**c**) chloromethyl-dichlorohydrofluorescein (reactive oxygen species), (**d**) monochlorobimane (glutathione), (**e**) Transmitted light image using Nomarski differential interference optics (Katey M. Lemar and Miguel A. Aon, unpublished 2006)

demonstrated in organelles isolated from cardiac muscle and other tissues (Mustafa et al., 1966; Packer et al., 1966) and (Gooch and Packer, 1971, 1974) are also responsible for *in vivo* mitochondrial respiratory oscillations in yeast.

3.4 Conclusions

In hepatocyte cultures circahoralian rhythms of protein synthesis induce a common rhythm throughout the population (spontaneous self-synchrony) by a process of direct cell-cell communication. Experiments clearly demonstrate that cellular co-operation involves gangliosides and/or catecholamines in a major part of all of the individuals, as after the initial triggering, the periodicity is retained for some time. Thus a medium-mediated co-ordination that supplements *in vivo* nervous and hormonal influences is physiologically effected by blood serum components. Agonist and antagonist administration indicates the pathway:signalling of gangliosides or other calcium agonists → efflux of Ca^{2+} from intracellular stores and elevation of Ca^{2+} cytosol → protein kinase activation → protein phosphorylation → synchronization of individual oscillations in protein synthesis rates → induction of a common rhythm throughout the population.

Recently, even wider implications for direct cell-cell communication have been uncovered extending its functions as a determinant of cell death. Thus protein kinase C that plays a key role in cell self-organization (Brodsky et al., 2007) can be a significant component of death receptor signalling (Harper et al., 2003). Activation of PKC targets death receptor signalling complex formation leading to inhibition of apoptosis in HeLa cells due to decreased capase-8 processing and incomplete activation of caspase-3. Disruption of contacts between epithelial cells along with damaged basement membrane results in apoptosis (Fouquet et al., 2004). E-cadherin, a key protein of cell adhesion, can control apoptosis as well. On the other hand, cell-cell contacts between fibroblasts, i.e. in the case of abnormality of direct contacts, trigger programmed necrosis (Bizik et al., 2004; Siren et al., 2006). Gangliosides, important signal factors of cell self-organization (Brodsky et al., 2000) can involve in cell death too (Bektas and Spiegel, 2004; Segnui et al., 2006).

Thus self-organization generated as a coherent output of multiple ultradian clock-driven and synchronized functions (Murray and Lloyd, 2006; Lloyd and Murray, 2007) and transmitted up the temporal heterarchy from the interactions of individual cells via the organized structure of tissues and organs is essential to the harmonious functioning of the whole organism. In addition to external hormones and neuronal controls, all the cells of a population can produce and release some signal molecules that are transferred between the cells so as to coordinate certain functions. This "third way" of functional organization may have been the first to have arisen in evolution (Fig. 3.8).

We see strong signs of this in experiments with yeast; that spontaneous generation of analogous coordination there is surprising, because our concept of *S. cerevisiae*

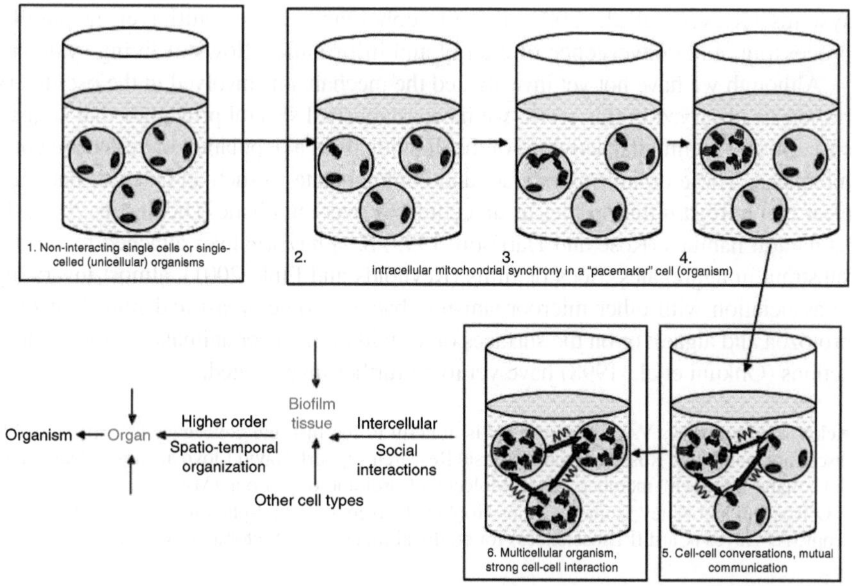

Fig. 3.8 The evolution of multicellular organisms has utilized ultradian rhythms for both intracellular coherence and intercellular signalling; this has led to population synchronization, social interactions and the development of tissues and organs leading to higher levels of complexity in metazoa and plants (Illustration by Dr. Victoria Gray)

as a typical unicellular organism would suggest independence and autonomy in each of the cells within a population.

Nevertheless, the discovery that yeast cells can communicate with one another can be traced back to 1969 when this was first proposed by Pye (1969) and further explored by Ghosh et al. (1971) and by Aon et al. (1992) in a search for the extracellular synchroniser of the glycolytic oscillator. The mechanism by which inter-organism communication occurs was proposed to involve acetaldehyde possibly along with other unidentified, suitably volatile, rapidly diffusing, low molecular weight, fermentation products. Extracellular Ca^{2+} also plays a role (Aldridge, 1976). Confirmation of the importance of acetaldehyde as a synchronising messenger on a longer time-scale has come with the construction of a phase response curve by addition of pulses of this compound to continuously cultured aerobic yeasts (Murray et al., 2003) at different times in the circa 40 min ultradian driven clock respiratory cycle. A second messenger substance, hydrogen sulphide, which shares with acetaldehyde another property, its ease of loss by further oxidation, is also implicated in the self-synchronising ability of *S. cerevisiae* (Sohn et al., 2001). These properties have been pivotal in elucidation of the "temporal landscape" of yeast growth (Lloyd and Murray, 2006) which finds extensive homologies in other eukaryotic metabolic and transcriptional control systems (Klevecz et al., 2004) and the coherence of the multilayering parallel processing, and convergence of energy and information flows of living systems.

Although we have not yet investigated the mechanisms involved in the oscillatory responses observed in this work, we have established several principles that suggest strongly that communal activities of the individuals in a population of yeasts are commonplace. These organisms interact and communicate to such an extent it becomes clear that to regard this species as unicellular is over-simplistic (Dickinson, 2005). In its natural habitats (Rose and Harrison, 1999) *S. cerevisiae* lives attached to various substrata in aggregates, flocs or films (Reynolds and Fink, 2001), almost invariably in association with other microorganisms (bacteria, other yeast and fungal species, protozoa and algae): or on the surfaces of, or within plants or animals. Its social interactions (Ohkuni et al., 1998) have yet to be further investigated.

Acknowledgements Vsevolod Brodsky is grateful to his Institute colleagues for very valuable discussion and to the Russian Fund for Basic Research support. David Lloyd acknowledges splendid collaboration with members of the Molecular Cardiobiology Group (MA Aon, S Cortassa and B O'Rourke) and Katey Lemar at Johns Hopkins University with 2-photon laser scanning microscopy, AJ Hayes (Cardiff Biosciences) for confocal images and Victoria Gray for Fig. 3.8.

References

Aldridge J (1976) Short-range intercellular communication, biochemical oscillations and circadian rhythms. In: Handbook of Engineering in Medicine and Biology (Fleming DG, Ternberg BN, eds) CRC, Cleveland OH, pp 55–147

Aon MA, Cortassa S (2006) Metabolic dynamics in cells viewed as multilayered, distributed, mass-energy-information networks. In: Encyclopedia of Genetics, Genomics, Protemics and Bioinformatics, Systems Biology, Vol. 6 (Dunn MY et al., eds) Wiley Interscience, New York

Aon MA, Cortassa S, Westerhoff HV, Van Dam K (1992) Synchrony and mutual stimulation of yeast cells during fast glycolytic oscillations. J Gen Microbiol 138:2219–2227

Aon MA, Cortassa S, Lloyd D (2000) Chaotic dynamics and fractal space in biochemistry: simplicity underlies complexity. Cell Biol Int 24:581–587

Aon MA, Cortassa S, Marbán E, O'Rourke B (2003) Synchronized whole cell oscillations in mitochondrial metabolism triggered by a local release of reactive oxygen species in cardiac myocytes. J Biol Chem 278:44735–44744

Aon MA, Cortassa SC, O'Rourke B (2006) The fundamental organization of cardiac mitochondria as a network of coupled oscillators. Biophys J 91:4317–4327

Aon MA, Cortassa S, Lemar KM, Hayes AJ, Lloyd D (2007) Single and cell population respiratory oscillations in yeast: a 2-photon scanning laser microscopy study. FEBS Lett 581:8–14

Barford D (1999) Structural studies of reversible protein phosphorylation and protein phosphatases. Biochem Soc Trans 27:751–766

Barnsley MF, Massopust P, Strickland H, Sloan AD (1987) Fractal modelling of biological structures. Ann NY Acad Sci 504:179–194

Bashford CL, Chance B, Lloyd D, Poole RK (1980) Oscillations of redox states in synchronous cultures of *Acanthamoeba castellanii* and *Schizosaccharomynces pombe*. J Biophys 29:1–11

Bektas M, Spiegel S (2004) Glycosphingolipids and cell death. Glycoconj J 20:39–47

Bennett M, Schatz MF, Rockwood H, Wisenfeld K (2002) Huygens's clocks. Proc Roy Soc A 458:563–579

Berridge MJ (1990) Calcium oscillations. J Biol Chem 265:9583–9586
Berridge MJ (1993) Inositol triphosphate and calcium signalling. Nature 361:315–325
Berridge MJ, Galione A (1988) Cytosolic calcium oscillators. FASEB J 2:3074–3082
Bergelson LD (1995) Serum gangliosides as endogenous immunomodulators. Immunol Today 16:483–486
Bizik J, Kanhuri E, Ristimaki A, Taieb A, Vaatalo H, Lubitz W, Vaheri A (2004) Cell-cell contacts trigger programmed necrosis and induce cyclooxygenase-2 expression. Cell Death Differ 11(2):183–195
Brodsky VY (1960) Quantitative studies of RNA and proteins in ganglion cells of retina. In 1st Conference on Cytochemistry, Moscow, pp 7–9
Brodsky VY (1975) Protein synthesis rhythm. J Theoret Biol 55:397–407
Brodsky VY (1992) Rhythm of protein synthesis and other circahoralian oscillations. The possible involvement of fractals. In: Ultradian rhythms in life processes (Lloyd D, Rossi EL, eds) Springer, London, pp 23–40
Brodsky VY (2006) Direct cell-cellcommunication: a new approach derived from recent data on the nature and self-organisation of ultradian (circahoralian intracellular rhythms. Biol Rev Camb Philos Soc 81:143–162
Brodsky VY, Dubovaja TK, Nechaeva NV, Fateeva VI, Novikova TE, Gvasava IG (1995) Protein synthesis rhythm in the denervated rat liver. Izv Akad Nauk Ser Biol 2:133–137
Brodsky VY, Nechaeva NV, Zvezdina ND, Novikova TE, Gvasava IG, Fateeva VI, Prokazova NV, Golovanova NK (2000) Ganglioside mediated metabolic synchronisation of protein synthesis activity in cultured hepatocyces. Cell Biol Int 24:211–222
Brodsky VY, Zvezdina ND, Nechaeva NV, Avdonin PV, Novgikova Te, Gvasava IG, Fateeva VI, Malchenko LA (2003a) Calcium ions as a factor of cell-cell cooperation in hepatocyte cultures. Cell Biol Int 27:965–976
Brodsky VY, Zvezdina ND, Nechaeva NV, Novikova TE, Gvasava IG, Fateeva VI, Gracheva HV (2003b) Loss of the hepatocyte co-operative activity after inhibition of ganglioside synthesis and shedding. Cell Biol Int 27:935–942
Brodsky VY, Nechaeva NV, Zvezdina ND, Novikova TE, Gvasava IG, Fateeva VI, Malchenko LA (2004) Small cooperative activity of old rat hepatocytes may depend on composition of the intercellular medium. Cell Biol Int 28:311–316
Brodsky VY, Zvezdina ND, Nechaeva NV, Novikova TE, Gvasava IG, Fateeva VI, Malchenko LA (2005) Single short-term signal that enhances cooperative activity of the old rat hepatocytes acts for several days. Cell Biol Int 29:971–975
Brodsky VY, Zvezdina ND, Fateeva VI, Malchenko LA (2007) Involvement of protein kinases in self-organization of the rhythm of protein synthesis by direct cell-cell communication. Cell Biol Int 31:65–73
Buck J (1988) Synchronous rhythmic flashing of fireflies. Quart Rev Biol 63:265–289
Buznikov GA (1990) Neurotransmitters in Embryogenesis. Harwood, New York/London
Carafoli E (2002) Calcium signalling: a fate for all seasons. Proc Natl Acad Sci USA 99:1169–1179
Chance B, Williamson G, Lee IY, Mela L, DeVault D, Ghoash A, Pye EK (1973) Synelvonization Pnenomena in Oscillations of Yeast Cells and Isolated Mitochondria. Academic, New York
Cortassa A, Aon MA, Winslow RL, O'Rourke B (2004) A mitochondrial oscillator dependent on reactive oxygen species. Biophys J 87:2060–2073
Dickinson JR (2005) Are yeasts free-living unicellular eukaryotes? Lett Appl Microbiol 41:445–447
Evans WH, DeVuyst E, Leybaert L (2006) The gap junction cellular intern: Connexin nemichannels enter the signalling limelight. Biochem J 397:1–14
Foquet S, Lugo-Martinez VN, Faussat AM, Renaud F, Cardot P (2004) Early loss of E-cadherin from cell-cell contacts is involved in the onset of Anoikis in enterocytes. J Biol Chem 279:43061–43069
Frame MK, DeFreijter AW (1997) Propagation mechanically induced intercellular calcium waves via gap junction and ATP receptors in rat liver cells. Exp Cell Res 230:197–207

Freestone PPE, Haigh RD, Lyte M (2007) Specificity of catecholamine-induced growth in *Escherichia coli, Salmonella enterica* and *Versinia enterocolitica.* Microbiol Lett 269:221–228

Gelfand IM, Tsetlin ML (1960) Continual models controlling systems. Dok Akad Nauk 131:1242 (in Russian)

Ghosh A, Chance B (1964) Oscillations of glycolytic intermediates in yeast cells. Biochem Biophys Res Commun 16:174–181

Ghosh AK, Chance B, Pye EK (1971) Metabolic coupling and synchronization of NADH oscillations in yeast cell populations. Arch Biochem Biophys 145:319–331

Gilbert DA, Visser G, Ferreira GMN, Hammond KD (2000) Transient chaos in intracellular dynamics. Biol Int 24:589–591

Glass L, Mackey MC (1988) From Clocks to Chaos: The Rhythms of Life. Princeton University Press, Princeton NJ

Goldberger AL, West B (1987) Chaos in Pathology. In: Chaos in Biological Systems (Degn H, Holden AV, Olsen LF, eds) Plenum, New York, pp 1–5

Gooch VD, Packer L (1971) Adenine nucleotide control of heart mitochondrial oscillations. Biochim Biophys Acta 245:17–20

Gooch VD, Packer L (1974) Oscillation systems in mitochondria. Biochim Biophys Acta 346:245–260

Gueron S, Levit-Gurevich K (1998) Computation of the internal forces in cilia: application to ciliary motion, the effects of viscosity and cilia interactions. Biophys J 74:1658–1676

Harper N, Hughes MA, Farrow SN, Cohen GM, MacFarlane M (2003) Protein kinase C modulates tumor-necrosis factor-related apoptosis-inducing ligand-induced apoptosis by targeting the apical events of death receptor signalling. J Biol Chem 278:4438–4497

Hastings JW, Greenberg EP (1999) Quorum sensing: the explanation of a curious phenomenon reveals a common characteristic of bacteria. J Bacteriol 181:2667–2668

Isakson BE, Seedort GJ, Libman RL, Evans WH, Boitano S (2003) Cell-cell communication in heterocellular cultures of alveolar epithelial cells. Am J Respir Cell Mol Biol 29:552–561

Jensen EW, Eldrup E, Kelbaek H, Nielsen SL, Christensen NJ (1993) Venous plasma noradrenaline increases with age: correlation with total blood volume and longterm smoking habits. Clin Phys 13:99–109

Josephson BD (1962) Possible new effects in superconductive tunnelling. Phys Lett 1:251–253

Josephson BD (1974) The discovery of tunnelling supercurrents. Science 184:527–530

Keulers M, Kuriyama H (1998) Acetaldehyde influence on metabolic oscillation of *Saccharomyces cerevisiae*. In: Information Processing in Cells and Tissues. Plenum, New York

Keulers M, Satroutdinov AD, Suzuki T, Kuriyama H (1996) Synchronization affector of autonomous short-period-sustained oscillation of *Saccharomyces cerevisiae*. Yeast 12:673–682

Klarreich E (2002) Huygens's clocks revisited. Am Sci 90:322–323

Klevecz RR, Bolen J, Forrest G, Murray DB (2004) A genomewide oscillation in transcription gates DNA replication and cell cycle. Proc Natl Acad Sci USA 101:1200–1205

Kwak W-J, Kwon G-S, Jin I, Kuriyama H, Sohn H-Y (2003) Involvement of oxidative stress in the regulation of H_2S production during ultradian metabolic oscillation of *Saccharomyces cerevisiae*. FEBS Microbiol Lett 219:99–104

Lee MW, Rees P, Shore KA, Ortin S, Pesquera L, Valle A (2005) Dynamic characterisation of a laser diode subject to double optical feedback for chaotic communications. IEEC Proc Optoelect 152:97–102

Lemar KM, Passa O, Aon MA, Cortassa S, Müller CT, Plummer S, O'Rourke B, Lloyd D (2005) Allyl alcohol and garlic (*Alium sativum*) extract produce oxidative stress in *Candida albicans*. Microbiology 151:3257–3265

Lemar KM, Aon MA, Cortassa S, O'Rourke B, Plummer S, Müller CT, Lloyd D (2007) Diallyldisulphide depleted glutathione in *Candida albicans*: oxidative stress mediated cell death studied by two-photon microscopy Yeast 24:698–706

Lloyd D (1992) Intracellular time-keeping: epigenetic oscillations reveal the functions of an ultradian clock. In: Ultradian rhythms in life processes (Lloyd D, Rossi EL, eds) Springer, London, pp 5–22

Lloyd D (2005) Systems dynamics of biology. J Appl Biomed 3:1–12
Lloyd AL, Lloyd D (1993) Hypothesis: the central oscillator of the circadian clock is a controlled chaotic attractor. BioSystems 29:77–85
Lloyd AL, Lloyd D (1995) Chaos: its significance and detection in biology. Biol Rhythm Res 26:233–252
Lloyd AL, May RM (1999) Synchronicity, chaos and population cycles: spatial coherence in an uncertain world. Trends Ecol Evol 14:417–418
Lloyd D, Murray DB (2005) Ultradian metronome: timekeeper for orchestration of cellular coherence. Trends Biochem Sci 30:373–377
Lloyd D, Murray DB (2006) The temporal architecture of eukaryotic growth. FEBS Lett 580:2830–2835
Lloyd D, Murray DB (2007) Redox rhythmicity: clocks at the core of temporal coherence. BioEssays 29:1–9
Lloyd D, Salgado LE, Turner MP, Suller MT, Murray D (2002a) Cycles of mitochondrial energization driven by the ultradian clock in a continuous culture of *Saccharomyces cerevisiae*. Microbiology 148:3715–3724
Lloyd D, Lemar KM, Salgado LEJ, Gould TM, Murray DB (2003) Respiratory oscillations in yeast: mitochondrial reactive oxygen species, apoptosis and time: a hypothesis. FEMS Yeast Res 3:333–339
Lloyd D, Salgado LEJ, Turner MP, Suller MT, Murray D (2002) Respiratory oscillations in yeast: clock driven mitochondrial cycles of energization. FEBS Lett 519:41–44
Lyte M, Ernst S (1992) Catecholamine induced growth of gram negative bacteria. Life Sci 50:203–212
May RM, Lloyd AL (2001) Infection dynamics on scale-free networks. Phys Rev E Stat Nonlin Soft Matter Phys 64: 066112
Michaels DC, Matyas EP, Jalife J (1987) Mechanisms of sinoatrial pacemaker synchronization: a new hypothesis. Circ Res 61:704–714
Morin LP (2007) SCN organization reconsidered. J Biol Rhythms 22:3–13
Murray DB (2004) On the temporal organization of *Saccharomyces cerevisiae*. Curr Genom 5:665–671
Murray DB, Lloyd D (2006)A tuneable attractor underlies yeast respiratory dynamics. Biosyctems, doi: 10.1016/j.biosystems.2006.09.032
Murray DB, Roller S, Kuriyama H, Lloyd D (2001) Clock control of ultradian respiratory oscillations found during yeast continuous culture. J Bacteriol 183:7253–7259
Murray DB, Klevecz RR, Lloyd D (2003) Generation and maintenance of synchrony in *Saccharomyces cerevisiae* continuous culture. Exp Cell Res 287:10–15
Mustafa MG, Utsumi K, Packer L (1966) Damped oscillatory control of mitochondrial respiration and volume. Biochem Biophys Res Commun 24:381–385
Nealson KH, Hastings JW (2006)Quorum sensing on a global scale massive numbers of bioluminescent bacteria make milky seas. Appl Environ Microbiol 72:2295–2297
Oleskin AV, Kirovskaya TA, Botvinko IV, Lysak LV (1998) Serotonin action on growth and differentiation of microorganisms. Rus J Microbiol 67:305–312 (in Russian)
Onkuni K, Hayashi M, Yamashita I (1998) Bicarbonate-mediated social communication stimulates meiosis and sporulation of *Saccharomyces cerevisiae*. Yeast 14:623–631
O'Rourke B, Cortassa S, Aon MA (2005) Mitochondrial ion channels: gatekeepers of life and death. Physiology (Bethesda) 20:303–315
Packer L, Utsumi R, Mustafa MG (1966) Oscillatory states of mitochondria. Damped oscillatory control of mitochondrial respiration and volume. Biochem Biophys Res Commun 24:381–385
Parekh AB, Putney JW (2005) Store-operated calcium channels. Physiol Rev 85:757–810
Pavlidis T (1973) Biological Oscillators. Their Mathematical Analysis. Academic, New York
Pikovsky A, Rosenblum M, Kurths J (2002) Synchronization: A Universal Concept in Nonlinear Science. Cambridge University Press, Cambridge
Prozorovskaya MP (1983) Age-related changes of adrenaline and noradrenaline in tissues of rats. Physiol J USSR 69:1244–1246 (in Russian)

Pye EK (1969) Biochemical mechanisms underlying the metabolic oscillations in yeast. Can J Bot 47:271–276

Pyragas K (2006) Delayed feedback control of chaos. Phil Trans Roy Soc A 364:2309–2334

Reynolds TB, Fink GR (2001) Baker's yeast, a model for fungal biofilm formation. Science 291:878–881

Romanovsky YM, Chernavsky DS (1972) Mutual synchronisation of many auto-oscillating systems contacted through diffusion. Thesis of 1V Biophysical Congress Moscow pp 51 (in Russian)

Rose AH, Harrison JS (eds) (1999) The Yeasts, Academic, London

Roussel MR, Lloyd D (2007) Observation of a chaotic multioscillatory metabolic attractor by real-time monitoring of a yeast continuous culture. FEBS J 274:1011–1018

Satroutdinov DV, Kuriyama H, Kobayashi H (1992) Oscillatory metabolism of Saccharomyces cerevisiae in continuous culture. FEMS Microbiol Lett 98:261–268

Segnui B, Andrieu-Abadie N, Jaffrezou JP, Benoist H, Levade T (2006) Sphingolipids as modulators of cancer cell death: potential therapeutic targets. Biochem Biophys Acta 1758(12): 2104–2120

Siren V, Salmenpera P, Kanhuri E, Bizik J, Sorsa T, Tervahartiala T, Vaheri A (2006) Direct cell-cell contacts between human dermal fibroblasts induce a novel form of cell activation leading to non-apoptotic programmed cell death. Ann Med 38:212–220

Skinner YE, Goldberger AL, Meyer-Kress G, Ideker RF (1990) Chaos in the heart: implications for clinical cardiology. Biotechnol 8:1018–1024

Sohn H, Kuriyama H (2001) Ultradian metabolic oscillation of *Saccharomyces cerevisiae* during aerobic continuous culture: hydrogen sulphide, a population synchronizer, is produced by sulphite reductase. Yeast 18:125–135

Sohn HY, Murray DB, Kuriyama H (2000) Ultradian oscillation of *Saccharomyces cerevisiae* during aerobic continuous culture: hydrogen sulphide mediates population synchrony. Yeast 16:1185–1190

Sohn HY, Kum EJ, Kwon GS, Jin I, Kuriyama H (2005a) Regulation of branched-charm and sulphur-containing amino acid metabolism by glutathione during ultradian metabolic oscillation of *Saccharomyces cerevisiae*. J Microbiol 43:375–380

Sohn HY, Kum EJ, Kwon GS, Jin I, Adams CA, Kuriyama H (2005b) GLR1 plays an essential role in the homeo dynamics of glutathione and the regulation of the H_2S production during respiratory oscillation of *Saccharomyces cerevisiae*. Biosci Biotechnol Biochem 69:2450–2454

Strogatz SH (2000) From Kuramoto Crawford: exploring the onset of chaos in populations of coupled oscillators. Physica D 143:1–20

Strogatz S (2003) Sync: The Emerging Science of Spontaneous Order. Hyperion, New York

Walters M, Sperandio V (2006) Autoinducer 3 and epinephrine signalling in the kinetics of locus of enterocyte gene expression in enterohemorragic *Escherichia coli*. Infect Immun 74:5445–5455

Wheatley D (2000) Ultradian rhythms in cells: questions regarding their existence and significance. Cell Biol Int 24:495–498

Wiley DA, Strogatz SH, Girvan M (2006) The size of the synch basin. Chaos 16:05103

Winfree AT (2001) The Geometry of Biological Time. Springer, New York

Woods NM, Cuthbertson KSR, Cobbold PH (1986) Repetitive rises in cytoplasmic free calcium in hormone-stimulated hepatocytes. Nature 319:600–602

Wuhl E, Hadstein C, Mehls O, Schaefer F (2005) Ultradian but not circadian blood pressure rhythms correlate with renal disfunction in children with chronic renal failure. J Am Soc Nephrol 16:746–754

Wurster B (1976) Oscillations in *Dictystelian discoideum*. Nature 260:703–704

Chapter 4
Phosphorylation Dynamics in Mammalian Cells

D.A. Gilbert[1] **and K.D. Hammond**[2]

Abstract Living cells are auto-dynamic because of control systems operating in their periodic mode. They comprise diverse regulatory networks and are thus multi-oscillators covering a wide range of characteristics. Phosphorylation reactions are involved in virtually all aspects of cell function. Here, we outline a range of our studies on ATP and protein phosphorylation in order to highlight certain features of ultradian dynamics not widely recognised nor appreciated. Our work in this field alone supports the multi-oscillator concept of the living cell and confirms its complexities not least with regard to the significance of temporal oganisation of dynamic processes. The findings support the view that the regulation of cell function, properties and behaviour is achieved through modulation of the dynamic characteristics and are consistent with our concepts of differentiation and cancer.

Key words Oscillations, cell dynamics, ATP, phosphorylation, differentiation

The living cell is most surely dynamic,
Its behaviour so very erratic.
Why is it so true
That all but a few
Insist on treating it static?

4.1 To Begin …

The nature of life determines all that happens in biology yet it is all too rarely taken into account when attempting to understand biological phenomena (Gilbert and Lloyd, 2000). The prime attribute of life is auto-dynamic behaviour and that

[1]113 Dellow Close, Ilford, Essex IG2 7ED, UK
Email: dgilbert92@googlemail.com

[2]Department of Molecular Medicine and Haematology, University of the Witwatersrand Medical School, York Road, Parktown, Johannesburg 2193, South Africa
Email: kate.hammond@wits.ac.za

is also the way we decide its existence. Of utmost significance to this feature are control systems, which, paradoxically, can favour both constancy of composition and also give rise to auto-dynamic behaviour (e.g. Gilbert, 1968). Cellular properties, function and behaviour thus depend on the quantitative aspects of cellular control systems and therefore organisation of processes in time, not least their coordination (Gilbert, 1968, 1969, 1974b, 1984). But the principles of dimensional analysis (e.g. Segel, 1980) show quite clearly that we can only explain quantitative and temporal features in terms of factors which, explicitly or implicitly, have mass and time embodied in them. This implies a total dependency of cellular processes on reaction rates, not gene structure. Here we merely focus attention on the existence of oscillatory behaviour in a number of cellular systems associated with phosphorylation. But we emphasise that the existence and characteristics of periodicities are governed entirely by the *sets* of reaction rates concerned. Thus, although we do not specifically mention the latter, we do acknowledge and imply their total significance.

Since predicting the cellular involvement of phosphorylation processes in theoretical studies on the nature of the cell cycle and cancer (Gilbert, 1974a; cf. also MacKinnon and Gilbert, 1992), we have been making experimental observations on the dynamics of such reactions. *In vitro* studies on mitosis by others have confirmed that several phosphorylated components are indeed so involved (e.g. Kirschner, 1992) but we have focused on reactions not directly concerned with mitosis and outline some studies here, in order to stress the complexity of cellular processes in time. We start with some of the 'simpler' issues.

Conventional approaches to cell biochemistry have indicated the importance of coenzymes and cofactors to function, properties and behaviour. This implies a dependence not only on the levels of such compounds but also on the way they change with time. In turn, those aspects reflect the synthesis and degradation (utilisation) and inter-conversion. There is still a detrimental focus of attention on the roles of genes but apart from other aspects, explanations for dynamic problems cannot come just from the identification of participating compounds nor is it adequate to interpret everything in terms of individual constituents when life is dependent on systems of components. Coenzymes and cofactors are dominant in cell function and behaviour for at least two reasons; firstly, they are involved in a wide range of reactions (which gives them the ability to coordinate processes) and secondly, those properties give rise to looped dependencies that provide control action. Without the cyclic sets of reactions, they can only act as regulators. Some problems associated with such studies have been considered by Gilbert and Ferreira (2000).

4.2 ATP and Related Aspects

We begin by considering one special phosphorylated compound of this kind. ATP is a component which is necessary for many of the reactions of interest and we briefly mention some of our findings because of this widespread significance. As mentioned above, by implication we are also concerned with the reactions involved in its synthesis, degradation and inter-conversion with ADP; one can expect it to be involved in several distinct control systems. We therefore briefly widen our remit slightly to include some comments on glycolysis and mitochondrial function even though those topics alone deserve a whole chapter and, no doubt, will receive attention elsewhere in this volume. The above considerations also provide reasons to expect oscillatory changes in the level of this component, ATP, and so is the case. Indeed, a number of authors have reported periodic variations so we focus on the complexity aspect not, we believe, hitherto considered by others although suggested by the complicated waveforms.

4.2.1 ATP

Figure 4.1A and B show Enright periodogram analyses of ATP oscillations observed in L cells and in murine erythroleukaemic (MEL) cells, respectively. In agreement with Gilbert and Lloyd (2000), these diagrams alone add further weight to the view

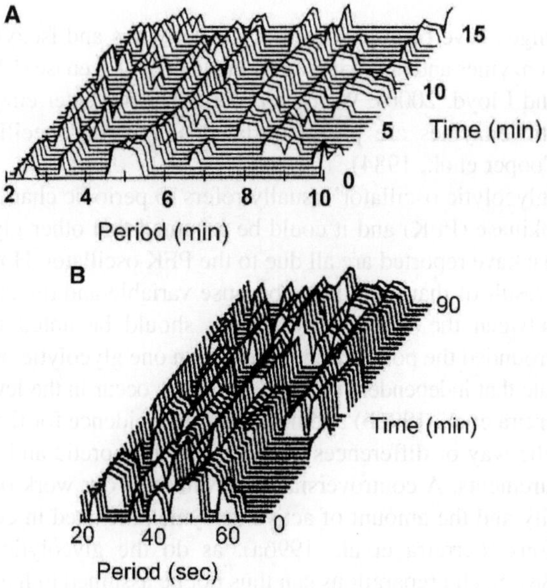

Fig. 4.1 ATP is involved in many cellular reactions and it is reasonable to expect many of them to form parts of distinct control systems that can oscillate in a relatively independent manner. That being so, the ATP level in a cell would reflect the summation of several rhythms. As can be seen in these Enright periodograms, time series analyses of the fluctuations in two mouse cell lines, L cells (Fig. 4.1A) and MEL cells (Fig. 4.1B), support that argument. Comparable results have also been obtained using a hamster kidney cell line

that cells are multi-oscillators. Each diagram indicates the existence in the data of several rhythms of different periods (frequencies) contributing to the total variation in the level of ATP in the indicated cells – and presumably all cells (we have observed similar results with HaK cells). The distinct bands of different periods strongly suggest the involvement of diverse metabolic processes although, because of the nature of the rhythmic processes, one cannot entirely rule out contributions from harmonics or morphological dynamics (Visser et al., 1990). However, there is an additional factor which cannot be dismissed, namely that we are dealing with summation of changes in sub-populations of cells which exhibit distinct periods of oscillation, as detected in studies on morphological periodicities (Visser et al., 1990). Not observed in these particular examples are drifts in the periods of some rhythms (e.g. Gilbert et al., 2000), seemingly due to gradual depletion of nutrients which also confirm theoretical arguments that the dynamic state is not rigidly fixed in characteristics.

Although some are not directly concerned with phosphorylation processes, it is pertinent to mention briefly oscillations associated with glycolysis and mitochondrial action which are responsible for ATP production.

4.2.2 Glycolysis

Rhythmic changes have been observed in the activities and isozyme patterns of a number of the enzymes and hexokinase and lactic dehydrogenase (LDH) in particular (see Gilbert and Lloyd, 2000). We note reports that the latter enzyme and certain other glycolytic enzymes are phosphorylated at tyrosine-specific sites (Cooper et al., 1983; Cooper et al., 1984).

The term 'glycolytic oscillator' usually refers to periodic changes generated by phosphofructokinase (PFK) and it could be assumed that other glycolytic enzyme periodicities we have reported are all due to the PFK oscillator. However, they cannot all be the result of that periodicity because variable and distinct timing differences exist between the various rhythms. It should be noted that much early discussion surrounded the possibility of more than one glycolytic oscillator. Studies on LDH indicate that independent periodic variations occur in the levels of individual isozymes (Ferreira et al., 1996b) and there is some evidence for the involvement of regulators in the way of differences between electrophoretic and extract pyruvate activity measurements. A controversial observation in our work on LDH was that both the activity and the amount of active isozyme oscillated in cell- and particle-free preparations (Ferreira et al., 1996a), as do the glycolytic substrates (see Edmunds, 1988). Such preparations can thus not be assumed to have constant composition, nor can the primary oscillations observed be attributed to interaction between cytoplasm and mitochondrion or to plasma membrane reactions. The behaviour, however, may relate to covalent modification of the enzyme protein by phosphorylation. In other studies, we have drawn attention to theoretical considerations on the efficiency of oscillating glycolysis, particularly in relation to the phenomenon of ageing (Gilbert, 1995).

4.2.3 Mitochondrial Action

Our studies on oscillations in the inter-conversion of NAD/NADH (the system redox state), and the effect thereon of insulin, have been described (e.g. Visser et al., 1990; Gilbert and Lloyd, 2000). This multi-frequency rhythm is usually attributed primarily to the PFK rhythm. By way of a change, here we include evidence for oscillatory changes in the redox state of FAD/FADH that can tentatively be associated with mitochondrial reactions and, hence, ATP synthesis (Fig. 4.2). The latter power spectrum analysis diagram again implicates the existence of a number of periods and also shows that the FAD oscillation is not rapidly stopped by suddenly dropping the temperature from 37°C down to 4°C; indeed the power is enhanced by the treatment (it should be pointed out that the data has been enhanced to emphasise the difference between the unaffected and cold shocked sections of the data, see Visser et al., 1990). Similar results have been noted with light scattering morphological studies – see for example Visser et al. (1990). It may be asked why such treatment should stimulate the rhythms. Elsewhere (Gilbert and Visser, 1993), we have shown that insulin stimulates cell surface vibrations and it was suggested that the non-specific actions of insulin on metabolism are due to effective stirring of the surrounding diffusion layer. And the FAD kind of effect discussed here could thus reflect a general attempt of cells to counteract a drop in metabolic rate by increasing the uptake of nutrients.

4.2.4 Phosphoamino Acid Phosphatases

Before the advent of suitable phosphorylated protein substrates, we developed a spectrophotometric assay for determining the hydrolysis of phosphotyrosine (Ferreira et al., 1996c). Figure 4.3, shows oscillations in this activity in MEL cells

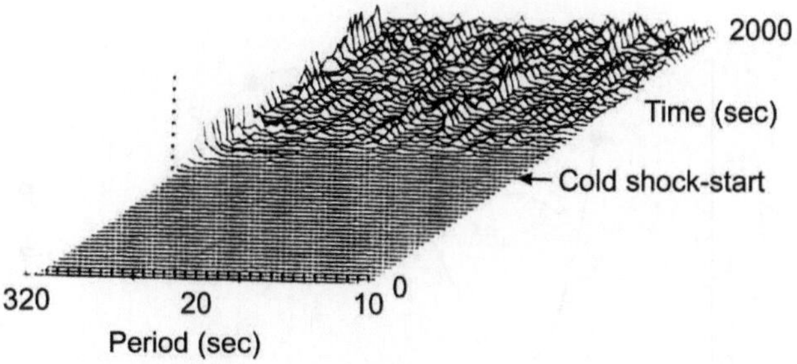

Fig. 4.2 We have used a difference fluorescence method (exciting the cells with light of one frequency and detecting the emitted light at another) to follow changes in the redox state of FAD in cells and here we show the stimulating effect of a sudden cold shock (from 37°C down to 4°C) on the oxidation state in a suspension culture of MEL cells using power spectrum analysis of the data. Although it appears that a range of oscillations occur only after the change in temperature, this is because we enhance the effect by only plotting values which are greater than an arbitrarily set threshold (Visser et al., 1990). Note the non-linear period scale of this method of analysis

and the effect thereon of insulin while the accompanying diagram, Fig. 4.4, also indicates that the hormone affects the rhythm and makes it evident that this activity is inhibited by lowering the temperature.

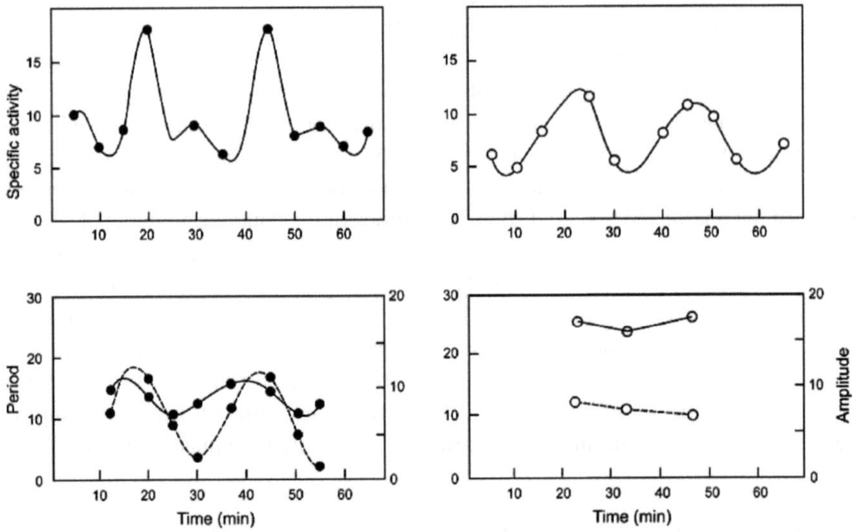

Fig. 4.3 The phosphotyrosine phosphatase (PTPase) activity in MEL cells has been found to oscillate and, as seen here, insulin affects this rhythm as it does for several cellular oscillations, consistent with the widespread effects of the hormone on cellular processes. It is not clear though if the latter acts directly on the PTPase periodicity or through another system which modulates the rhythm. Control cells are shown on the left and insulin-treated cells on the right. Period (———). Amplitude (...........)

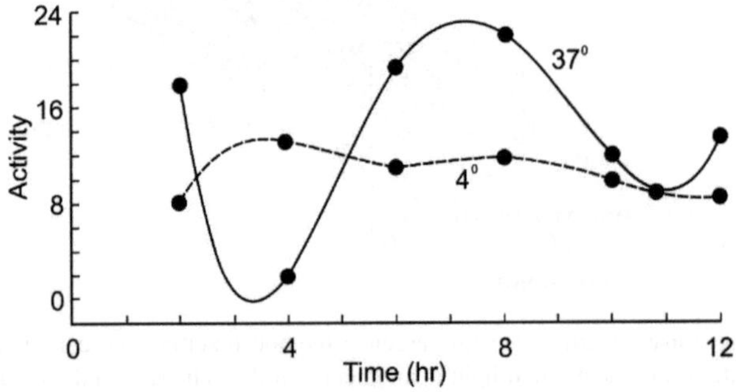

Fig. 4.4 Here it can be seen that the phosphotyrosine phosphatase oscillation in MEL cells is inhibited at low temperature. This diagram suggests an oscillation period of the order of hours, whereas the last figure seems to indicate a period of minutes. Both can be wrong but the former is certainly way off the mark. This is because of the problem of aliasing and hence a dependence on the sampling interval

4.2.5 Temporal Coordination

A widely ignored but major feature of great importance is the temporal co-ordination of cellular processes in time. The temporal relationship between two rhythms we thus study by using phase plane plots wherein the corresponding pairs of data values of interest, determined in the same cell extract, are plotted against each other (Gilbert, 1969). In these diagrams, time does not appear explicitly but as a movement of the point in the plane of the graph. In steady state situations the various data points fall on defined lines which are positive in slope where the two rhythms concerned are in-phase and negative in slope where they are out-of-phase, assuming that their periods are the same. Where an intermediate time relationship exists, the result is a closed curve under these conditions. More complex outcomes are obtained if the periods differ. In Fig. 4.5A and B, we give examples showing different phasings between the LDH and hexokinase isozyme rhythms in different cells. For LDH we have used the rates toward butyrate (B) as one substrate and pyruvate (P) for the alternate substrate to estimate isozyme pattern; on the other hand, for our

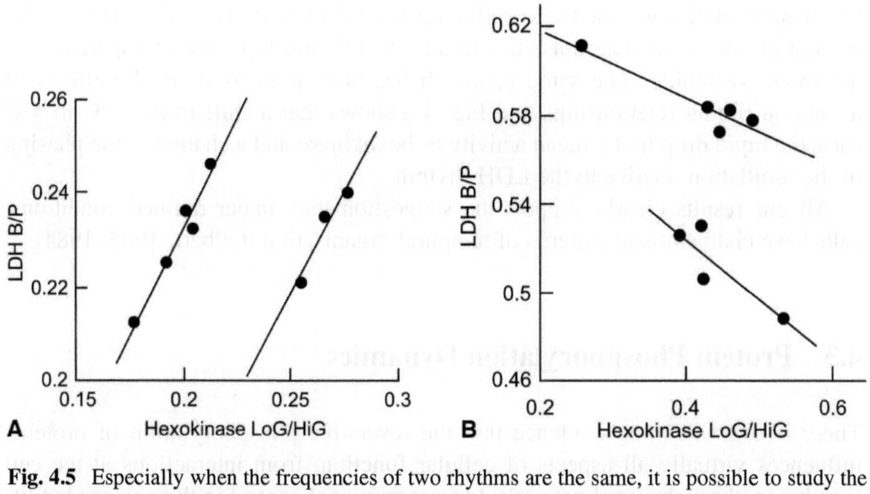

Fig. 4.5 Especially when the frequencies of two rhythms are the same, it is possible to study the phase relationship between them by plotting corresponding values, determined in the same extracts, against one another. This approach has been used in these two phase plane plot diagrams to observe the relative timings of periodic variations in the effective isozyme patterns for lactic dehydrogenase (B/P) and hexokinase (LoG/HiG) in HaK (Fig. 4.5A) and HeLa (Fig. 4.5B) cell lines. It can be seen that in both these cases, the data values fall on one or other of two straight lines, but in one the lines are of positive slope while of negative slope in the second. Our interpretation of this result is that (a) the two rhythms were in phase (positive slope) in one cell line but out of phase (negative slope) in the other and (b) metabolic switching was occurring in both cases giving rise to the pairs of lines. The different slopes thus support the view that cells can exhibit different patterns of temporal organisation and that the two periodicities are not due to the same rhythmic process. The highly linear nature of the lines indicates that both cells were essentially in a steady state (cf. Gilbert, 1974b, 1984)

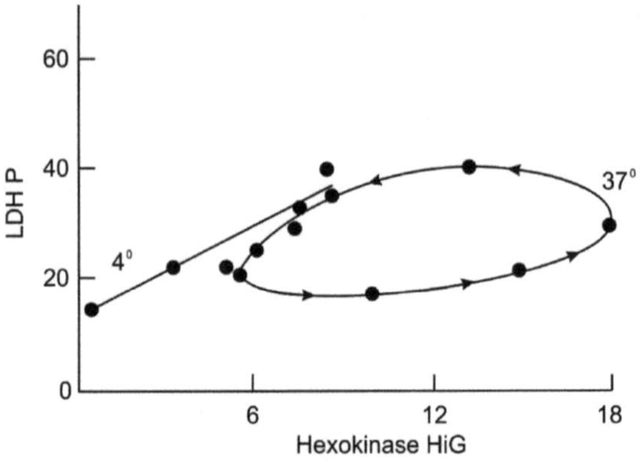

Fig. 4.6 With these rat hepatoma cells, the phase plane plot of the activities of LDH and hexokinase initially yielded a closed cyclic curve indicating that the two rhythms were in an intermediate phase relationship when the cells were at 37°C. However, when the temperature was rapidly decreased to 4°C, there was a rapid fall in the mean level of the hexokinase activity and the two rhythms then oscillated in phase, as shown by the linear plot

hexokinase studies we used a low glucose level (LoG) or high value (HiG). Also evident in both these diagrams are shifts in the relationships caused apparently by metabolic switching. The same approach has been used to study the effects of agents on timing relationships and Fig. 4.6 shows that a shift from 37°C to 4°C caused a rapid drop in the mean activity of hexokinase and a change in the phasing of the oscillation relative to the LDH rhythm.

All our results clearly support the suggestion that, under defined conditions, cells have characteristic patterns of temporal organisation (Gilbert, 1968, 1984).

4.3 Protein Phosphorylation Dynamics

There is overwhelming evidence that the reversible phosphorylation of proteins influences virtually all aspects of cellular function, from interactions at the cell membrane, through signal networks to transcriptional control in the nucleus but little attention has been paid to the temporal aspects of these processes, except with regard to mitosis. In the following sections, we consider the more general aspects but stress the cyclic behaviour relating to protein phosphorylation, highlighting the fact that cellular control of differentiation may be achieved through modulation of the rhythms (see Gilbert, 1984).

In the studies we describe, we have used the murine erythroleukaemic, MEL, and human acute promyelocytic leukaemic, HL60, cell lines. These cell lines, proliferating in culture, can be induced to differentiate and lose their malignant characteristics by a variety of agents; they thus provide valuable models for studying the molecular and cellular dynamics involved in differentiation, as well as the pathogenesis of cancer and its reversal (Rovera et al., 1979; Breitman et al., 1980; Reuben et al., 1980). The findings we discuss are consistent with our concept (Gilbert, 1984) that differentiation and cancer involve changes in the set of control systems dynamics, that is in the patterns of temporal organisation, by modification of the frequencies, amplitudes, means and phasings of cellular rhythms.

4.3.1 Phosphorylation Potential

The highly dynamic nature of protein phosphorylation processes, even greater than cell cycle studies seem to indicate (Norbury and Nurse, 1992), was illustrated in a study of changes in the phosphorylation potentials of certain specific proteins (Ferreira et al., 1994a). When MEL cell preparations were incubated with ^{32}P-ATP, and then subjected to SDS-PAGE, autoradiographic analysis showed a very intense band, a major component of the cells, designated protein X, corresponding to a molecular mass of 81 kDa; a fainter band, corresponding to a mass of 63 kDa and designated protein Y, was also seen. Remarkably, variations in phosphorylation of as much as 100-fold were found within an interval of 10 min. The identities of the X and Y proteins are not known and we are unable to define the nature of the underlying reactions, although we note a report describing the stimulatory effect of growth factors on the *in vivo* phosphorylation (possibly autophosphorylation) of an 80 kDa protein in 3T3 cells (Rozengurt et al., 1983).

Time plots revealed the irregular periodic nature of the changes for both protein X and protein Y despite the random selection processes involved; both the periods and amplitudes seemed to be rhythmically modulated, but apparently by different processes. Periodogram analysis of the data (Fig. 4.7) confirmed the complexity of the phosphorylation dynamics of the two proteins. Although there was some similarity between the X and Y protein oscillations, there was a lack of any distinct phase relationship between them; the ratio of intensities fluctuated in a pseudoperiodic manner with variations up to 50-fold. There may be several explanations for the complexity of the rhythms even if one assumes that the X and Y bands represent single proteins. Variations may be the result of changes in activities of kinases and phosphatases and the extents and multiplicity of phosphorylation (reflecting the involvement of several distinct kinases affecting different residues). As kinases are generally very specific, it seems probable that those catalysing the reactions for protein X and protein Y are different, irrespective of whether only one enzyme is involved in each case. If both activities vary with time in a periodic way, the observed complexities can be understood, especially if the frequencies are distinct. More than two oscillations seem to be involved for both proteins and this could account for the lack of a particular relationship in the phase plots, as might competition effects. However, the question arises as to how two, possibly more, kinase activities can be regulated independently in periodic manner. Changes in protein concentration may be another factor; oscillations are known to occur in protein synthesis

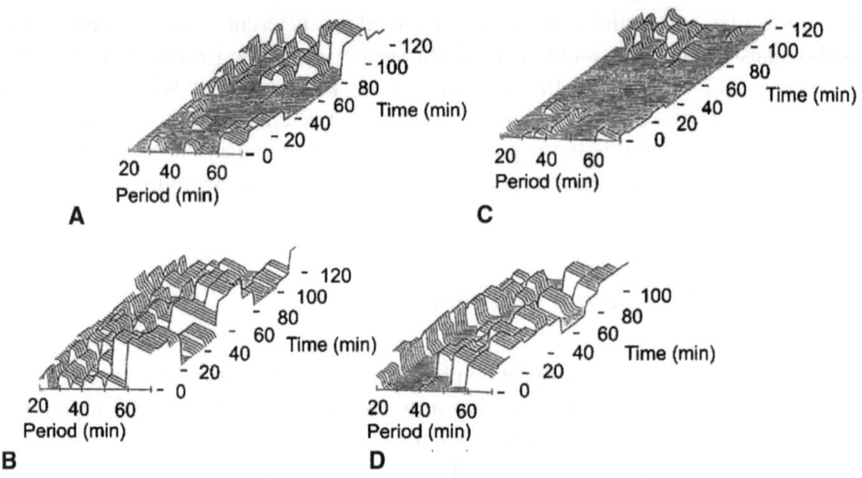

Fig. 4.7 These Enright periodograms show the phosphorylation dynamics of two proteins, protein X (**A** and **B**) and protein Y (**C** and **D**), in MEL cells. **A** and **C**: untreated control cells. **B** and **D**: cells treated with insulin. For the rhythms whose periods fall within the range shown, the magnitudes (the relative variance) tend to vary with time, but are more marked after addition of the hormone (Adapted from Hammond et al., 1998)

(although not dependent on ATP) and in the rate of incorporation of amino acids into protein (Brodsky et al., 1992).

Insulin was found to enhance the protein X and Y oscillations in a complex manner, in that it may affect the mean level of particular rhythms contributing to the overall behaviour as well as the dynamic characteristics of that contribution. The action of the hormone seemed to be general, but not all oscillations were affected similarly. Insulin stimulates rhythmic variations in cell morphology (Visser et al., 1990) and its effects on phosphorylation are in fact comparable. As with morphological changes, the apparent frequency dependent response to the hormone may be due to rapid, discrete change in the periods of certain oscillations rather than simple increases and decreases in their amplitudes. The complexity of the response to insulin is consistent with the multiplicity of its effects on cellular processes.

4.3.2 Protein Kinases and Phosphoprotein Phosphatases

Clearly, phosphorylation processes are highly dynamic, even under resting conditions, and it has become increasingly evident that the cyclic interconversion of the regulatory enzymes, the protein kinases and phosphoprotein phosphatases, represents the dynamic process in which the steady state equilibrium between active and inactive forms varies depending on the parameters determined by the metabolic state of the system. The emergent array of phosphoprotein phosphatases has shown that nature has created a diversity of structure and function far exceeding that of the protein kinases (Shi et al., 1998; Virshup, 2000; Janssens and Goris, 2001; Cohen, 2002; Sim and Ludowyke, 2002).

A possible role for serine/threonine phosphatases in regulating granulocytic differentiation of HL60 cells was suggested by Morita et al. (1992) and Tawara et al. (1993). Our studies of the serine/threonine phosphatases, protein phosphatase 1 (PP1) and protein phosphatase 2A (PP2A), and of protein tyrosine phosphatase 1B (PTP1B) enzyme protein expression during proliferation and differentiation of MEL and HL60 cells, which involved SDS-polyacrylamide gel electrophoretic analysis of cell extracts followed by immunoblotting using specific antibodies, showed marked temporal variations in the intensities of immunoreactive bands. The magnitude of the fluctuations for PP1 was as much as seven- to eightfold and for PP2A and PTP1B up to 10–12-fold (Hammond et al., 1998; Bhoola and Hammond, 2000; Bodalina et al., 2005).

The molecular mass of 38 kDa we observed for PP1 in HL60 cells corresponded closely to that given by Mumby and Walter (1993) for the catalytic subunit of the protein. Differentiation of HL60 cells along the granulocytic and monocytic pathways, induced by all-trans retinoic acid (ATRA) and phorbol myristate acetate (PMA), respectively, was associated with variations in the patterns for the PP1 band. With PMA, there was also a marked change in the mean expression of PP1. Representative results showing the effect of ATRA are given in Fig. 4.8. In MEL cells, we found a single immunospecific protein of mass 33 kDa. Temporal variations in the expression of this protein were seen and there were differences between the patterns in proliferating and hexamethylene bisacetamide (HMBA)-induced differentiating cells. In some cases, there were changes in amplitude and phasing, while in others there was a dampening or an overall increase in protein expression in HMBA-treated cells as compared with controls. An example is shown in Fig. 4.9. The difference plot effectively removes mean variations and provides further illustration of the dynamic nature of the system. Instantaneous period and amplitude both varied with

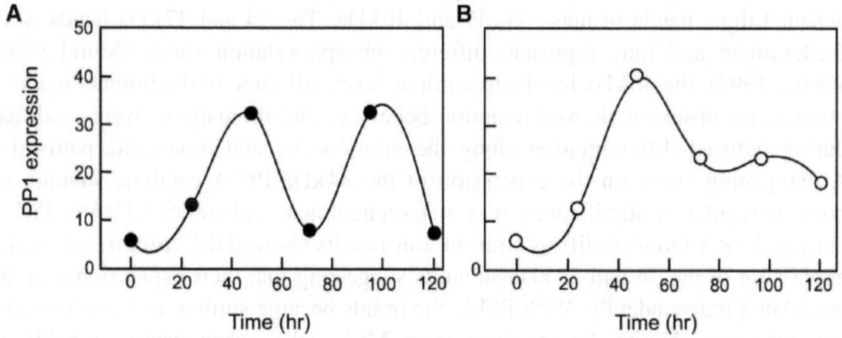

Fig. 4.8 The expression of PP1 in HL60 cells shows a rhythmic pattern of behaviour (**A**) that is modulated during ATRA-induced differentiation along the granulocytic pathway (**B**). The aliasing problem (see caption to Fig. 4.4) is obviously also relevant here, again because of the long sampling interval

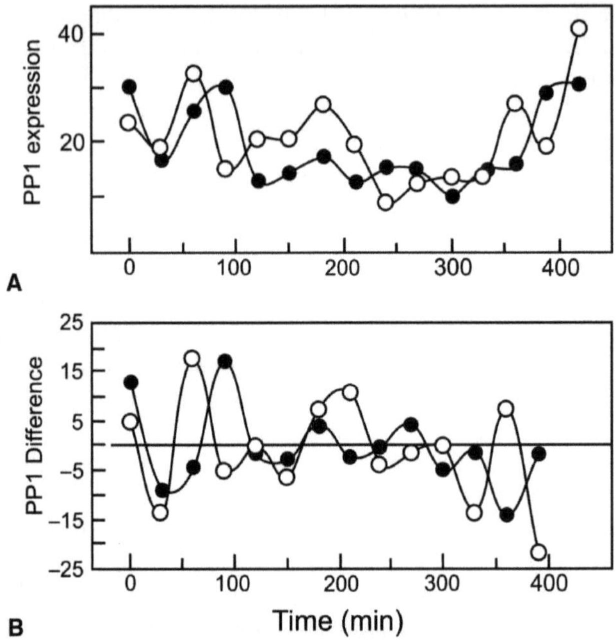

Fig. 4.9 Variations in the expression of PP1 can also be seen in untreated, proliferating (•) and HMBA-treated, differentiating (◦) MEL cells (**A**). The difference curves (**B**) provide further illustration of the dynamic nature of the system

time making it difficult to estimate mean period and amplitude; variations noted between the untreated and the HMBA-treated cells indicated that changes were taking place during cell differentiation.

In HL60 cells, with an antibody specific for the catalytic subunit of PP2A, we detected three bands of mass 34, 37 and 46 kDa. The 34 and 37 kDa bands were predominant and may represent different phosphorylation states (Mumby and Walter, 1993); the 46 kDa band was a minor form. All three of the immunoreactive proteins we observed showed temporal behaviour and the patterns were modified during induced differentiation along the granulocytic and monocytic pathways. A dampening effect on the expression of the 34 kDa PP2A catalytic subunit, of possible regulatory significance, was seen on treatment with either ATRA or PMA. During ATRA induced differentiation, our results showed different trends in the expression of the 34 and 37 kDa subunits suggesting that their expression may be modulated independently. With PMA, the trends became similar, and in phase, for the different subunits. Preparations from MEL cells, when probed for PP2A, showed two bands of masses 32 and 36 kDa; the 36 kDa band probably represented the catalytic subunit and the 32 kDa band a cleaved species of the catalytic subunit (see Tung et al., 1984, 1985). Time dependent variations in expression of the 32 and

36 kDa subunits were observed, using 15, 30 and 60 min and 12 h sampling times and HMBA modified the dynamics. Figure 4.10 shows ratios for the two subunits and a phase plane plot. Clearly, the two proteins were oscillating and there was some evidence that in certain instances they were doing so in phase with one another. This suggests a common underlying process, but on the other hand, there were differences that implied a certain degree of independence. As with PP1, the period and amplitude curves varied with time; mean periods covered a wide range. Observations again demonstrated that sampling frequency is an important factor in determining period and the true values may be much lower because of the problem of aliasing (Hammond et al., 1998; Gilbert and Ferreira, 2000; Calvert-Evers and Hammond, 2002).

We have also observed differentiation-associated changes in oscillatory behaviour for expression of PTP1B in HL60 cells. Our findings for expression of PTP1B, as for PP1 and PP2A, have indicated that the dynamic control mechanisms operating during induced differentiation along the granulocytic and monocytic pathways are different. The catalytic domain of PTP1B was detected as a 46 kDa band, which was usually the major form; in addition, a 42 kDa form was seen. During differentiation along the granulocytic pathway, induced by ATRA, there was an overall increase in

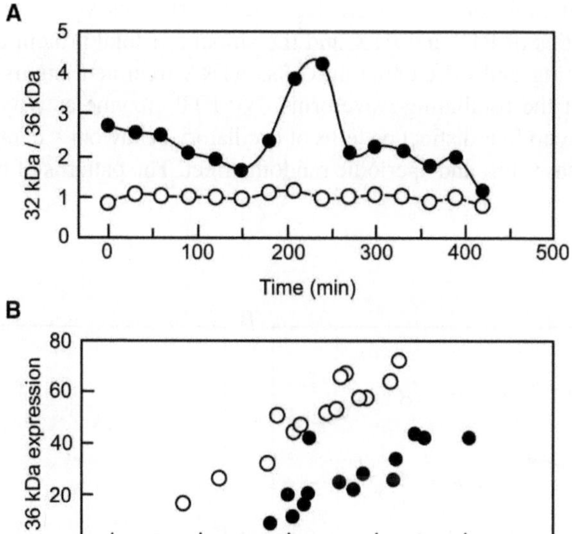

Fig. 4.10 The relationships between the 32 and 36 kDa PP2A components showed some variability; the two proteins were oscillating and in certain instances they were in phase with one another. In this experiment, their expression in untreated, proliferating (•) and HMBA-treated, differentiating (◦) MEL cells is plotted as ratios (**A**) and as a phase plane plot (**B**)

PTP1B expression and a dampening effect on the oscillation. For PMA-induced monocytic differentiation, there was an increase in the magnitude of the oscillation (Fig. 4.11). Further, the effect seen with PMA was remarkable with a complete change in distribution of the two forms in a time dependent manner and predominance of the 42 kDa subunit after 48 h. This change could have been the result of dephosphorylation, or alternatively, calpain catalysed cleavage of the protein may have caused a shift in the mobility (Frangioni et al., 1993). Studies have indicated that HeLa cells undergo phosphorylation of PTP1B during mitosis when the slower moving band is seen; in synchronous cells the faster band predominated (Schievella et al., 1993). In a study of the effects of different lysing buffers on PTP1B expression, we showed that when fluoride and phosphate were included, the protein was detected exclusively as the higher molecular mass form; in buffers from which these substances were omitted only the lower band was seen (Calvert-Evers and Hammond, 2000a).

Studies of the dynamics relating to protein concentrations and the activities of protein tyrosine phosphatase (PTP) and protein tyrosine kinase (PTK) have provided yet more evidence of oscillatory rhythms (Calvert-Evers and Hammond, 2000b, 2002, 2003). These involved examination of time-dependent variations during proliferation and differentiation, linear correlations and phasing relationships on a time basis and variations in the characteristics of the rhythms. Quantitative assessment of activity of PTPase and PTK in proliferating and ATRA-induced differentiating HL60 cells was by non-radioactive photometric enzyme linked immunoassay.

Temporal organisation and time-dependent variations were observed for the enzyme activities of PTP and PTK and the amount of total protein extracted from both proliferating and differentiating cells. ATRA treatment significantly altered the patterns of the oscillating waveforms. For PTP enzyme activity, the effect of ATRA gave rise to four distinct patterns of oscillatory behaviour – dampening, partial phase shift, phase shift and aperiodic random effect. The patterns of behaviour seen

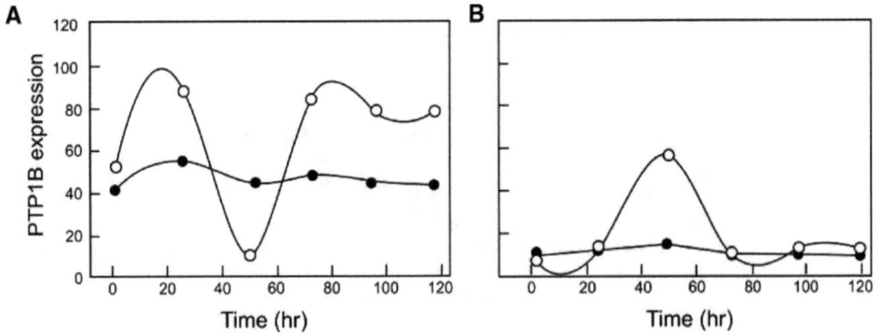

Fig. 4.11 Time dependent changes in the expression of the 42 (**A**) and 46 kDa (**B**) forms of PTP1B can be seen in HL60 cells. The patterns are enhanced during PMA-induced differentiation along the monocytic pathway. Untreated control cells (•); PMA-treated cells (◦)

for PTK activity on ATRA treatment included suppression or dampening, slight suppression with partial phase shift, increase in activity and a random irregular effect. Examples are shown in Figs. 4.12 and 4.13. No obvious relationship could be distinguished between the PTP or PTK enzyme activities and the protein concentration for either proliferating or differentiating cells; the poor degree of correlation suggested relative independence of these sets of oscillators. If temporal changes in PTP and PTK activities were caused simply by fluctuations in the amount of protein extracted from the cells, the timing relationship would not be apparent, as the enzyme activities should vary in unison with the protein changes. If the activities of the two enzymes were interdependent, phase plane plots should not have shown such marked deviation from linearity. As no particular relationships could be detected for these periodicities, it is possible that the frequencies of the rhythms are different, or that multiple interacting rhythms are involved. This is

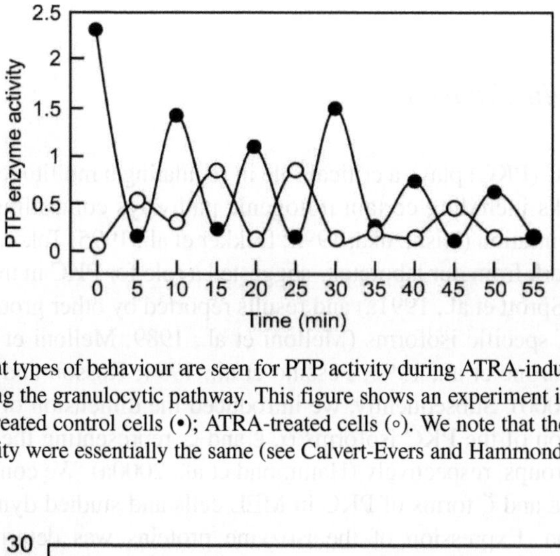

Fig. 4.12 Different types of behaviour are seen for PTP activity during ATRA-induced differentiation of HL60 cells along the granulocytic pathway. This figure shows an experiment in which there was a phase shift. Untreated control cells (•); ATRA-treated cells (○). We note that the patterns seen for PTP specific activity were essentially the same (see Calvert-Evers and Hammond, 2000b)

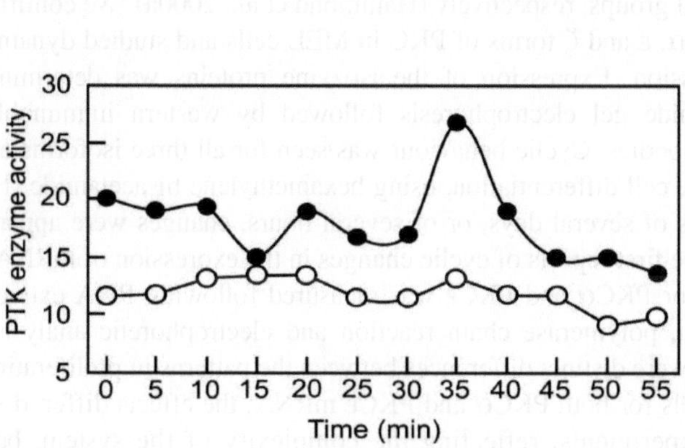

Fig. 4.13 Fluctuations in the activity of PTK in HL60 cells can be seen here. Differentiation along the granulocytic pathway induced by ATRA resulted in a dampening of the rhythms in this experiment. Untreated control cells (•); ATRA-treated cells (○)

consistent with our findings for other studies (Gilbert and Tsilimigras, 1981; Ferreira et al., 1994b). Analysis of the data showed that the period and amplitude of PTP and PTK enzyme activities in HL60 cells varied with time and that there were distinct changes in these characteristics during differentiation along the granulocytic pathway. The periods covered a wide range of values (5–26 min) and were of a similar order of magnitude in proliferating and differentiating cells, and for both enzymes. Differences in the amplitude values for PTP in untreated and induced cell were obvious. A very noticeable decrease was observed where there was dampening or suppression of activity during differentiation. This observation provides for an extra dimension of metabolic control through differential modulation of rhythmic characteristics such as amplitude.

The wide variety of modulatory effects following treatment of cells with differentiating agents reflects their dynamic nature and the complex time-dependent interactions between the individual regulatory processes within the cell, between the metabolic control systems of different cells and with the environment.

4.3.3 Protein Kinase C

Protein kinase C (PKC) plays a critical role in regulating a multitude of signal transducing networks including certain mitogenic pathways controlling cell proliferation and differentiation (Nishizuka, 1992; Dekker et al., 1995; Toker, 1998; Buchner, 2000). Early work from our laboratory suggested a role for PKC in the differentiation of MEL cells (Sprott et al., 1991a) and results reported by other groups indicated the involvement of specific isoforms (Melloni et al., 1989; Melloni et al., 1990; Leng et al., 1993; Patrone et al., 1994; Pessino et al., 1994; Rosson and O'Brien, 1995; Mallia et al., 2000). Subsequently, we introduced the dimension of time to a study of the expression of the PKC isoforms α, ε and ζ, representing the classical, novel and a typical groups, respectively (Hammond et al., 2000a). We confirmed the presence of the α, ε and ζ forms of PKC in MEL cells and studied dynamic aspects of their expression. Expression of the isozyme proteins was determined by SDS-polyacrylamide gel electrophoresis followed by western immunoblotting using specific antibodies. Cyclic behaviour was seen for all three isoforms and on induction of MEL cell differentiation, using hexamethylene bisacetamide (HMBA), over time periods of several days, or of several hours, changes were apparent. We also presented the first reports of cyclic changes in the expression of mRNA. Expression of mRNA for PKCα and PKCε was measured following RNA extraction, reverse transcription, polymerase chain reaction and electrophoretic analysis on agarose gels. There were distinct differences between the patterns in proliferating and differentiating cells for both PKCα and PKCε mRNA; the effects differed somewhat in different experiments, reflecting the complexity of the system, but in general there was a change in amplitude and phasing rather than frequency (Fig. 4.14). In these studies, for both mRNA and protein expression, we gave comparisons of

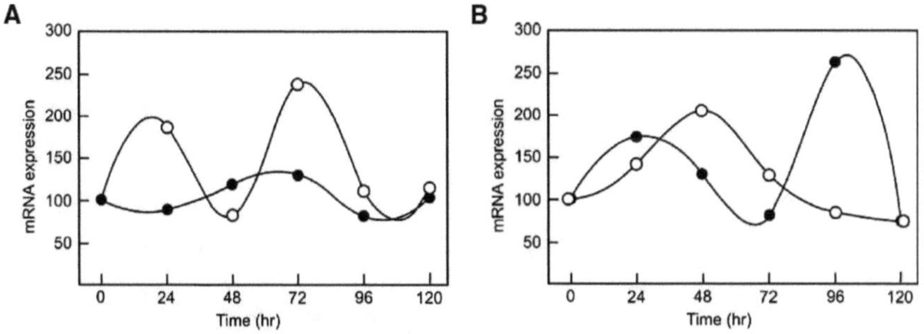

Fig. 4.14 Temporal variations in the expression of the mRNA for PKCα (**A**) and PKCε (**B**) seen in MEL cells and the effect of HMBA-induced differentiation are shown here. Untreated control cells (•); HMBA-treated cells (◦)

results for mean values with those corresponding to individual experiments. These were given particular mention as they demonstrated clearly the importance of examining the findings in a time-dependent manner for each of the data sets, in order to provide complete information without masking the dynamics. We suggested, based on our time studies, that modulation of the dynamics of the signals delivered by the different PKC isoforms is one of the molecular mechanisms involved in MEL cell differentiation. The rhythms of the different isoforms may vary in concert with each other as well as with those of transducing cascades, such as RAS/RAF (Kolch et al., 1993; Young et al., 1996; Van Dijk et al., 1997; Toker, 1998), in order to produce a coordinated response to extracellular effectors.

4.3.4 The RAS Signal Transducing Network

Another aspect of our work has involved studies of the *RAS* gene and its protein product in relation to induced differentiation of MEL cells. The *RAS* genes encode proteins that form part of a complex array of interacting networks regulated by kinases and phosphatases (Downward, 1990; Campbell et al., 1998; Chen et al., 1998; Malumbres and Pellicer, 1998). Several investigations suggested a role for the RAS protein in erythroid differentiation (Downward, 1990; Scheele et al., 1994; Ge et al., 1998; Matsumura et al., 1998); we showed changes in the activities of protein kinase A and protein kinase C, both of which interact with the RAS pathway and modulate its activity, during differentiation of MEL cells (Sprott et al., 1987, 1991a, b). We also saw temporal changes in expression of PKC isoforms in the study described above.

In our studies of *RAS,* we demonstrated cyclic changes in expression of H-*RAS* and N-*RAS* mRNA, as measured by northern blot analysis, and the protein product, as determined by SDS-PAGE and western immunoblotting (see Fig. 4.15)

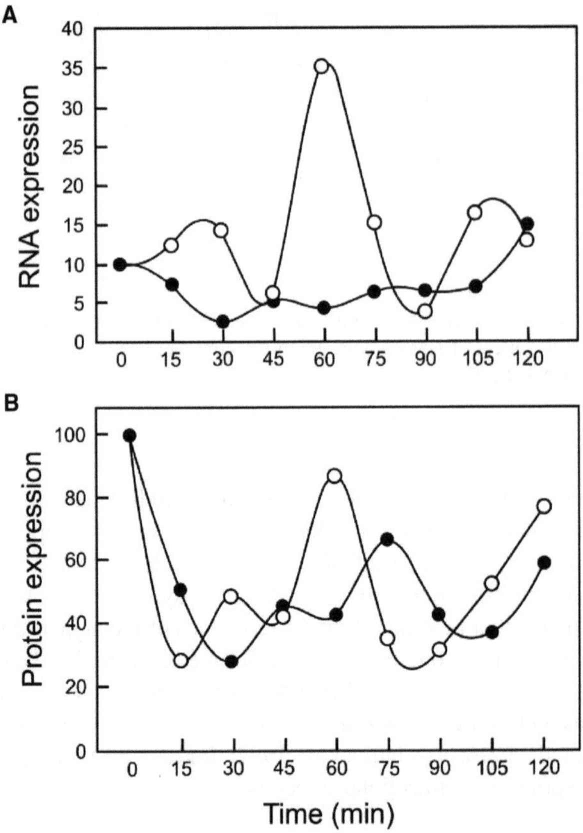

Fig. 4.15 The expression of the mRNA specific to H-*RAS* (**A**) and the expression of the RAS protein (**B**) in MEL cells show cyclic behaviour which is modified during induced differentiation. Untreated control cells (•); HMBA-treated cells (○)

(Hammond et al., 2000b). Once again, we showed that a range of results may be possible because of periodic behaviour and demonstrated that statistical comparisons based on single time point analyses provided an incomplete picture, which is not necessarily a reflection of the true situation and may be misleading. When mean values were assessed, increases in the expression of mRNA and protein were apparent after treatment with HMBA for 96 or 100h; these changes may be associated with activation of the RAS pathway leading to terminal differentiation, or they may be coincidental depending on the phasing of the rhythms. Changes in frequency and phasing of the patterns of expression on treatment with HMBA were seen throughout the time courses studied, suggesting involvement of RAS in MEL cell differentiation from an early stage. It was clear that that regulation of a critical signal transducing protein, RAS, like that of the other kinds of proteins we have studied, is not just a matter of a simple increase or decrease in concentration, it is more complex involving modulation of the pattern of temporal control.

4.3.5 *The P53 Protein*

The P53 protein is a nuclear phosphoprotein, which, like other proteins, is constantly adjusting to changes in the cell; it is a tumour suppressor protein that responds to DNA damage, abnormal cell proliferation and hypoxia by arresting cell proliferation or inducing cell death (Hollstein et al., 1991; Lane, 1992; Greenblatt et al., 1994; Vousden, 2002). The p53 system is of particular interest to us for examining our concepts of temporal organisation not least because the protein has been reported to oscillate. There are feedback loops in which P53 controls the expression of its own regulator, MDM2, and a simple mathematical model has been presented, which suggests that oscillations in P53 and MDM2 proteins can occur in response to a stress signal (Bar-Or et al., 2000; Monk, 2003; Lahav et al., 2004; Tyson, 2004; Harris and Levine, 2005). Oscillations of both proteins occur on exposure of various cell types to ionising radiation. These may be associated with repair of DNA, preventing continuous, excessive P53 activation. Other studies have shown that DNA damage elicits quantal pulses of P53 (Lahav et al., 2004; Tyson, 2004). Monk (2003) considers time delays resulting from transcription, transcription splicing and processing and protein synthesis, which in principle can result in oscillatory mRNA and protein expression; it can be shown mathematically that the observed oscillatory expression of proteins, including P53, which are components of short feedback inhibition loops appear to be driven by transcriptional delays.

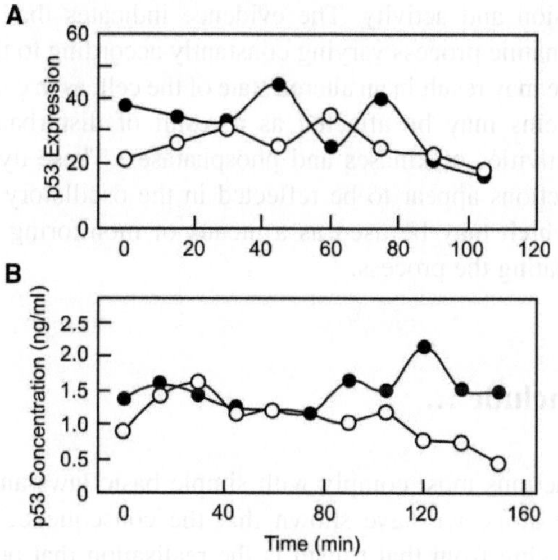

Fig. 4.16 Changes in the expression of the P53 protein in proliferating and differentiating MEL cells, measured by western blotting (**A**) or by ELISA (**B**). Untreated control cells (•); HMBA-treated cells (◦)

In our studies, aimed at exploring the contribution of P53 to the auto-dynamic nature of MEL cells, the protein was analysed both by western immunoblotting and by a more sensitive ELISA procedure; with both techniques, oscillatory behaviour, consistent with that seen in MCF7 cell populations (Lahav et al., 2004), was apparent (Bodalina et al., 2007). Our results were consistent with the view that HMBA influences P53 dynamics; (see Fig. 4.16) the period and amplitude curves differed in proliferating and differentiating cells. The period of the P53 rhythm was similar to that seen for PP1 and PP2A (30 min or less) in MEL cells (Bodalina et al., 2005). For PTK and PTPase in HL60 cells, the estimated period was lower and of the order of 10 min (Calvert-Evers and Hammond, 2002).

As in all our other investigations, the variations were complex as would be expected in the cellular environment where there are numerous interactions influencing protein expression. The models and experiments of Bar-Or et al. (2000) and Monk (2003), which revealed quantal pulses and harmonic oscillatory patterns, involved relatively uncomplicated situations in an ideal environment. Consistent behaviour is possible only in simple systems, that is in experiments where only a few isolated components are reacted together or in studies involving individual cells.

4.3.6 The Protein Phosphorylation Process

Support for the existence of numerous cellular periodicities associated with reversible protein phosphorylation is now considerable and includes rhythmic variations in expression of key phosphorylated and dephosphorylated proteins and enzyme protein expression and activity. The evidence indicates that phosphorylation of proteins is a dynamic process varying constantly according to the needs of the cell. A drastic change may result in an altered state of the cell, as in cancer. The expression of critical proteins may be affected as a result of disturbances to the balance between the activities of kinases and phosphatases. These dynamic changes and complex interactions appear to be reflected in the oscillatory patterns of cellular components, which may be used as a means of monitoring transformation and possibly modulating the process.

4.4 To Conclude ...

All cellular reactions must comply with simple basic laws and principles but we think that here alone we have shown that the consequences can be extremely complex. Following from that truism is the realisation that our ability to analyse and interpret actual data is incredibly challenged.

Acknowledgements Our thanks and appreciation to our colleagues and former colleagues for their invaluable contributions to these studies.

References

Bar-Or, R.L., Maya, R., Segel, L.A., Alon, U., Levine, A.J., and Oren, M. (2000). Generation of oscillations by the p53-Mdm2 feedback loop: a theoretical and experimental study. Proc Natl Acad Sci USA *97*, 11250–11255.

Bhoola, R., and Hammond, K. (2000). Modulation of the rhythmic patterns of expression of phosphoprotein phosphatases in human leukaemia cells. Cell Biol Int *24*, 539–547.

Bodalina, U., Hammond, K., and Gilbert, D. (2005). Temporal changes in the expression of protein phosphatase 1 and protein phosphatase 2A in proliferating and differentiating murine erythroleukaemia cells. Cell Biol Int *29*, 287–299.

Bodalina, U., Hammond, K., and Gilbert, D. (2007). Temporal variation in the expression of the p53 protein in proliferating and differentiating murine erythroleukaemia cells. Mol Cell Biochem *294*, 155–162.

Breitman, T., Selonik, S., and Collins, S. (1980). Induction of differentiation of the human promyelocytic leukaemia cell line (HL60) by retinoic acid. Proc Natl Acad Sci USA *77*, 2936–2940.

Brodsky, V., Bolkov, P., Nechaeva, N., Yurovtsky, Y., Novikova, T., Fateva, V., and Schevchenco, N. (1992). The rhythm of protein synthesis does not depend on an oscillation of ATP level. J Cell Sci *103*, 363–370.

Buchner, K. (2000). The role of protein kinase C in the regulation of cell growth and in signaling to the cell nucleus. J Cancer Res Clin Oncol *126*, 1–11.

Calvert-Evers, J., and Hammond, K. (2000a). The influence of buffer composition on the expression and activity of protein tyrosine phosphatase. Electrophoresis *21*, 2944–2946.

Calvert-Evers, J., and Hammond, K. (2000b). Temporal variations in protein tyrosine phosphatase activity during cell proliferation and differentiation. Cell Biol Int *24*, 559–567.

Calvert-Evers, J., and Hammond, K. (2002). Modification of oscillatory behaviour of protein tyrosine kinase and phosphatase during all-trans retinoic acid-induced differentiation of leukaemic cells. Cell Biol Int *26*, 1035–1042.

Calvert-Evers, J., and Hammond, K. (2003). Temporal variations in protein tyrosine kinase activity in leukaemic cells: response to all-trans retinoic acid. Mol Cell Biochem *245*, 23–30.

Campbell, S.L., Khosravi-Far, R., Rossman, K.L., Clark, G.J., and Der, C.J. (1998). Increasing complexity of Ras signaling. Oncogene *17*, 1395–1493.

Chen, C.Y., Liou, J., Forman, L.W., and Faller, D.V. (1998). Differential regulation of discrete apoptotic pathways by Ras. J Biol Chem *273*, 16700–16709.

Cohen, P. (2002). Protein phosphatases 1–targeted in many directions. J Cell Sci *115*, 241–256.

Cooper, J., Reiss, N., Schwartz, R., and Hunter, T. (1983). Three glycolytic enzymes are phosphorylated at tyrosine in cells transformed by Rous sarcoma virus. Nature *302*, 218–223.

Cooper, J., Esch, F., Taylor, S., and Hunter, T. (1984). Phosphorylation sites in enolase and lactate dehydrogenase utilized by tyrosine protein kinase in vivo and in vitro. J Biol Chem *259*, 7835–7841.

Dekker, L.V., Palmer, R.H., and Parker, P.J. (1995). The protein kinase C and protein kinase C related gene families. Curr Opin Struct Biol *5*, 396–402.

Downward, J. (1990). The ras superfamily of small GTP-binding proteins. TIBS *15*, 469–472.

Edmunds, L.N., Jr. (1988). Cellular and molecular bases of biological clocks. New York: Springer.

Ferreira, G., Hammond, K., and Gilbert, D. (1994a). Insulin stimulation of high frequency phosphorylation dynamics in murine erythroleukaemia cells. BioSystems *33*, 31–43.

Ferreira, G., Hammond, K., and Gilbert, D. (1994b). Oscillatory variations in the amount of protein extractable from murine erythroleukaemia cells: stimulation by insulin. BioSystems *32*, 183–190.

Ferreira, G., Hammond, K., and Gilbert, D. (1996a). Distinct, very high frequency oscillations in the activity and amount of active isozyme of lactate dehydrogenase in murine erythroleukaemic cells and a cell-free system. Cell Biol Int *20*, 625–633.

Ferreira, G., Hammond, K., and Gilbert, D. (1996b). Independent, high frequency oscillations in the amounts of individual isozymes of lactate dehydrogenase in HL60 cells. Cell Biol Int *20*, 607–611.

Ferreira, G., Wolfle, H., Hammond, K., and Gilbert, D. (1996c). High-frequency oscillations in the activity of phosphotyrosine phosphatase in murine erythroleukaemic cells: action of insulin and hexamethylene bisacetamide. Cell Biol Int *20*, 599–605.

Frangioni, J.V., Oda, A., Smith, M., Salsam, E.W., and Neel, B.G. (1993). Calpain-catalyzed cleavage and subcellular relocation of protein phosphotyrosine 1B (PTP-1B) in human platelets. EMBO J *12*, 4843–4856.

Ge, Y., Li, Z.H., Marshall, M.S., Bronmeyer, H.E., and Lu, I. (1998). Involvement of H-ras in erythroid differentiation of TF1 and human umbilical cord blood CD34 + + + cells. Blood Cells Mol Dis *24*, 124–136.

Gilbert, D., Visser, G., Ferreira, G., and Hammond, K. (2000). Transient chaos in intracellular dynamics. Cell Biol Int *24*, 589–591.

Gilbert, D.A. (1968). Differentiation, oncogenesis and cellular periodicities. J Theor Biol *21*, 113–122.

Gilbert, D.A. (1969). Phase plane analysis of periodic isozyme pattern changes in cultured cells. Biochem Biophys Res Commun *37*, 860–866.

Gilbert, D.A. (1974a). The nature of the cell cycle and the control of cell proliferation. BioSystems *5*, 197–206.

Gilbert, D.A. (1974b). The temporal response of the cell to disturbances and its possible relationship to differentiation and cancer. S Afr J Sci *70*, 234–244.

Gilbert, D.A. (1984). Temporal organisation, re-organisation and disorganisation in cells. In: Cell cycle clocks, ed. L.N. Jr. Edmunds. New York: Marcel Dekker, 5–25.

Gilbert, D.A. (1995). Ageing, oscillations and efficiency. BioSystems *36*, 1–5.

Gilbert, D.A., and Ferreira, G. (2000). Problems associated with the study of cellular oscillations. Cell Biol Int *24*, 501–514.

Gilbert, D.A., and Lloyd, D. (2000). The living cell: a complex autodynamic multioscillator system? Cell Biol Int *24*, 569–580.

Gilbert, D.A., and Tsilimigras, C.W.A. (1981). Cellular oscillations: relative independence of enzyme activity rhythms and periodic variations in the amount of extractable protein. S Afr J Sci *77*, 66–72.

Gilbert, D.A., and Visser, G. (1993). Insulin stimulation of morphological dynamics: nonspecific biophysical mechanisms for the generalized stimulation of metabolism. BioSystems *29*, 143–149.

Greenblatt, M.S., Bennett, W.P., Hollstein, M., and Harris, C. (1994). Mutations in the p53 tumor suppressor gene: clues to cancer etiology and molecular pathogenesis. Cancer Res *54*, 4855–4878.

Hammond, K., Bhoola, R., Bodalina, U., and Gilbert, D. (1998). Dynamic cells: temporal organisation and control of phosphorylation. Trends Comp Biochem Physiol *4*, 75–88.

Hammond, K., Savage, N., and Littlewood, M. (2000a). Protein kinase C in erythroleukaemia cells: temporal variations in the expression of the alpha, epsilon and zeta isoforms. Cell Biol Int *24*, 549–557.

Hammond, K., Savage, N., and Littlewood, M. (2000b). Rhythmic patterns in the expression of the *ras* oncogene in proliferating and differentiating erythroleukaemia cells. Cell Biol Int *24*, 529–537.

Harris, S.L., and Levine, A.J. (2005). The p53 pathway: positive and negative feedback loops. Oncogene *24*, 2899–2908.

Hollstein, M., Sidransky, D., Vogelstein, B., and Harris, C. (1991). p53 mutations in human cancers. Science *253*, 49–53.

Janssens, V., and Goris, J. (2001). Protein phosphatase 2A: a highly regulated family of serine/threonine phosphatases implicated in cell growth and signaling. J Biochem *353*, 417–439.

Kirschner, M. (1992). The cell cycle then and now. TIBS *17*, 281–285.

Kolch, W., Heidecker, G., Kochs, G., Hummel, R., Vahidi, H., Mischak, H., Finkenzeller, G., Marme, D., and Rapp, U.R. (1993). Protein kinase C alpha activates RAF-1 by direct phosphorylation. Nature *364*, 249–252.

Lahav, G., Rosenfeld, N., Sigal, A., Geva-Zatorsky, N., Levine, A.J., Elowitz, M.B., and Alon, U. (2004). Dynamics of the p53-Mdm2 feedback loop in individual cells. Nat Genet *36*, 147–150.

Lane, D.P. (1992). Cancer. p53 guardian of the genome. Nature *358*, 15–16.

Leng, L., Yu, F., Dong, L., Busquets, X., Osada, S., Richon, V.M., Marks, P.A., and Rifkind, R.A. (1993). Differential modulation of protein kinase C isoforms in erythroleukaemia during induced differentiation. Cancer Res *53*, 5554–5558.

MacKinnon, H., and Gilbert, D.A. (1992). To divide or not to divide. That is the question. Fundam Med Cell Biol *7*, 1–14.

Mallia, C.M., Aguirre, V., Mcgary, E., Tang, Y., Scandurro, A.B., Liu, C., Noguchi, C.T., and Beckman, B. (2000). Protein kinase C alpha is an effector of hexamethylene bisacetamide-induced differentiation of Friend erythroleukaemia cells. Exp Cell Res *246*, 348–354.

Malumbres, M., and Pellicer, A. (1998). RAS pathways to cell cycle control and cell transformation. Front Biosci *3*, 887–912.

Matsumura, I., Nakajima, K., Wakao, H., Hattori, S., Hashimoto, K., Sugahara, H., Kato, T., Miyazaki, H., Hirano, T., and Kanakura, Y. (1998). Involvement of prolonged ras activation in thrombopoietin-induced megakaryocytic differentiation of a human factor-dependent hematopoietic cell line. Mol Cell Biol *18*, 4282–4290.

Melloni, E., Pontremoli, S., Viotti, P.L., Marks, P.A., and Rifkind, R.A. (1989). Differential expression of protein kinase C isoforms and erythroleukaemia cell differentiation. J Biol Chem *264*, 18414–18418.

Melloni, E., Pontremoli, S., Sparatore, B., Patrone, M., Grossi, F., Marks, P.A., and Rifkind, R.A. (1990). Introduction of the beta-isozyme of protein kinase C accelerates induced differentiation of murine erythroleukaemia cells. Proc Natl Acad Sci USA *87*, 4417–4420.

Monk, N.A. (2003). Oscillatory expression of Hes1, p53, and NF-kappaB driven by transcriptional time delays. Curr Biol *13*, 1409–1413.

Morita, K., Nishikawa, M., Kobayashi, K., Deguchi, K., Ito, M., Nakano, T., Shimah, H., Nagao, M., Kuno, T., Tanaka, C., and Shirakawa, S. (1992). Augmentation of retinoic acid-induced granulocytic differentiation in HL60 leukaemia cells by serine/threonine protein phosphatase inhibitors. FEBS Lett *314*, 340–344.

Mumby, M.C., and Walter, G. (1993). Protein serine/threonine phosphatases: structure, regulation and functions in cell growth. Physiol Rev *73*, 673–699.

Nishizuka, Y. (1992). Intracellular signaling by hydrolysis of phospholipids and activation of protein kinase C. Science *258*, 607–614.

Norbury, C., and Nurse, P. (1992). Animal cell cycles and their control. Ann Rev Biochem *61*, 441–470.

Patrone, M., Pessino, A., Passalaqua, M., Sparatore, B., Melloni, E., and Pontremoli, S. (1994). Protein kinase C isoforms in murine erythroleukaemia cells and their involvement in the differentiation process. FEBS Lett *344*, 91–95.

Pessino, A., Sparatore, B., Patrone, M., Passalaqua, M., Melloni, E., and Pontremoli, S. (1994). Differential expression of protein kinase C isoform genes in three murine erythroleukaemia cell variants: implication for chemical induced differentiation. Biochem Biophys Res Comm *204*, 461–467.

Reuben, R., Rifkind, R., and Marks, P. (1980). Chemically-induced murine erythroleukaemic differentiation. Biochim Biophys Acta *605*, 325–346.

Rosson, B., and O'Brien, T.G. (1995). Expression and modulation of protein kinase C isoforms in differentiation-competent and differentiation-resistant erythroleukaemic cells. Biochem Biophys Res Commun *210*, 90–97.

Rovera, G., Santoli, D., and Damsky, C. (1979). Human promyelocytic leukaemia cells in culture differentiate into macrophage-like cells when treated with a phorbol diester. Proc Natl Acad Sci USA *76*, 2779–2783.

Rozengurt, E., Rodriguez-Pena, M., and Smith, K. (1983). Phorbol esters, phospholipase C and growth factors rapidly stimulate the phosphorylation of a M_r 80,000 protein in intact, quiescent 3T3 cells. Proc Natl Acad Sci USA *80*, 7244–7248.

Scheele, J.S., Pilz, R.B., Quilliam, L.A., and Boss, G.R. (1994). Identification of a ras-related protein in murine erythroleukemia cells that is a cyclic AMP-dependent protein kinase substrate and is phosphorylated during chemically induced differentiation. J Biol Chem *269*, 18599–18606.

Schievella, A.R., Paige, L.A., Johnson, K.A., Hill, D.E., and Erikson, R.L. (1993). Protein tyrosine phosphatase 1B undergoes mitosis specific phosphorylation on serine. Cell Growth Differ *4*, 239–246.

Segel, L.A. (1980). Dimensional analysis. In: Mathematical models in molecular and cellular biology. Cambridge: Cambridge University Press.

Shi, L., Potts, M., and Kennelly, P.J. (1998). The serine, threonine and/or tyrosine-specific protein kinases and protein phosphatases of eukaryotic organisms: a family portrait. FEMS Microbiol Rev *22*, 229–253.

Sim, A., and Ludowyke, R.I. (2002). The complex nature of protein phosphatases. IUBMB Life *53*, 283–286.

Sprott, S.C., Hammond, K., and Savage, N. (1987). Electrophoretic forms of protein kinases during differentiation of murine erythroleukaemic cells. Anticancer Res *7*, 521–525.

Sprott, S.C., Hammond, K., and Savage, N. (1991a). Protein kinases associated with proliferation and differentiation in murine erythroleukaemic cells. Int J Biochem *23*, 713–718.

Sprott, S.C., Hammond, K., and Savage, N. (1991b). Subcellular fractionation of murine erythroleukaemic cells: distribution of protein kinases. Anal Biochem *194*, 407–412.

Tawara, I., Nishikawa, M., Morita, K., Kobayashi, K., Toyoda, H., Omay, S.B., Shima, H., Nagao, M., Kuno, T., Tanaka, C., and Shirakawa, S. (1993). Down-regulation by retinoic acid of the catalytic subunit of protein phosphatase type 2A. FEBS Lett *321*, 224–228.

Toker, A. (1998). Signaling through protein kinase C. Front Biosci *3*, 1134–1147.

Tung, H.Y., Resnick, T.J., Hemmings, B.A., Shenolikar, S., and Cohen, P. (1984). The catalytic subunits of protein phosphatase-1 and protein phosphatase-2A are distinct gene products. Eur J Biochem *138*, 635–641.

Tung, H.Y., Alemany, S., and Cohen, P. (1985). The protein phosphatases involved in cellular regulation. 2. Purification, subunit structure and properties of protein phosphatases-2A0, 2A1, and 2A2 from rabbit skeletal muscle. Eur J Biochem *148*, 253–263.

Tyson, J.J. (2004). Monitoring p53's pulse. Nat Genet *36*, 113–114.

Van Dijk, M.C., Hilkmann, H., and Van Blitterswijk, W.J. (1997). Platelet-derived growth factor activation of mitogen-activated protein kinase depends on the sequential activation of phosphatidylcholine-specific phospholipase C, protein kinase C zeta and Raf-1. Biochem J *325*, 303–307.

Virshup, D.M. (2000). Protein phosphatase 2A: a panoply of enzymes. Curr Opin Cell Biol *12*, 180–185.

Visser, G., Reinten, C., Coplan, P., Gilbert, D., and Hammond, K. (1990). Oscillations in cell morphology and redox state. Biophys Chem *37*, 383–394.

Vousden, K.H. (2002). Activation of the p53 tumour suppressor protein. Biochim Biophys Acta *1602*, 47–59.

Young, S.W., Dickens, M., and Tavare, J.M. (1996). Activation of mitogen-activated protein kinase by protein kinase C isotypes alpha, beta I and gamma, but not epsilon. FEBS Lett *384*, 181–184.

Chapter 5
Is There a Mitochondrial Clock?

M.A. Aon*, S. Cortassa, and B. O'Rourke

Abstract A mitochondrial oscillator dependent on reactive oxygen species (ROS) was first described in heart cells. Available evidence now indicates that mitochondrial energetic variables oscillate autonomously as part of a network of coupled oscillators under both physiological and pathological conditions. Moreover, emerging experimental and theoretical evidence indicates that mitochondrial network oscillations exhibit a wide range of frequencies, from milliseconds to hours, instead of a dominant frequency. With metabolic stress, the frequency spectrum narrows and a dominant oscillatory frequency appears, indicating the transition from physiological to pathophysiological behavior.

Here we show that in the pathophysiological regime the mitochondrial oscillator of heart cells is temperature compensated within the range of 25–37°C with a Q_{10} = 1.13. At temperatures higher than 37°C, the oscillations stop after a few cycles, whereas at temperatures lower than 25°C the oscillations are asynchronous. Using our mitochondrial oscillator model we show that this temperature compensation can be explained by kinetic compensation. Furthermore, we show that in the physiological domain temperature compensation acts to preserve the broad range of frequencies exhibited by the network of coupled mitochondrial oscillators.

The results obtained indicate that the mitochondrial network behaves with the characteristics of a biological clock, giving rise to the intriguing hypothesis that it may function as an intracellular timekeeper across multiple time scales.

Keywords Mitochondrial oscillations, temperature compensation, biological clocks, membrane potential, reactive oxygen species, power spectral and relative dispersional analyses, fractal dynamics

---

The Johns Hopkins University, Institute of Molecular Cardiobiology, 720 Rutland Ave., 1059 Ross Bldg., Baltimore, MD 21205-2195

*Correspondence to: Email: maon1@jhmi.edu

5.1 Introduction

5.1.1 Biological Rhythms and Clocks

From an evolutionary perspective, the adaptation of an organism's behavior to its environment has depended on one of life's fundamental traits: biological rhythm generation. In virtually all light-sensitive organisms from cyanobacteria to humans, biological clocks adapt cyclic physiology to geophysical time with time-keeping properties in the circadian (24 h), ultradian (<24 h) and infradian (>24 h) domains (Edmunds, 1988; Lloyd, 1998; Lloyd et al., 2001; Lloyd and Murray, 2006; Lloyd, 2007; Pittendrigh, 1993; Sweeney and Hastings, 1960) By definition, all rhythms exhibit regular periodicities since they constitute a mechanism of timing. Timing exerted by oscillatory mechanisms are found throughout the biological world and their periods span a wide range from milliseconds, as in the action potential of neurons and the myocytes, to the slow evolutionary changes that require thousands of generations. In this context, to understand the synchronization of a population of coupled oscillators is an important problem for the dynamics of physiology in living systems (Aon et al., 2007a, b; Kuramoto, 1984; Strogatz, 2003; Winfree, 1967).

Circadian rhythms, the most intensively studied, are devoted to measuring daily 24 h cycles. A variety of physiological processes in a wide range of eukaryotic organisms display circadian rhythmicity which is characterized by the following major properties (Anderson et al., 1985; Edmunds, 1988): (i) stable, autonomous (self-sustaining) oscillations having a free-running period under constant environmental conditions of *ca.* 20–28 h; (ii) phase lability, reflected by the ability of the oscillation to reset its phase in response to an environmental stimulus (Zeitgeber); and (iii) conservation of the period (though not the amplitude) of the rhythm at different ambient temperatures within the physiological range. The latter property is termed temperature compensation; due to this trait, the free-running period of most circadian rhythms is quite stable over a wide range of constant temperatures.

However, the established view of the circadian rhythm as the dominating time frame of biological rhythms has been challenged. A distinct, emerging, view proposes to expand the 24 h time frame to include the ultradian domain (i.e. less than a day). Compellingly, this emerging framework argues that 24 h is a long time and, evidently, too narrow to account for the whole spectrum of time scales associated to crucial physiological functions exhibited by organisms and cells (Lloyd and Murray, 2007). In the past, ideas about time keeping of rhythms assumed the existence of a basic oscillator whose period will drive the rhythm. However, recent data seem to undermine this simplistic idea by showing that the whole genome is involved during respiratory oscillations in yeast (Klevecz et al., 2004). A very recent report (Roussel and Lloyd, 2007) describes that the yeast cell division cycle obtained in continuous cultures is only one among other various cyclic processes exhibited by yeast, among them the conspicuous circahoralian clock of ~40 min (Murray et al., 2001).

In this work we will subscribe to the expanded view of biological time keeping and elaborate on it based on the idea that biological clocks (i.e. temperature-compensated rhythms) function simultaneously across multiple time scales. Theoretically, our proposal is based on the idea that the interrelatedness between processes across multiple time scales is achieved through scaling so that what affects one time scale affects them all. This crucial property is captured by the concept of dynamic fractals. Furthermore, we will propose, and show supporting data, that the central core of the biological clock and its sheer dynamics is deeply nested in the mitochondrial network and associated processes. Very recent and rapidly accumulating evidence from yeast and heart mitochondria support this view.

5.2 Methods

Cardiomyocyte isolation. All experiments were carried out at 37°C on freshly isolated adult guinea pig ventricular myocytes prepared by enzymatic dispersion as previously described (Aon et al., 2003).

Fluorescent probes for two-photon laser scanning microscopy and image acquisition and analysis. The cationic potentiometric fluorescent dye TMRM (100 nM) was used to monitor changes in $\Delta\Psi_m$ as previously described (Aon et al., 2003, 2006). Images were recorded using a two photon laser scanning microscope (Bio-Rad MRC-1024MP) with excitation at 740 nm and the red emission of TMRM was collected at 605 ± 25 nm (Tsunami Ti:Sa laser, Spectra-Physics).

Analysis of TMRM time series. Extended time series of TMRM fluorescence (1,500 to 4,000 images) recorded at a maximal temporal resolution of 110 ms were subjected to Relative Dispersional Analysis (RDA) and Power Spectral Analysis (PSA) as previously described (Aon et al., 2006).

Computational model of the mitochondrial oscillator. An integrated model of mitochondrial energetics (Cortassa et al., 2003), which was extended to include a shunt of electrons of the respiratory chain towards the generation of O_2^-, a ROS scavenging system, and a ROS-activated anion efflux pathway across the inner membrane (Cortassa et al., 2004), was utilized for the simulations shown in Figs. 5.4 and 5.5, using the parametric conditions described in (Aon et al., 2006; Cortassa et al., 2003, 2004).

Statistical analysis. Estimation of the statistical significance of differences was performed with GraphPad Prism (Ver. 3.0; San Diego, CA). The results are presented as mean ± SEM (95% confidence interval) after a t-test (small samples, unpaired t-test with two tail P-values).

5.3 Results

Conservation of the period of a rhythm at different ambient temperatures within the physiological range is termed temperature compensation, and is one of the most conspicuous features that qualify a rhythm as a clock. The ability to compensate for fluctuations in temperature is precisely what one would anticipate of any clock mechanism in order to be a reliable timekeeper (Edmunds, 1988; Lloyd, 1998; Murray et al., 2001; Pittendrigh, 1993).

Elsewhere (Aon et al., 2006) we have shown that under "physiological" conditions mitochondrial $\Delta\Psi_m$ fluctuates in a non-random (correlated) manner at high frequency within a restricted amplitude range, implying depolarizations of only microvolts to a few millivolts. Our previous work (Aon et al., 2003, 2004) also showed that oxidative stress might elicit low-frequency high-amplitude oscillations that characterize the "pathophysiological" response. We asked whether both functional regimes of the cardiac mitochondrial network were temperature compensated.

5.3.1 Temperature Compensation in the Low-Frequency High Amplitude Regime

Whole cell mitochondrial oscillations, triggered by a local laser flash, as previously described (Aon et al., 2003), were analyzed with respect to their period at different temperatures in freshly isolated cardiomyocytes loaded with TMRM, a $\Delta\Psi_m$ probe. Figure 5.1A shows the mitochondrial oscillator's period at the two extremes of the temperature range, 25°C and 37°C, and beyond. It can be seen that the period of the oscillation remains relatively constant between 25°C and 37°C; the difference being not statistically significant. Moreover, the $\Delta\Psi_m$ oscillation period remained the same after a temperature jump from 25°C to 37°C within the same cell (Fig. 5.1B).

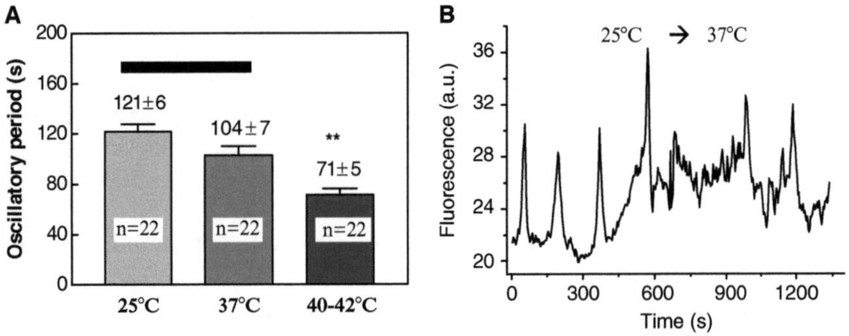

Fig. 5.1 Temperature compensation exhibited by the mitochondrial oscillator in the pathophysiological regime. Freshly isolated cardiomyocytes were loaded with 100 nM TMRM, a $\Delta\Psi_m$ probe, and imaged by two photon microscopy. Mitochondrial oscillations were triggered by a laser flash. The temperature at the stage of a Nikon E600FN upright microscope was thermostatically controlled with unrestricted access to atmospheric oxygen. (**A**) The oscillatory period ±SEM in seconds (three independent experiments) were analyzed at the indicated temperatures. The line on top of the two first bars indicates the temperature compensated range of the oscillations. (**B**) This panel shows a temperature jump between the two extremes (25–37°C) of the compensation range in a single cell. Notice the preservation of the oscillatory period. $^{**}p < 0.01$

Intermediate temperatures within the range of 25–37°C were also analyzed with similar results to those shown in Fig. 5.1A (not shown).

The constancy of the period of a clock is characterized by the temperature coefficient, Q_{10}. Generally speaking, the Q_{10} expresses that the rates of many physiological and biochemical processes differ by a factor of 2 or 3 at temperatures 10°C apart. Thus, when a biological rhythm is said to be temperature-compensated, the Q_{10} value is close to unity despite a wide range variation in the steady state ambient temperature. The temperature coefficient was calculated as follows (Anderson et al., 1985):

$$Q_{10} = \left(\frac{\tau_2}{\tau_1}\right)^{\frac{10}{(T1-T2)}}$$

where τ_1 and τ_2 are the mean periods, respectively, at temperatures T1 and T2 (with T1 > T2). We obtained a $Q_{10} = 1.13$ with $\tau_1 = 104$ s, $\tau_2 = 121$ s, $T_1 = 37$°C and $T_2 = 25$°C (Fig. 5.1A). Thus, the mitochondrial oscillations are temperature-compensated within the range 25–37°C with an average period of ~110 s for the whole range (Figs. 5.1A, 5.2B, C). Below that range (23°C) the synchronization mechanism fails showing only asynchronous oscillations (Fig. 5.2D) whereas above that (42°C), oscillations stop after a few cycles (Fig. 5.2A).

5.3.2 *Temperature Compensation in the High-Frequency Low Amplitude Regime*

We next determined the response to temperature of the physiological regime. Previously, we characterized this oscillatory regime of the cardiac mitochondrial network with Relative Dispersional (RDA) and Power Spectral (PSA) analysis (Aon et al., 2006).

In Fig. 5.3 are shown representative results obtained from heart cells loaded with TMRM, and imaged with two photon laser scanning microscopy. PSA detected a significant difference between cells analyzed outside (23°C) or within the temperature-compensated range (35°C). In fact, the power spectrum was characterized by several missing frequencies (i.e., shown as deep depressions) at 23°C as compared with 35°C. As shown in the insets of Fig. 5.3A and C, the spectral exponent, β, in the high frequency domain (frequencies > 0.3 Hz) was decreased together with the disappearance of certain frequencies at the temperature outside (β= 1.39) as compared to within (β = 1.54) the temperature-compensated range. The decrease in β in the high-frequency, low amplitude domain of the spectrum (Fig. 5.3A and C) was previously shown to predominantly result from a significant loss of correlation consistent with a weaker coupling between mitochondrial oscillators in the network (Aon et al., 2006). Nevertheless, the overall power spectrum exhibited similar β and the RDA analysis did not detect differences in the fractal dimension of the time

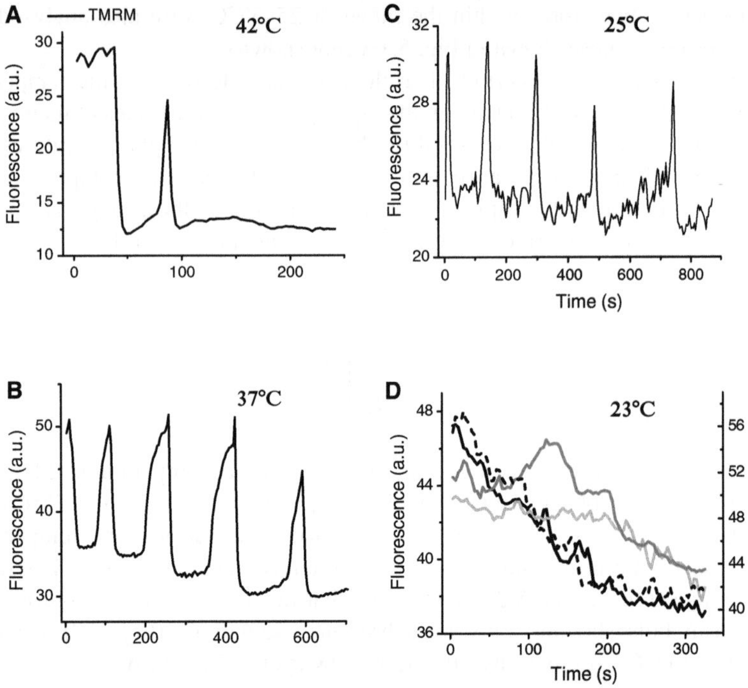

Fig. 5.2 Characteristics of mitochondrial oscillations within and outside the temperature-compensated range. Cardiomyocytes were isolated and mitochondrial oscillations triggered as described in the legend of Fig. 5.1. Shown are representative $\Delta\Psi_m$ oscillations obtained in living cardiomyocytes after a laser flash at the indicated temperatures. In panel **D**, the oscillations exhibited by four different cells at 23°C are shown

series (Fig. 5.3A–D). The latter effect indicates that, overall, and despite a decrease in correlation, the oscillators still exhibit long-term memory.

5.3.3 Temperature Compensation by Kinetic Compensation

We decided to test *in silico* the hypothesis of whether temperature compensation can be achieved through kinetic compensation. The 100s period $\Delta\Psi_m$ mitochondrial oscillations were simulated at two temperatures 10°C apart (27°C and 37°C) with our previously described model of the mitochondrial oscillator (Cortassa et al., 2004). The basic rationale of this *in silico* experiment is given by temperature dependence of a rate constant as described by the Arrhenius equation:

$$k_J = A_j e^{\frac{E_j}{RT}}$$

that after rearrangement becomes:

$$\ln k = \ln A - \frac{E}{RT}$$

with k, rate constant; A, Arrhenius constant; E, activation energy for the reaction; R, ideal gas constant; T, temperature.

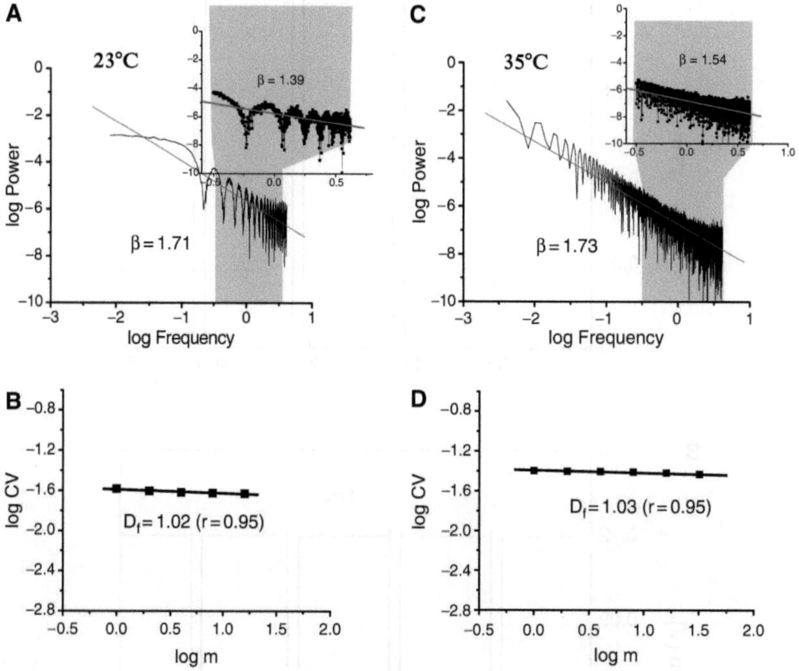

Fig. 5.3 Power Spectral (PSA) and Relative Dispersional (RDA) analyses within and outside the temperature-compensated range of $\Delta\Psi_m$ time series from living cardiomyocytes.
Freshly isolated cardiomyocytes were loaded with 100 nM TMRM and imaged by two photon microscopy at a frame interval of 110 ms as previously described (Aon et al., 2006). RDA and PSA were applied to time series composed of 2,000–4,000 time points (Aon et al., 2006). Depicted is the statistical analysis of $\Delta\Psi_m$ time series of the mitochondrial network of living cardiomyocytes loaded with TMRM outside (23°C) (**A**, **B**) or inside (35°C) (**C**, **D**) the temperature-compensated range, applying PSA (**A**, **C**) or RDA (**B**, **D**) analyses. In panels **A** and **C** the insets show in detail the spectrum in the high frequency domain (colored at >0.3 Hz). Notice the missing frequencies (seen as depressions) in the inset of panel A as compared to the richness of the inset in panel **C**. A major decrease in the spectral exponent, β, was determined at low temperatures.
Relative dispersion (or coefficient of variation) as a function of the aggregation parameter, m, gives a fractal dimension, D_f, close to 1.0 for myocytes showing large amplitude oscillations in $\Delta\Psi_m$ or those under physiological conditions, indicating a deterministic control process, while D_f is 1.5 for a completely random process (Aon et al., 2006). PSA of $\Delta\Psi_m$ time series after Fast Fourier Transform, also reveals a broad spectrum of oscillation in normally polarized mitochondria with a spectral exponent, β = 1.79 while a random process gives a β = 0 meaning that there is no relationship between the amplitude and the frequency in a random signal (Aon et al., 2006)

Activation energies for enzyme reactions are of the order 40–80 kJ mol^{-1}. As a rough guide this means that a 10°C rise in temperature between 20°C and 30°C will increase the rate of reaction by a factor of ~2. Thus, assuming that

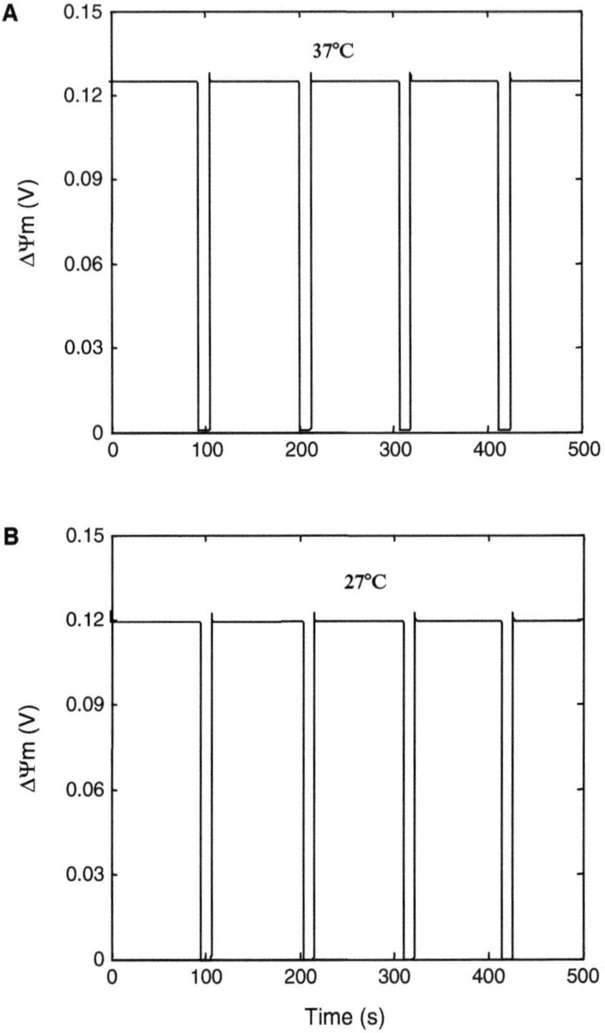

Fig. 5.4 Simulation of temperature compensation using the mitochondrial oscillator model. Simulation of the mitochondrial oscillations were performed utilizing the integrated model of mitochondrial energetics with incorporated IMAC, ROS production, and ROS scavenging previously described (Cortassa et al., 2004). The $\Delta\Psi_m$ oscillations shown correspond to the 100 s period observed experimentally in living cardiomyocytes under oxidative stress, pathophysiological, conditions (Aon et al., 2003). See Cortassa et al. (2004) and Cortassa et al. (2003) for parameters used in the simulations

the Arrhenius law applies to all reactions by the same factor, the rate constants of *all* processes at 27°C were halved with respect to the values utilized at 37°C to simulate the 100 s period oscillations in $\Delta\Psi_m$ (Fig. 5.4A) (Cortassa et al., 2004). Under these conditions, we obtain the same period at both temperatures (Figs. 5.4A and B).

In order to further demonstrate that there is kinetic compensation, and as a matter of control, we halved the rate constants of all processes with the exception of those related to the velocity of respiration, VO_2, ATPase, and the IMAC. Under these parametric conditions we obtained a 100 ms period instead of 100 s. In order to reestablish the 100 s period again, without decreasing VO_2, an almost twofold increase in the rate of ROS scavenging by the superoxide dismutase (Et_{SOD} from 1.87×10^{-3} to 3.48×10^{-3} mmol s^{-1}), had to be applied. This demonstrates that kinetic compensation is occurring since we need to increase the rate of ROS scavenging to compensate for higher rates of ROS production. Indeed, we obtain the 100 s period if, instead of increasing Et_{SOD}, a twofold decrease in the amount of electron carriers or enzymes or channels is imposed to account for the temperature jump through the rates of VO_2, ATPase, and IMAC, respectively.

5.4 Discussion

The main contribution of the present work is to show that the cardiac mitochondrial network exhibits temperature compensation within the range of 25–37°C. We recently described that cardiac mitochondria behave as a collection of weakly coupled oscillators with a broad frequency distribution under normal (physiological) conditions that can transition to an oscillatory state with a single dominant frequency under metabolic stressful conditions (Aon et al., 2006, 2007a). The results obtained show that temperature compensation affects both regimes although in a different manner. The low frequency high amplitude oscillations (pathophysiological) exhibit a Q_{10} of 1.13 meaning that the oscillation period remained constant even with temperatures as wide apart as 12°C. However, in the physiological domain, temperature compensation appears to keep the broad frequency distribution of the low amplitude oscillations. It seems that at temperatures lower than 25°C, outside the temperature-compensated range, the mitochondrial network losses "spectral reserve" by missing several frequencies. The term "loss of spectral reserve" (Goldberger et al., 1985) has been introduced in the context of cardiac electrical activity during the transition from normal to pathological states (Casolo et al., 1989; Ewing, 1991; Skinner et al., 1993).

What is the relevance of temperature compensation in a network that exhibits inverse power law behavior with a scale-free spectrum (as in cardiac mitochondria in the physiological domain)? We suggest that it is important for conserving the broad frequency distribution of oscillations exhibited by the network. For the sake of clarity, scale-free means that the mitochondrial network in the physiological domain does not exhibit a characteristic (oscillatory) frequency, but a whole range of them spanning at least three orders of magnitude. This is certainly a different situation in the pathophysiological regime, in which only one period predominates as compared with the physiological domain where several periods could be utilized as timers for a wide span of time scales. The fact that, according to RDA analysis, the overall correlation among oscillators is not lost, argues in favor of the idea that what matters in the physiological behavior of the mitochondrial network is the frequency distribution of the oscillators.

5.4.1 *Kinetic Compensation as a Mechanism of Temperature Compensation*

A complex web of reactions exhibiting oscillations is said to be temperature compensated when the rhythm's period remains constant despite temperature variation, and the fact that most of the individual reactions (enzymes, channels) are temperature-dependent. As a caveat, the Q_{10} of the IMAC has been shown to be dependent on pH and Mg^{2+} concentration in isolated mitochondria (Beavis and Powers, 2004), a dependence that is not accounted for in our mathematical model (Cortassa et al., 2004).

The kinetic compensation found in our model of the mitochondrial oscillator to simulate the temperature compensation described experimentally (Fig. 5.1), is related to the "antagonistic feedback" concept of Franck (Franck, 1980). This concepts states that the simultaneous occurrence of positive and negative feedbacks in oscillatory reaction networks, accounts for the positive and negative contributions to an oscillator's period. Considering that glycolytic oscillations are temperature-dependent (Betz and Chance, 1965; Mair et al., 2005), and accounting for an Arrhenius-type temperature dependence of rate constants, a similar approach was utilized by other authors to theoretically design a temperature-compensated glycolytic oscillator (Ruoff et al., 2003).

5.4.2 *Temperature Compensation and Mitochondrial Network Behavior*

For achieving temperature compensation, mitochondria should act as a network in order to simultaneously compensate for the kinetics of all processes involved after a 10°C temperature shift. In a highly correlated, coordinated, network of oscillators each mitochondrion affects the temporal behavior of its neighbors. This coordinated behavior is attained through ROS, the coupling factor between mitochondria, as suggested by the loss of correlation between oscillators induced by different interventions that decrease these intracellular messengers (Aon et al., 2006, 2007a). As a result, the dynamic behavior of the network corresponds to a statistically fractal, self-similar, process characterized by a large number of frequencies in multiple timescales.

Interestingly, when the network organization of mitochondria is lost, i.e. after isolation, the oscillations become temperature-dependent. The ionic oscillations described by (Chance and Yoshioka, 1966) in isolated mitochondria from pigeon heart exhibited temperature dependence. These authors detected an increase in frequency from 80 to 20 s in the temperature range of 26–35.5°C, respectively.

5.4.3 The Mitochondrial Clock and the Scaling Hypothesis

Herein we showed that the oscillations exhibited by the mitochondrial network and its net of interlinked reactions (TCA cycle, oxidative phosphorylation, Ca^{2+} dynamics, ROS-induced ROS release) is temperature compensated. Mitochondria act as metabolic "hubs" and as producers of ROS as signaling molecules with scale-free dynamics (Aon et al., 2007a). These findings strongly suggest that the intracellular timekeeping trait of mitochondria belongs to those of a biological clock.

In agreement with the self-similar, fractal, dynamics exhibited by the network of coupled mitochondrial oscillators (Aon et al., 2006), model simulations indicate that the oscillatory period can be modulated over a wide range of time scales (Aon et al., 2007b; Cortassa et al., 2004). This suggests that the mitochondrial oscillator may play a role as an intracellular timekeeper (Fig. 5.5). Under "physiological" conditions, mitochondrial $\Delta\Psi_m$ fluctuates at high frequency within a restricted amplitude range, implying depolarizations of only μV to a few mV (Fig. 5.5A). Oxidative stress may elicit low frequency, high amplitude oscillations that characterize the "pathophysiological" response (Fig. 5.5B). By just changing one parameter (the rate of superoxide dismutase) several periods across a wide range of time scales from milliseconds (Fig. 5.5A) to hours (Fig. 5.5B) can be obtained. Moreover, the experimentally found colored noise dynamics exhibited by mitochondria in the physiological regime, and its transition to pathophysiological behavior, could be simulated from deterministic (oscillatory) dynamics by mixing a certain number of frequencies from both domains (Figs. 5.5A and 5.5B). Thus, the scale-free behavior exhibited by the mitochondrial network dynamics allows us to propose that, as a timekeeper, the mitochondrial network functions with multiple outputs in several time scales simultaneously.

This proposal is in agreement with the scaling hypothesis introduced by West (1999). This hypothesis states that long-range correlations in the dynamic fluctuations of a system's variable (e.g., $\Delta\Psi_m$) are due to the existence of local instabilities in the system dynamics; otherwise the system retains its global dynamic stability. The longest and the shortest time scales exhibited by the dynamics of a process are tied together by means of the scaling, so what affects one time scale

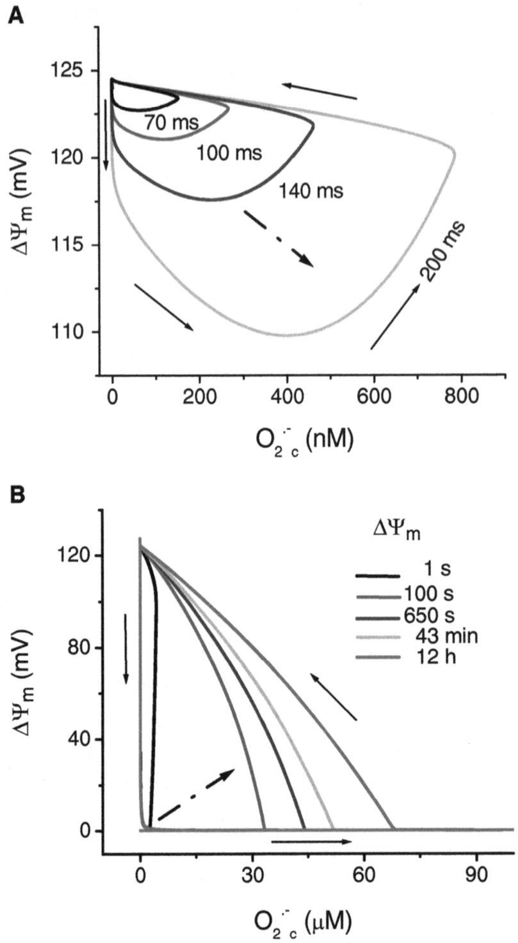

Fig. 5.5 Behavior of the mitochondrial model in the high and low frequency domain of oscillations. This figure shows the phase-plane relationship between $\Delta\Psi_m$ and periplasmic superoxide, $O_{2\ c}^{-}$, released from mitochondria, in the high (ms, **A**) and low (sec to h, **B**) frequency domain of the oscillatory behavior that correspond to stable limit cycles. Both panels show the wide excursions of ROS concentration associated with the extent of $\Delta\Psi_m$ depolarization and the increase in the oscillatory period. Outside arrows point to the sense of the cycling during the oscillation whereas inside (thick) ones denote the increase in period.
The different oscillatory periods were achieved, within the oscillatory domain, after changing the rate of superoxide dismutase (SOD) with (**A**) or without (**B**) slight modification of the shunt (defined as the fraction of the electron flow in the respiratory chain diverted to the generation of superoxide anion, O_2^-). See Cortassa et al. (2004) and Aon et al. (2006) for other parameters used in the simulations

affects them all. This kind of inter-relatedness of the data across multiple time scales is captured by the self-similar (fractal) statistics exhibited by the $\Delta\Psi_m$ time series of the cardiac mitochondrial network. That this kind of temporal fractal arises from the bifurcation of mitochondrial dynamics and its timekeeping properties across multiple time scales has already been shown (Fig. 5.5) (Aon et al., 2006, 2007a, b; Cortassa et al., 2004).

5.4.4 The Mitochondrial Clock, Aging and Disease

Harman (1956) introduced the hypothesis that aging would be the result of molecular damage induced by the ROS produced by mitochondria (Harman, 1956). In the summary of his seminal proposal Harman said: *"Aging and the degenerative diseases associated with it are attributed basically to the deleterious side attacks of free radicals on cell constituents and on the connective tissues. The free radicals probably arise largely through reactions involving molecular oxygen catalyzed in the cell by the oxidative enzymes and in the connective tissues by traces of metals such as iron, cobalt, and manganese ".*

Insightfully, Harman conjectured that a main target of ROS damage would be the DNA, a presumption that has largely been confirmed (Finkel and Holbrook, 2000; Lane, 2005; Wright et al., 2004). Another prescient observation was the association between aging and degenerative disease. In fact, the incidence of cardiovascular, cancer and neurodegenerative (e.g. Alzheimer, Parkinson) diseases increases exponentially within the period of life between 45 to 65 years, respectively (Finkel and Holbrook, 2000; Finkel, 2005).

The field of research on aging and, particularly, concerning the Harman's hypothesis, is full of controversy, contradictory results, and pitfalls (Lane, 2002, 2005). Although the general consensus is that Harman's hypothesis is, generally speaking, correct (Balaban et al., 2005), it is also known that, as originally formulated, it is too simplistic. The latter view has been substantiated by the results produced by intensive research on the effects of ROS on life span and disease. The present view is that mitochondria represent the main site of ROS production in the cell (Turrens, 2003), and oxidative stress is a major pathophysiological route to the collapse of $\Delta\Psi_m$. However, when kept under control, ROS serve as important signaling molecules (Cadenas, 2004; Droge, 2002; Haddad, 2004; Marin-Garcia and Goldenthal, 2004; Morel and Barouki, 1999) and their signaling of gene expression is highly dependent on the ability of cells to maintain a reductive intracellular environment (Cadenas, 2004; Droge, 2002). Another line of research on aging emphasizes other aspects than the "free-radical theory of aging" such as the approach led by Kirkwood and the "disposable soma theory" (Kirkwood, 1999, 2005). This approach connects with the association of aging and disease.

Several years later after formulating his hypothesis, Harman (1972), in an article entitled: *"The biologic clock: The mitochondria?"* proposed that the biological clock for the maximal life span of mammalians would be in the mitochondria (Harman, 1972). More precisely, he suggested that the life span is largely an expression of genetic control over the rate of oxygen utilization, a proposal that, in retrospect, appears also too simplistic. However, we would like to stress that at least in one aspect Harman was again correct: mitochondria have dynamic properties consistent with a biological clock.

5.4.5 *Concluding Remarks*

We conclude that the temperature compensation exhibited by the oscillatory behavior of the mitochondrial network qualifies it as a biological clock. Simulations studies performed with a mitochondrial oscillator dependent on ROS suggest that the temperature compensation arises as a result of kinetic compensation. Mitochondria in the network act as a hub nested in a web of interlinked reactions such as the TCA cycle, oxidative phosphorylation, Ca^{2+} dynamics, ROS production and ROS scavenging, and ROS-induced ROS release. The mitochondrial network timekeeping properties are expressed by the fractal dynamics exhibited by coupled oscillators with multiple and simultaneous outputs in several time scales.

References

Anderson, R. W., Laval-Martin, D. L. and Edmunds, L. N., Jr. (1985). Cell cycle oscillators. Temperature compensation of the circadian rhythm of cell division in Euglena. *Exp Cell Res* **157**, 144–158.

Aon, M. A., Cortassa, S., Marban, E. and O'Rourke, B. (2003). Synchronized whole cell oscillations in mitochondrial metabolism triggered by a local release of reactive oxygen species in cardiac myocytes. *J Biol Chem* **278**, 44735–44744.

Aon, M. A., Cortassa, S. and O'Rourke, B. (2004). Percolation and criticality in a mitochondrial network. *Proc Natl Acad Sci USA* **101**, 4447–4452.

Aon, M. A., Cortassa, S. and O'Rourke, B. (2006). The fundamental organization of cardiac mitochondria as a network of coupled oscillators. *Biophys J* **91**, 4317–4327.

Aon, M. A., Cortassa, S. and O'Rourke, B. (2007a). *On the network properties of mitochondria.* Wiley-VCH Verlag GmbH & Co. KGaA, Weinheim.

Aon, M. A., Cortassa, S. and O'Rourke, B. (2007b). *Mitochondrial oscillations in physiology and pathophysiology.* Austin, TX: Landes Bioscience.

Balaban, R. S., Nemoto, S. and Finkel, T. (2005). Mitochondria, oxidants, and aging. *Cell* **120**, 483–495.

Beavis, A. D. and Powers, M. (2004). Temperature dependence of the mitochondrial inner membrane anion channel: the relationship between temperature and inhibition by magnesium. *J Biol Chem* **279**, 4045–4050.

Betz, A. and Chance, B. (1965). Influence of Inhibitors and temperature on the oscillation of reduced pyridine nucleotides in yeast cells. *Arch Biochem Biophys* **109**, 579–584.

Cadenas, E. (2004). Mitochondrial free radical production and cell signaling. *Mol Aspects Med* **25**, 17–26.

Casolo, G., Balli, E., Taddei, T., Amuhasi, J. and Gori, C. (1989). Decreased spontaneous heart rate variability in congestive heart failure. *Am J Cardiol* **64**, 1162–1167.

Chance, B. and Yoshioka, T. (1966). Sustained oscillations of ionic constituents of mitochondria. *Arch Biochem Biophys* **117**, 451–465.

Cortassa, S., Aon, M. A., Marban, E., Winslow, R. L. and O'Rourke, B. (2003). An integrated model of cardiac mitochondrial energy metabolism and calcium dynamics. *Biophys J* **84**, 2734–2755.

Cortassa, S., Aon, M. A., Winslow, R. L. and O'Rourke, B. (2004). A mitochondrial oscillator dependent on reactive oxygen species. *Biophys J* **87**, 2060–2073.

Droge, W. (2002). Free radicals in the physiological control of cell function. *Physiol Rev* **82**, 47–95.

Edmunds, L. N., Jr. (1988). *Cellular and molecular basis of biological clocks: models and mechanisms for circadian timekeeping*. New York: Springer.

Ewing, D. J. (1991). Heart rate variability: an important new risk factor in patients following myocardial infarction. *Clin Cardiol* **14**, 683–685.

Finkel, T. and Holbrook, N. J. (2000). Oxidants, oxidative stress and the biology of ageing. *Nature* **408**, 239–247.

Finkel, T. (2005). Opinion: radical medicine: treating ageing to cure disease. *Nat Rev Mol Cell Biol* **6**, 971–976.

Franck, U. F. (1980). The Teorell membrane oscillator – a complete nerve model. *Ups J Med Sci* **85**, 265–282.

Goldberger, A. L., Bhargava, V., West, B. J. and Mandell, A. J. (1985). On a mechanism of cardiac electrical stability. The fractal hypothesis. *Biophys J* **48**, 525–528.

Haddad, J. J. (2004). Oxygen sensing and oxidant/redox-related pathways. *Biochem Biophys Res Commun* **316**, 969–977.

Harman, D. (1956). Aging: a theory based on free radical and radiation chemistry. *J Gerontol* **11**, 298–300.

Harman, D. (1972). The biologic clock: the mitochondria? *J Am Geriatr Soc* **20**, 145–147.

Kirkwood, T. (1999). *Time of our lives: the science of human aging*. New York: Oxford University Press.

Kirkwood, T. B. (2005). Understanding the odd science of aging. *Cell* **120**, 437–447.

Klevecz, R. R., Bolen, J., Forrest, G. and Murray, D. B. (2004). A genome wide oscillation in transcription gates DNA replication and cell cycle. *Proc Natl Acad Sci USA* **101**, 1200–1205.

Kuramoto, Y. (1984). *Chemical oscillations, waves, and turbulence*. Berlin: Springer.

Lane, N. (2002). *Oxygen: the molecule that made the world*. New York: Oxford University Press.

Lane, N. (2005). *Power, sex, suicide: mitochondria and the meaning of life*. Oxford: Oxford University Press.

Lloyd, D. (1998). Circadian and ultradian clock-controlled rhythms in unicellular microorganisms. *Adv Microb Physiol* **39**, 291–338.

Lloyd, D. (2007). *Respiratory oscillations in yeast*. Austin, TX: Landes Bioscience.

Lloyd, D. and Murray, D. B. (2006). The temporal architecture of eukaryotic growth. *FEBS Lett* **580**, 2830–2835.

Lloyd, D. and Murray, D. B. (2007). Redox rhythmicity: clocks at the core of temporal coherence. *BioEssays* **29**, 465–473.

Lloyd, D., Aon, M. A. and Cortassa, S. (2001). Why homeodynamics, not homeostasis? *ScientificWorldJournal* **1**, 133–145.

Mair, T., Warnke, C., Tsuji, K. and Muller, S. C. (2005). Control of glycolytic oscillations by temperature. *Biophys J* **88**, 639–646.

Marin-Garcia, J. and Goldenthal, M. J. (2004). Heart mitochondria signaling pathways: appraisal of an emerging field. *J Mol Med* **82**, 565–578.

Morel, Y. and Barouki, R. (1999). Repression of gene expression by oxidative stress. *Biochem J* **342(Pt 3)**, 481–496.

Murray, D. B., Roller, S., Kuriyama, H. and Lloyd, D. (2001). Clock control of ultradian respiratory oscillation found during yeast continuous culture. *J Bacteriol* **183**, 7253–7259.

Pittendrigh, C. S. (1993). Temporal organization: reflections of a Darwinian clock-watcher. *Annu Rev Physiol* **55**, 16–54.

Roussel, M. R. and Lloyd, D. (2007). Observation of a chaotic multioscillatory metabolic attractor by real-time monitoring of a yeast continuous culture. *FEBS J* **274**, 1011–1018.

Ruoff, P., Christensen, M. K., Wolf, J. and Heinrich, R. (2003). Temperature dependency and temperature compensation in a model of yeast glycolytic oscillations. *Biophys Chem* **106**, 179–192.

Skinner, J. E., Pratt, C. M. and Vybiral, T. (1993). A reduction in the correlation dimension of heartbeat intervals precedes imminent ventricular fibrillation in human subjects. *Am Heart J* **125**, 731–743.

Strogatz, S. H. (2003). *Sync: the emerging science of spontaneous order*. New York: Hyperion.

Sweeney, B. M. and Hastings, J. W. (1960). Effects of temperature upon diurnal rhythms. *Cold Spring Harb Symp Quant Biol* **25**, 87–104.

Turrens, J. F. (2003). Mitochondrial formation of reactive oxygen species. *J Physiol* **552**, 335–344.

West, B. J. (1999). *Physiology, promiscuity and prophecy at The Millennium: A tale of tails*. Singapore: World Scientific.

Winfree, A. T. (1967). Biological rhythms and the behavior of populations of coupled oscillators. *J Theor Biol* **16**, 15–42.

Wright, A. F., Jacobson, S. G., Cideciyan, A. V., Roman, A. J., Shu, X., Vlachantoni, D., McInnes, R. R. and Riemersma, R. A. (2004). Lifespan and mitochondrial control of neurodegeneration. *Nat Genet* **36**, 1153–1158.

Part II
Invertebrate Systems

Chapter 6
Ultradian and Circadian Rhythms: Experiments and Models

B. Fuentes-Pardo[1], C. Barriga-Montoya[1], M. Lara-Aparicio[2], and S. López de Medrano[3]

Abstract The importance of both ultradian and circadian rhythms in Nature is well known. In the literature it is possible to find many articles referring to the characteristics and properties of such rhythms and, in some cases, of their mathematical modeling.

However, as far as we know, there is not enough information relating these two types of rhythms either from a biological or mathematical point of view.

In this chapter we describe the main properties of the crayfish circadian rhythm structure and its mathematical simulation through different developmental stages and experimental situations. During the work we revealed the persistence of the ultradian rhythms even when they are masked by circadian rhythms.

Moreover, when transients were analyzed in both models, biological and mathematical, we were led to consider the hypothesis that ultradian rhythms could represent a regression to a primordial vital dynamic state.

Finally, we report on certain new experiments and their mathematical models, where we can observe the initial presence of a circadian rhythm and its subsequent regression to again an ultradian one by the influence of an external disturbance.

It seems that ultradian rhythms arise first in evolution and in ontogeny, preceding in both cases the appearance of circadian rhythms.

Keywords Ultradian rhythms, Circadian rhythms, Mathematical model

[1]Departamento de Fisiología, Facultad de Medicina, Universidad Nacional Autónoma de México, México, D.F., CP. 04510, México
Emails: bfpardo@servidor.unam.mx; carolina@mmc2.igeofcu.unam.mx

[2]Departamento de Matemáticas, Facultad de Ciencias, Universidad Nacional Autónoma de México, México, D.F., CP. 04510, México
Email: aparicio@servidor.unam.mx

[3]Instituto de Matemáticas, Universidad Nacional Autónoma de México, México, D.F., CP. 04510, México.
Email: santiago@matem.unam.mx

6.1 Biological Antecedents

Through the knowledge we obtained when we simulated the different stages and properties of a circadian rhythm in crayfish we realized the importance of the relationships between ultradian and circadian rhythms.

Circadian rhythms have been recognized for a long time as a fundamental feature in the organization of living systems (Aschoff, 1960; Bünning, 1973; Pittendrigh, 1974). Particularly in visual system of crayfish, rhythmic activity has been extensively studied (Fuentes-Pardo and Inclán-Rubio, 1981; Fuentes-Pardo and Ramos-Carvajal, 1982; Fanjul-Moles et al., 1987; Fuentes-Pardo et al., 1992, 1997; Fuentes-Pardo and Hernández-Falcón, 1993). When an intact animal is kept under complete darkness (except for test light pulses) and controlled ambient temperature, a clear circadian rhythm (with period of 23.4 h) is obtained in the amplitude of the visual photoreceptors' electrical response to light (electroretinogram, ERG). The ERG is just the sum of the activity of visual photoreceptors and its amplitude depends on both the probability that a photon reaches the retina and the photoreceptor cells excitability. The cells containing retinal shielding pigments, proximal and distal, are the structures in charge of regulating the transmission of light to visual photoreceptors. For many years it has been known (Welsh, 1941) that hormonal control dictates the retinal shielding pigments position; this is principally mediated by the pigment-dispersing hormone (PDH) and the red pigment-concentrating hormone (RPCH) (Garfias et al., 1995; Porras et al., 2001; Aréchiga and Rodríguez-Sosa, 2002). Recently, it has been proposed that besides the role of PDH in regulation of visual accessory pigments, this hormone influences the ERG circadian rhythm modifying the density and/or the sensitivity of membrane receptors involved in the phototransduction process, during the 24 h cycle (Verde et al., 2007).

At present, there are no conclusive studies on the origin of the ERG oscillation. One approach to this problem involves the study of the ontogeny of this circadian rhythm.

For the expression of a circadian pattern it is necessary that the anatomical structure reaches maturity and establishes the necessary structural and physiological relationships between its parts. During development, different structures and functions begin to show some temporal organization that eventually will acquire circadian characteristics. This implies that underlying the sense of time of the organism there are present a number of changes in its anatomical and functional organization. Indeed, the organism exhibits successive changes in period, relative amplitude and activity level of its circadian rhythms during all the ontogeny process. It can be assumed that variation in these parameters involves changes in the main elements of a circadian system, namely the structures implicated in the generation, expression and synchronization of the rhythm.

In very young crayfish (immediately after hatching), the ERG amplitude is very low (~4 μV) and shows clear ultradian periodical fluctuations with a period ranging from 15 min to 4 h. Four weeks old crayfish express higher ERG amplitude (~50 μV) and, for the first time, the presence of a circadian oscillation. The analysis of these recordings also reveals an ultradian rhythm similar to those recorded from the youngest crayfish, but now superimposed on an oscillation with circadian (although somewhat irregular) periods. When the young crayfish were recorded at different environmental temperatures the high frequency cycles did not change either frequency or amplitude. Older animals (around 5 months after hatching) show a progressive increment in ERG amplitude and period length, as well as a progressive disappearance of the high frequency cycles. From this moment on, amplitude and oscillation period show the characteristic values for this species (Fig. 6.1). These experimental results allow the establishment at maturity of a logarithmic relationship between the age of the animals and the voltage and period length of ERG oscillations (Fanjul-Moles et al., 1987).

It is tempting to propose that during ontogeny, before the complete maturation of the neuroendocrine system, there are only ultradian frequency oscillators, which progressively vanish until they disappear under the influence of some circadian neurosecretions released principally from the sinus gland (in the eyestalk). The appearance of a circadian rhythm coincides with the maturation of the neuroendocrine system. Supporting this idea is the well-established fact that in the first and second instar (from hatching to 28 days) the migration of screening pigments (which depends on the presence of at least two hormones, PDH and RPCH) is not observed. Advanced crayfish stages show a normal photomotor response, indicating that both screening pigmentary cells and neurosecretion release mechanism have attained a suitable maturation stage.

The 4 h pattern found in the youngest crayfish is similar to the pattern found in the ERG amplitude recorded from the isolated eyestalk of an adult crayfish (Sánchez and Fuentes-Pardo, 1976). In this experimental design, it could be observed that, superimposed on circadian variations in the ERG amplitude, there were high frequency cycles (with periods around 2–4 h) that seem to be correlated with the circadian time, since their amplitude depended on the phase of the circadian rhythm when

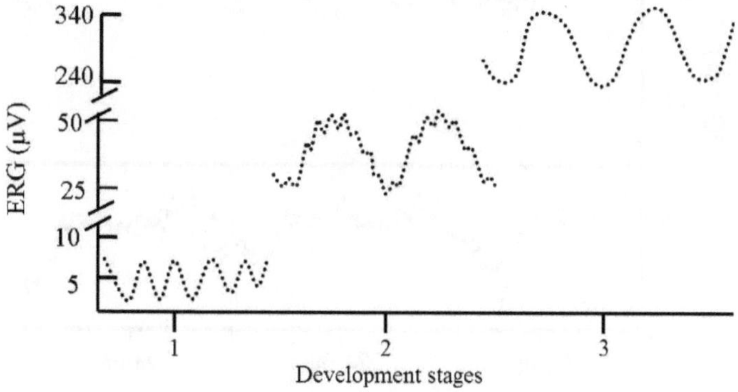

Fig. 6.1 Ontogeny of the ERG circadian rhythm. See text for explanation

they appeared (Fig. 6.2). It is remarkable that these periodical changes in ERG amplitude are present in the isolated eyestalk without changes in the locations of the accessory pigments.

Long term recordings of the ERG from crayfish deprived of one or both sinus glands show a very irregular circadian rhythm with superimposed ultradian cycles (Moreno-Sáenz et al., 1986, 1992): these were clearer in animals deprived of both sinus glands (Fig. 6.3). In these experiments, it seems that we are observing changes in the organization of a circadian rhythm produced as a consequence of

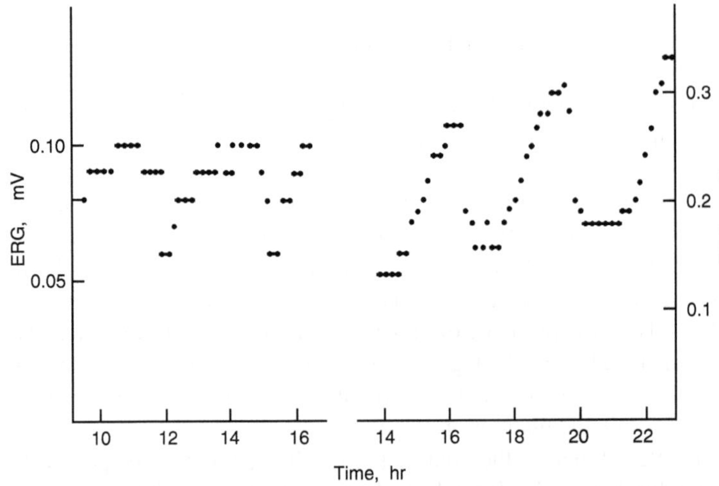

Fig. 6.2 ERG recording from an isolated eyestalk. High frequency oscillations, taken from the same experiment, at two different circadian times. Note that the waveform and amplitude are very different depending on the circadian time of observation

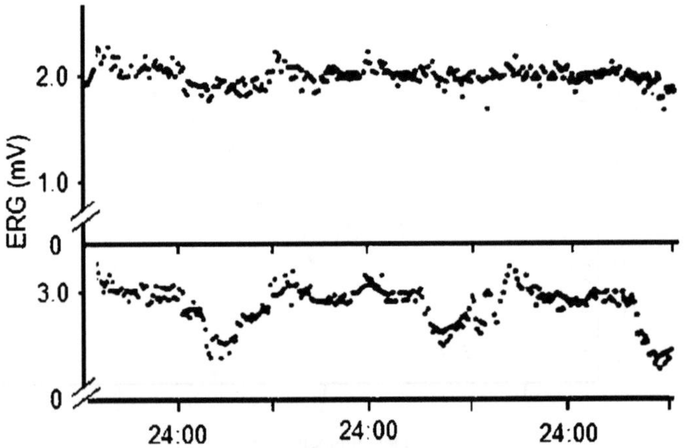

Fig. 6.3 Long-term simultaneous ERG recordings from both eyes of crayfish with sinus gland ablation. Note the presence of high frequency cycles superimposed on the circadian activity

the loss of an oscillator. This oscillator would be an output of the sinus gland which, in normal conditions, blocks the ultradian cycles and imposes circadian changes in the photoreceptors' light sensitivity through the circadian release of PDH (Verde et al., 2007).

More difficult to explain are the results obtained with experiments using deuterium oxide (D_2O); this was employed at different doses in long term recordings of the ERG (Fuentes-Pardo and Moreno-Sáenz, 1988). In this work the authors analyzed the effect of deuterium oxide on the circadian oscillations obtained from both intact animal and isolated eyestalk, and upon the ultradian oscillations obtained from the isolated eyestalk. The results showed a lengthening of both circadian and ultradian periods. Moreover, there was a direct relationship between the D_2O dose and the length of period in both circadian and ultradian cycles. There was also a direct relationship between the lengthening of the ultradian and circadian cycles observed with each dose of D_2O employed. It was proposed that the lengthening of the circadian period is due to the effect of D_2O upon the high frequency oscillations. The isolated eyestalk experiments reinforce this interpretation, since the group of oscillators (cells) that were partially uncoupled due to the eyestalk ablation, also shows an increment in period in the presence of D_2O, since it causes a lengthening of the ultradian cycles superimposed on the circadian cycles (Fig. 6.4). The lengthening produced by D_2O could result from the diffusion of this substance to all the cells, tissues and organs of the organism, particularly to the cells involved in the generation and expression of the ERG oscillatory activity. These cells would actually be the oscillators that reduce their oscillation frequency produced by the physicochemical changes of D_2O. The results allow us to propose that D_2O affects the oscillatory machinery by its well-established general property of slowing biochemical reaction kinetics due to the "heavy isotope effects".

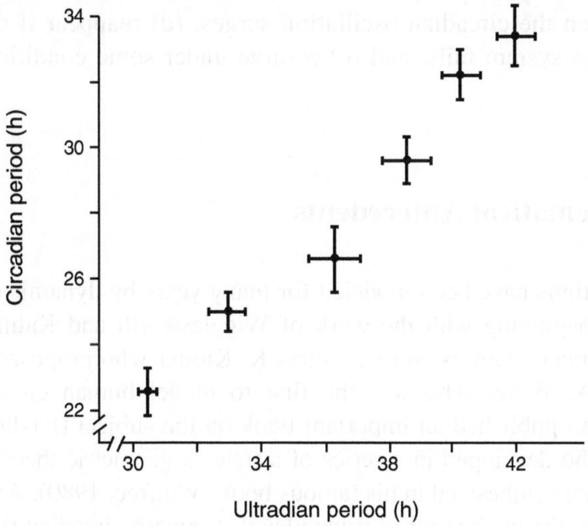

Fig. 6.4 Relation between ultradian and circadian cycles in the isolated eyestalk obtained with different D_2O doses (1.25%, 2.5%, 5.0%, 10.0%, 20.0% and 30.0%)

When the ERG circadian rhythm recorded from a crayfish kept under constant darkness is perturbed by the presence of a light stimulus, this induces a phase change, an advance or a delay, that can be detected once the rhythm returns to a steady state. The sense and magnitude of change depend on the circadian time of stimulus application. It is worthwhile noting that immediately after the light application, the ERG circadian rhythm shows a transitory stage characterized by the presence of irregular high frequency (ultradian) oscillations; the characteristics of these, particularly phase as well as the circadian moment when the stimulus was applied, seem to determine whether the rhythm will show advance or delay in the steady state.

It can be proposed that the ultradian rhythms observed in each and every one of the above experimental scenarios (ontogeny, sectioned eyestalk, sinus gland deprivation, D_2O and application of a light stimulus) appear as a consequence of the failure (or decreased velocity in the case of D_2O) of the periodical release of some hormone, particularly PDH.

Unfortunately, we don't have a conclusive explanation of the physiological role of the ultradian cycles in the ERG circadian rhythm. It is tempting to propose that in ontogeny, before the complete maturation of the neuroendocrine system, there are only ultradian oscillators, and that in adult animals their resurgence occurs during some exogenous perturbations.

In summary, and in accordance with many authors working diverse species (Goldbetter, 1991; Veldhuis, 1992; Lloyd and Kippert, 1993; Lloyd et al., 2002; Lloyd and Murray, 2005), the high frequency cycles: (a) posses similar properties to circadian cycles i.e., they show temperature compensation mechanisms and are affected by the same chemicals affecting the circadian ones. Somewhat similar observations have been made by authors working with diverse biological systems; (b) are a very early expression of the cellular activity in the ontogeny; (c) disappear when the circadian oscillation surges; (d) reappear if the neuroendocrine circadian system fails; and (e) resurge under some conditions of external perturbation.

6.2 Mathematical Antecedents

Circadian rhythms have been modeled for many years by dynamical systems with limit cycles, beginning with the work of Wigglesworth and Kalmus (1960) and continued by many authors, among others K. Klotter who proposed a van der Pol oscillator, R.A. Wever who was the first to model human circadian rhythms, T. Pavlidis who published an important book on the subject (Pavlidis, 1973), and A. Winfree who developed in a series of articles a geometric theory of biological clocks which he synthesized in his famous book (Winfree, 1980). An important list of problems on the mathematical principles that govern circadian rhythmicity was also set by Winfree (1979). For a detailed account of these developments see Lara-Aparicio et al. (2006).

Ultradian rhythms can be modeled in the same way as circadian ones. In fact it has been observed that both rhythms share many of their properties and, up to now, ultradian rhythms have not posed new problems for modeling.

More interesting has been the problem of modeling systems involving both circadian and ultradian rhythms in interaction. Most of the work on this aspect is related to the question of the origin of the circadian rhythms. Since most rhythms at the cellular level are ultradian it is natural to imagine that the global circadian ones are the result of the coupling of a large number of local ultradian oscillators. An early attempt to construct a model on this hypothesis was made by Winfree (1967) who coupled physically a large number of oscillators. From that time on, many different mathematical models have been suggested. One of the more recent ones (Paetkau et al., 2006) has the feature of being based on models of actual ultradian oscillators in molecular biology. Nevertheless, in this case, the circadian oscillator is not given a concrete interpretation based on real biologically-derived data, and the assumed linearity of the oscillators involved does not reflect all the features of the biological ones. Many of the other proposed models just try to model the abstract possibility of such couplings with no concrete interpretation of either oscillator.

An interesting such mathematical model was constructed by Pavlidis (1969) by coupling a large number of oscillators of the same frequency in such a way that the resulting summated output has a considerably smaller frequency.

A limitation of the Pavlidis model was pointed out in Barrio et al. (1997). The frequency of the resultant of the ultradian oscillators in the model depends strongly on the number of them, in contrast with the observed behavior of circadian oscillators that preserve their frequency even when a large portion of the tissue supporting it are removed. In this interesting article it is shown how to improve Pavlidis' model by a hierarchical coupling of the oscillators in such a way that the resulting oscillator's frequency does not depend greatly on the number of ultradian oscillators above a certain quantity. Other interesting models are also discussed in Barrio's article.

We encountered in our modeling a very different situation. In the ontogenesis of a circadian rhythm the experiments showed a global ultradian rhythm in very young individuals that evolves into a global circadian rhythm with an intermediate stage in which both rhythms are superposed. How could one construct a model of this based on the coupling of many ultradian oscillators?

The coupling strength of the oscillators should vary during the development of the rhythm: In the early stages, the coupling should be weak enough to preserve the ultradian character of the resulting output, but strong enough to synchronize all the oscillators (otherwise the result would be a practically constant signal, being the sum of many unsynchronized clocks). At the other end of the process, in the adult stage, the strength of the coupling should be able to produce a global circadian result, as in the models described above. And it should also reproduce the intermediate step where both ultradian and circadian rhythms coexist!

For our purpose we have taken a different option, without denying that it would be interesting to search for a simple model whose individual oscillators and modes of coupling evolve naturally in the above manner. Our approach consists in assuming the existence of a well-defined ultradian rhythm from the beginning of the ontogenesis and of a circadian rhythm that emerges gradually, but whose circadian character is well defined from its first manifestation. And we concentrate on modeling the relations between the two. This does not exclude the possibility that the circadian oscillator might be the result of the coupling of many ultradian oscillators. Whether the ultradian oscillator that we see from the beginning is one of them or not is irrelevant for the modeling. Nevertheless, the biological interpretation points very emphatically to the fact that each of the oscillators exists in different anatomical structures.

For some time we have been working with a mathematical model based on the experimental work described above, first developed to simulate the ontogeny of the circadian rhythm in crayfish and then confronted with other experimental situations where it was subject just to minor adaptations. This model is based on the consideration that a certain anatomical structure has the capacity to oscillate with an ultradian frequency while another one is capable of oscillating with a circadian frequency.

Let us explain how this mathematical model works in descriptive terms:

We worked in agreement with a long tradition in the subject (going at least as far back as Kalmus and Wigglesworth, 1960), which associates a limit cycle to a circadian rhythm (see Lara-Aparicio et al., 1993), but we used a simple two dimensional oscillator with a circular limit cycle instead of the usual van der Pol oscillator.

Basically, the model is constituted by two coupled limit cycle oscillators of that kind, one of them of a relative high frequency (ultradian), and the other one with a relative low frequency (circadian) which governs to the above oscillator.

A complete description of the mathematical model appears in many published articles (See the literature at the end of the chapter.)

For the time being we want to emphasize several important facts that interrelate ultradian and circadian rhythms, since one can check through the behavior of the model the persistence of the ultradian cycles during the complete life of the crayfish.

The ultradian rhythm is seen in the electroretinogram (ERG) of the very young crayfish, and even during its youth; it only disappears in the adult age (Figs. 6.1 and 6.5).

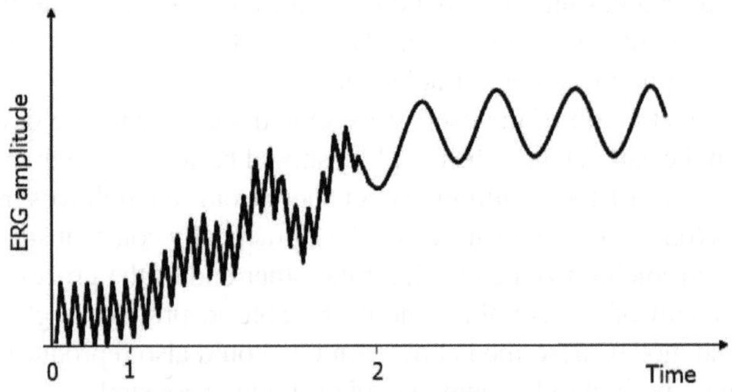

Fig. 6.5 ERG amplitude curve generated by the mathematical model. It can be seen the ultradian (0→1), superimposed (1→2) and circadian (2→) stages

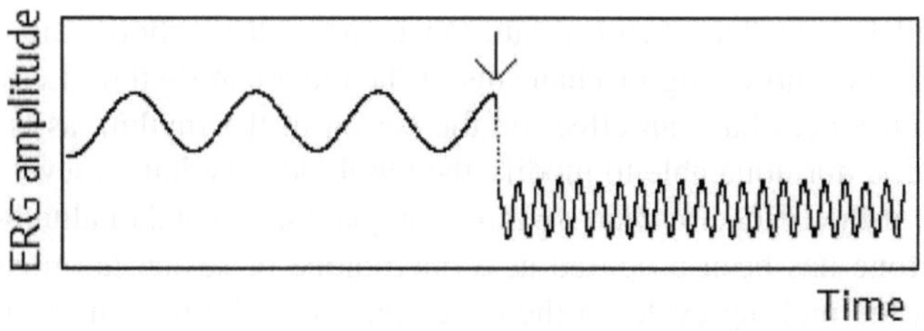

Fig. 6.6 Mathematical modelling of the suppression of the governing circadian rhythm

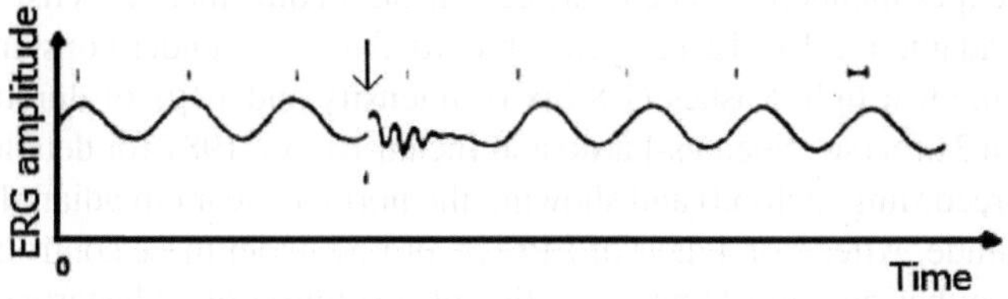

Fig. 6.7 Effect of a single light pulse on the ERG circadian rhythm. The arrow indicates zeitgeber application

The existence of a latent ultradian rhythm in the adult stage is a basic fact of the model that can be observed in different experimental situations:

1. When the governing circadian rhythm is suppressed, the ultradian rhythm reappears (Fig. 6.6).
   If one perturbs the eye of the crayfish with a single signal light pulse, one observes the reappearance of the ultradian rhythm in the response curve during the short transient phase that precedes the restitution of the circadian rhythm with an advance or delay phase shift (Fig. 6.7). These transient dynamics are of a great evolutionary importance according to A. Winfree who wrote (Winfree, 1979): *The dynamics of transients have been almost universally ignored. But to my mind it is the essence of temporal adaptation and the foremost target of evolutionary change....*
2. More recently, in the experiments described in the following section, when a male crayfish is perturbed by the presence of a female one, the emergence of the ultradian rhythm is observed again, this time superimposed on a circadian one.

6.3 Action of Sexual Hormones on ERG Circadian Rhythm

The circadian rhythm ERG amplitude in crayfish is well characterized. With the aim to obtain (and sometimes to predict) modifications in its period, amplitude and phase many different external signals, both photic and non-photic, have been applied. These manipulations permit a better understanding of the physiological principles that underlie the rhythm.

It is well known that chemical substances as well as photic stimulus greatly influence the synchronizing mechanisms of the circadian rhythm. Less frequently, chemical substances have an effect on the period of the rhythm, as is the case of ecdysterone, a hormone able to modify the amplitude, excitation level, period and noise level of the ERG circadian rhythm in a quantitative and qualitative manner.

Ecdysterone has been proposed as a pheromone or sexual hormone. It is also involved in the molting cycle, in the development of the nervous system, in programmed neuron death and in many other functions. In crayfish, its synthesis takes place in the Y-organ and is regulated by the molting-inhibiting hormone, a polypeptide secreted by the X-organ-sinus gland complex in the eyestalk.

Recent experiments have been carried out on a adult male crayfish (*P. clarkii*) previously adapted to LD 12:12 cycles, recorded at 16°C under constant darkness, except for the test light flashes (1.8 lux of intensity and 10 μs of duration) sent to the eye each 3 min (see Fuentes-Pardo and Inclán-Rubio, 1981 for detailed explanation of the recording method) and showing the normal, clear circadian rhythm in its ERG amplitude. After 3 or 4 days of ERG recording under these controlled ambient conditions, either an adult female crayfish or a solution of ecdysterone was introduced into the bath during 20 min. Compared with the control experiment before the arrow, when we go back to the controlled conditions, important qualitative changes can be observed in this rhythm (Fig. 6.8).

The analysis of the time series before and after the disturbance shows that time series are not stationary: shortening of the length of the circadian period (typically half an hour) is observed, as well as the appearance of high frequency (ultradian) oscillations with periods of about 16, 12, 6 and 3 h superimposed on the circadian oscillation. It is crucial to understand too that these disturbances are irreversible during the whole duration of the experimental recording (about 20 days).

As a control for our experiment we put a male crayfish in the bath (instead of the female one). During the ERG recording we cannot observe any change in the characteristics of ERG circadian rhythm. The same happens if the observed male

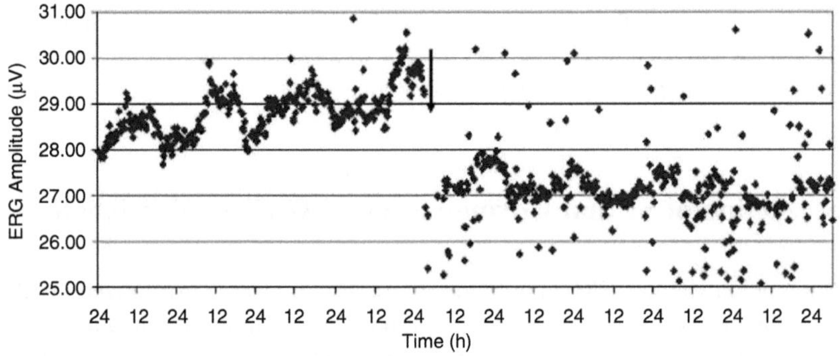

Fig. 6.8 ERG circadian rhythm of an adult, male crayfish recorded in presence of a female (indicated by the arrow) crayfish

crayfish's deutocerebrum is injured or eliminated before the ERG recording: the presence of either a female, or a male, or of a solution of ecdysterone in the bath does not produce the disturbances observed in the intact crayfish in presence of a female.

It is known that ecdysterone (and more generally any chemical external signal) is detected by chemical receptors located in the base of antennulas and that this information is processed by the deutocerebrum (a probable site for circadian activity generation).

We propose that deutocerebrum sends a circadian signal to sinus gland (via protocerebrum) masking the initial sinus gland circadian oscillation and expressing a new circadian oscillation with high frequency cycles superimposed on it.

6.4 Mathematical Modeling of Sexual Hormones Experiments

Considering that the model we initially used to simulate the ontogeny of the circadian rhythm (Lara-Aparicio et al., 1993) has proved useful for the modeling of several characteristics of the circadian rhythm, among them those in paragraphs (1) and (2) in Section 6.2 (Fuentes-Pardo et al., 1995, 2001; Lara-Aparicio et al., 2006), we decided to employ it, with pertinent modifications, to simulate the rhythm of the perturbation of the male crayfish by the presence of a female (Fig. 6.9).

In the experimental system, our biologists observed that the original circadian frequency shrinks while another circadian frequency appears (with a value slightly different from the original circadian oscillation) on which is superimposed a relative high frequency oscillation (ultradian).

From the biological evidence we proceeded to modify our previous model through the addition of a third oscillator, assumed to be modelling the one located

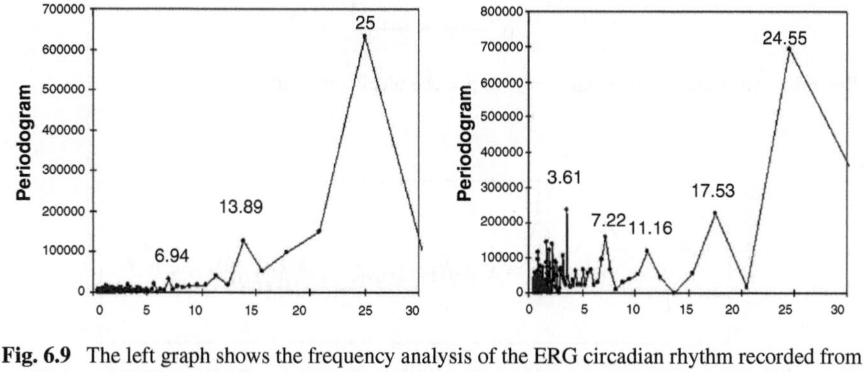

Fig. 6.9 The left graph shows the frequency analysis of the ERG circadian rhythm recorded from a male crayfish during the first 4 days of recording. The right graph shows the frequency analysis of the same rhythm during and after the presence of a female crayfish. Note the presence of high frequency cycles and the slight change in frequency of the circadian oscillation

$$
\begin{aligned}
\dot{x}_1 &= -kx_2 + e\left(x_1 + y_1 - \frac{3}{2}y_3 - x_3\right)\left(a - \frac{1}{5}x_3 y_1 - \left(x_1 + y_1 - \frac{3}{2}y_3 - x_3\right)^2 - x_2^2\right)\\
\dot{x}_2 &= k\left(x_1 + y_1 - \frac{3}{2}y_3 - x_3\right) + ex_2\left(a - \frac{1}{5}x_3 y_1 - \left(x_1 + y_1 - \frac{3}{2}y_3 - x_3\right)^2 - x_2^2\right)\\
\dot{x}_3 &= c_1 x_3 - c_2 x_3^2\\
\dot{y}_1 &= -\left(l + \frac{z_1}{4}\right)y_2 + f\left(y_1 - \frac{3}{4}y_3 - \frac{z_1}{9}\right)\left(\frac{3}{2}y_1 y_3 - \frac{z_1}{4} - \frac{y_3^2}{2} - \left(y_1 - \frac{z_1}{9}\right)^2 - y_2^2\right)\\
\dot{y}_2 &= \left(l + \frac{z_1}{4}\right)\left(y_1 - \frac{3}{4}y_3 - \frac{z_1}{9}\right) + fy_2\left(\frac{3}{2}y_1 y_3 - \frac{z_1}{4} - \frac{y_3^2}{2} - \left(y_1 - \frac{z_1}{9}\right)^2 - y_2^2\right)\\
\dot{y}_3 &= d_1 y_3 - d_2 y_3^2\\
\dot{z}_1 &= -mz_2 + g(z_1 + c)\left(s - (z_1 + c)^2 - z_2^2\right)\\
\dot{z}_2 &= (mz_1 + c) + gz_2\left(s - (z_1 + c)^2 - z_2^2\right)
\end{aligned}
$$

Where $k = 0.6$; $e = 25$; $l = 0.1$; $f = 10$; $g = 10$; $a = 0.5$; $c =$ alleatory variable; $c_1 = 0.04$; $c_2 = 0.02$; $d_1 = 0.04$; $d_2 = 0.02$; $m = 0.085$; $s = 0$ or 1

Fig. 6.10 System of differential equations of the mathematical model

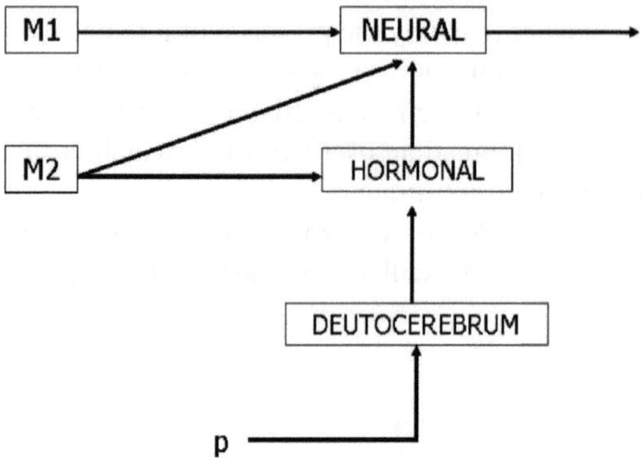

Fig. 6.11 Mi = maturation stage; i = 1, 2; p = chemical perturbation

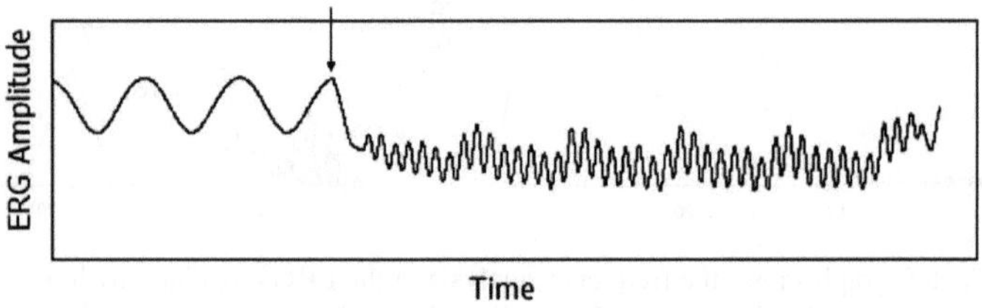

Fig. 6.12 Mathematical simulation of ERG circadian rhythm of an adult, male crayfish recorded in presence of a female (indicated by the arrow) crayfish

in the deutocerebrum. It is assumed to be circadian with a frequency slightly different of the one of original circadian oscillation and it mainly affects the original hormonal circadian oscillator which in turn governs the neural ultradian one. The final mathematical model is given by the above set of equations (see Fig. 6.10).

In Fig. 6.11 we present a block diagram of the relation between the different oscillators. The type of coupling employed is the same as in the previous models (Lara-Aparicio et al., 1993).

Finally, Fig. 6.12 shows the simulated curve.

6.5 Conclusion

The experimental and modeling work presented here supports the thesis that ultradian and circadian rhythms coexist and together participate in the organization of the periodical changes (adaptive) in light sensitivity characteristic of the crayfish visual system. We propose that ultradian activity initiates earlier during but is masked by the circadian oscillations when the crayfish has reached a certain maturation degree. Also, it resurges when some exogenous perturbations affect the circadian system and so is omnipresent.

Is phylogeny repeated in ontogeny? This would mean that in our experiments we might be also revealing the masked ultradian rhythms that were present during the evolutionary succession.

The authors thank professor David Lloyd for commenting their manuscript and for sharing with them his insights on this subject.

Supported by IN200207 DGAPA grant.

References

Aréchiga H., Rodríguez-Sosa L. (2002). Distributed circadian rhythmicity in the crustacean nervous system. The crustacean nervous system (ed Wise K.). Springer, Berlin.

Aschoff J. (1960). Exogenous and endogenous components in circadian rhythms. Biological clocks. Cold Spring Harbor Symp. Quant. Biol. 25:11–27.

Barrio R. A., Zhang L., Maini P. (1997). Hierarchically coupled ultradian oscillators generating robust circadian rhythms. Bull. Math. Biol. 59(3):517–532.

Bünning E. (1973). The physiological clock. Circadian rhythms and chronometry. Springer, New York, 258 pp.

Fanjul-Moles M. L., Moreno-Sáenz E., Villalobos-Hiriart N., Fuentes-Pardo B. (1987). ERG circadian rhythm in the course of ontogeny in crayfish. Comp. Biochem. Physiol. 88A(2):213–219.

Fuentes-Pardo B., Hernández-Falcón J. (1993). Neurobiology of the circadian clock of crayfish. Trans. Com. Biochem. Physiol. 1:635–673.

Fuentes-Pardo B., Inclán-Rubio V. (1981). Correlation between motor and electroretinographic circadian rhythm in the crayfish *Procambarus bouvieri* (Ortmann). Com. Biochem. Physiol. 68A:477–485.

Fuentes-Pardo B., Moreno-Sáenz E. (1988). Action of deuterium oxide upon the ERG circadian rhythm in crayfish *Procambarus bouvieri*. Comp. Biochem. Physiol. 90A(3):435–440.

Fuentes-Pardo B., Ramos-Carvajal J. (1982). The phase response curve of electroretinographic circadian rhythm of crayfish. Comp. Biochem. Physiol. 74A (3):711–714.

Fuentes-Pardo B., Fanjul-Moles M., Moreno-Sáenz E. (1992). Synchronization by light of the ERG circadian rhythm during ontogeny in the crayfish. J. Interdiscip. Cycle Res. 23:81–91.

Fuentes-Pardo B., Lara-Aparicio M., López de Medrano S. (1995). Perturbation of a circadian rhythm by single and periodic signals and its mathematical simulation. Bull. Math. Biol. 57(2):175–189.

Fuentes-Pardo B., Solórzano-García S., De la O Martínez A. (1997). Effects of complete and skeleton photoperiods on the circadian rhythm of the electroretinogram of the crayfish. Biol. Rhythm Res. 28:69–84.

Fuentes-Pardo B., Lara-Aparicio M., López de Medrano S. (2001). On the ontogeny of the motor circadian rhythm in crayfish. Bull. Math. Biol. 63(2):353–369.

Garfias A., Rodríguez-Sosa L., Aréchiga H. (1995). Modulation of crayfisf retinal function by red pigment concentrating hormone. J. Exp. Biol. 198:1447–1454.

Goldbetter A. (1991). A minimal cascade model for the mitotic oscillator involving cycline and *cdc*2 kinase. Proc. Natl. Acad. Sci. USA 88:9107–9111.

Kalmus H., Wigglesworth L. A. (1960). Shock excited systems as models for biological rhythms. Biological clocks. Cold Spring Harbor Symp. Quant. Biol. 25:211–216.

Lara-Aparicio M., López de Medrano S., Fuentes-Pardo B., Moreno-Sáenz E. (1993). A qualitative mathematical model of a circadian rhythm in crayfish. Bull. Math. Biol. 55(1):97–110.

Lara-Aparicio M., Barriga-Montoya C., Fuentes-Pardo B. (2006). A brief biomathematical history of circadian rhythms: from Wigglesworth to Winfree. Scientiae Mathematicae Japanicae 64(2):357–370.

Lloyd D. and Kippert F. (1993). Intracellular coordination by the internal clock. Cell Biol. Int. 17(12):1047–1052.

Lloyd D. and Murray D. B. (2005). Ultradian metronome: timekeeper for orchestration of cellular coherence. Trends Biochem. Sci. 30(7):373–377.

Lloyd D., Eshantha L., Salgado J., Turner M. P., Murray D. B. (2002). Respiratory oscillations in yeast: clock-driven mitochondrial cycles of energization. FEBS Lett. 519:41–44.

Moreno-Sáenz E., Hernández-Falcón J., Fuentes-Pardo B. (1986). Role of the sinus gland in crayfish circadian rhythmicity-II. ERG circadian rhythm. Comp. Biochem. Physiol. Part A 87A(1):119–125.

Moreno-Sáenz E., Fuentes-Pardo B., Hernández-Falcón J. (1992). Photoentrainment of the circadian rhythmin the electroretinogram of the crayfish and its dependence on the sinus gland. J. Exp. Zool. 264:144–152.

Paetkau V., Edwards R., Illner R. (2006). A model for generating circadian rhythm by coupling ultradian oscillators. Theor. Biol. Med. Model. 3(12):1–10.

Pavlidis T. (1969). Populations of interacting oscillators and circadian rhythms. J. Theor. Biol. 22:418–436.

Pavlidis T. (1973). Biological oscillators: their mathematical analysis. Academic, New York.

Pittendrigh C. S. (1960). Circadian rhythms and the circadian organization of living systems. Cold Spring Harbor Symp. Quant. Biol. 25:159–184.

Pittendrigh, C. S. (1974). Circadian organization in cells and the circadian organization of the multicellular system. Neurosciences Third Study Program. (eds Schmitt S. O. and Worden F. G.). MIT, Cambridge.

Porras M. G., López-Colomé A. M., Aréchiga H. (2001). Red pigment-concentrating hormone induces a calcium-mediated retraction of distal retinal pigments in the crayfish. J. Comp. Physiol. 187:349–357.

Sánchez J. A., Fuentes-Pardo B. (1976). Circadian rhythm in the amplitude of the electroretinogram in the isolated eyestalk of the crayfish. Comp. Biochem. Physiol. 56A:601–605.

Veldhuis J. D. (1992). A parsimonious model of amplitude and frequency modulation of episodic of hormone secretory bursts of a mechanism for ultradian signaling for endocrine glands. In:

Ultradian rhythms in life processes (eds Lloyd D. and Rossi E. L.) Springer, London, pp. 139–172.
Verde M. A., Barriga-Montoya C., Fuentes-Pardo B. (2007). Pigment dispersing hormone generates a circadian response to light in the crayfish, *Procambarus clarkii*. Comp. Biochem. Physiol. Part A 147:983–992.
Welsh J. H. (1941). The sinus gland and 24-hours cycle of retinal pigment migration in the crayfish. J. Exp. Zool. 186:35–49.
Winfree A. T. (1967). Biological rhythms and the behavior of populations of coupled oscillators. J. Theor. Biol. 16, 15–42.
Winfree A. T. (1979). 24 hard problems about 24-hour rhythms. Nonlinear oscillations in biology. American mathematical Society. Lectures in Applied Mathematics (ed Hoppensteadt F. C.), Vol. 17, pp. 93–126.
Winfree A. T. (1980). The geometry of biological time, xiii, Springer, New York, 530 p.

Chapter 7
Ultradian Lovesong Rhythms in Drosophila

C.P. Kyriacou

Abstract Ultradian courtship song cycle appear to be a feature of male courtship songs among the *melanogaster* subgroup of species as well as some other more distantly related species. Species-specific differences in the song cycle between *D. melangaster* and *D. simulans* map onto the Threonine-Glycine repetitive region encoded by the clock gene *period*. Mutations in the *period* gene alter in a parallel fashion both circadian and lovesong cycles, but female mutants do not prefer the song characteristics of the corresponding mutant male, so there is no 'genetic coupling' between the male and female communication systems with respect to *period*. The song cycle appears to have a neural basis in the thorax, possibly within glia that express PER protein. Recent finding suggest that plastic changes in the thoracic nervous system that are mediated by the *period* and *timeless* genes may play a role in the neurogenesis of the courtship song cycle

Keywords Circadian, lovesong, *period, timeless,* species-specific, behaviour

It has been more than 15 years since the first edition of this volume and the chapter on the lovesong cycle of Drosophila. This ultradian behavioural rhythm was first identified by Kyriacou and Hall (1980) and stimulated considerable interest and some controversy throughout that decade. Rather than repeat the whole story that has been reviewed many times, I intend to focus on the developments that have occurred since the writing of that last contribution in 1991–1992. However, this requires weaving the new material in with the old, so I hope the reader will forgive some repetitions.

Department of Genetics, University of Leicester, Leicester LE1 7RH, UK
Email: cpk@leicester.ac.uk

7.1 The Courtship Song Rhythm: A Species-Specific Behaviour

Fruitflies, like many higher organisms, indulge in complex, innate courtship behaviours before mating. In *Drosophila melanogaster*, males do the active courting, while females do the rejecting, at least until they finally relent (Billeter et al., 2006). The most eye-catching part of the display is the male's wing vibration, which produces a patterned acoustic output. This can be captured using a sensitive microphone, and consists of a train of pulses, plus a humming sound (Fig. 7.1). The pulses are separated by interpulse intervals (IPIs) of 10–200 ms, depending on the species, but these intervals are not invariant, as initially believed, but oscillate, with a period of 30–80 s, again depending on the species, or the genotype within the species (Demetriades et al., 1999; Kyriacou and Hall, 1980, 1986). In *D. melanogaster*, this song cycle has a period of between 50–65 s, superimposed on an IPI that varies between 15 to 100 ms (Fig. 7.1). An average IPI for this species is between 30 to 40 ms, whereas that for the sympatric sibling species *D. simulans*, is between 40 and 70 ms, depending on the strain (Kyriacou and Hall, 1986). In *D. simulans* however, the song cycle is shorter, about 35–40 s. Playback experiments in which the two song characteristics, namely the average species IPI and the song cycle are mixed and

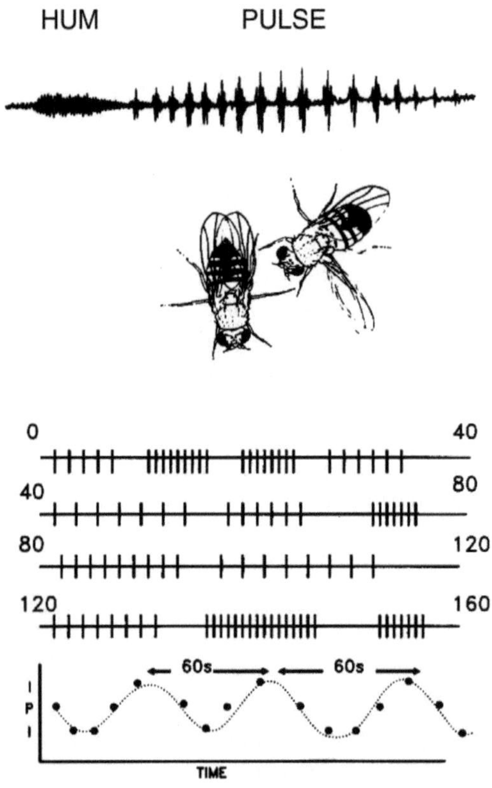

Fig. 7.1 Top: a burst of courtship song that is produced by the male's wing vibration is shown, which consists of hums and pulses. Individual interpulse intervals (IPI) can lie anywhere between 15–100 ms, in any single burst of wing vibration. The average IPI for *D. melanogaster* males is 30–40 ms. Lower: over the 160 s of real time shown, the average IPIs oscillate with a period of ~60 s

matched between species, reveal that the combination of the two species-specific characteristics enhances the female's receptivity (Kyriacou and Hall, 1982). Thus an artificial song containing 55 s cycles and 35 ms average IPIs is highly effective in stimulating *D. melanogaster* females, whereas 48 ms IPIs and 35 s cycles are preferred by *D. simulans* females (Kyriacou and Hall, 1982, 1986). Any other mixture of IPI and cycle is not much more effective than silence.

An interesting modification of this theme was also applied by presenting to females, the same individual IPIs that would be presented in a coherent *melanogaster* song cycle, but presenting them at random although within normal length song bursts (Ritchie et al., 1999). Under these conditions, females preferred the cycling songs, reinforcing the functional value of this type of patterned input. Why do females prefer these rhythmic songs? One idea was that each individual female has a preferred IPI, and she needs to hear a number of these before she mates. A male, in a natural situation, may encounter several females, and by systematically changing his IPI, this may allow him to 'scan' the preferred IPIs of several females, giving him additional opportunities to mate. An attempt was made to bidirectionally select for females that preferred higher or lower IPIs, and the results from the first few generations were encouraging. However, the obligatory inbreeding that is a feature of such selection experiments finally got the better of the experimental design (Greenacre, 1990; Kyriacou et al., 1992).

While songs can be presented to females in the presence of courting but mute (wingless) males, another paradigm is to pre-expose females with songs, then add normal winged males, and observe whether the priming stimulus has had any effect on female receptivity (Kyriacou and Hall, 1984). Using this method, it was observed that females apparently 'remember' the previous rhythmic pulse song exposure for up to 3–4 min, in that they mate faster having heard it. Hum song is also an effective prestimulus, but pre-exposing females to a patterned artificial song with both a rhythmic pulse song and a hum song component significantly potentiates the effect on receptivity (Griffith et al., 1993). The relationship of this type of acoustic sensitisation to learning and memory was revealed when it was observed that females carrying the classic mutations *dunce*, *amnesiac* and *rutabaga*, showed a disrupted response to song pre-exposure (Kyriacou and Hall, 1984).

7.2 Genetics of the Song Rhythm

Genetic crosses between *D. melanogaster* and *D. simulans* have revealed that the *X* chromosome is important for specifying the species-specific song rhythm (Kyriacou and Hall, 1986). Thus a hybrid male with a *melanogaster X* (from his mother), sings with a *melanogaster* like, ~60 s cycle, whereas the reciprocal male carrying the *simulans X*, sings with shorter IPI cycles characteristic of the maternal species. Female hybrids carrying both species *X* chromosomes have a complex set of preferences depending upon whether the average IPI that is presented to them is *melanogaster*-like (35 ms average), or *simulans*-like (48 ms average), or intermediate (42 ms). Basically on a *melanogaster* IPI background, they prefer 55 s (*melanogaster*) cycles, but they switch to 35 s cycles when offered a *simulans* IPI background. Remarkably, when given an intermediate IPI, they prefer an intermediate rhythm of 45 s (Kyriacou and Hall, 1986). Consequently, IPI and song rhythm appear to be inextricably linked within the female's brain.

Genes on the *X* chromosome were therefore implicated in the control of the species-specific song rhythm. The only candidate gene on offer at the time was *period* (*per*), the sex-linked circadian clock locus (Konopka and Benzer, 1971).Using chemical mutagenesis, three *per* mutants had been isolated, which lengthened, shortened or obliterated the fly's circadian rhythm of locomotor activity. When the song rhythm of these mutants was examined, they reflected what was occurring in the circadian domain, so that the per^{s} mutation that had a 19 h free-running behavioural cycle, showed a short, 40 s song rhythm. The per^{L} mutation was long on both phenotypes (29 h and 80 s), while the arrhythmic per^{01} mutant showed little evidence for a coherent song rhythm (Kyriacou and Hall, 1980) although a later study revealed that half of per^{01} males showed very fast but weak ~30 s cycles (Kyriacou and Hall, 1989). Thus a gene product was common to two timing systems whose periodicities were orders of magnitude apart!

This unexpected connection between ultradian and circadian phenotypes, was also later revealed in the *tau* mutant hamster, whose shorter circadian cycle was reflected in a reduced frequency of ultradian ~30 min bursts of LH secretion (Loudon et al., 1994). For completeness, the *Drosophila per* mutants also affected infradian phenotypes, such as the duration of the different stages of growth and development (Kyriacou et al., 1990a). These early results have been extended by several other groups, particularly Sharma and colleagues who have studied the relationship between the clock, early and late eclosion and developmental time (e.g. Paranjpe et al., 2005). The relationship between circadian and cell cycles, on which development is based (Hunt and Sassone-Corsi, 2007), has also provided a fertile research field since it was first discovered that key cell cycle genes showed circadian cycling in gene expression (Akhtar et al., 2002).

One interesting question that arose from the *per* mutants was whether, for example, a per^{s} female would prefer a per^{s} song rhythm? If a *per*-mutant female was 'tuned' in to the corresponding mutant male song, this would support the view that in animal communication systems, there might be a genetic coupling of the transmitter (the male in this case), with the receiver (the female). Initial playback experiments revealed in fact that per^{s} females actually preferred 40 s per^{s}-like songs (Greenacre et al., 1993). However, this experiment had used a homozygous stock of per^{s} females that had been maintained in the laboratory for about 20 years. When a rigorous system of backcrosses was implemented, so that the per^{s} and wild-type alleles were placed on a congenic background, the normal 55 s song rhythm preference of per^{s} females was restored. The 'coupling' effect obtained with the long-standing per^{s} stock was evidently due to secondary selection at other loci. The males of that same stock had maintained their per^{s} 40 s cycle over two decades suggesting that female preference was more evolutionarily flexible than the male character (Greenacre et al., 1993). Thus those females who had preferred 40 s song cycles (these cycles were the

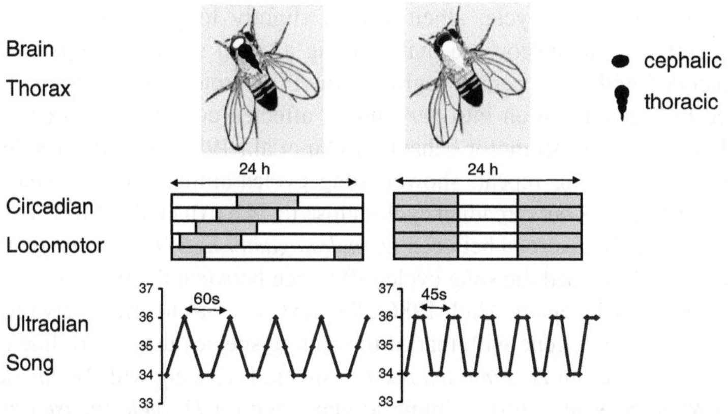

Fig. 7.2 Diagrammatic representation of results of Konopka et al. (1996). The *per*^{s} allele is semi-dominant in both circadian and song characters and shorter periods in both phenotypes can be distinguished from those of wild-type (Konopka and Benzer, 1971; Kyriacou and Hall, 1980). When the cephalic ganglion of a mosaic was heterozygous (white) but the thoracic nervous system was wild-type (black), circadian behaviour of that mosaic showed a short semi-dominant periodicity (~20–22 h), as shown in the actogram (5 days of activity are illustrated), but the song rhythm (Y-axis is IPI in ms, X-axis is time) gave a wild-type periodicity ~60 s. The reciprocal mosaic (brain wild-type-black, but thorax heterozygous-white), gave wild-type circadian behaviour but short ~40–45 s song cycle

only ones on offer in that stock) must have been at a selective advantage, in that they would have mated faster, and laid their progeny in the food medium first. Consequently it is not difficult to imagine that their song preference changed over many generations, providing a compelling example of microevolution.

A neurogenetic analysis of the lovesong rhythm was also informative using all-male mosaics, that carried both *per*$^+$ and *per*^{s}/*per*$^+$ tissue. As *per*^{s} has a dominant song phenotype as well as a semi-dominant circadian phenotype, it was possible to simultaneously score the circadian and song behaviour of each fly, before determining the internal distribution of *per*$^+$ and *per*^{s}/*per*$^+$ tissue using a benign enzyme marker. It was observed that circadian phenotype mapped to the brain, and song phenotype to the thoracic ganglion, revealing a developmental independence of the two oscillators (Fig. 7.2; Konopka et al., 1996). As nervous system PER expression in the thorax appears to be restricted to the glia (see Hall, 2003) it would seem that this cell type plays an interesting and unexpected role in the generation of song cycles.

7.3 Molecular Basis of Song Rhythms

The molecular basis of song rhythms was initiated with the basic cloning of *per* DNA fragments, and their transformation into *per*^{01} mutants (Hamblen et al., 1986; Zehring et al., 1984). The same DNA fragments that restored circadian rhythmicity also rescued the song cycle, albeit with a slightly longer than normal period. Subsequently, it was discovered that deleting a small stretch or repetitive amino acids encoded within the central region of the PER protein by *in vitro* mutageneis, followed by transformation into *per*^{01} hosts, affected courtship song cycles much more than circadian locomotor behaviour (Yu et al., 1987). In fact flies deleted of this Thr-Gly encoding repeat, showed song cycles similar to *D. simulans*, with periods of about 40 s but circadian cycles close to 24 h (Yu et al., 1987). This region of PER was highly diverged between *D. melanogaster* and *D. simulans*, and as previous work had mapped the song cycle difference between the two species to the *X* chromosome (Kyriacou and Hall, 1986), the next step was to effectively replace the *D. melanogaster per* gene with that of the sibling species using germline transformations. The resulting *D. simulans per* transgenic flies sang with *D. simulans* 40 s cycles (Wheeler et al., 1991). Chimeric genes with a *D. melanogaster per* gene whose Thr-Gly region was replaced with that of *D. simulans*, produced transgenic flies which sang with *D. simulans* characteristics. The converse, a *D. simulans per* with a *D. melanogaster* Thr-Gly, switched the transgenic flies songs back to *D. melanogaster* 60 s cycles (Wheeler et al., 1993).

This remarkable study had revealed that a small DNA fragment, which carried at most, 12 amino acid differences between the two species, was responsible for encoding a species-specific behaviour. This was something of a shock, especially to evolutionary biologists, who had come to believe that adaptive species-specific characters accumulated over thousands of generation by small changes in hundreds of genes (Coyne, 1992; Orr, 2005). Evidently this was not the case with *per* and song rhythms, nor was it with species-specific circadian locomotor or mating patterns, both of which were also similarly encoded by *per* (Petersen et al., 1988; Tauber et al., 2003). These experiments signalled a change in the view, held for many decades, on the 'infinitesimal' view of adaptations, originally championed by Fisher (1930).

7.4 Song Rhythms: Real or Statistical Artefacts?

At the time of the last chapter however, storm clouds had gathered over this song rhythm work, in that two studies had been published in 1988 that questioned the existence of these song cycles, suggesting that they were, in fact, statistical artefacts. It thus also followed that there could be no significant differences between the songs of *per* mutants (Crossley, 1988; Ewing, 1988). The key difference between the approaches of Kyriacou and Hall in their studies, and the critics, Crossley on one hand and Ewing on the other, was in the way they treated missing song data i.e. times when the male was silent, which is inherent in most courtships. Crossley and Ewing would necessarily estimate song data that was 'missing' to apply their 'off the shelf' spectral analysis, whereas Kyriacou and Hall had avoided this potential artefact, initially using regression analysis, then later applying spectral analysis where missing data was treated as exactly that, *missing*. An important turning point in the debate was when the song records of Crossley in which there were no missing data (i.e. her spectral analysis was appropriate for these) were inspected (Kyriacou and Hall, 1989). It turned out that seven out of nine of these records showed ~60 s spectral periods, just as expected, and the probability of this occurring by chance was tiny.

While Kyriacou and Hall had spent considerable effort in reporting their methods, to show how these experiments 'should' be done and analysed (Kyriacou and Hall, 1989; Kyriacou et al., 1990b), an independent group, more or less ignored all of these instructions, and went ahead in an attempt to resolve the debate using their own behavioural, song recording and statistical methods (Alt et al., 1998). It was thus a pleasant surprise that they came up with similar results as Kyriacou and Hall, namely that the four *per* alleles had different song cycles, suggesting that the song rhythm is more robust than originally believed (Alt et al., 1998). Taken together with the replications and extensions of the playback work by the independent group mentioned earlier (Ritchie et al., 1999), it would seem that the debate has been finally laid to rest.

Others have now discovered song rhythms in other Drosophila species. In a study of *D. pseudoobscura* and *D. persimilis* songs, a robust short-cycle 1.23 s rhythm was detected in *D. persimilis* songs (Noor and Aquadro, 1998). This rhythm was temperature independent, reminiscent of the IPI cycles of *D. melanogaster* (Kyriacou and Hall, 1980). An interesting observation was also made on the songs of the pest species, the melon fly, *Bactrocera cucurbitae*. The males have courtship songs with a very dense generation of pulses in each train, and with a regular interval between these trains (Miyatake and Kanmiya, 2004). Two circadian mutant strains were studied, one with a short 22 h, the other with a long 30 h free running circadian period. The pulse train intervals were significantly shorter for both young and old males in the 22 h strain compared to the 30 h strain (Miyatake and Kanmiya, 2004). These results again reveal a genetic connection between circadian clocks and courtship songs.

7.5 How Is the Song Cycle Generated?

The canonical clock gene, *period*, is a negative autoregulator of its own transcription, and generates 24 h cycles of its own mRNA and protein (Hall, 2003). One might therefore legitimately ask how such a dynamic phenomenon might relate to song cycles in the time domains of 30–80 s. Clearly the song rhythm cannot be transcriptional within this limited time frame. The answer to this conundrum probably lies in the observations made using microarrays with fly mRNA. Many genes show transcriptional cycles of 24 h in a wild-type genetic background (Ceriani et al., 2002; Claridge-Chang et al., 2001; Lin et al., 2002; McDonald and Rosbash, 2001; Ueda et al., 2002), and most of these disappear in a *per*^{01} mutant background (Claridge-Chang et al., 2001; Lin et al., 2002). However, what is interesting and relevant, is that several hundred other genes, that are not normally rhythmically expressed, are either up or downregulated in *per*^{01} mutants (Claridge-Chang et al.,

2001; Lin et al., 2002). These are downstream effectors of PER, and control many different physiological processes. Thus it could be that whatever transcription factor that is involved in the pathway to the song cycle, is either up or down regulated to different extents in the classic *per* mutants.

We would expect, based on gene dosage experiments with *per* and the song cycle, that per^s represents a hypermorphic type of mutation, because stacking up extra copies of per^+ gives shorter periods (Hamblen et al., 1986). Conversely, per^L would represent a hypomorphic allele. The null mutant, per^{01}, under this scheme, would have a super long song phenotype, yet it does not – half the flies are arrhythmic, the other half have super short 30 s cycles that are rather weak (Kyriacou and Hall, 1989; Kyriacou et al., 1990a). This latter result is consistent with short-period locomotor activity cycles observed in per^{01} mutants (Dowse et al., 1987; Dowse and Ringo, 1987) and experiments in which heterozygous rhythmic *per* mutant diplo-*X* females, that have been transformed to pseudomales using the *transformer* sex determination mutation, show a semi-dominant shortening effect on the song cycle in any combination with per^{01} (Kyriacou and Hall, 1980). Clearly, the route between PER as a transcriptional regulator of downstream song factors is complex. Yet we can surmise that whatever neuronal network generates the song cycle, and we presume it is neurogenic rather than myogenic based on experiments that reversibly shut down neuronal transmission (Kyriacou and Hall, 1985), mutations in *per* tweak this cycle in the right direction via some as yet undetermined downstream factors, probably expressed in thoracic glia (Konopka et al., 1996; Hall, 2003).

Recently, the role of *per*, and its partner molecule, *timeless* (*tim*), has been documented in a different role. The dorsolongitudinal muscles control the movements of the fly's wings and it has been observed that bouton size on the neuromuscular terminals has a circadian rhythm that is maintained in constant conditions and is disrupted by null mutations in the *per* and *tim* genes (Mehnert et al., 2007). In addition, neuronal branching at these terminals is non-rhythmic, but it is dramatically affected by tim^{01} null mutants, which show hyperbranching, and by per^{01} mutants that show a modest reduction in branching (Mehnert et al., 2007). Thus PER and TIM have opposite effects on this branching phenotype, with the former acting as a growth factor, and the latter as a repressor, suggesting that the balance between them is important. These are novel and interesting results, showing additional non-rhythmic phenotypes determined by *per* and *tim*. Yet they are difficult to understand if we consider that in tim^{01} mutants, very little PER remains because PER requires TIM for stability (Price et al., 1995). Thus in terms of PER, a fly that is tim^{01} or per^{01} gives a similar very low to non-existent level of PER product, so the levels of PER alone cannot explain the opposing phenotypes. Perhaps these effects may be mediated by different regulatory targets of PER and TIM, rather than simply be an indirect effect of the tim^{01} mutation on PER levels? Yet might there be functional implications of these results for song rhythms?

Many years ago, we noted that tim^{01} mutants which are circadian arrhythmic, are nevertheless robustly rhythmic in their song cycles, but their periods are superlong, much longer than per^L, and opposite to the short cycle per^{01} song cycles (H. Roe and C.P. Kyriacou, unpublished observations (1977)), providing a correlation with the branching phenotype observed in the two clock mutants (Mehnert et al., 2007). Could it be that this hyperbranching phenotype at the neuromuscular junction in *tim*^{01} mutants lengthens the period of the song cycle whereas the hypobranching in *per*^{01} shortens or obliterates the cycle? It's a tantalising thought. However, in contrast, one might also rationalise the song results as follows, again using the effect of *tim*^{01} on PER levels. The *tim*^{01} mutants are extremely hypomorphic for PER expression (Price et al., 1995) whereas *per*^{L} mutants are hypomorphs in a genetic context for both circadian and song rhythms (Hamblen et al., 1986; Smith and Konopka, 1982). Thus the longer song cycle of *tim*^{01} compared to *per*^{L} may reflect its more 'extreme' hypomorphic nature. However this latter hypothesis would suggest that the opposite neuronal branching phenotypes observed for *per*^{01} and *tim*^{01}, are simply incidental.

7.6 Summary

In conclusion, the next few years may see two lines of further enquiry into fly song rhythms. One might be evolutionary, with a description of song rhythms in other closely related Drosophilid species, particularly those of the *melanogaster* subgroup. Do all these species have song rhythms, or is it just limited to *D. melanogaster*, *simulans* and *yakuba*, the latter of which has a long 70 s cycle (Demetriades et al., 1999). If song rhythms are more wide ranging, might the periods show any consistent phylogenetic patterns? A second approach could be to use the Gal4/UAS binary misexpression system (Brand and Perrimon, 1993) in order to probe the underlying neurogenetic basis for the song cycle. This could be done in a number of ways, for example, an immediate experiment that suggests itself might be to drive PER expression in the thorax via a glial promoter to see whether such restricted expression restored song rhythmicity to *per*^{01} males. These types of experiments would possibly represent the best way forward for attempting to understand the elusive biological nature of the fly song rhythm.

Acknowledgements CPK thanks the Royal Society for a Wolfson Research Merit Award.

References

Akhtar, R.A., Reddy, A.B., Maywood, E.S., Clayton, J.D., King, V.M., Smith, A.G., Gant, T.W., Hastings, M.H. and Kyriacou, C.P. 2002. Circadian cycling of the mouse liver transcriptome, as revealed by cDNA microarray, is driven by the suprachiasmatic nucleus. Curr Biol 12: 540–550.

Alt, S., Ringo, J., Talyn, B., Bray, W. and Dowse, H. 1998. The *period* gene controls courtship song cycles in *Drosophila melanogaster*. Anim Behav 56: 87–97.

Billeter, J.C., Rideout, E.J., Dornan, A.J. and Goodwin, S.F. 2006. Control of male sexual behavior in drosophila by the sex determination pathway. Curr Biol 16: R766–R776.

Brand, A.H. and Perrimon, N. 1993. Targeted gene-expression as a means of altering cell fates and generating dominant phenotypes. Development 118: 401–415.

Ceriani, M.F., Hogenesch, J.B., Yanovsky, M., Panda, S., Straume, M. and Kay, S.A. 2002. Genome-wide expression analysis in Drosophila reveals genes controlling circadian behavior. J Neurosci 22: 9305–9319.

Claridge-Chang, A., Wijnen, H., Naef, F., Boothroyd, C., Rajewsky, N. and Young, M.W. 2001. Circadian regulation of gene expression systems in the Drosophila head. Neuron 32: 657–671.

Coyne, J.A. 1992. Genetics and speciation. Nature 355: 511–515.

Crossley, S.A. 1988. Failure to confirm rhythms in Drosophila courtship song. Anim Behav 36: 1098–1109.

Demetriades, M.C., Thackeray, J.R. and Kyriacou, C.P. 1999. Courtship song rhythms in *Drosophila yakuba*. Anim Behav 57: 379–386.

Dowse, H.B. and Ringo, J.M. 1987. Further evidence that the circadian clock in Drosophila is a population of coupled ultradian oscillators. J Biol Rhythms 2: 65–76.

Dowse, H.B., Hall, J.C. and Ringo, J.M. 1987. Circadian and ultradian rhythms in *period* mutants of *Drosophila melanogaster*. Behav Genet 17: 19–35.

Ewing, A.W. 1988. Cycles in the courtship song of male *Drosophila melanogaster* have not been detected. Anim Behav 26: 1091–1097.

Fisher, R.A. 1930. The genetical theory of natural selection. Oxford University Press, Oxford.

Greenacre, M.L. 1990. Genetic analysis of courtship behaviour and biological rhythms in Drosophila. Ph.D. thesis, University of Leicester.

Greenacre, M.L., Ritchie, M.G., Byrne, B.C. and Kyriacou, C.P. 1993. Female song preference and the *period* gene in *Drosophila*. Behav Genet 23: 85–90.

Griffith, L.C., Verselis, L.M., Aitken, K.M., Kyriacou, C.P., Danho, W. and Greenspan, R.J. 1993. Inhibition of calcium/calmodulin-dependent protein kinase in Drosophila disrupts behavioral plasticity. Neuron 10: 501–509.

Hall, J.C. 2003. Genetics and molecular biology of rhythms in Drosophila and other insects. Adv Genet 48: 1–280.

Hamblen, M., Zehring, W.A., Kyriacou, C.P., Reddy, P., Yu, Q., Wheeler, D.A., Zwiebel, L.J., Konopka, R.J., Rosbash, M. and Hall, J.C. 1986. Germ-line transformation involving DNA from the *period* locus in *Drosophila melanogaster*: Overlapping genomic fragments that restore circadian and ultradian rhythmicity to per^0 and per^- mutants. J Neurogenet 3: 249–291.

Hunt, T. and Sassone-Corsi, P. 2007. Riding tandem: Circadian clocks and the cell cycle. Cell 129: 461–464.

Konopka, R.J. and Benzer, S. 1971. Clock mutants of *Drosophila melanogaster*. Proc Natl Acad Sci USA 68: 2112–2116.

Konopka, R.J., Kyriacou, C.P. and Hall, J.C. 1996. Mosaic analysis in the Drosophila CNS of circadian and courtship-song rhythms affected by a *period* clock mutation. J Neurogenet 11: 117–139.

Kyriacou, C.P. and Hall, J.C. 1980. Circadian rhythm mutations in *Drosophila melanogaster* affect short-term fluctuations in the male's courtship song. Proc Natl Acad Sci USA 77: 6729–6733.

Kyriacou, C.P. and Hall, J.C. 1982. The function of courtship song rhythms in Drosophila. Anim Behav 30: 794–801.

Kyriacou, C.P. and Hall, J.C. 1984. Learning and memory mutations impair acoustic priming of mating behaviour in Drosophila. Nature 308: 62–65.

Kyriacou, C.P. and Hall, J.C. 1985. Action potential mutations stop a biological clock in Drosophila. Nature 314: 171–173.

Kyriacou, C.P. and Hall, J.C. 1986. Interspecific genetic control of courtship song production and reception in Drosophila. Science 232: 494–497.

Kyriacou, C.P. and Hall, J.C. 1989. Spectral analysis of Drosophila courtship song rhythms. Anim Behav 37: 850–859.

Kyriacou, C.P., Oldroyd, M., Wood, J., Sharp, M. and Hill, M. 1990a. Clock mutations alter developmental timing in Drosophila. Heredity 64: 395–401.

Kyriacou, C.P., van den Berg, M.J. and Hall, J.C. 1990b. Drosophila courtship song cycles in normal and *period* mutant males revisited. Behav Genet 20: 617–644.

Kyriacou, C.P., Ritchie, M.G. and Greenacre, M.L. 1992. Genetics of sexual selection in Drosophila. Amer Zool 32: 31–39.

Lin, Y., Han, M., Shimada, B., Wang, L., Gibler, T.M., Amarakone, A., Awad, T.A., Stormo, G.D., Van Gelder, R.N. and Taghert, P.H. 2002. Influence of the period-dependent circadian clock on diurnal, circadian, and aperiodic gene expression in *Drosophila melanogaster*. Proc Natl Acad Sci USA 99: 9562–9567.

Loudon, A.S., Wayne, N.L., Krieg, R., Iranmanesh, A., Veldhuis, J.D. and Menaker, M. 1994. Ultradian endocrine rhythms are altered by a circadian mutation in the Syrian hamster. Endocrinology 135: 712–718.

McDonald, M.J. and Rosbash, M. 2001. Microarray analysis and organization of circadian gene expression in Drosophila. Cell 107: 567–578.

Mehnert, K.I., Beramendi, A., Elghazali, F., Negro, P., Kyriacou, C.P. and Cantera, R. 2007. Circadian changes in Drosophila motor terminals. Dev Neurobiol 67: 415–421.

Miyatake, T. and Kanmiya, K. 2004. Male courtship song in circadian rhythm mutants of *Bactrocera cucurbitae* (tephritidae: Diptera). J Insect Physiol 50: 85–91.

Noor, M.A.F. and Aquadro, C.F. 1998. Courtship songs of *Drosophila pseudoobscura* and *D. persimilis*: Analysis of variation. Anim Behav 56: 115–125.

Orr, H.A. 2005. The genetical theory of adaptation: A brief history. Nat Rev Genet 6: 119–127.

Paranjpe, D.A., Anitha, D., Chandrashekaran, M.K., Joshi, A. and Sharma, V.K. 2005. Possible role of eclosion rhythm in mediating the effects of light-dark environments on pre-adult development in *Drosophila melanogaster*. BMC Dev Biol 5: 5.

Petersen, G., Hall, J.C. and Rosbash, M. 1988. The *period* gene of Drosophila carries species-specific behavioral instructions. EMBO J 7: 3939–3947.

Price, J.L., Dembinska, M.E., Young, M.W. and Rosbash, M. 1995. Suppression of PERIOD protein abundance and circadian cycling by the Drosophila clock mutation *timeless*. EMBO J 14: 4044–4049.

Ritchie, M.G., Halsey, E.J. and Gleason, J.M. 1999. Drosophila song as a species-specific mating signal and the behavioural importance of Kyriacou & Hall cycles in *D. melanogaster* song. Anim Behav 58: 649–657.

Smith, R.F. and Konopka, R. 1982. Effects of dosage alterations at the *per* locus on the circadian clock of Drosophila. Mol Gen Genet 183: 30–36.

Tauber, E., Roe, H., Costa, R., Hennessy, J.M. and Kyriacou, C.P. 2003. Temporal mating isolation driven by a behavioral gene in Drosophila. Curr Biol 13: 140–145.

Ueda, H.R., Matsumoto, A., Kawamura, M., Iino, M., Tanimura, T. and Hashimoto, S. 2002. Genome-wide transcriptional orchestration of circadian rhythms in *Drosophila*. J Biol Chem 277: 14048–14052.

Wheeler, D.A., Kyriacou, C.P., Greenacre, M.L., Yu, Q., Rutila, J.E., Rosbash, M. and Hall, J.C. 1991. Molecular transfer of a species-specific behavior from *Drosophila simulans* to *Drosophila melanogaster*. Science 251: 1082–1085.

Wheeler, D.A., Hamblen-Coyle, M.J., Dushay, M.S. and Hall, J.C. 1993. Behavior in light-dark cycles of Drosophila mutants that are arrhythmic, blind, or both. J Biol Rhythms 8: 67–94.

Yu, Q., Colot, H.V., Kyriacou, C.P., Hall, J.C. and Rosbash, M. 1987. Behaviour modification by *in vitro* mutagenesis of a variable region within the *period* gene of *Drosophila*. Nature 326: 765–769.

Zehring, W.A., Wheeler, D.A., Reddy, P., Konopka, R.J., Kyriacou, C.P., Rosbash, M. and Hall, J.C. 1984. P-element transformation with *period* locus DNA restores rhythmicity to mutant, arrhythmic *Drosophila melanogaster*. Cell 39: 369–376.

Chapter 8
Mid-range Ultradian Rhythms in *Drosophila* and the Circadian Clock Problem

H.B. Dowse

Abstract The current molecular model of the circadian timer presumes a 24-h escapement based on a transcription/translation oscillator (TTO) as the ultimate frequency source. The output of a very accurate mammalian circadian clock functioning both in free-run and LD-entrained mode is here analyzed using clock-theory methodology. The results are used to ask whether this postulated molecular oscillator is consistent in theory with the precision observed. The outcome suggests that the TTO is found wanting in this regard and a number of reasons why this may be so are discussed. In contrast, mid-range ultradian oscillations ($\tau \sim 1$-18 h) provide likely evidence for even faster oscillations, and these very high frequency ultradian oscillations in turn offer an alternative to the TTO as the biological clock's escapement. The well-studied TTO would act as the "integrator" of this inherently more precise timing information translating it into a relevant time frame. Two-part timing systems such as this are universal in all man-made clocks.

Keywords Circadian, clock, *Drosophila*, timekeeper theory, ultradian

Rhythmicity in living organisms has been known for millennia. It was probably first noticed in plants because it was remarkable that something lacking a nervous system could have a "sense" of the time of day. The phenomenon came under scientific scrutiny in the 18th century (e.g. DeMairan, 1729). None other than Charles Darwin (Darwin and Darwin, 1880) took a close look. But careful study with modern equipment awaited the rapid advances of physiology in the 20th century, with the early work of Bünning, among others, laying the foundation (review: Bünning, 1964). It was clear that in some way, organisms could tell time accurately and some living horologe was hypothesized to underlie the observed phenomena. The work in these halcyon days was largely delineating the "formal properties" of the rhythms. The mechanisms remained black boxes. It was hoped

School of Biology and Ecology, 5751 Murray Hall, University of Maine, Orono, ME 04469-5751
E-mail: dowse@maine.edu

that the overt behavior of the black boxes would yield clues about their internal workings. In the process, a massive amount of information on rhythms and their behaviors was compiled from a wide variety of organisms under multiple experimental paradigms. Temperature effects, entrainment behavior, phasic and tonic effects of light and other zeitgebers, response or lack thereof to many toxins and drugs, all were carefully recorded. The landmark 1960 Cold Spring Harbor Symposium summarized the bulk of this classical work and remains a difficult-to-obtain classic worth owning (Chovnik, 1960).

It was not until the 1960s that concerted efforts to figure out the clockwork itself began in earnest (review: Edmunds, 1988). A series of hypothetical clock models systematically incorporated breakthroughs in cell and molecular biology as they appeared. Among these, famously, was the chronon (Ehret and Trucco, 1967), which postulated a cyclical chain of sequential transcriptional and translational events. Not long after the fluid mosaic model of the membrane was established, an escapement model based on membrane diffusion was proposed (Njus et al., 1974). Oscillations in proton gradients on mitochondria were proposed (Chay, 1981). There has been no lack of proposals for 24-h escapements.

8.1 The Molecular Clock Dogma

In the late 1960s, Ronald Konopka decided to take a forward genetic approach to the problem using the most logical organism to work with at the time, or perhaps any time, namely *Drosophila melanogaster*. The fly's contribution to biology across the board has been and remains titanic and the molecular and genetic tools available still lead the pack (see, e.g. Hall, 2003a). Konopka's breakthrough discovery was the *period* (*per*) gene (Konopka and Benzer, 1971). Here was one gene with three alleles in hand that yielded short (per^S), long (per^L) periods, and nulls (per^0 and per^-) that were arrhythmic. By some criteria, this was formal evidence that a "true gear" of the clock had been found. The logic is, if you can change period or eliminate the timer by changing a single protein, then it must, perforce, be a part of an oscillator with a 24-h period (Dunlap, 1999).

Starting with this toehold, the rest of the molecular machinery that now comprises the almost universally accepted theory of biological timekeeping was revealed. These have been reviewed extensively elsewhere on multiple occasions, and we shall not do so here other than to set forth the basics of the mechanism sufficient to our purpose (review: Dunlap, 1999). Essentially, there is a subset of "clock genes" that produce proteins through transcription and translation. This is the "positive" part of the loop. These clock proteins then interact in various ways ultimately feeding back negatively on their own transcription in concert with other clock gene products. The result is reduced titers of the initial proteins. The cycle continues as the negative elements decline in the system and the upswing begins again. This is the core Transcription-Translation Oscillator (TTO) (Dunlap, 1999). The specific genes and proteins and their roles vary across organisms, but the principle remains the same. In the fly, the *per* and *timeless* genes are transcribed in the positive phase, and the PER and TIM proteins initiate negative feedback (review: Hall, 2003b).

When the molecular and genetic labs grabbed the *per* findings and ran, the imposing body of information compiled by the early workers, most of them careful, innovative individuals, seems to have been largely forgotten. But there always comes a reckoning. This chapter will start by revisiting an old result and setting it up for careful analysis as a "gold standard" clock. We shall apply these numbers to the current generic molecular TTO clock model to critique its performance. We shall then look at an alternative model and the concrete evidence for it found in fly rhythms.

8.2 Lessons from the Heirs of William Harrison

To analyze our test clock's performance we will employ the standard statistical techniques of the science of timekeeping (Allan et al., 1997). Here, of course, we are dealing with an inherently *inaccurate* clock. It is *circa*dian, so it does not measure the length of a day particularly well. But a given circadian clock can still be very *precise*, in that it may maintain its own innate period well, and by relaxing the requirement that it measure a day accurately, we can compare it statistically to other clocks.

Almost all known clocks of human devising consist of two distinct parts (Allan et al., 1997). The first is the time source oscillator or *frequency standard*. In times long past, this consisted of observation of celestial processes such as the earth's rotation measured against the sun, moon, or named stars. This progressed to pendulum clocks with energy sources and escapement mechanisms to release that energy in regulated aliquots. Continued need for ever more accurate clocks led to employing piezoelectric quartz crystal vibrations and thence to atomic clocks using the so-called fine structure of quantum absorption and emission lines. The Cesium atomic clock is an example and provides the current standard for the UTC second used in the GPS time distribution network and in almost all current time standard laboratories. The second part of a clock is a *counter* or "*integrator*" that turns the output of the fundamental oscillator into useful time information. The relatively high frequency periodicity of the pendulum, e.g., was connected to a mechanism that converted the input oscillation into useful time measurement. Both parts of the mechanism will have inherent inaccuracies that arise from causes unique to their operating mechanics (review: Allan et al., 1997).

To test maximum performance, one needs a benchmark. If one specific version of a system achieves a given performance benchmark, then it must be accepted that the system and its relatives are, in general, capable of at least similar performance even if the standard is not always reached in all individual cases. We will take an extremely well-behaved and precise circadian timer as our benchmark. In the above mentioned 1960 Cold Spring Harbor Symposium volume (Chovnik,

1960), records were published for the flying squirrel, *Glaucomys volans* by DeCoursey (1960). Figure 2 of that paper, in particular, deserves attention. The animal was put in DD at 20° and its wheel running activity was recorded for 25 days. Feeding was done as needed and unobtrusively in complete darkness. The precision of the periodicity is striking, and this is not atypical of the performance of which mammals are capable.

Figure 2 from the DeCoursey (1960) paper was scanned (HP Scanjet 4850®) and the result was rendered as a jpg image (Fig. 8.1). To extract information on timing, a single event in the animal's phase was chosen, namely the onset of running wheel activity. This was done by working with the scanned image in Microsoft Photo Editor® which provides the position of the cursor in pixels. The 24-h scale in the scanned image was 529 pixels in length. The onset of activity during each objective 24-h day could thus be estimated to the limit of 2 min 42 s, the width of a single pixel. Pixel number was converted to hours and fractions thereof of the 24-h day (ordinate) with hour of the experiment as the abscissa. These data were subjected to regression analysis (Statistical Analysis System, GLM procedure). As might be expected, the linear fit is excellent: P = 0.0001, P = 0.0001, $r^2 = 0.99945$. The intercept is 14.49 ± 0.028 h (SEM) and the slope is −0.017 ± 0.00008337 (SEM). From this regression line, we estimate the period (τ) of the rhythm, at 23.586 h in the standard manner.

Normally, analysis of timekeeping accuracy is done by determining how well a clock estimates the length of the standard second over some interval. This is an

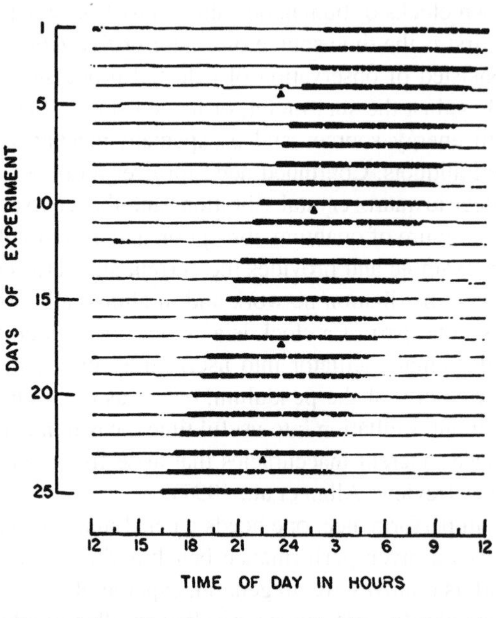

Fig. 8.1 Twenty-five days of data taken by event recorder for wheel running activity of a flying squirrel, *Glaucomys volans*, housed at 20° in constant darkness. (From DeCoursey, 1960.) Hours of onset were digitized for analysis as described in text

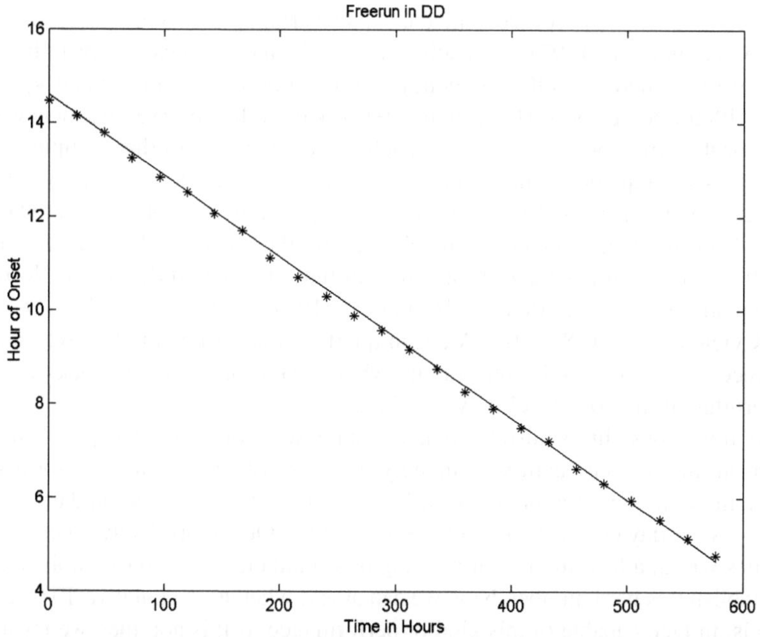

Fig. 8.2 Times of onset of activity taken from Fig. 8.1 are represented by "*". The solid line is the regression of hour of onset of activity on time (in hours)

interval established by the oscillation of a Cs atom (Quinn, 1991). Essential measurements are frequency accuracy and stability and time accuracy and stability. We shall make only one assumption to work the squirrel data into this format, namely that the animal's clock is attempting to measure a standard second by counting up 84,909.6 s per circadian cycle, i.e. from activity onset to the next onset. This number comes from the period calculated from the regression line as the actual period of the cycle, but given that the linear model is a near perfect fit, this is warranted. Initially, for clarity and continuity with the actogram plot and common practice in the literature, hours will be used as the fundamental time interval instead of the second.

We first look at the error by plotting the difference between the points calculated from the regression equation and each day's actual onset. This yields Fig. 8.2. Analyzing these errors, we get $\sigma^2 = 0.2292$ h and $\sigma = 0.0957$ h. So our squirrel can measure 23.586 h, or 84,909.6 s to ± 0.0957 h or ± 344.5 s, nearly within the limits imposed by the pixel size in the scanned image. This further translates to measuring an hour to within ± 14.3 s and the standard UTC second to $\pm 3.99 \times 10^{-3}$s, or roughly 4 ms. Here is the biological clock's benchmark.

There are more mathematically sophisticated ways to analyze these data which we will employ for completeness and to allow comparison with other timekeepers. Standard measures of variance are not well behaved when applied to timers. They tend to increase without bound as the length of the sample interval is extended and a better statistic was worked out, the Allan variance σ_τ^2 (Sullivan et al., 1990; review: Allan et al., 1997). For each test, we take the frequency error to be y(t) = $(\nu_1 - \nu_0)/\nu_0$, where ν_0 is the frequency of the calibrating oscillator and ν_1 that of the oscillator being tested Howe et al., 1981). We set the observed frequency as the reciprocal of the observed period of each cycle compared to that computed from the regression equation. The frequency errors for each cycle $y_1, y_2, \ldots y_N$ are differenced to produce N – 1 readings: $y_1 = e_2 - e_1$, and the sum of the squared differences from the mean is calculated followed by the mean/2. This statistic clearly has the form of variance and its square root that of standard deviation. Normally, this square root σ_τ is reported out (Howe et al., 1981). Doing this for the *Glaucomys* clock yields $\sigma_\tau = 7.118 \times 10^{-4}$. A cheap quartz crystal watch might have an Allan variance of 10^{-7} over this time range, while a Hydrogen-maser clock would be performing at the 10^{-14} level (Clynch, 2002).

So how does this squirrel clock compare with other timekeepers? William Harrison, in the 18th century, ultimately won a substantial cash prize for a ship's chronometer to measure longitude sufficient to meet the Crown's standard of accuracy – 3 s per day (Allan et al., 1997; Sobel, 1995). Our squirrel's living clock actually lies within a few minutes of meeting this standard and we have published data to establish this hard number. Now we can assess whether the putative TTO mechanism is, in fact, capable of this kind of performance. If it is not, then we must, perforce, revisit the fundamental question of what the timer actually IS. We continue our analysis.

8.3 Sources of Accuracy and Inaccuracy in TT Oscillators

We will consider the biological clock in light of this fundamental two part clock generalization. Whatever the fundamental frequency of the underlying oscillator is, it must be coupled to biologically meaningful output. The instant of onset of running is one phase point commanded by the output system, but is not necessarily directly connected to the frequency standard. The animal is prompted to enter the wheel by some signal from the CNS, then begins running. This must be a fairly noisy sequential chain of events in and of itself. Part of the error in free run is from the oscillator's inherent inaccuracy (imprecision, strictly speaking), part from the counter or coupler and now we must add an error from the neural mechanism coupling the clock to actual behavior.

There is a way to partition the error if sufficient data are in hand. On the same page of DeCoursey's (1960) article, there is a record of another squirrel entrained to an LD cycle (her Fig. 4). Here, we have a clock that is both precise *and* accurate. The output period of the clock/integrator/coupler is almost exactly 24 h judging by the running onset phase point, and the frequency standard timing information from the mechanical timing clock can be considered perfect in relative terms. If we look at the error of the entrained clock in this animal, we can estimate how much variance is a result of the "counting/behavioral output" half of the clock. This is the part

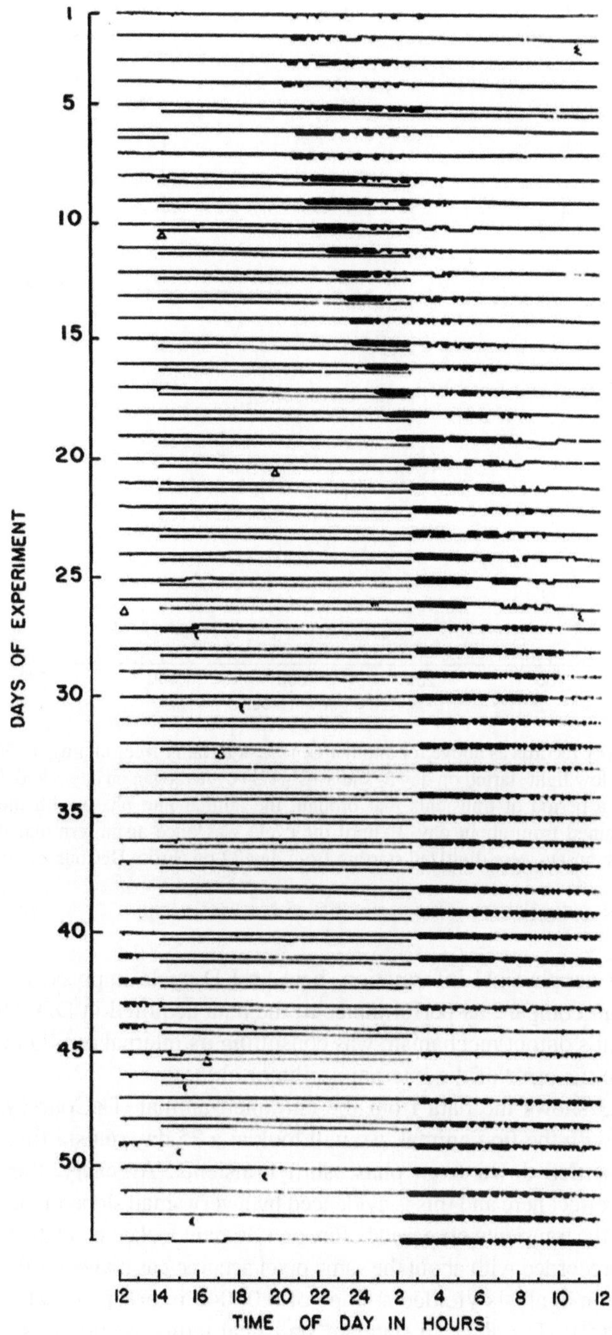

Fig. 8.3 Fifty-three days of running wheel activity from a flying squirrel. Light-on during LD phase indicated by solid lines. (From DeCoursey 1960, Fig. 4.)

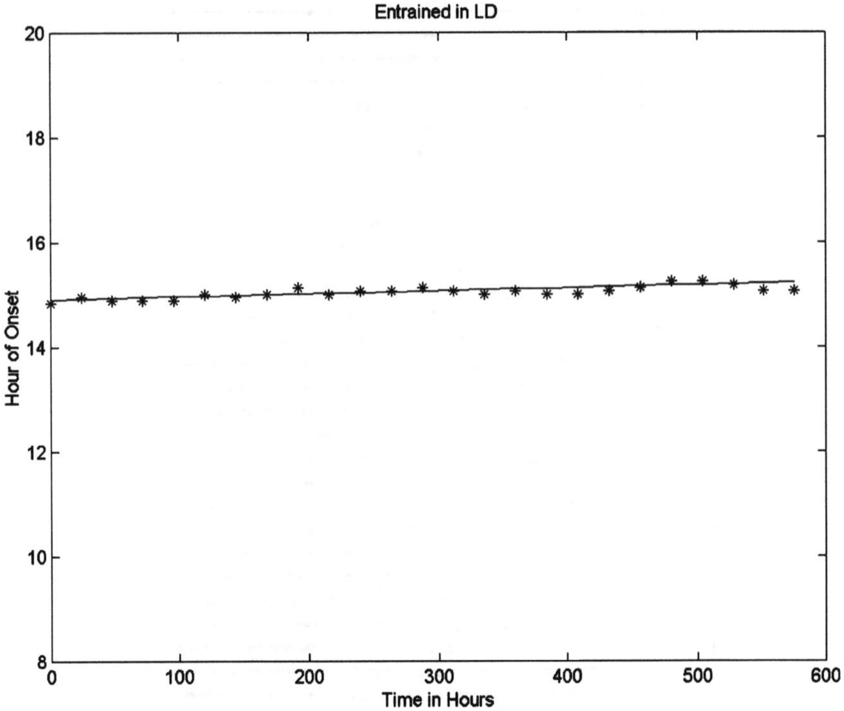

Fig. 8.4 Activity records as above for another *G. volans* initially free-running in DD through day 5. Exposure to low light started on day 5, and a 12L:12D cycle began on day 8 and continued until day 48. After a period of transients that brought the animal into phase with the LD cycle, it remained entrained from about day 21 until the cycle was once again terminated. Twenty-five days of activity onsets were digitized starting from day 22 on (From DeCoursey, 1960)

that is using geophysical information, here an LD cycle imposed by an accurate clock. We can compare its performance to the data acquired in DD when that particular animal's output mechanism was consulting its internal oscillator. The difference must be the error of the internal oscillator alone.

Figure 8.3 shows the data from the entrained animal (DeCoursey, 1960). For consistency with the first animal, we will look at a 25 day run starting 3 days after the animal settled down from phase shift transients. Arguably, there may be a residual aftereffect here and this is evidenced by a very small slope in the line with the same sign as the transients observed before entrainment locked in. Data were similarly scanned and recorded with about the same pixel accuracy as above for the DD animal. The regression analysis yielded a slope of 0.00046 intercept 14.91 h. $\tau = 24.01$ h, with $p = 0.0001$, (Fig. 8.4). In standard statistical terms, accuracy is ±2.57 min or 154 s where we are comparing NOT to the animal's internal clock, which is not putting out exactly a 24 h period, but to the zeitgeber period of exactly 24.0 h as this is what it is ultimately consulting. Allan statistic $\sigma_\tau = 8.98 \times 10^{-5}$, or about an order of magnitude smaller than the clock in free run in DD. Simply subtracting this "integrator" standard deviation from the value for the animal in DD yields ± 190 s, or less than 2 ms error in measuring a standard second.

8.4 The TTO Objectively Considered as a Frequency Standard

In the case of our squirrel's putative TTO-based clock, three *per* genes are transcribed non-synchronously in response to a positive signal from a heterodimer, CLOCK::BMAL, consisting of products of the *clock* and *Bmal* genes. These form a complex in the cytoplasm, potentially with other players, thence returning to the nucleus where they operate negatively on CLOCK::BMAL to shut down transcription of the *per* genes. The PER complex diminishes in the cytoplasm, likely owing to phosphorylation, and the negative feedback ceases as the complex subsequently diminishes in the nucleus, freeing CLOCK::BMAL to ramp up transcription on the three *per* genes once again to renew the cycle (review: Dunlap, 1999). We ask if such a sequence of events, with broad peaks of abundance of the transcripts and proteins on the order of 8–12 h, is capable of the kind of precision we see in the clock we have just analyzed. The onset of running MUST be timed by the onboard clock with a known error that is very small compared to the long events of the oscillation.

The TTO family of oscillators has been modeled extensively, and mathematical models can be helpful in addressing the precision issue, but there are problems. It has been argued (Endy and Brent, 2001) that modeling of cellular processes is problematic in that few examples have led to predictions of future behaviour. The initial *Drosophila* clock model of Goldbeter (1995) is an example worth looking at in this light. It is deterministic with free parameters that must be chosen carefully and arbitrarily. It has only a narrow range of parameter values that can yield a solution that oscillates, although these oscillations are robust. Beyond this window lie damped oscillations that degenerate to a steady state. Discovery of further molecular components rendered it moot rapidly, but it served as a basis for further work.

Importantly, this was a deterministic model, i.e. based on differential equations solved numerically (Goldbeter, 1995). The problem is, the crucial reactions in the TTO clock scheme presented above are stochastic, and subject to random error. CLOCK::BMAL must find its binding sites. The rate of this depends critically on the number of these heterodimers in the nucleus (Riggs et al., 1970; Berg et al., 1981). Similarly, to create the complex necessary for transport out of the nucleus through the nuclear pore, stochastic binding events to transport proteins must occur (Alberts et al., 2002). In the cytosol, the PER complex must form as a result of diffusion and random enocunters. And so on at every step. With this in mind, a stochastic model was developed that took these concerns into account (Barkai and Leibler, 2001). Using Monte Carlo simulations, a very simple clock model was tested. Initially, their model produced widely fluctuating periods and amplitudes. However, with refinement, this problem was cleared up. Still, the best it could do was to produce periods with errors somewhat less than 10%, i.e. 2.4 h (Barkai and Leibler, 2001). This is far from the roughly 2 min the squirrel can produce. A further deterministic model for the mammalian clock was developed that took later molecular information into account (Leloup et al., 2003), as well as a stochastic version based on a simplified model (Gonze et al., 2002). In this latter instance, the period produced was ±2.8 h. Still far off the mark.

Starting with the known precision of the exemplary clock above, and given the unknown, but arguably moderate number of steps in any conceivable real-world TTO, the error term for each step would have to be tiny once it is divided up, certainly on the order of just a few seconds to yield 2–3 min errors at the end of each cycle. This is especially daunting when the resulting time information must be integrated to put the squirrel in its running wheel at just the right time, a process with its own inherent errors to be added in.

8.5 An Ultradian Frequency Standard for the Circadian Clock

Is this the only way a cell could tell time, the only model left standing? A family of competing hypotheses holds that the observable 24-h TTO molecular system is NOT the frequency standard, rather that it comprises the other half of the clock, the counter or integrator/output side that couples whatever the actual oscillator is to the organism's physiology. The salient assumptions are that much higher-frequency ultradian cellular oscillators provide the frequency standard, and that there are multiple oscillators in a single cell acting in ensemble. The oscillators are independent of the systems being timed to prevent inaccuracies owing to potentially chaotic feedback. The reason for hypothesizing this scenario is NOT that it draws on human-engineered timepieces for inspiration, rather the opposite. From an evolutionary perspective, nature tends to evolve the most effective configurations by natural selection. Later on, engineers find that nature has beaten them to the punch. A case in point: the clock system that programs events during the cell cycle provides some excellent examples of gating logic that are used in digital computers (Alberts et al., 2002). This is a clear case of convergent evolution. No human-designed clock has had a 24-h frequency standard since sundials and astronomical occultation tables were abandoned (Allan et al., 1997).

The argument put forward, and touted as ironclad, that *per* must encode or actually be a gear, is that the allelic series for the gene shows short, long, and null periodicities in the circadian range as described in more detail below (Dunlap, 1999). By analogy, by changing the length of or destroying a pendulum in a mechanical clock, one would realize similar outcomes. But consider the following: The input to a common electric wall clock is the sinusoidal 60 Hz line voltage. This is the frequency standard. The second half of this two-part clock is a geared mechanism that transforms this high frequency input to seconds, minutes and hours. If you change the diameter or destroy a single gear of this integrator mechanism, you will also change the output predictably. But you will *not* have changed the frequency standard at all.

An ultradian oscillator as frequency standard has several very attractive components. First, the most accurate clocks in current use have the highest frequency oscillators as their base (Itano and Ramsey, 1993). For a celestial clock, the earth's rotation is a fair 24-h time base, but with tides, etc. not as good as one might think. Certainly it is far off the money compared to modern atomic clocks (Allan et al., 1997). Second, given that even the most accurate clocks have inherent errors, the published time standard is not based on a single unit sitting somewhere, but on the combined output of an ensemble of clocks. All official timing centers use this paradigm (Allan et al., 1997).

The statistics of why it works is straightforward. Consider an oscillator operating independently. Each cycle will have an error that is independent of any previous error, but the errors will still be cumulative. If our oscillator has a deviation of 1 ms in 1 s (based on calculating one standard deviation unit from the data), then after 100 s, it would be off by $1 \times \sqrt{100}$, or 10 ms. But if we have, for example, 100 of these oscillators operating in an ensemble, then we take $10/\sqrt{100}$, or only 1 ms (worked example: Allan et al., 1997). By using an ensemble of high frequency oscillators within the cell itself, the pacemaker could easily achieve the high standard set by our squirrel's clock. This has been demonstrated at the macroscopic level. When individual neurons from the mammal's master clock in the Suprachiasmatic Nucleus (SCN) are tested individually, they produce inaccurate rhythms. As an ensemble, however, the accuracy improves markedly (Herzog et al., 2004).

Ultradian rhythms are nearly as ubiquitous as circadian rhythms. Here, we define ultradian periods as those between about 1 and 18 h and label them mid-range to distinguish them from very high frequencies: faster oscillations reflect metabolic processes (Lloyd and Edwards, 1984; Edmunds, 1988), and 19 h is approximately the mean period shown by the circadian clock mutant *per*^{S} in *Drosophila* (Konopka and Benzer, 1971; Hall and Kyriacou, 1990). Ultradian oscillations have been found in organisms as disparate as unicells (Lloyd et al., 1982; Michel and Hardeland, 1985) and mammals (Daan and Aschoff, 1981). They occur in both the presence and the absence of normal circadian rhythms, and are found also in animals that lack circadian rhythms but do exhibit tidal or lunar periodicity (Dowse and Palmer, 1990, 1992).

8.6 Ultradian Rhythms in *Drosophila*

We now turn to ultradian rhythms in the fly, and focus on three fundamental questions. First, are ultradian rhythms evidence of underlying oscillations or are they simply epiphenomena owing their existence to interactions among circadian clocks or "random noise"? Second, what is the function of these high frequency clocks if any? Third, is there any functional relationship between circadian and ultradian rhythms? To address these questions, we review experiments, primarily those using *Drosophila* as a model system.

Ultradian rhythms in *Drosophila* locomotor activity were first reported by Dowse et al. (1987) in mutant animals lacking a functional allele of the *period (per)* gene. Mutations of this sex-linked gene affect rhythmicity in *Drosophila* in striking ways (Konopka and Benzer, 1971; Hall and Kyriacou, 1990). *per^{Long}* *(per^L)* slows the clock to a period of about 29 h, while *per^{Short}* *(per^S)* accelerates the clock to a period of about 19 h. The non-functional *(null)* allele mentioned above, *per^0*, along with heterozygous overlapping deletions that eliminate the locus entirely (here *per^-*) (Smith and Konopka, 1981; Bargiello and Young, 1984; Reddy et al., 1984) also eliminate normal circadian rhythms, while revealing a wealth of higher-frequency periodicities (Dowse et al., 1987).

When per^o or per^- flies are raised in a light-dark (LD) cycle and their movement is recorded in constant darkness (DD), there is no consistent organization of the activity into clear bouts (Konopka and Benzer, 1971; Smith and Konopka, 1981). This contrasts with wild-type, per^+, flies, which exhibit modulation of the amplitude of activity into obvious cycles of about 24 h (Konopka and Benzer, 1971). Simple inspection of event recorder data, and the use of grossly insensitive digital signal analysis techniques such as the "periodogram" (Enright, 1965, 1990) are usually inadequate to reveal the underlying patterns (see Dowse and Ringo, 1991), however these rhythms are actually sufficiently robust that even this analytical technique does reveal ultradian periodicities in about 30% of per^{01} flies tested (Tomioka et al., 1998).

We (Dowse et al., 1987) used a sensitive spectral analysis technique, MESA (Maximum Entropy Spectral Analysis; Dowse and Ringo, 1989a, 1991; Levine et al., 2002), coupled with a robust test for significance, the correlogram (Chatfield, 1989; Dowse and Ringo, 1989a), and discovered that more than half of the per^o and per^- animals tested had clear-cut ultradian rhythms in the range 4–18 h. Usually, the spectral analysis revealed more than one periodicity. The periods of these rhythms were seldom simple multiples (e.g. 10.0 and 17.5 h), and thus could not have arisen from harmonics of some longer period. The presence of ultradian rhythms is not limited to per^0 and per^- mutants. We found that per^L flies also displayed strong ultradian rhythms in spectra of their activity; wild type did as well, although to a lesser extent (Dowse and Ringo, 1987). per^S animals display much less power in the ultradian range (Dowse and Ringo, 1987); alternating current signal power is equivalent to the variance in the data with the mean (direct current) component removed and comes from an electronic signal analogy (Beauchamp and Yuen, 1979).

When wild-type *Drosophila* are reared in constant darkness and are then tested as adults in DD, having never been exposed to light let alone any cyclic illumination, they frequently resemble per^0 in being either arrhythmic or in displaying multiple ultradian rhythms (Dowse and Ringo, 1989b; Power et al., 1995a). Some of these DD-reared animals have residual wild-type circadian rhythmicity, some have per^L-like rhythms, and some are unusual in displaying short, weak circadian rhythms (τ = 20–22 h; Dowse and Ringo, 1989b; Power et al., 1995a). Rearing animals in LD or constant light (LL) followed by testing in LL also produces per^o-like behaviour, with multiple ultradian periodicities (Fig. 8.5) (Power et al., 1995a).

Other workers have reported that DD-rearing of wild-type flies does not produce the ultradian rhythmicity we have observed (Sehgal et al., 1992; Tomioka et al.,

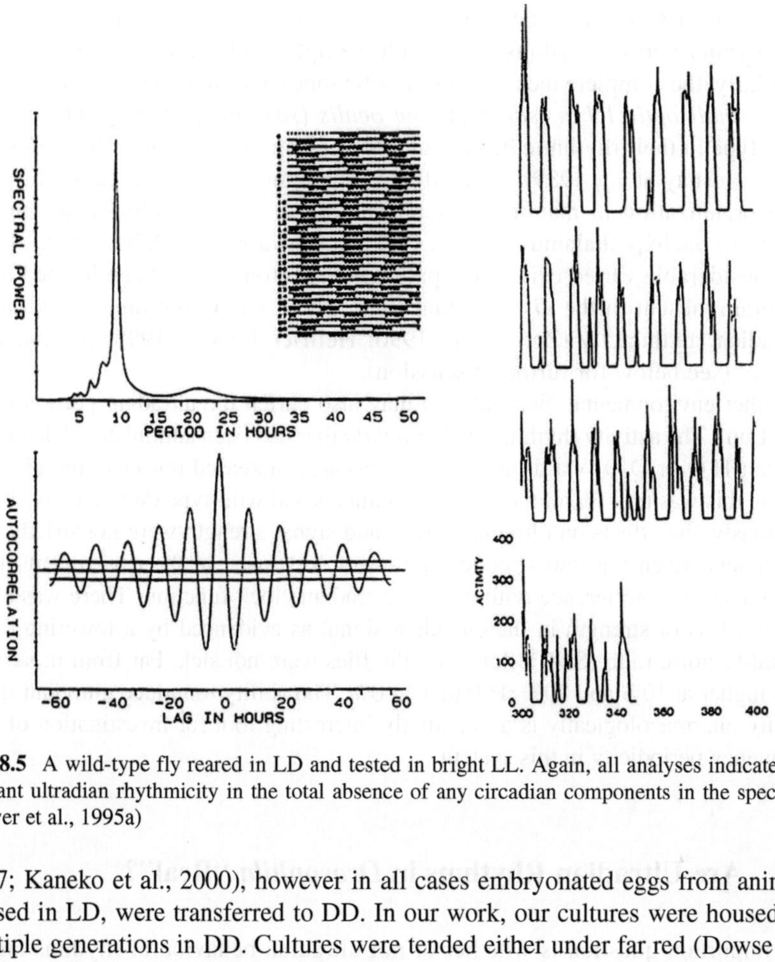

Fig. 8.5 A wild-type fly reared in LD and tested in bright LL. Again, all analyses indicate significant ultradian rhythmicity in the total absence of any circadian components in the spectrum (Power et al., 1995a)

1997; Kaneko et al., 2000), however in all cases embryonated eggs from animals housed in LD, were transferred to DD. In our work, our cultures were housed for multiple generations in DD. Cultures were tended either under far red (Dowse and Ringo, 1989b) or under infrared light (Power et al., 1995a). At no time were they exposed to any other light.

It is difficult to switch flies over from ultradian mode to circadian mode. Flies rendered rhythmic in the ultradian mode by proper DD rearing can be switched over by exposing them to either an hour pulse or to two full cycles of LD as adults (Power et al., 1995b). When returned to DD, *per*^{S} animals were restored to their normal rhythmic patterns with appropriate short periodicity, while wild-type flies often failed to convert to their near 24-h periodicity. Exposure of embryos or larvae had no effect on the adult patterns (Power et al., 1995b).

disconnected (disco) mutants possess eyes that fail to connect to the brain developmentally, and are missing other portions of the nervous system, including portions of the optic lobes (Steller et al., 1987). These animals are blind, and are *per*^{o}-like in rhythmic behaviour (Dushay et al., 1989). They also display strong ultradian rhythms (Dowse et al., 1989). Blindness alone is insufficient to block normal rhythmicity, as the mutant *no receptor potential A (norpA)* blocks retinal light transmission everywhere, not just in the compound eyes (Pak, 1975, 1979), yet permits normal rhythms entrainable to light cycles (Dushay et al., 1989). Similarly, the complete lack of eyes or optic lobes does not yield the *disco* result, since *small optic lobes (sol)* and *sine oculis (so)* mutations are very close to wild-type in their rhythmic behaviour (Helfrich and Engelmann, 1983; Helfrich, 1986; Dushay et al., 1989). This effect is analogous to the destruction of circadian organization in mammals after lesioning of the suprachiasmatic nucleus (SCN) of the hypothalamus (Rusak, 1977; Rosenwasser and Adler, 1986). Based on considerable varied evidence, "pacemaker neurons" have been located in the fly brain, absent in the *disco* mutant flies, which seem absolutely essential for circadian rhythmicity (Zerr et al., 1990; Helfrich-Förster, 1998; review: Hall, 2003b) (see below for further discussion).

Other environmental manipulation may also enrich the ultradian portion of the spectrum. The activity rhythms of *Drosophila* that had been administered deuterium oxide (^2H_2O or D_2O) via drinking water revealed increased power in the ultradian range; this was true for all the *per* mutant alleles and wild-type (White et al., 1991). Curiously, the effects on ultradian power and signal strength were not strictly dose dependent. Even the lowest concentration of $^2H_20$ used in the experiment (10%) caused severe interference with normal circadian clock function. There was also a general loss of strength in the circadian signal as evidenced by a lowering of the signal-to-noise ratio (SNR). And, yet, the flies were not sick. Far from it: survival was higher at 10% and 20% $^2H_20$ than at 0%. The ability to induce ultradian rhythmicity pharmacologically is a potentially interesting tool for investigation of high frequency periodicity in this system.

8.7 Are Ultradian Rhythms in *Drosophila* "Real"?

An important question is whether or not ultradian behavioural rhythms are (1) produced by interactions of circadian timekeepers, (2) pure random artifact, or (3) overt expressions of very high frequency cellular clocks. Ultradian periodicities are well documented in mammalian systems. For example, they occur naturally in the form of bouts of rapid eye movement (REM) sleep, replaced with periodicity in the state of awareness and psychomotor proficiency during waking periods (Webb and Dube, 1981). In rats, the extent of ultradian rhythmicity in the activity record is under genetic control, with some strains showing very pronounced high frequency bouts of activity superimposed on the circadian rhythm (Buttner and Wollnik, 1984; Wollnik et al., 1987; Schwartz and Zimmermann, 1990). Ultradian rhythmicity can also be induced surgically. In test animals, the circadian system typically breaks down after SCN lesions, leaving only ultradian rhythms (Rusak, 1977; Rosenwasser and Adler, 1986). Surgery on the SCN may also affect ultradian rhythmicity in strains in which it occurs naturally, since SCN lesions abolish ultradian as well as circadian rhythms in rats (Wollnik and Turek, 1989).

For this system, it has been proposed that these ultradian periods are a result of uncoupling multiple circadian oscillators (Rosenwasser and Adler, 1986). Spectral analysis of the activity of mammals may indicate ultradian rhythms, but inspection of the raw data plotted in actograms indicates multiple circadian rhythms still at work. How could multiple oscillations interact to produce other, distinct rhythms? In a simple example, if two circadian clocks act in phase to produce a single peak, and subsequently uncouple, two things may happen. The two rhythms may remain uncoupled and drift apart. Assuming no destructive interference occurs, a sufficiently long record with a high resolution analysis would show both periodicities in the spectrum, neither ultradian. A spurious long period would also appear at the "beat" frequency, the point at which mutual reinforcement occurs. Beat frequencies must always be longer than both contributing cycles (Tyson et al., 1976). Alternatively, the two oscillators may recouple at 180° out of phase. Spectral analysis would show an "ultradian" 12-h peak, but the fundamental oscillations would be circadian. If three oscillators are at work, they might produce a trimodal pattern with an 8-h peak in the spectrum.

It is more difficult to explain ultradian rhythms in *Drosophila* as the result of interacting circadian rhythms. Actograms from individuals displaying ultradian rhythms, e.g. *per^o*, do not show multiple circadian components and peaks in spectra are not submultiples of 24 h. Continuing the example above to explain these results, assuming the coupling is not 180°, but perhaps 105°, then two intervals will occur during each 24-h period: 7 and 17 h. But it does not follow that spectral analysis will result in two spectral lines at 7 and 17 h. Only half the number of peaks are present needed for 7- and 17-h rhythms, as every other one is "missing". Practically speaking, when analysing *Drosophila* records with two activity peaks per *ca* 24-h day aligned as described above, routinely we see only a single spectral line at the circadian period (H. B. Dowse, 1989). To explain multiple periods resulting from interacting circadian periods thus requires "Ptolemaic reasoning", multiplying the number of components to ridiculous complexity. A 7-h period would need four 28-h clocks coupled 7-h out of phase, and there is no way to explain the 17-h period other than two 34-h clocks 180° antiphase. We cannot conclude from this, however, that the rhythms we record actually reflect the direct output of oscillators with the same period.

We can exclude another explanation offered by Enright (1990), namely that these rhythms are short-lived phenomena that appear in the data record for just one or two cycles. These could either be short-lived transient ripples in a steady state, or truly spurious random variations devoid of any biological significance. When actograms from records rich in ultradian rhythms are plotted modulo 24 h – the normal procedure for circadian rhythms-ultradians are seldom obvious. If, however, ultradian periodicities are detected by spectral analysis, and the data are plotted modulo τ for a strong ultradian rhythm period, that ultradian period becomes obvious in the actogram (Dowse et al., 1987). Ultradian rhythms plotted in this way appear as strong and regular as circadian rhythms, and, critically, they continue unchanged in period or phase for the entire duration of some records. The

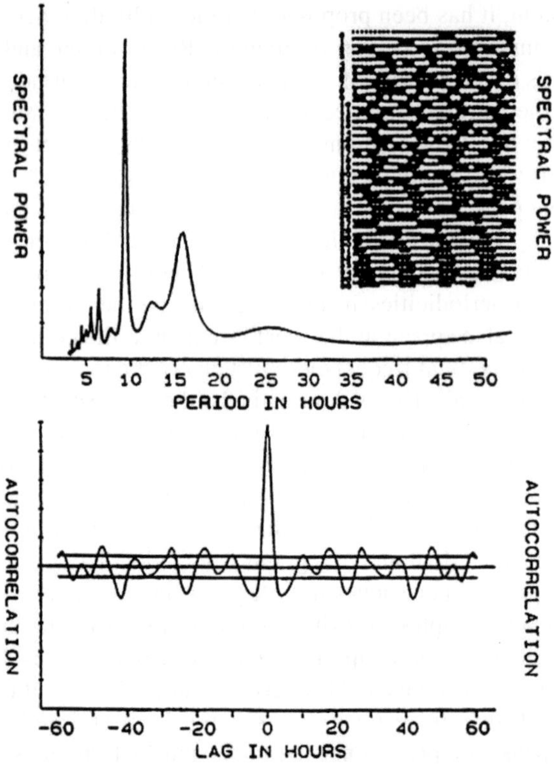

Fig. 8.6 Data from a *Drosophila melanogaster* bearing the *per*^{0} mutation that was reared and tested in strong LL. There is no evidence whatsoever of rhythmicity in the circadian range, however the ultradian periodicity is very clear. Maximum Entropy Spectral analysis (MESA) gives a sharp peak at 11 h, and this is strongly corroborated by the autocorellogram. The raw data are plotted as an array analysis actogram (inset) modulo the primary ultradian periodicity(inset), and a simple plot of activity as a function of time is given to the right (From Power et al., 1995a)

record from a *per*^{0} fly both reared and tested in strong LL is shown in Fig. 8.6. Either the LL conditions or the genotype should have resulted in a profoundly arrhythmic animal individually, let alone together, but the ultradian oscillations are robust and long-lived (Power et al., 1995a). This is scarcely what would be expected if only one or two cycles were present, as Enright (1990) has suggested. This also further cuts into any suggestion that interacting circadian rhythms are producing the output, as under this combination of conditions, no such rhythms should be present at all.

It should be noted that, if, as we argue below, the circadian clock itself uses an ensemble of high frequency ultradian oscillators for its time base, it may be that the observed mid-range ultradian rhythms in *Drosophila* behaviour records reflect artifactual outputs produced or rendered visible by a damaged or absent integrator system. This would account for the high variability that we have seen in the periods of these rhythms. There is no favoured ultradian period in *Drosophila* – values are spread fairly uniformly through the 4- to 18-h region of the spectrum. This leaves open the question of whether very high frequency quantal oscillators may yet be found that can account for both the longer observed ultradian periods and the circadian clock itself.

8.8 Do Mid-range Ultradian Rhythms Have a Function?

It is next appropriate to ask what purpose these ultradian rhythms (oscillators?) serve in *Drosophila*, if any. There are no obvious or even cryptic relevant geophysical correlates; further, the overt rhythms are characterized by their wide variability (Dowse et al., 1987), making a search for a single phase-locked environmental variable fruitless. Behaviourally, a 24-h timer is clearly useful. For example constraining eclosion and adult activity to the cool moist times near dawn and dusk is critical, since *Drosophila* are highly susceptible to desiccation (Pittendrigh, 1958). It seems unlikely that organization of *Drosophila* activity into discrete short bouts could have the same adaptive value along the lines postulated for vertebrates (Daan, 1981) when there is such a high interindividual variation both in period and in level of expression.

8.9 Is There a Relationship Between Ultradian and Circadian Rhythms?

The final question to be addressed concerns the relationships among the hypothetical quantal oscillators, the observed mid-range behavioural ultradian rhythms, and the circadian system. If the observed mid-range ultradian rhythms lack any demonstrable function in *Drosophila,* and yet if they are real (i.e. not "random noise"), then they are most likely to be by-products of the system providing temporal order, and thus afford a window into the physiological processes of this system. We have argued at length (Dowse and Ringo, 1992) that the circadian clock that we observe is actually the coupling mechanism, i.e. the frequency standard is NOT an oscillator with a 24-h escapement. We have hypothesized (Dowse et al., 1987; Dowse and Ringo, 1992), in company with many other investigators (e.g. Pavlidis, 1971; Winfree, 1980; Klevecz et al., 1984; Lloyd and Edwards, 1987), that the circadian clock is built upon many escapements in ensemble. The time constants of these components in many of these proposed systems are closer to those of intermediate metabolic processes about two to three orders of magnitude faster than any 24-h process (Lloyd and Edwards, 1987; Edmunds, 1988). Similarly, the phenomena of temperature compensation, precision and resistance to pharmacological meddling may be easier to explain in populations of coupled oscillators. The models have taken many forms. Some are concerned with ensembles of neurons that, as noted above, can behave more accurately in the circadian range than one individual cell (Herzog et al., 2004).

Multiple TTOs have been proposed as ultradian units coupled to form the circadian timer (Paetkau et al., 2006). Multiple ultradian units coupled within the SCN of the mammalian clock have been proposed as the timing source, and the relationship between the number of participating units and the output period has been investigated numerically (Barrio et al., 1997). A way to validate such hypotheses has been proposed by Edwards et al. (2007), namely by spectral analysis of rhythmic records, and the work done in our lab (Dowse and Ringo, 1987) is cited as "First steps in this direction ...". These workers have produce a realistic, hence stochastic, model showing how ultradian periodicities might yield circadian rhythms.

We specifically hypothesized earlier that *per* in *Drosophila* and *per*-like analogues in other organisms are required for assembly of the ultradian components we discovered into the circadian system Dowse and Ringo (1992). We postulated a population of oscillators coupled by weak, non-linear, inhibitory phasic links. The role of the PER protein in this scheme was to facilitate coupling among the oscillators.

This hypothesis is no longer tenable. The role of PER in the system, at least as far as its role in the 24-h TTO is concerned, is now well worked out and there seems to be no possibility of it functioning in this manner. Further, it has been argued cogently by Hall (2003b) that the very ubiquity of the ultradian oscillators belies a role for *per* in coupling them. If PER protein is responsible for coupling rhythms, either intra- or inter-cellularly, the manifestation of this coupling or lack thereof would be seen in the output of the neurons identified as the "master clock" in the fly. He points out that if the neurons are missing entirely, as in *disco* flies, then animals bearing this mutation should be profoundly arrhythmic, even to the level of ultradians. Yet *disco* flies have high frequency rhythms in about the same proportion as is seen in *per*^{0} (Dowse et al., 1987, 1989).

But this scarcely dooms the concept of ultradian oscillators as frequency standards. On the contrary, the structure of the TTO elucidated in the intervening years lends itself ideally to its reinterpretation as the counter in a two part clock structure as delineated above. As the above analysis shows, it is very difficult to support the idea that such an inherently sloppy system could measure time with the observed precision. One cannot time an event to the level of seconds well with a frequency standard orders of magnitude longer. Subdividing such a long-period cycle into thousands of subunits seems beyond the capacity of the cell to effect accurately. We thus hypothesize a dual mechanism clock, as above. An ensemble of high frequency oscillators provides the frequency standard and the TTO is the counter/time dispersal subunit.

The robustness of this counter is of primary importance to the countenance of the resulting observable rhythms. Normally, the system has an approximate 24-h output. This will appear in virtually every aspect of the organism's metabolism at all levels. By altering one gear of the output mechanism, longer, shorter, or a lack of periods would be observed as has been argued above. This is evident in the *per* allelic series, and there is circumstantial evidence to support the claim. The proportion of ultradian rhythmicity in flies homozygous for each of four alleles of *per:* wild-type, *Long, short,* and *null*. We found that the *null* allele, *per*^{0}, has almost exclusively ultradian rhythms, *per*^{L} has fewer, but they are still numerous and are found in the presence of the characteristic long period, *per*$^+$ has a few ultradians, and *per*^{S} has virtually no ultradian periods at all in the spectrum (Dowse and Ringo, 1987).

Thus period relates to the amount of ultradian power in a simple way. Robustness of coupling of the frequency standard half of the clock to the counter half would permit varying power in the ultradian range to appear in the locomotor activity rhythm. This was formally quantified as SNR, calculated using a waveform-independent technique of our own devising (Dowse and Ringo, 1989a). In light of the bipartite clock hypothesized here, the correlation between putative oscillator/counter coupling strength and period, the loglinear dependence of frequency on SNR reported in this work makes even more sense. It is predictable from the regression equation that, if *per*^{o} were to have a circadian rhythm at all, it would appear in the 35- to 40-h range (D and R). This was confirmed dramatically by the analysis of the *per*^{o} allele, *per*^{04}. This is a hypomorphic mutation, in contrast to the original amorph, *per*^{01} (there are no *per*^{02}. or *per*^{03} alleles). Periods of approximately 40h were uncovered with a spectral analysis technique specifically set up to look for such long cycles (Hamblen-Coyle et al., 1989).

It was also hypothesized (Dowse and Ringo, 1989b) that in the *Drosophila* system light or some other phasically active stimulus is required for proper initialization of the circadian clock, in this case either internal coupling of the multiple component frequency standard oscillators, or coupling of this population to its TTO counter, or both. DD-reared animals display a wide range of periodicities in their activity, as has been noted above, and random and partial assembly of subunits into various levels of organization is the only currently tenable hypothesis for the wide range of results observed. Note that *per*^{S} flies reared in DD can actually show *per*^{L}-like rhythms (Fig. 8.7). It is puzzling how this might be possible if the molecular machinery is geared to produce a short period. However, if the TTO is an integrator, its connection to the underlying ultradians could be highly malleable, especially in DD-reared animals, easily explaining the broad distribution of periods observed across genotypes under these circumstances (Power et al., 1995a). The oscillator/counter hypothesis is consistent with all the evidence. It suffers from the problem that the underlying high frequency oscillators must be identified.

8.10 Is There Evidence for Very High Frequency Candidate Ultradian Oscillators?

Bünning (1964) hypothesized long ago that maintenance of internal temporal order was a primary duty of any cellular clock: mutually incompatible cellular processes that cannot be separated in space must be separated in time. It is necessary for products of the cell's metabolism to be at the right place at the right time. Oatley and Goodwin (1971) also noted that oscillations around a steady state are inevitable, and that these oscillations may be acted upon by natural selection to produce

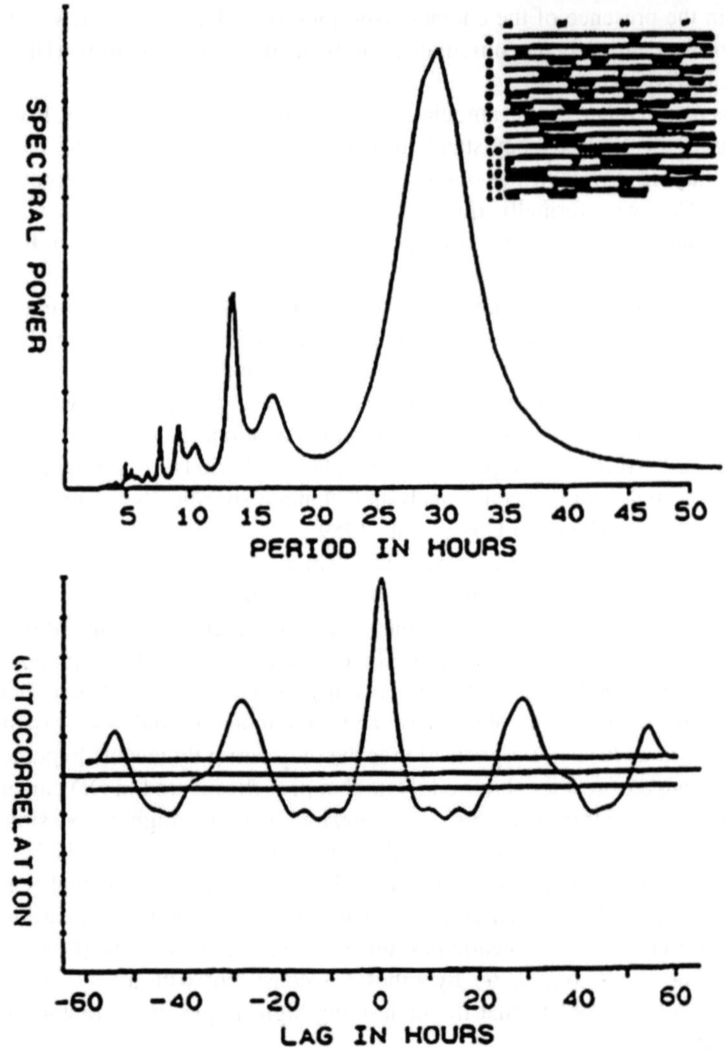

Fig. 8.7 This fly, tested in DD, was from a stock of animals bearing the *per*^{S} mutation that was kept for many generations in DD with no exposure to light other than infra red. It was not exposed to visible light at any time. Remarkably, it's record is indistinguishable from those produced by flies carrying *per*^{L} mutations that were reared normally in LD and tested in DD (From Power et al., 1995a)

any period of use. It is proposed, then, that the multiple mid-range ultradian oscillations we see in *Drosophila* are manifestations of the higher frequency internal cellular-housekeeping clocks. Their appearance as mid-range periodicities in the activity record is an artifact of the disruption by whatever means of the intracellular integrator system responsible for delivering circadian timing information to the upper behavioural centres in the CNS. The ultimate quantal clocks would presumably have a much shorter period, and would be found in every cell of the body. A cell lacking these timekeepers would die.

Extensive evidence has accumulated on long-lived, temperature-compensated oscillations in a number of unicells (Lloyd and Edwards, 1984). The cell cycle in these organisms is temperature dependent, and thus seems not to be a candidate for putative control by these higher-frequency oscillators. But if they are quantal in nature, like building blocks, they might be added or subtracted to lengthen or shorten the period of the controlled cycle as the temperature changes, and the discrepancy would be resolved (Lloyd and Edwards, 1987; Lloyd and Kippert, 1987). Using the same logic, Lloyd and Edwards (1987) have extended this argument to the circadian clock.

There are well-characterized high frequency oscillations in the cell that are strong candidates for the quantal oscillators underlying such a clock (Klevecz et al., 1984; Klevecz and Dowse, 2000; Klevecz et al., 2004; Li and Klevecz, 2006; Lloyd and Murray, 2007). Looking for further evidence of such oscillators and elucidating their molecular mechanisms and relationships to circadian rhythms in *Drosophila* seems to offer the best chance of unraveling the 24-h clock in this organism.

8.11 Summary

It is argued that the precision of exemplary circadian clocks is numerically inconsistent with the 24-h Transcription/Translation Oscillator (TTO) hypothesis of biological timing. Mid-range ultradian rhythms (having periods in the range 4–18 h) have been observed in the locomotor activity of *Drosophila melanogaster*. These ultradian activity rhythms are especially prominent when strong circadian rhythms are absent, which is characteristic of two unrelated clock mutants *per*^{o} and *disco,* as well as of individuals from populations of flies that have never been exposed to light over generations or are raised and tested in LL. Ultradian activity rhythms occur also in flies with normal circadian rhythms, e.g. LD-reared, wild-type flies tested in DD; in these normal flies, though, ultradian rhythms are usually weak and are present only in some individuals. In general, the strength of the ultradian rhythms of individuals varies inversely with the strength of their circadian rhythms. In many of the genotypes on which we have reported thus far, the periods of these rhythms are distributed evenly throughout the range 4–18 h. Often, they are strong and long-lived, but, since no geophysical correlate is evident, their function in regulating behaviour is problematic. It is hypothesized that these rhythms reflect the activity of yet higher-frequency cellular oscillators, and that these oscillators underlie the circadian timing system by acting as frequency standards in ensemble. Circumstantial evidence favours this hypothesis.

Acknowledgements Dr. John Ringo contributed extensively to this chapter in the first edition of this book. I thank Dr. Joel Levine for comments. This work was made possible in part by sale of family heirlooms

References

Alberts B, Johnson A, Lewis J, Raff M, Roberts K, Walter P (2002) Molecular biology of the cell. Garland, New York

Allan DW, Ashby N, Hodge CC (1997) The science of timekeeping. Hewlett Packard Application Note 1289

Bargiello TA, Young MW (1984) Molecular genetics of a biological clock in *Drosophila*. Proc Natl Acad Sci USA 81: 2142–2146

Barkai N, Leibler S (2001) Circadian clocks limited by noise. Nature 403: 267–268

Barrio RA, Zhang L, Maini PK (1997) Hierarchically coupled ultradian oscillators generating robust circadian rhythms. Bull Math Biol 59: 517–532

Beauchamp K, Yuen C (1979) Digital methods for signal analysis. George, Allen, Unwin, Boston

Berg OG, Winter RB, von Hippel PH (1981) Diffusion-driven mechanism of protein translocation on nucleic acids. 1. Models and theory. Biochemistry 20: 6929–6948

Bünning E (1964) The physiological clock. Academic Press, New York

Buttner D, Wollnik F (1984) Strain differentiated circadian and ultradian rhythms in locomotor activity of the laboratory rat. Behav Genet 14: 137–151

Chatfield C (1989) The analysis of time series. Chapman and Hall, London

Chay TR (1981) A model for biological oscillations. Proc Natl Acad Sci USA 78: 2204–2207

Chovnik A (ed) (1960) Cold Spring Harbor Symposia on Quantitative Biology XXV Waverly, Baltimore MD

Clynch JR (2002) Precise time and time interval clocks, time frames and frequency. http://www.gmat.unsw.edu.au/snap/gps/clynch_pdfs/pttinote.pdf

Daan S (1981) Adaptive daily strategies in behavior. In: Aschoff J (ed) Biological rhythms. Plenum, New York, pp 275–298 (Handbook of behavioral neurobiology, vol 4)

Daan S, Aschoff J (1981) Short-term rhythms in activity. In: Aschoff J (ed) Biological rhythms. Plenum, New York, pp 491–498 (Handbook of behavioral neurobiology, vol 4)

Darwin C, Darwin F (1880) The power of movement in plants. John Murray, London

DeCoursey P (1960) Phase control of activity in a rodent. In: Chovnik A (ed) Cold Spring Harbor Symposia on Quantitative Biology XXV Waverly, Baltimore MD, pp 49–55

DeMairan JB (1729) Observation botanique. In: Histoire de l'Academie Royale des Sciences. Paris, p 35

Dowse H, Palmer J (1990) Evidence for ultradian rhythmicity in an intertidal crab. In: Hayes DK, Pauly JP, Reiter RJ (eds) Chronobiology: its role in clinical medicine, general biology, and agriculture, Part B. Wiley-Liss, New York, pp 691–697

Dowse HB, Palmer JD (1992)Comparative studies of tidal rhythms. XI. Ultradian and circalunidian rhythmicity in four species of semiterrestrial, intertidal crabs. Mar Behav Physiol (21: 105–119)

Dowse H, Ringo J (1987) Further evidence that the circadian clock in *Drosophila* is a population of coupled ultradian oscillators. J Biol Rhythms 2: 65–76

Dowse H, Ringo J (1989a) The search for hidden periodicities in biological time series revisited. J Theor Biol 139: 65–76

Dowse H, Ringo J (1989b) Rearing *Drosophila* in constant darkness produces phenocopies of *period* circadian clock mutants. Physiol Zool 62: 785–803

Dowse HB, Ringo JM (1991) Comparisons between "periodograms" and spectral analysis: Apples are apples after all. J Theoret Biol 148: 139–144.

Dowse H, Ringo J (1992) Is the circadian clock a "metaoscillator?" Evidence from studies of ultradian rhythms in *Drosophila*. In: Young M (ed) Molecular approaches to circadian clocks. Marcel Dekker, New York, pp 195–220

Dowse H, Hall JC, Ringo J (1987) Circadian and ultradian rhythms in *period* mutants of *Drosophila melanogaster*. Behav Genet 17: 19–35

Dowse H, Dushay M, Hall JC, Ringo J (1989) High-resolution analysis of locomotor activity rhythms in *disconnected,* a visual system mutant of *Drosophila melanogaster*. Behav Genet 19: 529–542

Dunlap J (1999) Molecular bases for circadian clocks. Cell 96: 271–290

Dushay MS, Rosbash M, Hall JC (1989) The *disconnected* visual system mutations of *Drosophila melanogaster* drastically disrupt circadian rhythms. J Bio Rhythms 4: 1–27

Edmunds LN Jr (1988) Cellular and molecular bases of biological clocks. Springer, Berlin/Heidelberg/New York

Edwards R, Gibson R, Illner R, Paetkau V (2007) A stochastic model for circadian rhythms from coupled ultradian oscillators. Theor Biol Model 4: 1–11

Ehret CF, Trucco E (1967) Molecular models for the circadian clock. I. The chronon concept. J Theor Biol 15: 240–262

Endy D, Brent R (2001) Modelling cellular behaviour. Nature 409: 391–395

Enright JT (1965) The search for rhythmicity in biological time-series. J Theor Biol 8: 662–666

Enright JT (1990) A comparison of periodograms and spectral analysis: don't expect apples to taste like oranges. J Theor Biol 143: 425–430

Goldbeter A (1995) A model for circadian oscillations in the *Drosophila* period protein (PER). Proc R Soc Lond B 261: 319–324

Gonze D, Halloy J, Golbeter A (2002) Robustness of circadian rhythms with respect to molecular noise. Proc Natl Acad Sci USA 99: 673–678

Hall JC (2003a) A neurogeneticist's manifesto. J Neurogenet 17: 1–90

Hall JC (2003b) Genetics and molecular biology of rhythms in *Drosophila* and other insects. Advances in genetics, vol. 48. Academic, Boston MA

Hall JC, Kyriacou CP (1990) Genetics of biological rhythms in *Drosophila.* Adv Insect Physiol 22: 221–298

Hamblen-Coyle M, Konopka RJ, Zwiebel LJ, Colot HV, Dowse HB, Rosbash M, Hall JC (1989) A new mutation at the *period* locus of *Drosophila melanogaster* with some novel effects on circadian rhythms. J Neurogenet 5: 229–256

Helfrich C (1986) Role of the optic lobes in the regulation of the locomotor activity rhythm of *Drosophila melanogaster*. Behavioral analysis of neural mutants. J Neurogenet 3: 321–343

Helfrich C, Engelmann W (1983) Circadian rhythm of the locomotor activity in *Drosophila melanogaster* mutants sine oculis and small optic lobes. Physiol Entomol 8: 257–272

Helfrich-Förster C (1998) Robust circadian rhythmicity of *Drosophila melanogaster* requires the presence of lateral neurons: a brain-behavioral study of *disconnected* mutants. J Comp Phsyiol A 182: 435–453

Herzog ED, Aton SJ, Numano R, Sakaki Y, Tei H (2004) Temporal precision in the mammalian circadian system: a reliable clock from less reliable neurons. J Biol Rhythms 19: 35–46

Howe DA, Allan DW, Barnes JA (1981) Properties of signal sources and measurement methods. Proc. 35th Annual Symposium on Frequency Control. In: Sullivan DB, Allan DW, Howe DA, Walls FL (eds) Characterization of clocks and oscillators, NIST technical note 1337, United States Department of Commerce, National Institute of Standards and Technology.

Itano WM, Ramsey NF (1993) Accurate measurement of time. Sci Am 269: 56–65

Kaneko M, Hamblen MJ, Hall JC (2000) Involvement of the *period* gene in developmental time-memory: effect of the per^{Short} mutation on phase shifts induced by light pulses delivered to *Drosophila* larvae. J Biol Rhythms 15: 13–30

Klevecz RR, Kaufmann SA, Shymko RM (1984) Cellular clocks and oscillators. Int Rev Cytol 5: 97–126

Klevecz RR, Dowse HB (2000) Tuning the transcriptome: basins of attraction in the yeast cell cycle. Cell Proliferation 33: 209–218

Klevecz RR, Pilliod J, Bolen J (1991) Autogenous formation of spiral waves by coupled chaotic attractors. Chronobiol Int 8: 6–13

Klevecz RR, Bolen J, Forrest G, Murray D (2004) A genomewide oscillation in transcription gates DNA replication and cell cycle. PNAS 101: 1200–1205

Konopka RJ, Benzer S (1971) Clock mutants of *Drosophila melanogaster*. Proc Natl Acad Sci USA 68: 2112–2116

Leloup J-C, Gonze D, Goldbeter A (2003) Limit cycle models for circadian rhythms based on transcriptional regulation in *Drosophila* and *Neurospora*. J Biol Rhythms 14: 433–448

Levine J, Funes P, Dowse H, Hall J (2002) Signal analysis of behavioral and molecular cycles. Biomed Central Neurosci 3: 1
Li CM, Klevecz RR (2006) A rapid genome-scale response of the transcriptional oscillator to perturbation reveals a period-doubling path to phenotypic change. PNAS 103: 16254–16259
Lloyd D, Edwards S (1984) Epigenetic oscillations during the cell cycles of lower eukaryotes are coupled to a clock. In: Edmunds LN Jr (ed) Cell cycle clocks. Marcel Dekker, New York, pp 27–46
Lloyd D, Edwards S (1987) Temperature-compensated ultradian rhythms in lower eukaryotes: timers for cell cycles and circadian events? In: Pauly JE, Scheving LE (eds) Advances in chronobiology, Part A, Alan R Liss, New York, pp 131–151
Lloyd D, Kippert F (1987) A temperature-compensated ultradian clock explains temperature dependent quantal cell cycle times. In: Bowler K, Fuller BJ (eds) Temperature and animal cells. Cambridge University Press, Cambridge, pp 135–155 (Symp Soc Exp Biol 41)
Lloyd D, Edwards SW, Fry JC (1982) Temperature compensated oscillations in respiration and cellular protein content in synchronous cultures of *Acanthamoeba castellanii*. Proc Natl Acad Sci USA 79: 3785–3788
Lloyd D, Murray DB (2007) Redox rhythmicity: clocks at the core of temporal coherence. BioEssays 29: 465–473
Michel U, Hardeland R (1985) On the chronobiology of *Tetrahymena*. III. Temperature compensation and temperature dependence in the ultradian oscillation of tyrosine aminotransferase. J Interdiscipl Cycle Res 16: 17–23
Njus D, Sulzman FM, Hastings JW (1974) Membrane model for the circadian clock. Nature 248: 116–120
Oatley K, Goodwin BC (1971) The explanation and investigation of biological rhythms. In: Colquhoun WP (ed) Biological rhythms and human performance. Academic, New York, pp 1–38
Paetkau V, Edwards R, Illner R (2006) A model for generating circadian rhythm by coupling ultradian oscillators. Theor Biol Med Model 3: 12
Pak WL (1975) Mutants affecting the vision of *Drosophila melanogaster*. In: King RC (ed) Handbook of genetics, vol 3. Plenum, New York, pp 703–733
Pak WL (1979) Study of photoreceptor function using *Drosophila* mutants. In: Breakfield XO (ed) Neurogenetics: genetic approaches to the nervous system. Elsevier/North Holland, New York, pp 67–79
Pavlidis T (1971) Populations of biochemical oscillators as circadian clocks. J Theor Biol 33: 319–338
Pittendrigh CS (1958) Adaptation, natural selection, and behavior. In: Roe A, Simpson GG (eds) Behavior and evolution. Yale University Press, New Haven, CT, pp 390–416
Power J, Ringo J, Dowse H (1995a) The effects of *period* mutations and light on the activity rhythms of *Drosophila melanogaster*. J Biol Rhythms 10: 267–280
Power J, Ringo J, Dowse H (1995b) The role of light in the initiation of circadian activity rhythms of adult *Drosophila melanogaster*. J Neurogenet 9: 227–238
Quinn TJ (1991) The BIPM and the accurate measurement of time. Proc of the IEEE: Special issue on time and frequency 79: 894–905
Reddy P, Zehring WA, Wheeler DA, Pirotta V, Hadfield C, Hall JC, Rosbash M (1984) Molecular analysis of the *period* locus in D. *melanogaster* and identification of a transcript involved in biological rhythms. Cell 38: 701–710
Riggs AD, Bourgeois S, Cohn M (1970) The lac repressor-operator interaction. 3. Kinetic studies. J Mol Biol 53: 401–417
Rosenwasser A, Adler N (1986) Structure and function in circadian timing systems: evidence for multiple coupled circadian oscillators. Neurosci Biobehav Rev 10: 413–448
Rusak B (1977) The role of the suprachiasmatic nucleus in the generation of circadian rhythms in the golden hamster, *Mesocricetus auratus*. J Comp Physiol 118: 145–164
Schwartz W, Zimmermann P (1990) Circadian timekeeping in *BALB/c* and C57BL/6 inbred mouse strains. J Neurosci 10: 3685–3694

Sehgal A, Price J, Young M (1992) Ontogeny of a biological clock in *Drosophila melanogaster*. PNAS 89: 1423–1427

Smith RF, Konopka RJ (1981) Circadian clock phenotypes of chromosome aberration with a breakpoint at the *per* locus. Mol Gen Genet 185: 243–251

Sobel D (1995) Longitude. Walker, New York

Steller H, Fischbach KF, Rubin G (1987) Disconnected: a locus required for neuronal pathway formation in the visual system of *Drosophila*. Cell 50: 1139–1153

Sullivan DB, Allan DW, Howe DA, Walls FL (1990) Characterization of clocks and oscillators. NIST technical note 1337. US Department of Commerce, National Institute of Standards and Technology

Tomioka K, Uwozumi K, Matsumoto N (1997) Light cycles given during development affect freerunning period of circadian locomotor rhythm of *period* mutants in *Drosophila melanogaster*. J Insect Physiol 43: 297–305

Tomioka K, Sakamoto M, Harui Y, Matsumoto N, Matsumoto A (1998) Light and temperature cooperate to regulate the circadian locomotor rhythm of wild type and *period* mutants of *Drosophila melanogaster*. J Insect Physiol 43: 297–305

Tyson JJ, Alivisatos SGA, Richter O, Grün F, Schneider FW, Pavlidis T (1976) Mathematical background report. In: Hastings JW, Schweiger H-G (eds) The molecular basis of circadian rhythms. Dahlem Konferenzen, Berlin, pp 85–108

Webb W, Dube M (1981) Temporal characteristics of sleep. In: Aschoff J (ed) Biological rhythms. Plenum, New York, pp 499–522 (Handbook of behavioral neurobiology, vol 4)

White L, Ringo J, Dowse H (1991) A circadian clock of *Drosophila:* effects of deuterium oxide and mutations at the *period* locus. Chronobiol Int 9: 250–259

Winfree AT (1980) The geometry of biological time. Springer, Berlin/Heidelberg/New York

Wollnik F, Turek F (1989) SCN lesions abolish ultradian and circadian components of activity rhythms in LEW/Ztm rats. Am J Physiol 25: R1027–R1039

Wollnik F, Gartner K, Buttner D (1987) Genetic analysis of circadian and ultradian locomotor activity rhythms in laboratory rats. Behav Genet 17: 167–178

Zerr DM, Hall JC, Rosbash M Siwicki KK (1990) Circadian fluctuations of *period* protein immunoreactivity in the CNS and the visual system of *Drosophila*. J Neurosci 10: 2749–2762

Chapter 9
Tidal Rhythms

M.K. Chandrashekaran and V.K. Sharma

Abstract Several marine organisms living in the inter-tidal zone show tidal rhythms. The endogenous nature of the rhythm enables organisms to determine *time* accurately even when the moon is hidden for two to three months because of perpetually overcast sky. While the basic characteristics of the rhythms are endogenously determined, it is fine-tuned by high and low tides caused by gravitational pull exerted by sun and moon. Hydrostatic pressure, turbulence of the waves and resultant mechanical agitation have all been proposed to be time cues. The fundamental nature of the timing mechanisms of intertidal organisms has been controversial for a long time. Empirical evidence strongly suggests a circadian component in activity appearing superimposed on a tidal component. A close relationship between tidal rhythms and circadian rhythms has also been inferred from the fact that tidal rhythms are also, to a large extent, refractory to temperature. While in most cases the ecological importance of the lunar cyclic phenomena is quite obvious, results of a few early studies suggest that tidal, circadian and lunar rhythms are intricately connected conferring an adaptive value to organisms inhabiting the edges of the oceans. We review here some of the major findings from early studies that may be relevant to understanding the basic nature of tidal rhythms. We critically review these studies in order to bring some clarity in the way we view tidal rhythms, and to understand their relationship with circadian rhythms.

Keywords Tidal rhythms, lunar, marine, crab, zeitgeber, circadian, circatidal.

9.1 The Intertidal Environment

The intertidal area, also called the littoral zone, is the relatively narrow strip of land, which lies between the high and low tide marks. Rich in nutrients and oxygen, this zone is a constantly changing and inhospitable environment inundated twice daily

Chronobiology Laboratory, Evolutionary and Organismal Biology Unit,
Jawaharlal Nehru Centre for Advanced Scientific Research, Bangalore–560 064, India
Email: mkc@jncasr.ac.in

by the high tide and exposed by the receding tide. Organisms living in the littoral zone have adapted to the huge changes in moisture, temperature, turbulence and salinity. Many intertidal animals burrow into the sand like clams, mole crabs, and copepods, live under rocks, or attach themselves to rocks such as mussels and barnacles. Four vertical zones characterize the littoral strip of land:

1. The spray or splash zone is dry much of the time and is sprayed with salt water during high tides and is home to barnacles, isopods, lichens, woodlice, limpets and snails.
2. The high tide zone, which is flooded only during high tide, is home to sea anemones, barnacles, sea stars, brittle stars, chitons, a variety of crabs, green algae, isopods, limpets, mussels, snails and some marine vegetation.
3. The middle tidal zone is an area of near perpetual turbulence covered and uncovered twice a day with salt water from the tides and is inhabited by much the same animals as are found in the high tide zone and additionally by sponges.
4. The low tide zone is perpetually under water being exposed when the tide is unusually low. This zone is home to abalone, brown seaweed, crabs, hydroids, sea cucumber, sea lettuce, sea palm, surf grass and tube worms.

Since many of the littoral organisms named above are known to undertake tidal migrations they may be found in all four zones at appropriate tidal phases. Only barnacles and mussels, which attach themselves firmly to the rocks at the high tide zone, do not indulge in the tidal migrations. Littoral animals have a wide variety of predators such as fish, gulls, crows and kites, who eat them.

9.1.1 Ocean Tides

The ocean is in a constant predictable state of flux. The sun and moon exercise a gravitational pull towards them on the surface of the earth, and this force particularly affects oceanic waters. The gravitational pull on the oceans causes the water to bulge towards the orbiting moon creating the high tide. In a semi-diurnal tidal environment, between two high tides occur the low tides. As the moon rotates around the earth each day, water is displaced from the side of the earth facing the moon and also from the side of the earth facing *away* from the moon (Fig. 9.1). The bulging of the waters on both sides of the earth occurs in response to the moon's orbit. Because each lunar orbit takes 24 h and 50 min, there are usually two high tides and two low tides a day. However, because each orbit takes slightly longer than 24 h, the timing of high and low tides shift by about *ideally* 25 min. When the sun and the moon are aligned – during full moon and new moon – the gravitational pull is greater causing very high tides and very low tides called *spring tides*. By contrast, when the moon and the sun are not aligned – in the first and third quarters of the lunar cycle – their gravitational forces partially cancel each other out, and the tidal range or amplitude is less. These tides are called *neap tides*. Over most of the Atlantic coast and the rest of the world, the average difference of water levels

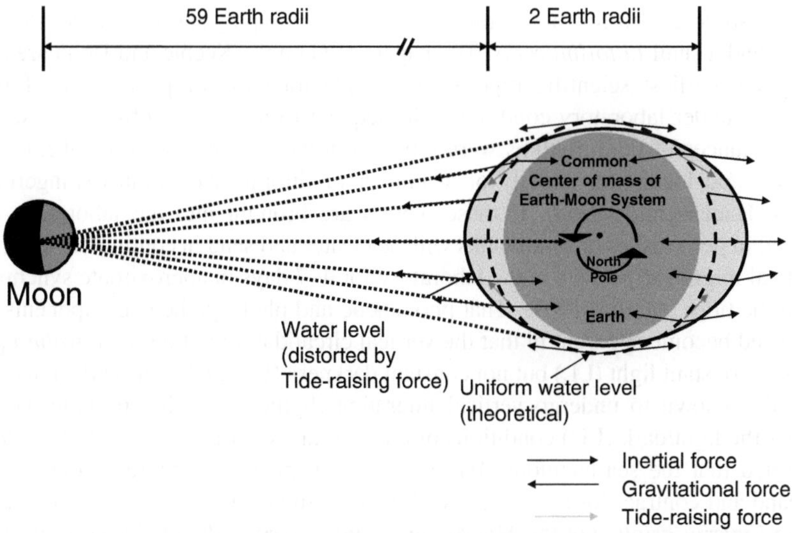

Fig. 9.1 The primary factor that controls the rhythm and amplitude of the tides is the moon. The gravitational attraction of the moon on the surface of the earth causes two tidal bulges on the side closest to the moon and on the opposite side. When the sun and the moon are aligned – during full moon and new moon – the gravitational pull is greater causing very high tides and very low tides called *spring tides*. By contrast, when the moon and the sun are not aligned – in the first and third quarters of the lunar cycle – their gravitational forces partially cancel each other out, and the tidal range or amplitude is less. These tides are called *neap tides*

at high tide and low tide – tidal amplitude – is 2 m. The greatest tidal amplitude in the world occurs in the Bay of Fundy – between Nova Scotia and New Brunswick reaching values of 16 m.

9.2 Earlier Studies on Tidal Rhythms

Several organisms living in the inter-tidal zone of the sea show tidal rhythms. In the Welsh coast near Cardiff and Swansea and in the beaches in Brittany, in a narrow zone just below the high tide mark of neap tides, lives a small acoelous turbellarian worm, *Convoluta roscoffensis*. Over the hours of darkness of the nights and during the tides, these zoo-xanthellae-bearing worms live buried in the sand but emerge on to the surface during daytime low tides. Thus, exposed to the sun the symbiotic algae within the beautifully sinuous body of the planarians photosynthesise, giving the worms a grass-green sheen. Since these worms occur in great density of thousands per square meter, large expanses of the beaches at low tide appear spinach-green. When the sea rises and covers the *convoluta* patches, the worms burrow into the sand to avoid being washed up on to the beach. They re-emerge with the advent of low tide (Gamble and Keeble, 1903; Bohn, 1903, 1904). These movements of the turbellarians are a type of *vertical migration*. Tidal rhythms were also described for the swimming activity of the polychaete *Hedistes diversicolor* and a snail *Littorina rudis*. The papers of Gamble, Keeble and Bohn are possibly the earliest scientific reports on the phenomenon of persistence of tidal rhythms under laboratory conditions. Subsequent to these early observations, several instances of tidal rhythms were reported in the rate of oxygen uptake, colour change and locomotor activity of a number of littoral invertebrates (Fingerman, 1960; Hauenschild, 1960). If these worms are brought into the laboratory and placed in petri dishes or small tanks with sea water and sand in continuous light, the vertical migration persists in the laboratory for 4–7 days in approximate synchrony with the tides (Martin, 1907). That phototactic and photosynthetic components are involved become apparent in that the vertical circatidal migration of *convoluta* persists in constant light (LL) but not constant darkness (DD). Diatoms and euglenoids are also known to undergo vertical migration rhythms, which reportedly persist under the light/dark (LD) conditions of the laboratory for a few days. Bohn (1906) reported that the sea anemone *Actinia equina,* kept in an aquarium continues to expand and contract for up to 8 days. However, studies with the inter-tidal invertebrate *Emerita asiatica* in the Madras coast and *Emerita talpoida* in the beaches in Wilmington, North Carolina revealed that the number of days over which tidal rhythms persist may depend on the environmental conditions such as ambient temperature and oxygen partial pressure of the sea water as well as on the species.

Brown et al. (1953) were among the earliest to report tidal and diurnal rhythms in colour change, activity, and oxygen uptake in the fiddler crab *Uca pugnax*. Brown unfortunately used abstruse statistical methods to 'extract' tidal rhythms from long time series data. He additionally confounded his readers by tirelessly expounding on how electrostatic and magnetic field variations arising from the rotation of the earth provided organisms with 24 h cues even under the constant conditions of the laboratory. Cole (1957) in a hard-hitting paper, using Brown's controversial statistics demonstrated a strictly 24 h biological rhythm in that mythical animal, the unicorn. Thereafter Brown fought a lone and losing battle and his influence in chronobiology became weak. Fortunately this did not act as a deterrent or set-back for the progress of the field.

Buried in the upper reaches of the beaches of southern California lives the amphipod *Synchelidium*. Its habitat is submerged during high tides when the amphipods come out, and swim about and feed in the wave wash. Two or 3 h later when the sea ebbs, the amphipods burrow into the moist sand to safety. Enright (1963) investigated the persistence of tidal rhythmicity in the swimming activity of this crustacean under the constant conditions of laboratory. He placed large populations of *Synchelidium* in cuvette-like glass troughs with seawater and sand, and followed with time-lapse photography the number of amphipods swimming. There was an impressive tidal rhythm in the swimming activity of these animals (Fig. 9.2), which however damped out after about 3 days in the laboratory. The first two peaks faithfully mirrored the phase, height and semidiurnal inequalities of the tides of the Californian coast. Three other crustaceans occurring in the same beach, an anomuran mole crab *Emerita,* a mysid *Archaeomysis* and an isopod *Excirolana chiltoni* all possessed similar tidal rhythms as *Synchelidium* (Enright, 1963).

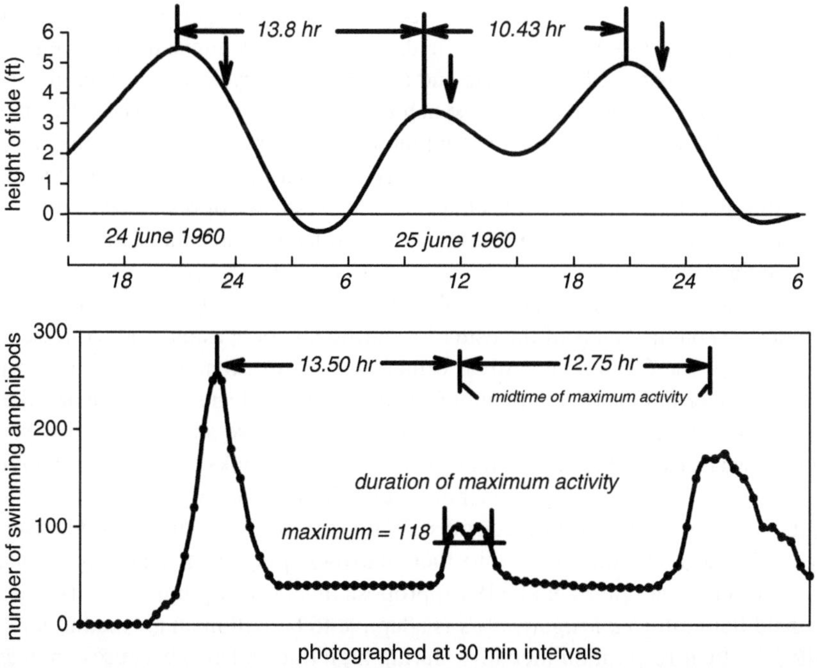

Fig. 9.2 Tidal rhythm in the swimming activity of freshly collected *Synchelidium* sp., upper graph: height of tide from U.S. Coast and Geodetic Survey predictions for La Jolla, California. MLLW: zero tide reference level = mean lower low water; lower graph: counts of number of swimming amphipods during laboratory observations of two photographs per hour. Center lines of activity peaks determined as midtime of maximum activity (Modified from Enright, 1963)

The isopod *Excirolana chiltoni* which also occurs in the same beaches in California with its complex patterns of persistent tidal rhythms was the object of Klapow's (1972) study. Under the constant conditions of the laboratory, the patterns of swimming activity of this burrowing crustacean mirror the amplitude and phase of the tides, which were themselves quite complex. There was, however, substantial amount of variability in the persistent tidal rhythms of *Excirolana.* There were especially tenacious specimens of *Excirolana* (Enright, 1972) (the virtuoso isopod) in which the tidal rhythm in the swimming activity persisted up to 65 days – a record, all the while mimicking the changing features and inequalities of the tides. The unanswered question is how even without reinforcing Zeitgeber inputs there could be such intricate and long-lasting synchronization of the activity patterns to the actual height of the tides.

Honnegger (1973) worked in California on the fiddler crab *Uca crenulata.* He exposed his crabs to artificial tides and reported that most specimens could be synchronized by the tides. But the locomotor activity was merely exogenously driven and the apparent entrainment might have only been a "masking" effect of the tides. Under constant conditions, after tidal entrainment, bimodal or unimodal activity patterns, were displayed. It must be noted that Honegger collected his fiddler crabs from a tidal lagoon where only hydrostatic changes minus the turbulence are experienced. When a 24 h LD cycle and a 12.35 h tidal cycle were imposed simultaneously, their combined effect in synchronizing the crabs' activity was variable. It is likely that crabs, which exhibit a unimodal activity pattern, are sensitive to light stimuli, whereas those with a bimodal activity pattern respond to light as well as tidal stimuli. Honegger concludes that since the response of his crabs to tides and LD cycles were similar, the basic oscillator for tidal and circadian activity rhythms may be the same.

In India one may spot fiddler crab males of *Uca annulipes* wave their claws (chelicerae) on the banks of the estuaries during low tide, as may be witnessed on the banks of the Cooum and Adyar estuaries in Madras and on the banks of the Mandovi river in Goa. The waving of the enlarged chelicerae is very marked and in great synchrony when low tide coincides with sunrise and are reminiscent of dhobis (washermen) beating clothes on stones on the banks of rivers – a common sight in India before washing machines came into vogue. So the older literature of Indian origin refers to fiddler crabs as "dhobi-crabs". When the water rises at high tide the crabs disappear into their burrows, plug them and stay virtually under water. de la Iglesia et al. (1994) propose that burrow plugging behaviour of their fiddler crab *Uca uruguayensis* is adaptive to be within an air medium, more suited for their respiration modality, during high tide and also because it prevents burrow collapse. Fiddler crabs do not directly inhabit the intertidal zone as do other species of brachyuran crabs such as *Sesarma quadratum* and the ghost crab *Ocipoda albicans*.

The inter-tidal fish *Blennius pholis* shows clear-cut tidal rhythm in its swimming activity when observed under laboratory conditions for at least 5 days (Gibson, 1967). This is a littoral fish which lives under loose stones in pools between the high and low tide zones and migrates between feeding and resting grounds with each flood tide. Gibson studied the locomotor activity behaviour of this fish in the laboratory in an actograph that was designed on the knowledge that the fish sank and rested at the bottom of the sea between bouts of the activity coinciding with the high tide. Recording made in LL and DD at constant temperature and away from the tides in the laboratory revealed impressive overt tidal rhythmicity in the swimming activity, which, however, waned and damped out after 4–5 days (Fig. 9.3). Gibson has since reported (1973) that the young plaice (*Pleuronectes platessa* L.) caught in the intertidal zone exhibit a short-lived tidal rhythm when their swimming activity is recorded in darkness of the laboratory. This tidal rhythm rapidly changes to one of circadian frequency both in darkness and in light-dark cycles. Gibson concludes that the basic rhythm is circadian in nature, which can be entrained to be in phase with the tides by some, as yet unknown, Zeitgeber present under tidal conditions. Gibson's (1967) paper on circatidal rhythms in the locomotor activity of the intertidal fish *Beennius pholis* was the first report to describe such rhythms in a vertebrate. Daan and Koene (1981) reported that the oystercatcher birds *Haematopus ostralegus* synchronized their foraging flights such that they arrived during low tides on tidal mudflats.

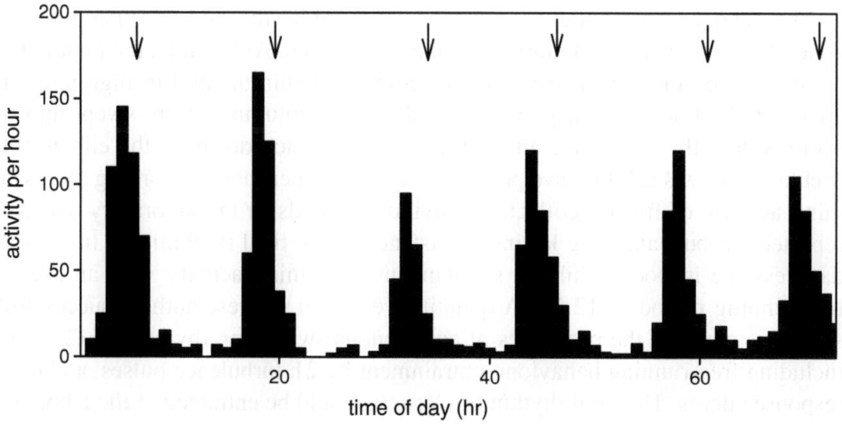

Fig. 9.3 The persistent tidal activity of the shanny *Blennius pholis* in continuous darkness (Modified from Gibson, 1965)

9.3 A Few Recent Researches on Tidal Rhythms

Wikelski and Hau (1995) observed for 3 years on two islands the Galapagos marine iguanas, a herbivorous reptile that graze on algae in the intertidal areas during low tide. These reptiles walked to the intertidal foraging grounds from far off resting areas in anticipation of the low tide. The degree of anticipation depended on environmental parameters such as wave action and food supply. They postulate an endogenous tidal rhythm to explain this behaviour. The iguanas assembled in good numbers on intertidal rocks during the neap tide nights, which may indicate that the iguanas possessed information on semi-monthly rhythms in tide heights. Enclosure experiments showed that the bitidal foraging rhythms of the iguanas might free-run in the absence of direct cues of their habitat. A tidal rhythm of gaping of valves under constant conditions of the laboratory coinciding with high tide has been reported (Williams and Pilditch, 1997) for the intertidal, burrowing bivalve *Austrovenus stutchburyi, which* also wanes after several days. This tidal rhythm in gaping could be reinstated by the authors by offering them pulses of algae made available at a tidal frequency. The authors state that the opening of the valves is not solely a response to the arrival of the food; usually, their opening anticipates the arrival of the food pulses. The tidal pattern seen during the pulsed food treatment persists for two cycles when food is withheld indicating the involvement of an internal timing system. *Helicion pectunculus* is a high-shore, crevice dwelling limpet active during nocturnal low tides and daytime low tides. Gray and Hodgson (1999) performed laboratory experiments recording their locomotor activity. Maximum entropy spectral analysis (MESA) revealed that the limpets possess a free-running endogenous rhythm of locomotor activity with both circadian (period = 28.1 h) and circatidal (period = 13.8 h) components.

The sand beach isopod *Cirolana cookii* inhabits the mid-to-upper intertidal zone of sheltered shores of northeastern New Zealand. At low tide, the populations are distributed in distinct bands in the sediment within the mid-to-high intertidal zone. As the tide rises, they swim into the water column and are swept into the shore, where they scavenge and are then washed back down by the ebbing tide. Mehta and Lewis (2000) have performed critical experiments recording the swimming activity of freshly collected individual isopods in the laboratory and have reported a robust and long lasting persistence of the tidal rhythmicity. In constant darkness the isopods exhibited spontaneous swimming activity with an average free-running period of 12.5 h. A quantitative model of these authors successfully simulates many of the properties of endogenous swimming rhythms of *C. cookii,* including free-running behaviour, entrainment by 2 h turbulence pulses, and phase response curves. The tidal rhythm of *C. cookii* could be entrained in the laboratory by turbulence cycles 1:11 (1 h pulse every 12 h) created by vigorous aeration of the sea water. Activity onset locks on to 1 h interval of agitation after a few days of exposure. Mehta and Lewis (2000) interpret the two peaks of activity coinciding with high tides as being circatidal in nature and as temporal adaptations to a complex intertidal environment. Circadian component was absent in the activity patterns of *C. cookii.*

The fundamental nature of the timing mechanisms of intertidal organisms has been controversial for a long time. Recently two major hypotheses have been formulated to explain the underlying control of tidal rhythms that show two components per lunar day. The first hypothesis was proposed for the clock control of the tidal rhythms of the Mediteranean shore crab *Carcinus maenas* (Reid and Naylor, 1989). This hypothesis is an explanation of the independent behaviour of the two clock components, solar day (24 h) and tidal (12.4 h) versions. However, an alternate hypothesis states that each component of a tidal rhythm is controlled by its own clock, which can function independently of its counterpart. Each clock has a fundamental lunidian (24.8 h) periodicity and is coupled antiphase to its partner (Palmer and Williams, 1986; Palmer, 1989, 1995). Under natural conditions, the two activity phases are maintained at a fixed distance from each other and display overt tidal (12.4 h) rhythmicity.

Mehta and Lewis (2000) conclude for the tidal rhythms in *Cirolana cookii* that: "The behaviour of our population of isopods and the quantitative model of their rhythmic timing system ... in general terms support the circalunidian model first proposed by Palmer and Williams (1986). Our data do not support Naylor's interacting circadian-circatidal system for the shore crab *Carcinus maenas* (Naylor, 1996). This crab shows a strong circadian component (Palmer, 1997)." It is relevant to mention here that Blume et al. (1962) had worked on the circadian and tidal rhythms of the shore crab *Carcinus maenas* and subjected their data to mathematical analyses, which revealed a periodicity of about 12.4 h besides the normal 24 h periodicity.

Observations on the rhythmic activity of juvenile specimens of the intertidal blenny *Zooarces viviparous* (Cummings and Morgan, 2001) revealed endogenous patterns of swimming at three different periodicities: the most obvious and most frequently observed pattern was a predominantly circatidal (12.4 h) period. Activity peaks are phased initially to the hours of the expected high tides of the habitat but gradually drift out of phase. Circadian modulation of the swimming activity becomes evident when some of the long-term data are plotted as actograms. Most fish studied became inactive within a week or so of collection, but in eight fish the level of activity remained sufficiently high to permit analyses over a long term. The overall activity fluctuated with a semilunar period of the activity maximum coinciding with the spring tides. Embryos attached to the female crab *Sesarma haematocheir* hatch synchronously within 1 h. Hatching is also synchronized near the time of the nocturnal high tide. These events are governed by a single circatidal clock in the female crab (Saigusa, 2002).

The cumacean *Dimorphostylis asiatica* (Crustacea) exhibits a circatidal swimming activity rhythm (Akiyama, 2004). The animals were exposed to a 12.5 h sinusoidal change of hydrostatic pressure of 0.3 atm amplitude in the laboratory. Under constant dark conditions, most of the specimens were entrained to a daily bimodal swimming activity rhythm by the hydrostatic pressure cycle. A marked feature was a flexible phase relationship between the entrained bimodal rhythm and the hydrostatic pressure cycles suggesting a week entrainment effect of the hydrostatic pressure.

Chabot et al. (2004) report circatidal and circadian rhythms of locomotion in the horseshoe crab *Limulus polyphemus*. The nocturnal increase in the sensitivity of the lateral eye of this crab found along the Atlantic coast, have been firmly established as being controlled by an endogenous circadian clock located in the brain (Barlow, 1983). Not much is known, about the control of the animal's tidal behavioural rhythms of mating and spawning that are observed in the intertidal zone during high tides late in spring. The semi-diurnal sandy beach serves as a breeding site for marine animals as phylogenetically diverse as the Pacific grunion *Leuresthes tenuis*, several species of sea turtles and the horseshoe crab *Limulus polyphemus* (Rudloe, 1980).

9.4 Tidal and Circadian Rhythms in *Emerita asiatica* (M.-Edw.)

The results are based on extensive experiments performed by one of the authors in the period 1961–1963 (Chandrashekaran, 1963, 1965). The anomuran mole crab *Emerita asiatica* (M.-Edw.) lives in populous colonies in the narrow intertidal strip of the beach in Madras (13° 4′ N, 80° 15′ E), which is only 8–12 m broad and quite steep. The colonies live close to the water edge and move in great synchrony up and down the beach during high and low tide, thus always remaining under a few cm of water. Committed to filtering its food through its elaborately plumose antennae the crabs may be seen to strain particulate matter and small sand-living copepods from the receding washes at high tide. In doing so the crabs face seawards and hop out of the sand and rebury themselves with the advent of every breaker.

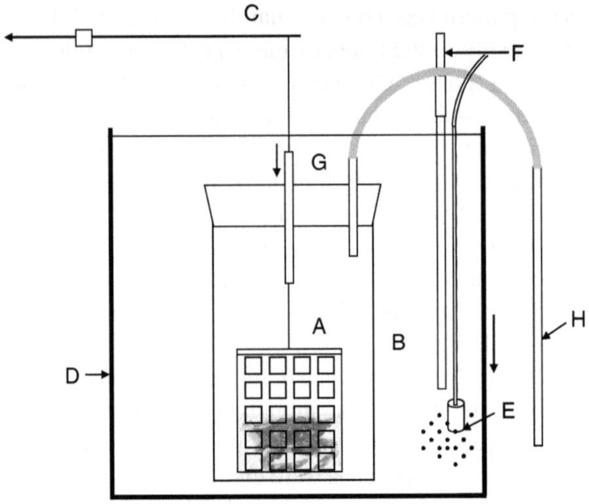

Fig. 9.4 The simple plastic cage device used to record the swimming activity of the crab *Emerita asiatica* with the continuous flow sea water arrangement to enable simultaneous oxygen consumption estimations. (A) activity cage (B) animal chamber (C) marking lever (D) constant water level trough (not drawn to scale) (G) inlet (H) outlet. When the crab swam up into the water away from the bottom, the cage rose inside the chamber. Such bouts of activity registered themselves through the free end of the writing lever on a slowly revolving kymograph drum (Modified after Chandrashekaran, 1965)

Experiments were performed in the laboratory in order to examine if the swimming activity rhythms in *Emerita* would persist in the constant conditions of the laboratory in the absence of the habitat tides. Using a simple continuous flow and activity cage set-up (Fig. 9.4) the oxygen consumption at 2 h intervals and continuous locomotor activity were monitored. Since the tidal rhythms in activity do not persist beyond 4–5 days the crabs were placed in the actograph within hours of capture from the beach which was a mere 200 m from the aquarium and the laboratory. Figs. 9.5–9.8 describe the rate of oxygen consumption and swimming activity measured on four female crabs for 24 h periods individually. We had not succeeded in finding male *Emerita* and it was not known then that there was an impressive sexual dimorphism in this mole crab. The males are considerably smaller than the females and are often transported in vast numbers under the folded telson of the females. They were even mistaken for juvenile crabs. Furthermore, the non-ovigerous females in the intermoult stages were sturdier and of ideal size for activity recordings. It is clear from the figures that there is a persistent tidal rhythmicity in the metabolic rate and swimming activity.

A closer examination of Figs. 9.5–9.8 reveals that the height of the high tide occurring during hours of darkness in the Madras coast was higher on all 4 days than that coinciding with daylight hours. Since daylength does not significantly vary in this latitude, one of the high tides invariably occurred during night hours. Interestingly it also turns out that all four crabs were more active during the nightly high tide. Therefore it is not clear if the exaggerated activity reflected a circadian

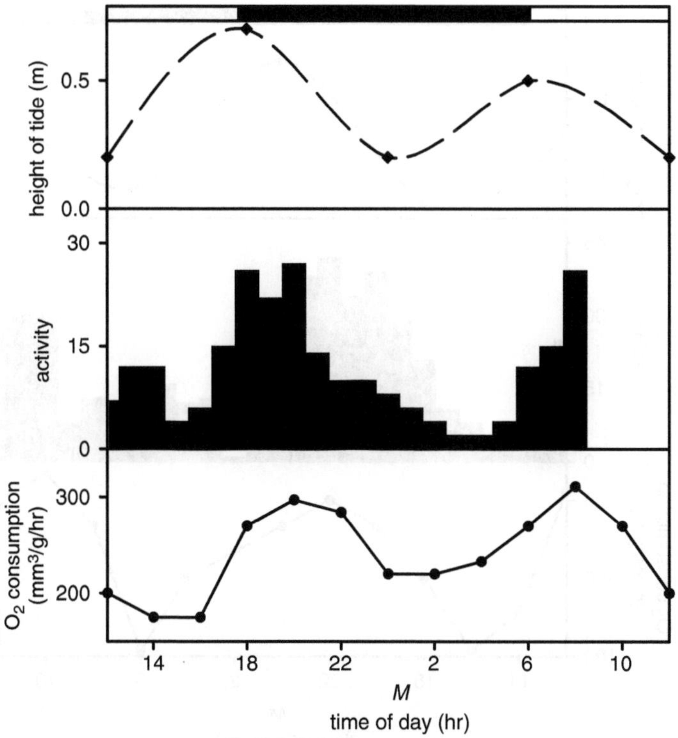

Fig. 9.5 The rate of oxygen consumption and spontaneous swimming activity estimated simultaneously on a non-ovigerous female *Emerita asiatica* crab in relation to the tidal stages in the habitat. Tidal data were obtained from the Tide Tables for the Madras Coast (13° 4′ N lat; 80° 15′ E long) published by the Geodetical Survey of India. Shaded bars indicate hours of darkness in the habitat (Modified after Chandrashekaran, 1965)

component being added on to the tidal activity component or if the elevated swimming activity was a function of the height of the high tides. Predation pressure is not as high during hours of darkness and the crabs could be more active. The higher the tide the more land area is under water and consequently more food is available. A critical experiment to tease apart circadian super-position on the one hand and the activity occurring as the function of the height of the high tide on the other hand, would be to perform experiments on days when the height of the nightly high tide would be the same as the height of the daytime high tide. Such days were very rare for unknown geophysical reasons – at least unknown then to the experimenter. For the coastline of India data on time of expected high and low tides are given in the Tide Tables published annually by the Geodetical Survey of the Government of India.

A scrutiny of the Tide Tables revealed that between 12 and 14 February 1963 the high tide at night was predicted to be of the same magnitude as that during daylight (0.9 m). An experiment was carried out during this period and the results are presented in Fig. 9.9. It was found that the crab showed a peak activity level

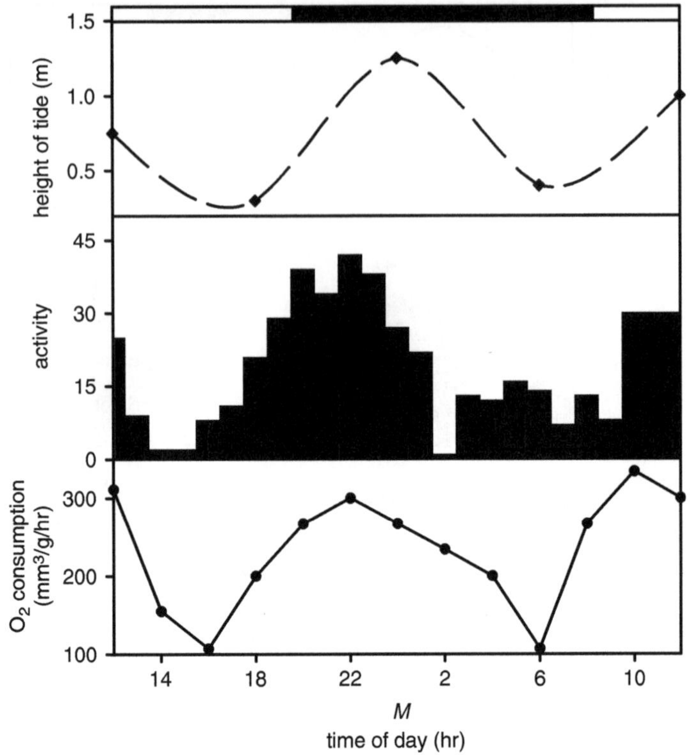

Fig. 9.6 The rate of oxygen consumption and spontaneous swimming activity estimated simultaneously on a non-ovigerous female *Emerita asiatica* crab in relation to the tidal stages in the habitat. Other details as in Fig. 9.5 (Modified after Chandrashekaran, 1965)

(39 cage-tilts/h) clearly higher at night than peak activity during the earlier daytime high tide (29 cage-tilts/h) at the peaks. Even the overall magnitude of activity was greater at nighttime.

Another experiment was performed when the noontime high tide was *higher* than that at both nights before and after. Such an inverted tidal regime was considered to be an ideal situation to clarify this complex issue further. The locomotor activity of a freshly captured crab was recorded for a continuous period of 39 h from 9 p.m. 2 March to 12 a.m. 5 March 1963. It was again noticed (Fig. 9.10) that the activity was more intense during the high tide coinciding with the first (B_1) and second nights (B_2) (26 cage-tilts/h and 24 cage-tilts/h respectively). The activity coinciding with the daytime high tide (A), which was much higher in the coastline than tides B_1 and B_2, was much more subdued No quantitative correlation between the amplitude of the high tide and magnitude of locomotor activity could be discerned. The results strongly suggest a circadian component in activity appearing superimposed on a tidal component, accounting for elevated activity levels during the nocturnal high tides.

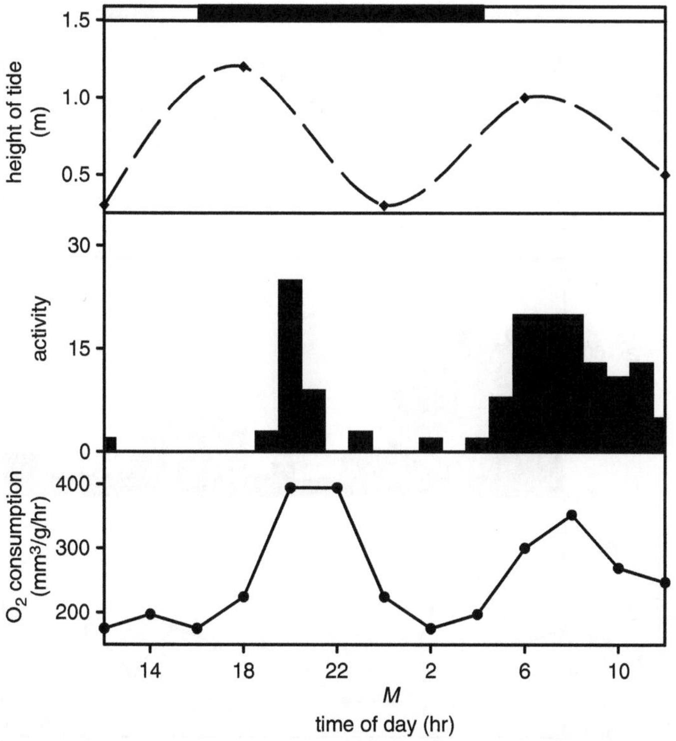

Fig. 9.7 The rate of oxygen consumption and spontaneous swimming activity estimated simultaneously on a non-ovigerous female *Emerita asiatica* crab in relation to the tidal stages in the habitat. Other details as in Fig. 9.5 (Modified after Chandrashekaran, 1965)

9.4.1 The Waning of the Tidal Rhythms of Emerita in the Laboratory

The swimming activity of a solitary *Emerita* was kymographically traced through 6 consecutive days and nights. The crab, a non-berried female in the intermoult stage, was in a healthy condition on release from the activity cage at the close of the experiment. The data obtained in this experiment are presented in Fig. 9.11. It is obvious that the amplitude and phase of the activity rhythms are in clear agreement with the amplitude and phases of the tidal cycles out in the habitat for the first 72 h of the experiment. On the 4th, 5th and 6th days there is a marked decrease in the level of activity. The amplitude and phases of the activity rhythm are so staggered in relation to the tidal cycle on these days that no clear-cut rhythmicity can be seen. About the latter part of the 6th day what was continuous activity earlier appears to have lapsed into spasmodic efforts. This is a clear case of waning of rhythms in the laboratory in the absence of tidal reinforcement. The elevated levels of locomotor activity during night time high tides for the first 3 days are impressive

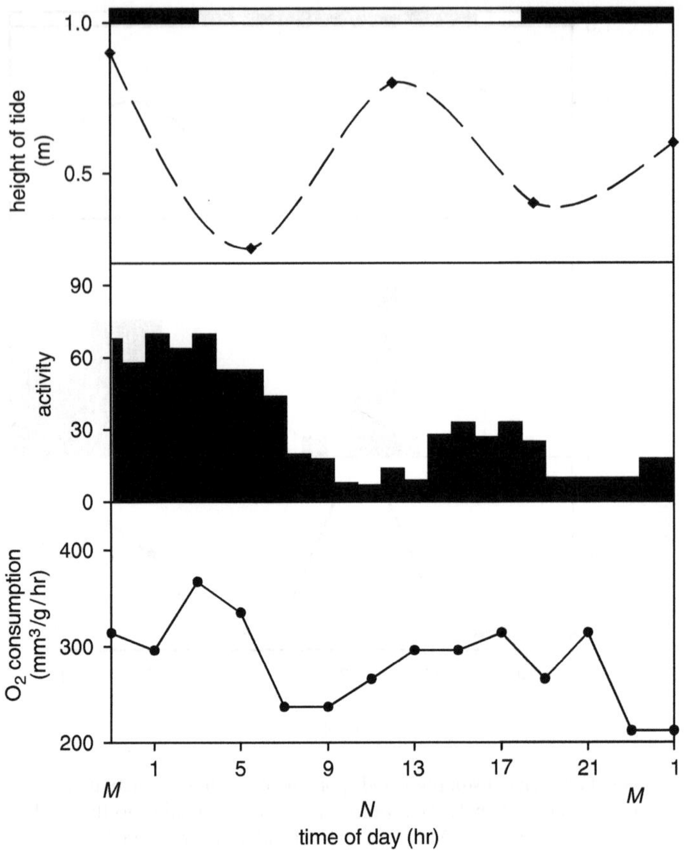

Fig. 9.8 The rate of oxygen consumption and spontaneous swimming activity estimated simultaneously on a non-ovigerous female *Emerita asiatica* crab in relation to the tidal stages in the habitat. Other details as in Fig. 9.5. Experiments illustrated in Figs. 9.5–9.8 were performed on different days (Modified after Chandrashekaran, 1965)

and a further confirmation of circadian (nocturnal) superposition of activity on tidal rhythms. This interpretation would also explain the findings of Enright (1963) on the elevated activity levels in *Synchelidium* since the nighttime high tides on the Californian coast where also the height of the nightly high tide is clearly higher than the height of the high tide of daylight hours.

Since the ambient seawater temperature of the water troughs in which *Emerita* were maintained was in the range of 26–28°C, it was difficult to keep the crabs in good health for longer than 3–4 weeks. The water was also vigorously aerated to keep the oxygen content reasonably high. One crab was examined for the profile of its swimming activity patterns 10 days after capture. Figure 9.12 illustrates the data and it is apparent that only the circadian component had survived the constant conditions of the laboratory.

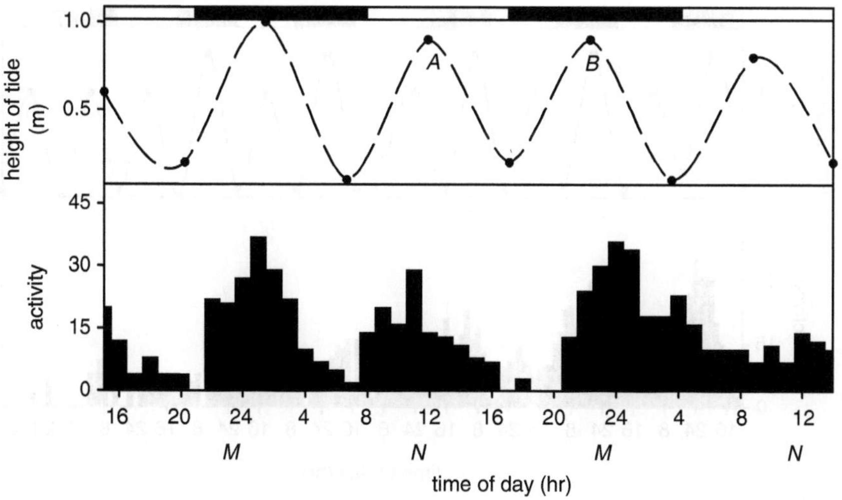

Fig. 9.9 Illustration of the tidal and circadian rhythms of locomotor activity of a solitary female *Emerita* over a period of 48 h when day (A) and night (B) high tides were of equal magnitude (0.9 m). Shaded bars indicate hours of darkness in the habitat. Details in text (Modified after Chandrashekaran, 1965)

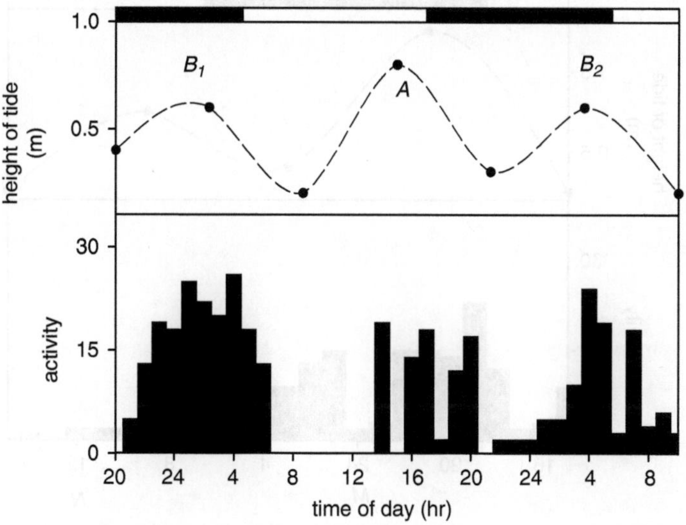

Fig. 9.10 Illustration of the tidal and circadian rhythms in the locomotor activity of a female specimen of *Emerita* recorded through 39 h when the daytime high tide (A – 0.8 m) was higher than the night time high tide high tide (B_1 and B_2 – 0.6 m) in the habitat Shaded bars indicate hours of darkness in the habitat. Details in text (Modified after Chandrashekaran, 1965)

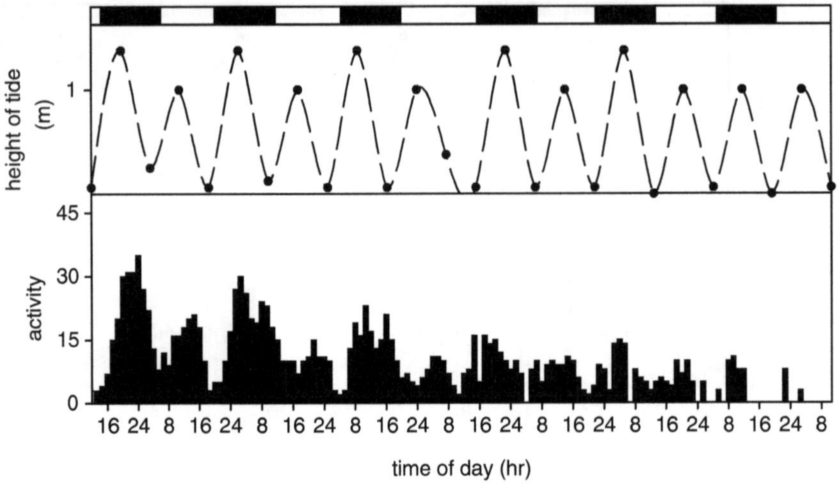

Fig. 9.11 The waning of the tidal rhythm in the swimming of a female *Emerita* in the course of the 6-day experimental recording. Shaded bars indicate hours of darkness outside (Modified after Chandrashekaran, 1965)

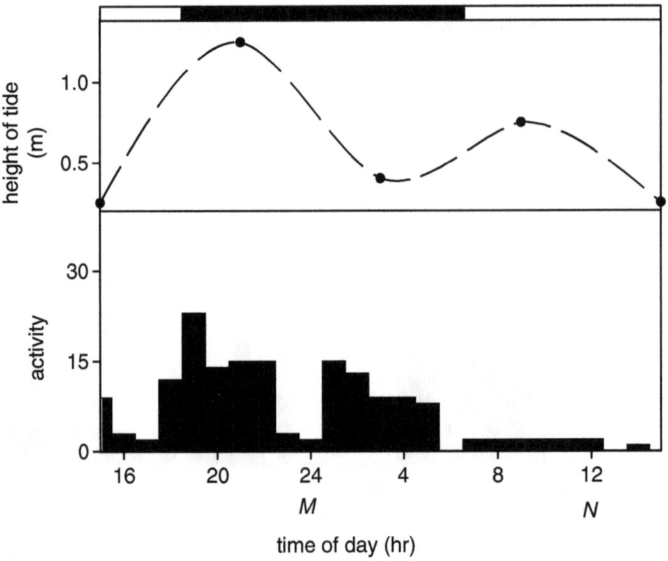

Fig. 9.12 The pattern of the spontaneous locomotor activity of a female *Emerita* 10 days after capture from the habitat. Shaded bar indicates hours of darkness outside (Modified after Chandrashekaran, 1965)

9.5 Zeitgeber(s)

The zeitgebers (entraining cues) of tidal rhythms are really not known. Hydrostatic pressure, turbulence of the waves and resultant mechanical agitation have all been claimed to be entraining cues. Enright (1961) found that an amphipod from the tidal zone can react to pressure changes of less than 0.01 atmospheres. It may be mentioned here that during the experiments and maintenance of crabs in seawater troughs that *Emerita asiatica* was extremely sensitive to hydrostatic pressure. They lived comfortably if sand was strewn on the bottom of the troughs and the seawater column was just 3–4 cm. When the bigger crabs scampered around their slate grey backs were exposed and churned and aerated the surface water. Enright (1963) also observed that the amphipod *Synchelidium* from La Jolla, California responds primarily to the turbulence of the streaming water as the decisive synchronizer of tidal rhythms. This holds also for the amphipod *Excirolana chiltoni* (Klapow) Variations in the hydrostatic pressure and wave action are synchronizers for the sand crab *Eurydice pulchra* (Jones and Naylor, 1970). It is reasonable to conclude that the intertidal invertebrates of the Madras coast were also additionally responding in nature to the turbulence of seawater, as did *Synchelidium, Excirolana chiltoni* and *Eurydice pulchra.*

One of us had (MKC) observed that in addition to *Emerita asiatica,* another anomuran burying crab *Albunea* sp. and a clam *Donax cuneatus* showed twice-daily tidal migrations in the Madras coast. These animals are washed up the shore during high tide and 3 h later down the shore during low tide, by so-called breakers. Breakers are big tidal waves, which occur once every 10–15 waves and end with a big swishing noise and a lot of foam thereby richly oxygenating the water. The mystery was how these animals "knew" a breaker was rolling towards them, especially in the case of the clam *Donax cuneatus* with their white and sturdy calcareous shells. The clams literally popped up from the sand each time a breaker washed up the beach. It was difficult to see live clams for they promptly reburied themselves in the sand and disappeared from sight. Turner and Belding (1957) suggested that the tidal migratory limits of the bivalve *Donax variabilis* occurring in the east coast of the USA were probably set by the "acoustic shocks" generated by the breakers. To prove this they stomped on the wet beach and saw that the bivalves just popped out. Chandrashekaran (1963) performed an experiment in the laboratory to investigate if *Donax cuneatus* responded to these "acoustic shocks" using a Philips audio-frequency generator (GM 2308/01) with a 20 W amplifier and a 10 W diaphragm loudspeaker for the purpose and discovered that *Donax cuneatus* responded to acoustic bombardments of frequencies 50, 100 and 150 Hz with increasingly rapid ventilating shell movements and protrusion of the foot and siphons. This interestingly also happens to be the range of acoustic/mechanical vibrations the breakers generate in the beach. If one lay close to the water-mark with an ear to the ground, one could hear the breakers rumbling with these typical frequencies. The clams did not respond to higher frequencies of 500, 1,000, 1,500, 2,000 and 2,500 Hz, which was in the nature of a buzz, which obviously do not have any environmental relevance to them. The results of these experiments confirm that the acoustic bombardment generated by the breakers trigger the popping up of organisms like *Emerita*, *Albunea* and *Donax cuneatus* from underneath the sand, to be washed up and down the beach (Chandrashekaran, 1963).

Even though the sensitivity of several species of *Donax* to sound or vibration had been reported as early as in 1938 (Mori, 1938) quantitative data on the sounds that elicited the popping up response of the clams were not available literally for the next 50 years until 1995 (Ellers, 1995a, b). Ellers (1995b) noted that on a beach close to North Carolina the *Donax variabilis* clams 'jumped' out of the sand and rode waves, migrating shoreward with the rising tide and seaward during the falling tide. This method of locomotion has been named 'swash riding' (Ellers, 1987, 1988). He elaborates "On dissipative beaches where the clams live, a breaking wave becomes a bore (a traveling cliff-like structure of tumbling water) and the bore becomes a swash, which the clams (*Donax spp.*) ride". Ellers postulates that sound is a reasonable cue that enables clams to anticipate waves since sound travels faster than, and in front of, the bore, thus announcing its arrival. The rumbling noise of an approaching bore consisted mostly of low frequencies ranging from 40 to 300 Hz, with the largest amplitudes being in the range of 60–100 Hz. The *Donax variabilis* were freshly captured and brought to the laboratory and placed on a sandy terrain between the front of the speaker of a hydrophone and the far end of the aquarium. Sounds were produced in different ways. Sound was recorded from waves on the beach or generated by a sine wave generator circuit. Pulsed sound waves of 22 s duration were presented to the clams starting 40 min after and ending 3 h and 15 min after high tide. Clams 4–16 cm away from the loudspeaker jumped out of the sand in response to sound pulses. Clams were responsive to low frequency sounds similar to those produced on a beach by waves and they were more responsive to louder sounds. Responsiveness to sound stimuli in the laboratory had an impressive endogenous tidal rhythm with high activity coinciding with hours of high tide and no activity at all during low tide (Fig. 9.13). Like all tidal rhythms reported this endogenous sensitivity to sound also persists over only three to five tidal cycles before waning. A careful scrutiny of Fig. 9.13 reveals an elevated sensitivity of the clams during high tides coinciding with hours of darkness in the habitat. We would like to believe that the elevated activity might be indicative of the superposition of a nocturnal circadian component on the nightly high tide component. Since the experiments were not performed beyond the fifth tidal cycle, and since the author has also not given the details of the actual height of the high tides, it is difficult to be assertive on this point. This is the first report demonstrating use of flow-induced sounds by an invertebrate.

Wave surge most likely represents a releasing as well as an orienting stimulus for movement of animals to breeding grounds and during their twice daily tidal migrations up and down the beach. Tidal rhythms wane rapidly, they are impervious to LD entrainment, compensated for temperature changes and the zeitgebers are not quite known. To an extent they are ephemeral phenomena to the rhythm researchers. The precision of tidal rhythms are also in no way comparable to that of circadian rhythms, which is not surprising, since in nature the light/dark cycles engendered by sunrise and sunset entrain them. Sunrise and sunset in any latitude are a highly dependable and noise free zeitgeber (Bünning, 1969; Chandrashekaran

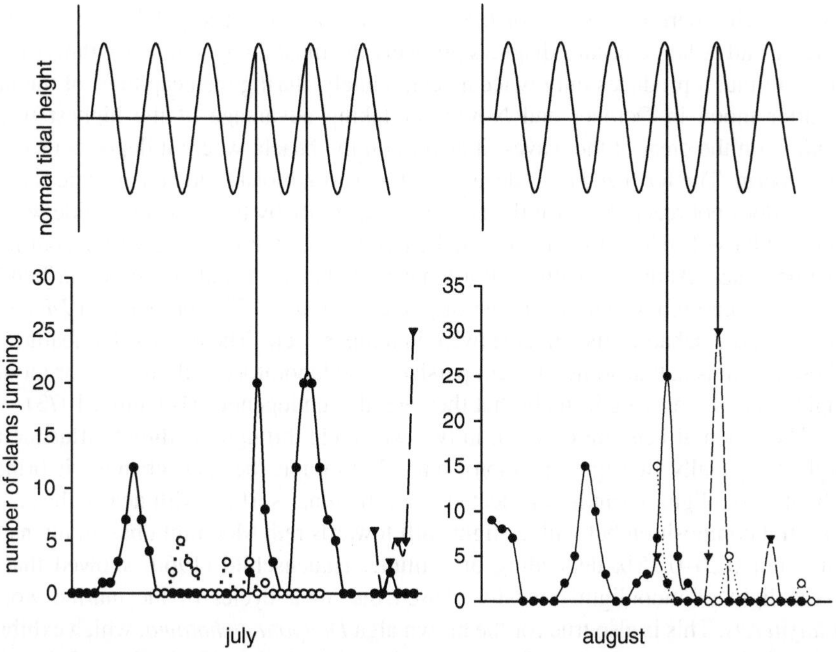

Fig. 9.13 Number of *Donax variabilis* in the laboratory jumping relative to the time of day and tide. Noon is N; midnight is M. Clams were most responsive around the time of high tide and showed no response around the time of low tide. The pattern shown is a compilation of the results of several experiments. Experiments were done in continuous light (Modified after Ellers, 1995b)

and Loher, 1969). Lunar/tidal rhythm researchers, for practical reasons, can work only in marine biological institutions situated close to the sea. These factors may account for the relative paucity of literature on tidal rhythms.

Textbooks inform us that a lunar day is 24 h 50 min long. Two high tides must therefore occur 12 h 25 min apart and so on. But a careful study of the tide table will make it clear that the situation never really happens in the real world outside owing to the topography of the tidal strip and meteorological conditions. Nature obviously cannot programme to accommodate for noise, which may be why tidal rhythms lack the precision of circadian rhythms. Lability of the period may be an asset in a constantly changing environment like the inter-tidal environment. It is also not known if tidal/lunar rhythms have a genetic basis like the circadian rhythms.

9.6 Lunar and Semi-Lunar Rhythms

Several marine organisms are known to have lunar breeding cycles with their physiological activities coinciding with neap or spring tides. They either have the lunar-monthly period of 29 days or semilunar periods of 14–15 days (Korringa, 1957).

We describe here only some of the better-known cases. The palolo worm of the Pacific and Atlantic oceans displays probably the most famous lunar rhythms of all. This animal reproduces only twice a year, namely, during the neap tides of the last quarter moon in October and November taking advantage of the highest tides. Riding on the crest of the waves, it arrives on the beach, where it deposits its eggs and sperm. The fertilized eggs develop in the warm, moist sand and are safe for the water does not reach them for the next 2 weeks. Only by the next spring tide would the fish have developed sufficiently to be freed from their eggs and washed out into the open sea. Activities with such a lunar periodicity have also been described in mussels, sea urchins and discharge of gametes in algae. The brown alga *Dictyota dichotoma* discharges its gametes twice in a lunar cycle. The remarkable feature of these rhythms is that some of them persist in the laboratory without tidal or moonlight exposure making it likely that they are also endogenous (Bünning, 1973).

The moon affects life on earth in two ways: (1) through the direct influence of light (especially during full moon), and (2) through the tidal changes it brings about. Moonlight is sunlight reflected from the lunar surface with nearly the same spectral composition but with a slight shift towards red. Moonlight has an intensity of about 0.20–0.25 lx depending on latitude. Hauenschild (1960) showed that a weak light of moonlight intensity phased the lunar cycles of the marine worm *Platynereis*. This is also true for the brown alga *Dictyota dichotoma*, which exhibits this relationship even more clearly. Bünning and Müller (1962) showed that the maximum discharge of eggs in *Dictyota dichotoma* takes place 9 days after exposure to moonlight. The next maximum then follows after an interval of 15–16 days. Midges of genus *Clunio* are found in the inter-tidal zones of the Atlantic and Pacific shores from the temperate areas to the Arctic. They have very curious and complicated life cycles. The *Clunio marinus* of the Atlantic coast of western Europe, for example, lives in the *lowest* parts of the inter-tidal zone exposed only during spring low water. It is only during these times that the insects are able to emerge (Neumann, 1963). Both sexes are extremely short-lived (ca. 2 h) and therefore copulation and oviposition must occur before the tide rises and covers the larval site. On the Normandy coast, low tide occurs twice a day at intervals of 12.4 h; consequently, low and high tides are about 50 min later each day and it takes a period of about 15 days before the times of low and high tides come full circle. Superimposed on this semi-diurnal tidal cycle is a semi-lunar tidal range. Eclosion of *Clunio marinus* in the Normandy coast is restricted to evening low tide (circadian component) and semi-lunar (lunar component) rhythms.

Neumann (1963, 1966a, b) analysed eclosion rhythms in *C. marinus* and showed it to be regulated by the super-position of the circadian rhythms governing eclosion and the semi-lunar rhythms governing initiation of pupation. In populations of *C. marinus* reared in the laboratory in LD cycles of 16:8 h, eclosion occurs towards the end of the photoperiod. The LL cultures show arrhythmic eclosion but a transfer to LD or an exposure to a single dark period re-initiates a circadian rhythm with a near 24 h period. Neumann's results show that, a circadian clock controls the *C. marinus* eclosion rhythm. The semi-lunar rhythm of pupation, which is superimposed on the circadian cycle, was entrained by natural or artificial moonlight (Neumann, 1966b).

Cultures of *C. marinus* from Normandy were raised in LD 12:12 h or LD 16:8 h and then exposed to pulses of weak light (0.4 lx) during the dark period for 4–6 days at intervals of 30 days. This treatment initiated and entrained a semi-lunar rhythmicity, which was absent from control populations without moonlight. In LD cycles of various periods, a strong synchronization was only possible in LD 12:12 h and LD 11:11 h, whereas in LD 10:10 h and LD 15:15 h the synchronization by the 30 day moonlight cycle was weak or absent. Neumann (1989) concluded that an LD cycle with a period close to 24 h is an essential zeitgeber for semilunar and lunar timing in this marine insect.

Considerable differences are known to occur between populations of these midges in different localities and different species of *Clunio*. These differences have been shown to be genotypic (Neumann, 1966a, b) and polygenic. Lunar rhythms of activity or eclosion (with a period of ca. 28 days) have been described under field conditions for a number of insects (Saunders, 1976). In at least two examples, in the mayfly *Povilla adusta* and ant lion *Myrmeleon obscurus* (Hartland-Rowe, 1958; Youthed and Moran, 1969a), these rhythms have been shown to be endogenous. Hartland-Rowe (1955, 1958) showed that the mayfly *Povilla adusta* emerged from the waters of Lake Victoria in its greatest numbers just after the full moon. This rhythm was maintained even after the nymphs had been kept in the dark for 10 days and in two individuals for 6 weeks. The second case concerns the rhythm of pit-building activity by larvae of the ant lion *Myrmeleon obscurus*. Using mean pit volume as a measure of activity, Youthed and Moran (1969b) showed that the maximum activity occurred at full moon. There was also a clear lunar day (24.8 h) rhythm with a peak in activity about 4 h after moonrise. The authors demonstrated that the observed lunar rhythm (period about 28 days) was a combination of this lunar day and the solar day (circadian) rhythm (Youthed and Moran, 1969a), which produces a peak in activity shortly after dark. The functional significance of these rhythms is still unclear.

Gwinner (1973) established that the nocturnal restlessness of robins and redstars increased during the migratory season in conjunction with the nocturnal illumination from the moon. In the night monkey *Aotus trivigatus*, activity increased with moonlight but in the phyllostomatid bat *Artibeus literatus,* activity decreased with moonlight (Erkert, 1974). Lunar monthly rhythms have been reported from light-trap studies in the abundance of the number of insect species, most of them from tropical latitudes. In a few species of microchiropteran bats in Madurai we found that there was very little overt foraging activity during full moon nights and enhanced activity during new moon nights. Morrison (1978) reported that the fruit bat *Artibeus jamaicensis* returned to their roosts between 01.00 and 07.00 h on full moon nights even though the sky was overcast. On new moon nights they continued to forage outside throughout the night. Morrison postulated that the activity cycle in the fruits bats is in some manner locked to the phases of the lunar cycle and called the phenomenon lunar phobia.

We at Madurai Kamaraj University (9° 58′ N lat; 78° 10′ E long) were out making bat counts of a few species of insectivorous bats on the night of 13 March 1979 when, unknown to us, a lunar eclipse occurred from 01.15 to 04.30 a.m. The bat counts made were mainly on *Pipistrellus* spp., *Rhinopoma hardwickei* and *Hipposideros speoris* or *H. bicolor*. The observers worked in two groups in two sites (sites A and B) at eastern and western extremes of the university campus. We found that the bat foraging activity (bat counts) suddenly increased during the eclipse. Figure 9.14 sets out the data on bat counts and insect abundance at the two sites for the night of 13 March 1979. The black areas of the horizontal bars on top indicate the time and duration of the lunar eclipse during this full moon night. The graphs at the bottom show control measurement at the same sites on a full moon night a

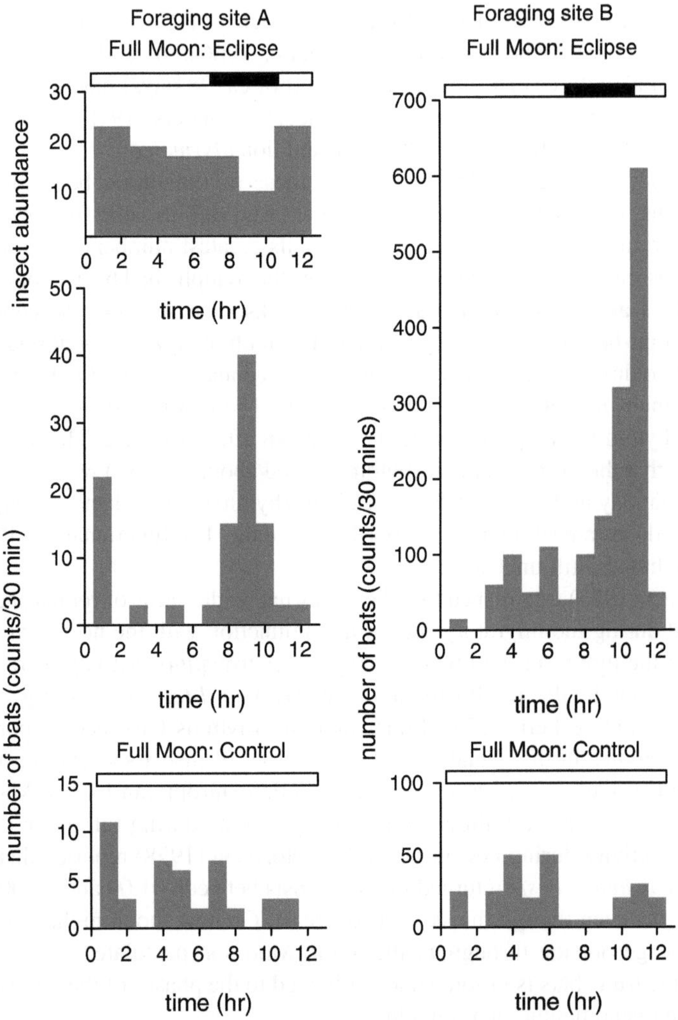

Fig. 9.14 Flight activity patterns of several species of microchiropteran bats on the night of 13 March 1979 recorded simultaneously by two groups of researchers at two foraging sites ca. 6 km apart in Madurai (9° 58′ N lat; 78° 10′ E long) (Modified after Usman et al., 1980)

lunar month later (Usman et al., 1980). We interpret the bat activity data to mean that full moon light is far too bright as it increases the risk of predatory attacks on these echolocating bats by owls that rely on eyesight. Therefore the bats start to fly under canopy cover. The only environmental factor we had noticed that can suppress the activity of these bats and drive them to their daytime roosts are gusty winds and torrential rains. We find no necessity to invoke an endogenous lunar rhythm to explain the activity of the bats on full moon nights. In spite of the early reports that moonlight acts as a zeitgeber to lunar cycles, more careful experiments need to be done to clinch the issue.

9.7 Similarities Between Circatidal, Circadian and Lunar Rhythms

A close relationship of the tidal rhythms with circadian rhythms can be inferred from the fact that tidal rhythms are also, to a large extent, independent of temperature (Bünning, 1973; Chandrashekaran, 2005). Even more striking is the fact that shore crabs, such as *Carcinus maenas* show only the circadian component of activity. In *Emerita asiatica* the circadian component is the more resilient and persists for weeks in the constant conditions of the laboratory. The findings of Honnegger (1973) for *Uca crenulata* also support strongly the inter-relatedness of tidal and circadian rhythms. Results of Neumann's experiments (1963, 1966a, b) are the most convincing support for the hypothesis that tidal, circadian and lunar rhythms are intricately interdependent conferring an adaptive value to organisms inhabiting the edges of the oceans.

9.8 Ecological Importance

In most cases the ecological importance of the lunar cyclic phenomena is quite obvious. Reproduction could be attuned to the environmental conditions of spring and neap tides. In some cases, the release of male and female gametes is synchronized and the chances of fertilization considerably enhanced. In addition reproduction being restricted to a few days in the lunar cycle, it is often restricted to a certain time of year and even to a certain time of the day. This confines the reproductive processes in marine organisms, both plants and animals, sometimes to even a few hours in an entire year as in the case of the grunion fish *Leuresthes tenuis* of the California coast and he palolo worm of the Pacific and Atlantic oceans, which reproduces only twice a year, as mentioned earlier. This holds, for example, also for the semilunar and lunar cycles of spore release in the alga *Dictyota*. By such "gating" of gamete release, the chances of male and female gametes uniting are increased a 1,000-fold over the chances when liberation is at random (Bünning, 1973). The endogenous nature of the rhythmicity enables the organism to find the appropriate time in nature, even when the moon is hidden for 2 or 3 months because of a perpetually overcast sky.

References

Akiyama, T. 2004. Entrainment of the circatidal swimming activity rhythm in the cumacean *Dimorphostylis asiatica* (Crustacea) to 12.5-hour hydrostatic pressure cycles. *Zool. Sci.* 21: 29–38.

Barlow, R. B. 1983. Circadian rhythms in the *Limulus* visual system. *J. Neurosci.* 13: 31–41.

Blume, J., Bünning, E. and Müller, D. 1962. Periodenanalyse von Aktivitätsrhythmen bei *Carcinus maenas. Biol. Z.* 81: 569–575.

Bohn, G. 1903. Sur les mouvements oscillatoires des *Convoluta roascoffensis. C. R. Acad. Sci.* (Paris) 137: 576–578.

Bohn, G. 1904. Periodicité vitale des animaux soumis aux oscillations du niveau des mers. *C.R. Acad. Sci.* (Paris) 139: 610–611.

Bohn, G. 1906. As cited in Bünning, 1973.

Brown, F. A., Bennett, M. F. and Webb, H. M. 1953. Endogenously regulated diurnal and tidal rhythms in metabolic rate in *Uca pugnax. Biol. Bull.* 105: 371.

Bünning, E. 1969. Common features of photoperiodism in plants and animals. *Photochem. Photobiol.* 9: 219–228.

Bünning, E. 1973. The Physiological Clock. 3rd rev Engl ed. pp. 258. Berlin/Heidelberg/New York: Springer.

Bünning, E. and Müller, D. 1962. Wie Messen Organismen lunare Zyklen? *Z. Naturforsch.* 16b: 391–395.

Chabot, C. C., Kent, J. and Watson, W. H. III. 2004. Circatidal and circadian rhythms of locomotion in *Limulus polyphemus. Biol. Bull.* 207: 72–75.

Chandrashekaran, M. K. 1963. Studies on the basal metabolism of tropical poikilotherms. Unpublished Ph.D. thesis. University of Madras, Madras, India.

Chandrashekaran, M. K. 1965. Persistent tidal and diurnal rhythms of locomotor activity and oxygen consumption in *Emerita asiatica. (M-Edw). Z. vergl. Physiol.* 50: 137–150.

Chandrashekaran, M. K. 2005. *Time in the Living World.* pp. 197. India: University Press.

Chandrashekaran, M. K. and Loher, W. 1969. The relationship between the intensity of light pulses and the extent of phase shifts of circadian rhythms in the eclosion rate of *Drosophila pseudoobscura. J. Exp. Zool.* 172: 147–152.

Cole, L. C. 1957. Biological clock in the unicorn. *Science.* 125: 874–876.

Cummings, S. M. and Morgan, E. 2001. Time-keeping system of the eel pout, *Zoarces viviparous. Chronobiol. Int.* 18: 27–46.

Daan, S. and Koene, P. 1981. On the timing of foraging flights by oystercatchers *Haematopus ostralegus* on tidal mudflats. *Neth. J. Sea Res.* 15: 1–22.

de la Iglesia H., Rodrguez, E. M. and Dezi, R. E. 1994. Burrow plugging in the crab *Uca uruguayensis* and its synchronization with photoperiod and tides. *Physiol. Behav.* 55: 913–919.

Ellers, O. 1987. Passive orientation of benthic animals in flow. In: *Signposts in the Sea. Proceedings of a Multidisciplinary Workshop on Marine Animal Orientation and Migration.* W. F. Herrenkind and A. B. Thistle, eds, pp. 45–68. Tallahassee, FL: Department of Biological Sciences, Florida State University.

Ellers, O. 1988. Locomotion via swash-riding in the clam *Donax varaiablis.* Ph.D. Dissertation, Duke University, Durham, NC.

Ellers, O. 1995a. Behavioral control of swash-riding in the clam *Donax variabilis. Biol. Bull.* 189: 120–127.

Ellers, O. 1995b. Discrimination among wave-generated sounds by a swash-riding clam. *Biol. Bull.* 189: 128–137.

Enright, J. T. 1961. Pressure sensitivity of an amphipod. *Science* 133: 758–760.

Enright, J. T. 1963. The tidal rhythm of activity of a sand-beach amphipod. *Z. vergl. Physiol.* 46: 276–313.

Enright, J. T. 1972. A virtuoso isopod: circa-lunar rhythms and their fine structure. *J. Comp. Physiol.* 77: 141–162.

Erkert, H. G. 1974. Einfluss des Mondlichtes auf die Aktivitätsperiodik nachtaktiver Säugetiere. *Oecologia* (Berlin) 14: 269–287.

Fingerman, M. 1960. Tidal rhythmicity in marine organisms. *Cold Spring Harb. Symp. Quant. Biol.* 25: 481–489.

Gamble, F. W. and Keeble, F. 1903. The bionomics of *Convoluta roscoffensis*, with special reference to its green cells. *Proc. Roy Soc.* B. 72: 93–98.

Gibson, R. N. 1965. Rhythmic activity in littoral fish. Nature 207: 544–545.

Gibson, R. N. 1967. Experiments on tidal rhythms of *Blennius pholis. J. Mar. Biol. Assoc. UK* 47: 97–111.

Gibson, R. N. 1973. Tidal and circadian activity rhythms in juvenile plaice. *Pleuronectes platessa. Mr. Biol.* 22: 379–386.

Gray, D. R. and Hodgson, A. N. 1999. Endogenous rhythms of locomotor activity in the high-shore limpet *Helcion pectunculus (Patellogastropoda). Anim. Behav.* 57: 387–391.

Gwinner, E. 1973. Circannual rhythms in birds: their interaction with circadian rhythms and environmental photoperiod. *J. Repod. Fertil. (Suppl)* 19: 51–65.

Hartland-Rowe, R. 1955. Lunar rhythm in the emergence of an *Ephemeropteran. Nature* (London) 176: 657.

Hartland-Rowe, R. 1958. The biology of a tropical mayfly *Povilla adusta Navas* (*Ephemeroptera, Polymitarcidae*) with special reference to the lunar rhythm in emergence. *Revue. Zool. Bot. Afr.* 58: 185–202.

Hauenschild, C. 1960. Lunar periodicity. *Cold Spring Harb. Symp. Quant. Biol.* 25: 491–497.

Honnegger, H. -W. 1973. Rhythmic motor activity responses of the California fiddler crab *Uca crenulata* to artificial light conditions. *Mar. Biol.* 18: 19–31.

Klapow, L. A. 1972. Natural and artificial rephasing of a tidal rhythm. *J. Comp. Physiol.* 79: 233–258.

Korringa, P. 1957. Lunar periodicity. *Mem. Geol. Soc. Am.* 67: 917–934.

Martin, L. 1907. La memoire chez *Convoluta roscoffensis. C. R. Acad. Sci.* (Paris) 145: 555–557.

Mehta, T. S. and Lewis, R. D. 2000. Quantitative tests of a dual circalunidian clock model for tidal rhythmicity in the sand beach isopod *Cirolana cookii. Chronobiol. Int.* 17: 29–41.

Mori, S. 1938. *Donax semigranosus Dkr* and the experimental analysis of its behaviour on the flood tide. *Zool. Mag*. Tokyo 50: 1–12.

Morrison, D. W. 1978. Lunar phobia in a neotropical bat *Artibeus jamaicensis. Anim. Behav.* 26: 852–855.

Naylor, E. 1996. Crab clockwork: the case for interactive circatidal and circadian oscillators controlling rhythmic motor activity of *Carcinus maenas. Chronobiol. Int.* 13: 153–161.

Neumann, D. 1963. Über die Steuerung der lunaren Schwärmperiode der Mücke *Clunio marinus. Verh. Dt. Zool. Ges. Wien.* 1962: 275–285.

Neumann, D. 1966a. Die intraspezifische variabilität der lunaren und täglichen Sclüpfzeiten von *Clunio marinus* (Diptera: Chironomidae). *Verh. Dt. Zool. Ges. Jena.* 1965: 223–233.

Neumann, D. 1966b. Der lunare und tägliche Sclüpfperiodik der Mücke *Clunio* steuerung und abstimmung auf die Gezeitenperiodik. *Z. Vergl. Physiol.* 53: 1–61.

Neumann, D. 1989. Circadian components of semilunar and lunar timing mechanisms. *J. Biol. Rhythms* 4: 285–294.

Palmer, J. D. 1989. Comparative studies of tidal rhythms. VIII. A translocation experiment involving circalunidian rhythms. *Mar. Behav. Physiol.* 14: 231–243.

Palmer, J. D. 1995. Review of the dual-clock control of tidal rhythms and the hypothesis that the same clock governs both circatidal and circadian rhythms. *Chronobiol. Int.* 12: 299–310.

Palmer, J. D. 1997. Dueling hypothesis: circatidal versus circalunidian battle basics. *Chronobiol. Int.* 14: 337–346.

Palmer, J. D. and Williams, B. G. 1986. Comparative study of tidal rhythms: the dual clock control of the locomotor rhythms of two decapod crustaceans. *Mar. Behav. Physiol.* 12: 269–278.

Reid, D. G. and Naylor, E. 1989. Are there separate circatidal and circadian clocks in the shore crab *Carcinus maenas*? *Mar. Ecol. Prog*. 52: 1–6.

Rudloe, A. 1980. The breeding behaviour and patterns of movement of horseshoe crabs. *Limulus polyphemus,* in the vicinity of breeding beaches in Apalachee Bay, FL. *Estuaries* 3: 177–183.

Saigusa, M. 2002. Hatching controlled by the circadian clock and the role of the medulla terminalis in the optic peduncle of the eyestalk, in an estuarine crab *Sesarma haematocheir. J. Exp. Biol.* 205: 3487–3504.

Saunders, D. S. 1976. *Insect Clocks*. pp. 279. Oxford: Pergamon.

Turner, H. J. and Belding, D. L. 1957. The tidal migration of Donax variables. Contribution No. 886. Woodshole, MA.

Usman, K., Habersetzer, J., Subbaraj, R., Gopalakrishnaswamy, G. and Paramanandam, K. 1980. Behaviour of bats during a lunar eclipse. *Behav. Ecol. Sociobiol.* 7: 79–81.

Youthed, G. J. and Moran, V. C. 1969a. The solar-day activity rhythm of Myrmeleontid larvae. *J. Insect Physiol.* 15: 1103–1116.

Youthed, G. J. and Moran, V. C. 1969b. The lunar-day activity rhythm of Myrmeleontid larvae. *J. Insect Physiol.* 15: 1259–1271.

Wikelski, M. and Hau, M. 1995. Is there an endogenous tidal rhythm in marine iguanas? *J. Biol. Rhythms* 10: 335–350.

Williams, B. G. and Pilditch, C. A. 1997. The entrainment of persistent tidal rhythmicity in a filter-feeding bivalve using cycles of food availability. *J. Biol. Rhythms* 12: 173–181.

Part III
The Neuroendocrineal and Developmental Level

Chapter 10
Pulsatile Hormone Secretion: Mechanisms, Significance and Evaluation

J.D. Veldhuis

Abstract Secretion of anterior and posterior pituitary hormones, adrenal glucocorticoids, mineralocorticoids and catecholamines, gonadal sex steroids, parathormone, insulin and glucagon is pulsatile (burst-like or episodic). Neuronal inputs, cellular excitability and feedback with time delays constitute proximate mechanisms driving recurrent pulses. Both the amplitude and frequency of gonadotropin, thyrotropin, prolactin and sex-steroid pulses determine their mean concentrations. In contradistinction, primarily the amplitude of growth hormone, adrenocorticotropin, cortisol, parathormone and insulin pulses controls their average values. Gonadotropin-releasing hormone and growth-hormone pulses convey unique signaling information to target tissues. Evaluation of the mechanisms that govern pulsatile hormone secretion requires simultaneous quantification of the number, size and shape of secretory bursts, underlying nonpulsatile (basal) secretion and associated elimination kinetics. The necessary methodology is termed deconvolution analysis. More complex ensemble models are used to interlink neurohormone signals by estimating endogenous dose-response curves noninvasively. The approximate entropy statistic is a specific and sensitive measure of feedback fidelity within an ensemble system. Implications of neurohormone pulsatility include regulation of somatic growth, reproduction, muscle and bone mass, visceral fat burden, glucose metabolism, parturition, and sodium and water balance.

Keywords Pulsatile, hormone, analysis, secretion, model, deconvolution, signal, endocrine

---

Endocrine Research Unit, Mayo Medical and Graduate School, Center for Translational Science Mayo Clinic, Rochester, MN 55905, USA
Email: Veldhuis.Johannes@mayo.edu

10.1 Introduction

The intermittency of hormone secretion is endowed by the embryonic origin of endocrine cells from primitive neuroectoderm and the anatomic interface between neurons and endocrine cells in the adult. A pulse is definable as a punctuated time-delimited event marked by a sequential increase and decrease in the neurohormone secretion rate. Pulses serve as signals mediating homeostatic control. Multiple regulatory inputs are necessary to generate recurrent pulses. Physiological mechanisms that engender pulsatility, albeit not fully understood, are altered in aging, regulated by development and disrupted by stress, illness and disease. Pathologically flawed pulse-generating mechanisms both reflect and give rise to overt disease, such as hypogonadotropic hypogonadism when gonadotropin-releasing hormone (GnRH) pulses are abolished (Urban et al., 1988), and growth failure when growth-hormone (GH) releasing-hormone (GHRH) receptor function is abrogated (Giustina and Veldhuis, 1998; Veldhuis et al., 2006). The theme of hormone pulsatility embraces both the mechanisms of pulse generation and the target-tissue actions of pulses. In both circumstances the clinician and investigator must be able to quantify pulses objectively and accurately. Thus, this chapter addresses mechanisms of pulse generation, physiological significance and analytical quantification.

10.2 Mechanisms of Pulse Generation

10.2.1 Prototypical Neuroendocrine Axes

A functional neurohormone axis requires the existence and operation of two or more regulatory signals, e.g. a driving (feedforward) signal emanating from the hormone-secreting gland, and an inhibitory (feedback) signal secreted by cells acted upon by the hormone. An axis may be viewed at various levels of complexity. One simple axis comprises parathormone (PTH) and calcium ions, which interact reciprocally (Fig. 10.1). In particular, PTH elevates calcium concentrations, and calcium concentrations repress PTH secretion. In most axes, a third regulatory signal confers more precise modulation. In the calcium-PTH system, the third effector is Vitamin D. At higher levels of regulatory complexity, numerous other signals also participate. For calcium/PTH regulation, collateral effectors include phosphorous, magnesium, phosphotonin and other modulators.

10.2.2 PTH and Calcium Pulses

Parathormone (PTH) secretion is pulsatile. However, the fundamental mechanisms of PTH pulse generation have not been elucidated. The frequency of PTH pulses (one pulse every 7 min) is much higher than that of anterior-pituitary hormones

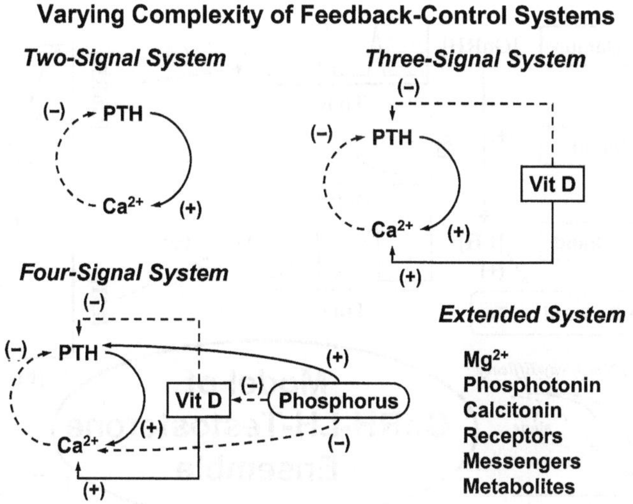

Fig. 10.1 Models of a prototypical endocrine axis comprising PTH and Ca^{2+} (*top left*), PTH, Ca^{2+} and vitamin D (*top right*), and the same three signals along with phosphorus (*bottom left*) and other comodulators (*bottom right*)

(about one pulse every 60 min). In addition, at least 65% of PTH secretion is basal (nonpulsatile), compared with only 5–15% for most pituitary hormones (Veldhuis et al., 1990). PTH and Ca^{2+} concentrations vary inversely with a 2-min lag time, denoting tight reciprocal coupling. Brief induction of hypercalcemia or hypocalcemia respectively decrease and increase PTH pulse size without disrupting frequency. Analogously, pathophysiologies like end-stage renal failure and primary hyperparathyroidism augment the mass of PTH released per burst without altering burst frequency (Goodman et al., 1998). Therefore, PTH secretion is under strong amplitude-specific control.

10.2.3 Gonadotropic Axis

In the reproductive axis, brain gonadotropin-releasing hormone (GnRH) pulses trigger pituitary luteinizing hormone (LH) pulses, which in turn stimulate gonadal estrogen and testosterone (Te) pulses (Evans et al., 1992; Urban et al., 1988). The set of signals (GnRH, LH and sex steroids) constitutes the core axis, given that loss of any single effector abolishes overall function. Feedback (inhibitory) and feedforward (stimulatory) signal exchange maintains the concentrations of GnRH, LH and sex steroids within relatively narrow, age-, gender- and species-specific ranges (Keenan and Veldhuis, 2003a) (Fig. 10.2). The dose-response properties of the interactions change during intrauterine development, infancy, childhood, puberty and aging, as well as in response to internal and external stress or overt disease. Recent

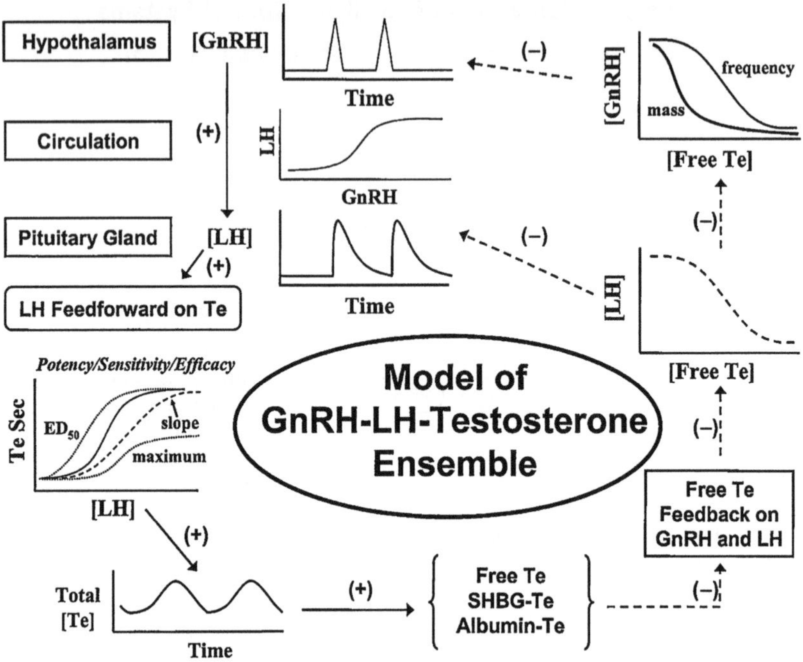

Fig. 10.2 Simplified schema of the male GnRH-LH-testosterone [Te] (hypothalamo-pituitary-Leydig cell) axis with key feedback (–, inhibitory) and feedforward (+, stimulatory) interactions. Interactions are mediated by implicit *in vivo* dose-response interfaces, shown by the descending and ascending curves

analyses of the male gonadal axis indicate that feedback and feedforward (homeostatic) dynamics unfold in a manner that reflects not only cause-and-effect (deterministic) but also unpredictable (random, stochastic) inputs (Keenan et al., 2004). For example, stochastic features typify timing of the GnRH pulsing mechanism, the size of successive LH pulses and burst-by-burst dose-response properties defining LH → Te feedforward (Keenan et al., 2006).

In the gonadal axis, all three signals (GnRH, LH and Te or estradiol) are vividly pulsatile when measured at or near their sites of secretion (Veldhuis et al., 2004). In the monkey, horse, sheep and rat, GnRH and LH pulses coincide one-to-one when quantified together in hypothalamopituitary portal blood or intercavernous-sinus blood exiting the pituitary gland (Clarke and Cummins, 1982). Although invasive sampling of pituitary blood is not possible in the human, injected GnRH pulses elicit LH pulses in a dose-dependent manner in GnRH-primed patients with GnRH deficiency (Conn and Crowley, 1991; Veldhuis et al., 1989b).

Genetic and pharmacological experiments establish that hypothalamic GnRH output supervises pulsatility of LH and thereby Te secretion. Whereas neither LH nor Te secretion is pulsatile *in vitro*, LH and Te pulses coincide closely *in vivo* (Foresta et al., 1997). These events are not independent of the hypothalamus, since

GnRH-receptor antagonists abolish both LH and Te pulsatility. In addition, Te and its metabolites repress the number and size of GnRH pulses via negative feedback, thereby framing a testis-brain inhibitory loop. Moreover, Te reduces the efficacy of GnRH in stimulating gonadotropes, therein creating a testis-pituitary feedback arc (Urban et al., 1988; Veldhuis et al., 2004). Thus, the overall dynamics arise from two feedforward (GnRH → LH and LH → Te) and two feedback (Te → GnRH and Te → [GnRH → LH]) linkages (Fig. 10.2). Precisely how all four regulatory loops modulate dynamic pulsatility has not been established experimentally. However, biomathematical models can satisfactorily relate the combined pathways via time-delayed reciprocal interactions (Keenan et al., 2004, 2006).

10.2.4 Somatotropic Axis

The secretion of each of GH, GH-releasing hormone (GHRH) and somatostatin (SS) is pulsatile. A third peptidyl regulator is the GH-releasing peptide (GHRP), ghrelin, which functions as an amplifier of GHRH drive and an antagonist of SS restraint (Veldhuis et al., 2006). Whether SS, GHRH or ghrelin neurons are able to generate coherent secretory bursts *ex vivo* remains uncertain. Accordingly, recent biomathematical constructs formulate GH pulsatility as the consequence of time-delayed reciprocal interactions between GHRH (stimulatory) and SS (inhibitory) neurons in the hypothalamus and between GHRH and SS peptides' directing the synthesis and release of GH by the pituitary (Farhy et al., 2007; Farhy and Veldhuis, 2004) (Fig. 10.3). Interneuronal

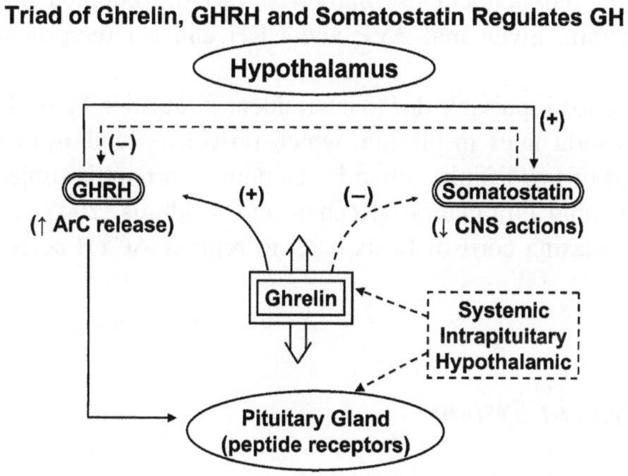

Fig. 10.3 Connections (– stimulatory, and—inhibitory) among three primary peptidyl regulators that jointly control GH secretion. GHRH = GH-releasing hormone, ArC = arcuate nucleus, and CNS = central nervous system. Interactions among the signals, rather than the effects of any one signal, control the pituitary secretion of GH

synapses, rather than GHRH peptide per se, seem to regulate SS and thereby GH pulses, since rare patients with deletional mutations of the GHRH receptor maintain GH pulses that are 30-fold smaller but of normal timing (Roelfsema et al., 2000). Alternatively, periventricular SS neurons may function as oscillators which are driven by time-delayed negative feedback via autoreceptors (SSTR-1). A primary SS pulse generator would be consistent with preservation of normal GH pulse frequency in volunteers with GHRH-receptor mutations and patients given single and combined continuous intravenous infusions of GHRH and GHRP [a synthetic ghrelin analog] (Roelfsema et al., 2000; Vance et al., 1985; Veldhuis et al., 2002). Whereas different simulation models are able to create expected patterns of GH, GHRH and SS pulsatility, direct experimental discrimination among the models is not yet possible. This deficiency reflects the enormous difficulty in studying *in vivo* mechanisms of pulse generation. Interdisciplinary efforts are needed to successfully combine molecular, cellular, neurophysiological and mathematical methods.

10.2.5 Stress-Responsive Adrenal Axis

The corticotropic (ACTH) axis is regulated primarily by two hypothalmic feedforward signals, corticotropin-releasing hormone (CRH) and arginine vasopressin (AVP), and two adrenally derived feedback signals, cortisol and aldosterone (Keenan et al., 2001) (Fig. 10.4A). The proximate site of pulse generation is hypothalamic, since both CRH and AVP are secreted from the brain in distinct bursts (Engler et al., 1989; Redekopp et al., 1986). Moreover, neither ACTH nor cortisol secretion is pulsatile *in vitro*. Whether CRH and AVP neurons are strictly independent oscillators or variably coupled pulse generators also is not certain, given that AVP and CRH can act reciprocally within the hypothalamus.

Cortisol secretion is pulsatile due to intermittent stimulation by ACTH (Fig. 10.4B). Cortisol has several fates in plasma, which reflect its binding to a high-affinity globulin (CBG) or low-affinity albumin. Unbound cortisol is subject to diffusion, advection and rapid elimination (Keenan and Veldhuis, 2003a). Whether free, bound or total plasma cortisol feeds back to repress ACTH secretion is not yet clear.

10.2.6 Oxytocin System

Oxytocin neurons have been studied more extensively. Whereas the precise molecular basis for pulse generation is undisclosed, autofeedback by oxytocin onto cognate neurons appears to be important (Moos et al., 2004). Selective genetic deletion of brain oxytocin receptors will be needed to verify this inference.

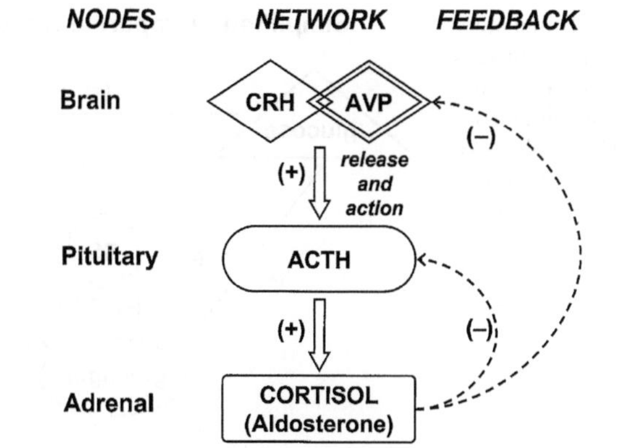

Fig. 10.4 ***Panel A***. Dual hypothalamic feedforward by CRH and AVP onto ACTH secretion, and dual feedback by cortisol and aldosterone onto CRH, AVP and ACTH secretion. ***Panel B***. ACTH concentrations drive cortisol secretion (*top section*). Free and bound cortisol concentrations (*bottom section*) are determined by the concentrations of total cortisol, albumin and cortisol binding-globulin (CBG). Cortisol may undergo the five fates indicated by the subscripted rate constants (k's)

10.2.7 Single-Gland Pulsatility

Pancreatic islets generate insulin pulses every 4–8 min *in vivo* and *in vitro*. Glucose selectively augments the size of insulin secretory bursts in the human without altering their frequency (Ritzel et al., 2003). Peaks are partially censored immediately after secretion, because the liver extracts 50–85% of inflowing insulin, and the large systemic distribution volume dilutes insulin markedly (Meier et al., 2005).

Individual beta cells maintain rapid rhythmic oscillations (every 12 s and every 3 min) in intracellular Ca^{2+} concentrations and membrane capacitance, the latter

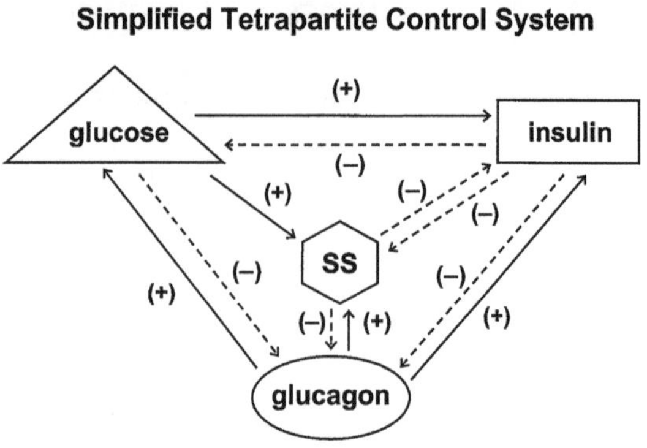

Fig. 10.5 Hypothesized stimulatory (+) and inhibitory (–) linkages among insulin, glucagon and somatostatin (SS) in regulating glucose concentrations. The collective interactions rather than the effect of any single hormone putatively interlink glucose, insulin, SS and glucagon pulses

reflecting exocytotic vesicle fusion (Porksen et al., 2002). What remains unknown is how secretion by individual beta cells within an islet and among islets is synchronized to yield rapid bursts. Diffusible substances within the islet (e.g., excitatory and inhibitory neurotransmitters, ATP, amylin, somatostatin, C-peptide, glucagon, zinc, adenosine *inter alia*) may be relevant. In addition, an intrapancreatic neuronal network originating in the celiac plexus and insulinotropic enteric peptides like GLP (glucagon-like peptide) might confer larger-scale synchrony among islet clusters. A plausible integrative model for insulin secretion, albeit unproven, is that intraislet (a) insulin stimulates glucagon and somatostatin release; (b) somatostatin inhibits secretion of both other peptides; and (c) glucagon induces somatostatin and represses insulin secretion (Fig. 10.5). Indirect evidence for this notion is the time-delayed inverse correlation between insulin and glucagon secretion, and the capability of somatostatin to quench the release of both peptides (Porksen et al., 1996).

10.3 Implications of Pulsatile Hormone Secretion

10.3.1 Reproductive Axis

A compelling example of the requirement for pulsatility is GnRH's intermittent stimulation of LH secretion (Veldhuis et al., 2004). GnRH neurons in the arcuate nucleus of the primate hypothalamus exhibit recurrent bursts of multiunit electrophysiological activity, which correlate one-to-one with LH and putatively GnRH pulses (Knobil, 1990). Direct sampling of median-eminence tissue fluids, hypothalamo-pituitary portal blood and pituitary (cavernous-sinus) blood has demonstrated punctuated bursts of GnRH outflow (Evans et al., 1992; Urban et al., 1988). With rare possible exceptions, such as end-stage renal failure, GnRH pulses coincide one-to-one with LH pulses sampled in jugular or peripheral blood (Schaefer et al., 1994). Destruction of the arcuate nucleus, surgical interruption of the hypothalamo-pituitary stalk and genetic GnRH deficiency abolish large-amplitude LH pulses, whereas injected GnRH pulses restore concordant LH pulses (Conn and Crowley, 1991).

Continuous delivery of GnRH rapidly downregulates (causes pharmacological tachyphylaxis of) LH secretion (Conn and Crowley, 1991). Downregulation is also imposed by higher frequencies of GnRH pulses, typically exceeding one pulse every 40 min (Knobil, 1990). The mechanisms of downregulation comprise repression of the LH-specific beta-subunit gene, depletion of membrane-associated GnRH receptors, and impaired intracellular signaling to secretory granules. For reasons that are not so clear, higher GnRH pulse frequencies upregulate FSH-specific beta-subunit gene expression. The clinical implication of these dynamics is that potent long-acting synthetic GnRH-receptor agonists suppress gonadotrope LH (and to a lesser degree FSH) secretion, and thereby decrease gonadal sex-steroid concentrations (Conn and Crowley, 1991). GnRH agonists are used to treat prostatic cancer, endometriosis, precocious puberty and other disorders defined by gonadal sex-steroid dependence.

In contrast to the pituitary gland's requirement for GnRH pulses, whether maximal gonadal sex-hormone secretion requires pulsatile LH stimulation has not been ascertained in the human. However, pulsatile and constant LH administration have comparable efficacy in stimulating Te secretion in the rat and sheep (Chase et al., 1988; Gibson-Berry and Chase, 1990).

10.3.2 Growth-Hormone Axis

GHRH induces homologous-receptor downregulation *in vitro* (Giustina and Veldhuis, 1998; Veldhuis et al., 2006). In contradistinction, constant infusion of GHRH for 2 weeks *in vivo* augments pulsatile GH secretion without altering the timing of GH pulses (Vance et al., 1989). Indeed, rare neuroectodermal tumors that secrete massive amounts of GHRH in a continuous fashion drive high-amplitude GH pulses (Veldhuis et al., 2006). Thus, physiological intermittency of GHRH release is not obligatory to maintain pulsatile GH secretion.

The endogenous GHRP, ghrelin, is minimally pulsatile compared with GHRH or GnRH. Whether continuous ghrelin exposure maintains pulsatile GH secretion is not known. The consideration is likely, because constant infusion of the ghrelin analog, GHRP-2, amplifies pulsatile GH output by twofold for at least 90 days (Bowers et al., 2004).

Hypothalamic SS secretion is pulsatile. Nonetheless, continuous administration of a SS analog (octreotide or lanreotide) suppresses the amplitude of GH pulses by >85% in healthy individuals and by >50% in some patients with GH-secreting pituitary tumors (Biermasz et al., 2004).

Intermittent and continuous GH injections differ in their efficacies for inducing or repressing certain genes (Jaffe et al., 2002; Shapiro et al., 1989). Large infrequent GH pulses (≤6 pulses/24 h) typical of the adult male rat induce hepatic CYP2C11 and muscle IGF-I gene expression more effectively than small frequent

GH pulses (≥7 pulses/day) characteristic of the female animal (Veldhuis et al., 2001b). The feminine GH pulse pattern especially induces hepatic expression of CYP2C12, IGF-I, GH-binding protein and the LDL receptor (Giustina and Veldhuis, 1998). Clinical studies indicate that a single pulse of GH is more effective than the same dose of GH infused continuously in stimulating lipolysis, and that the IGF-I response in estrogen-sufficient women is less responsive GH than that in testosterone-sufficient men (Veldhuis et al., 2006). In longer-term studies, pulsatile and continuous GH delivery in the human exert distinguishable effects on hepatic drug metabolism, LDL cholesterol and IGF-I concentrations (Jaffe et al., 2002). The reproducibility and significance of these findings is not yet clear.

Molecular analyses indicate that a single GH pulse can activate the STAT-5b (signal transducer and activator of transcription) signaling pathway. Rare genetic mutations and transgenic silencing of STAT-5b further demonstrate that this intracellular signal mediates GH's growth-promoting actions (Hwa et al., 2005). Continuous GH exposure downregulates STAT-5b activity by inducing the feedback repressor, SOCS (suppressor of cytokine signaling), and concomitantly upregulates GH-responsive genes via STAT-5a, HNF (hepatic nuclear factor) and other transcriptional proteins. Estrogen also attenuates GH action by stimulating SOCS (Leung et al., 2003). Accordingly, the time-mode of GH delivery to target cells directs distinct signaling pathways that converge on specific genes. Pulsatility patterns in the rat endow strong sexual dimorphism of body size, hepatic drug-detoxifying enzymes, lipoprotein receptors and tissue-specific IGF-I synthesis. Although similar gender-dependent differences may exist in the human, clinical studies have been few, conflicting, and imperfectly designed.

10.3.3 Parathormone (PTH)

PTH secretion occurs in frequent bursts (one burst every 7 min) superimposed upon high (>65%) basal hormone release. No infusion studies have reproduced the high-frequency pattern in the human. However, on a time-scale of hour rather than minute, intermittent PTH delivery is more effective than continuous PTH administration in maintaining bone mass (Morley, 2005). This important observation, albeit incompletely parsed mechanistically, has been translated clinically into once-daily injection of biosynthetic PTH (teriparatide) in the treatment of osteoporosis.

10.3.4 Insulin and Glucagon

Insulin secretion is at least 70% pulsatile, as inferred by frequent direct sampling of pancreatico-portal venous blood in the dog and human (Porksen et al., 1997). Several, but not all, clinical studies indicate that pulsatile is more effective then continuous insulin infusion over the short term (hour) in suppressing hepatic gluconeogenesis

(Porksen et al., 2002). Conversely, pulsatile glucagon infusion may stimulate hepatic glucose output more than continuous infusion. Nonetheless, clinical studies necessarily deliver insulin peripherally rather than portally. What remains to be tested is whether portal-vein insulin pulses regulate hepatic glucose metabolism more powerfully than systemic insulin pulses. The rationale for this postulate is that pancreatic islets release insulin almost directly into the liver; portal-venous insulin pulses are six-fold larger than their systemic counterparts; and transhepatic insulin flux removes 50–85% of the hormone contained in pulses (Meier et al., 2005; Porksen et al., 1997).

10.3.5 *Other Hormones*

Testosterone, estradiol, progesterone, aldosterone, cortisol, oxytocin and antidiuretic hormone are also secreted in distinct pulses (Evans et al., 1992; Siragy et al., 1995; Urban et al., 1988; Veldhuis et al., 1989a). Except for oxytocin, the target-tissue impact of pulses of these hormones is not known. Clinical studies suggest that pulsatile injections of oxytocin require less hormone than continuous infusion to stimulate effectual uterine contractions in late gestation (Salamalekis et al., 2000).

10.4 Quantification of Hormone Pulsatility

10.4.1 *Pulsatile Secretion*

Episodic hormone release is difficult to quantify in an objective, reproducible, precise and valid manner. Salient obstacles include sparse data series, procedural and assay variability, missing observations, unequal elimination kinetics in different individuals, unknown secretory-burst shape, size and number, concomitant nonpulsatile secretion, and time-of-day dependencies. Significant technical developments in the last decade have begun to address these issues. In this context, a major conceptual advance is to quantify episodic hormone secretion using composite analytical models. Composite analytical constructs include parameters for flexibly shaped secretory bursts, biexponential elimination kinetics, basal release and random effects (Keenan et al., 1998, 2005). To ensue mathematical validity, all parameters must be estimated simultaneously.

A recent maximum-likelihood estimation method entails first creating multiple sets of objectively estimated candidate pulse-onset times (Keenan et al., 2005) (Fig. 10.6). A procedure termed deconvolution ('disentangling, unrolling') is then applied to each of the potential pulse-time sets to calculate of pulsatile secretion, biexponential elimination, basal secretion and random variability. The optimal pulse-time set is then chosen using statistical model-selection criteria (such as the Akaike or Bayesian information criterion). This strategy has been validated empirically by frequent (5 min) and extended (4–12 h) sampling of hypothalamo-pituitary

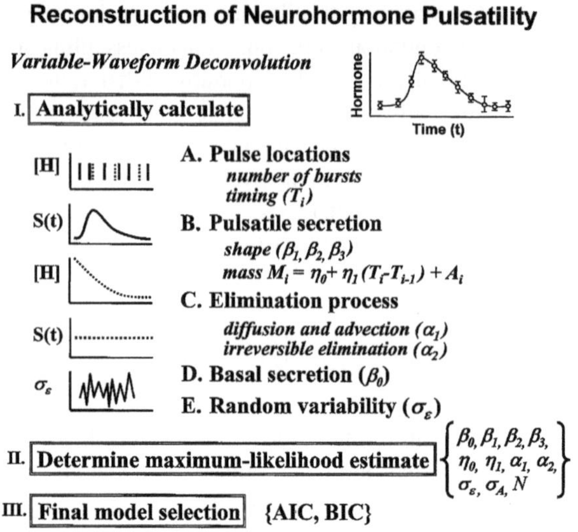

Fig. 10.6 Basic concept of deconvolution analysis to reconstruct hormone pulsatility. Deconvolution signifies to unravel or decompose a hormone-concentration pulse (*top right*) into its contributing parts (*middle*) marked A through E. Candidate pulse-time sets are determined first by incremental smoothing (A). Then, for each candidate set, the analytical algorithm calculates maximum-likelihood parameters for pulsatile secretion (B), hormone elimination (C) and basal secretion (D), while accounting for random variability (E). The final step is statistical model selection, which defines the optimal pulse-time set using the Akaike or Bayesian information criterion (AIC, BIC). Greek symbols denote parameters

portal and jugular-venous blood in the conscious sheep and horse (Keenan et al., 2001, 2004). The model framework has been verified by direct mathematical proof [Ph.D. theses of W. Sun (1995), R. Yang (1997) and S. Chattopadhyay (2001)].

10.4.2 Ensemble Model-Based Analyses

Suitably framed ensemble (multihormone) models permit one to reconstruct interactions among several signals. The ideal model would have both analytical and simulation capabilities. Examples are biomathematical formalisms embodying time-delayed, dose-responsive interactions among (i) GnRH, LH and testosterone; and (ii) CRH, AVP, ACTH and cortisol (Keenan et al., 2001, 2004) (Figs. 10.2 and 10.4). The analytical strength of ensemble analysis of the male gonadal axis is illustrated by the inference that aging in healthy men increases the number and decreases the size of unobserved (secreted) GnRH pulses (Keenan et al., 2006). In the corticotropic axis, noninvasive dose-response reconstruction indicates that consecutive ACTH secretory bursts have variable efficacy in driving cortisol secretion, consistent with possible pulse-by-pulse desensitization and resensitization of ACTH action on adrenal steroidogenesis (Fig. 10.7). A comparable inference applies to LH $\rightarrow$ Te feedforward (Fig. 10.8).

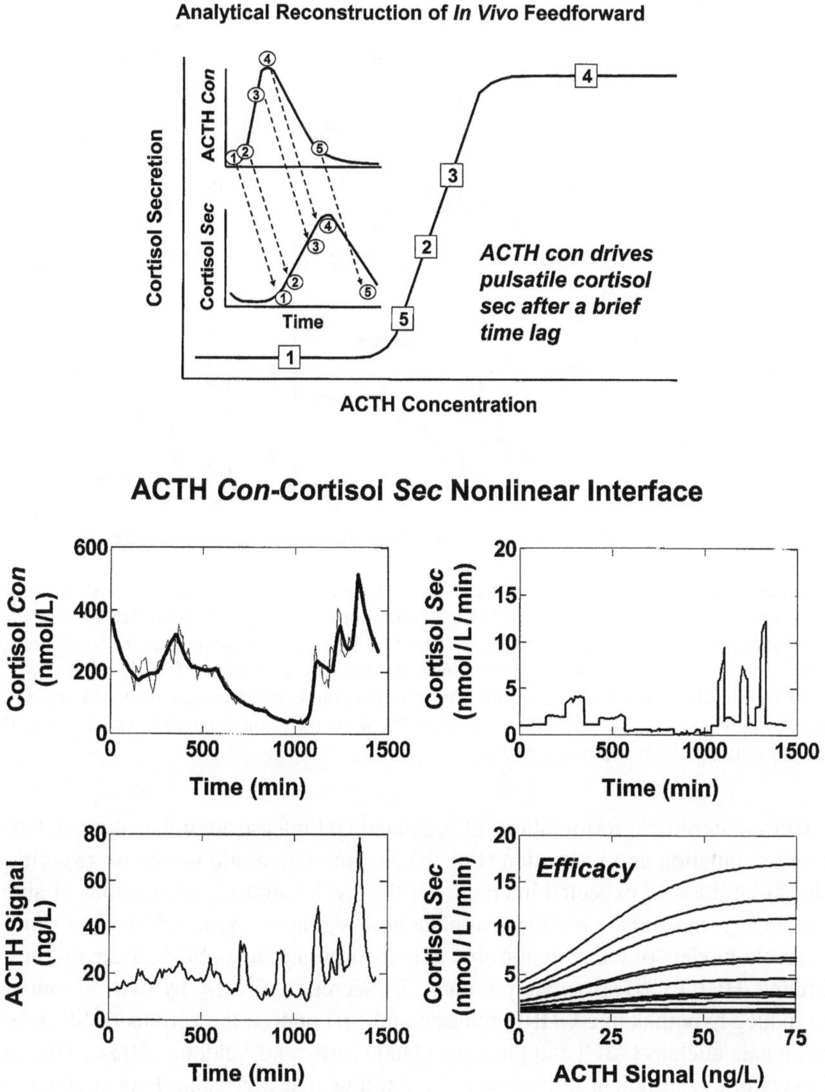

Fig. 10.7 ***Panel A***. Reconstructing (analytically estimating) a feedforward dose-response function from time-delayed pairs (interrupted lines) of serially measured concentrations of a pulsatile input (here, the agonist is ACTH) and pulsatile output (here, cortisol secretion). The nonlinear dose-response function (continuous ascending curve with numbers in boxes) is reconstructed mathematically. Outcomes are the stimulus potency (one-half maximally effective stimulus concentration), stimulus efficacy (maximal response), and response sensitivity (maximal positive slope of the dose-response curve). ***Panel B***. Cortisol and ACTH concentrations (Con and Signal) [*left column*], calculated cortisol secretion rate (Sec) [*top right*] and interlinking nonlinear dose-response functions calculated on a pulse-by-pulse basis (*bottom right*). The efficacy of ACTH → cortisol drive was allowed to vary from pulse to pulse over the 24-h sampling session. Variously steep and flat dose-response curves reflect momentary enhancement and repression of adrenal responses to ACTH pulses

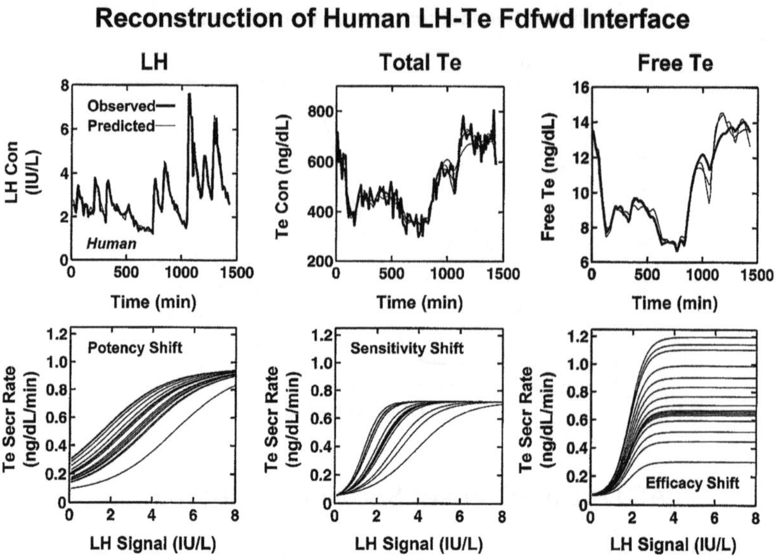

Fig. 10.8 Analytical reconstruction of LH-Te feedforward dose-response relationships *in vivo*. Estimation was based on measuring LH concentrations (*top left*) and total Te concentrations (*top middle*) in a healthy young man every 10 min for 24 h and calculating free Te concentrations (*top right*). The outcome comprises curvilinear dose-response functions for LH → Te (*bottom, left-to-right*). The statistical models allowed possible random effects on response sensitivity, stimulus potency and stimulus efficacy. Estimates were made on a pulse-by-pulse basis, thus yielding multiple dose-response curves

Current ensemble formulations of regulated GH release are not analytical, but do permit simulation and prediction (Fig. 10.3). Simulation allows one to: (a) critique clinical intuition of expected interactions; (b) test integrative implications of alterations in any one signal; and (c) formulate new regulatory hypotheses. For example, ensemble models of the somatotropic axis incorporate the published capabilities of ghrelin/GHRP to: (i) stimulate pituitary GH secretion directly by two- to fourfold; (ii) induce hypothalamic GHRH release; and (iii) oppose the actions of SS in both the arcuate nucleus [ArC] and pituitary gland (Farhy and Veldhuis, 2004). This simple network is sufficient to explicate the findings that transgenic knockdown (50% silencing) of brain ghrelin receptors reduces pulsatile GH secretion and IGF-I concentrations by 35% in the female (but not male) mouse (Shuto et al., 2002).

10.4.3 Approximate Entropy to Measure Ensemble Interactions

Approximate entropy (ApEn) is a statistical regularity measure, which quantifies pattern consistency in time series (Pincus, 2000). The statistic discriminates subtle differences in serial regularity or orderliness of subpatterns. Disorderly hormone secretion is evoked experimentally by interrupting negative feedback or superimposing unvarying feedforward (Veldhuis et al., 2001c). Mathematical models establish that uncoupling of feedback connections within a system increases randomness. Accordingly, elevated ApEn quantifies disruption of interactive control within a network (Pincus et al., 1996).

ApEn is calculated as a single number between zero [denoting perfectly ordered output] and the natural logarithm of 10 (2.3026) [signifying maximally random output]. Figure 10.9A illustrates orderliness of a perfect sine wave (ApEn = 0) compared with progressively corrupted versions of the same (increased ApEn). ApEn has high specificity and sensitivity (both > 90%) in discriminating between normal and disorderly time series (Hartman et al., 1994; Pincus et al., 1999). In particular, autonomous tumors maintain highly irregular or irreproducible (high-ApEn) secretion patterns, as documented in acromegaly (GH), Cushing's disease (ACTH), prolactinoma, hyperparathyroidism, aldosteronoma and adrenal cortical adenoma (Goodman et al., 1998; Hartman et al., 1994; Roelfsema et al., 1998; Siragy et al., 1995). Puberty also elevates ApEn of LH and GH secretion signifying reduced inhibitory control (Pincus et al., 2000; Veldhuis et al., 2001a). ApEn of insulin secretion rises in early noninsulin-dependent (type II) diabetes mellitus, thus defining impaired beta-cell regulation before overt disease (Schmitz et al., 1997). Increased ApEn is also typical of LH secretion in older individuals (Pincus et al., 1996) (Fig. 10.9B (*top*)).

A related statistic, cross-ApEn, quantifies the relative synchrony between two time series. The utility of cross-ApEn lies in identifying impaired pathway-dependent or ensemble-level regulation. Cross-ApEn values of LH-Te and ACTH-cortisol secretion pairs are high in aging people, denoting deterioration of coordinate control of the linked signals (Pincus et al., 1996). Figure 10.9B (*bottom*) shows that aging elevates cross-ApEn of LH secretion and nocturnal penile tumescence (NPT), a marker of neurohormone outflow. High cross-ApEn in this context quantifies significant erosion of synchrony between two central-neural processes,

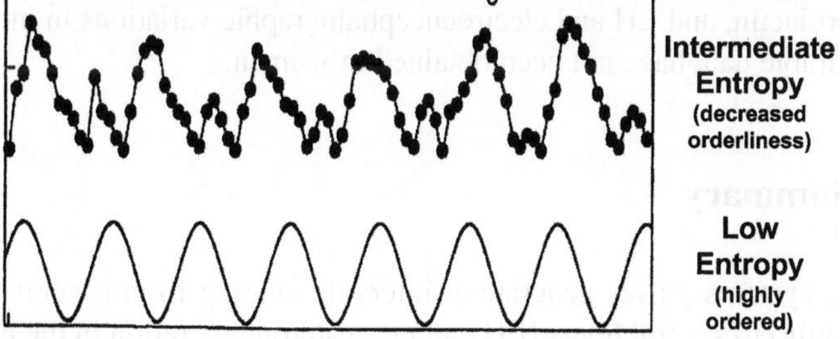

Fig. 10.9 (continued)

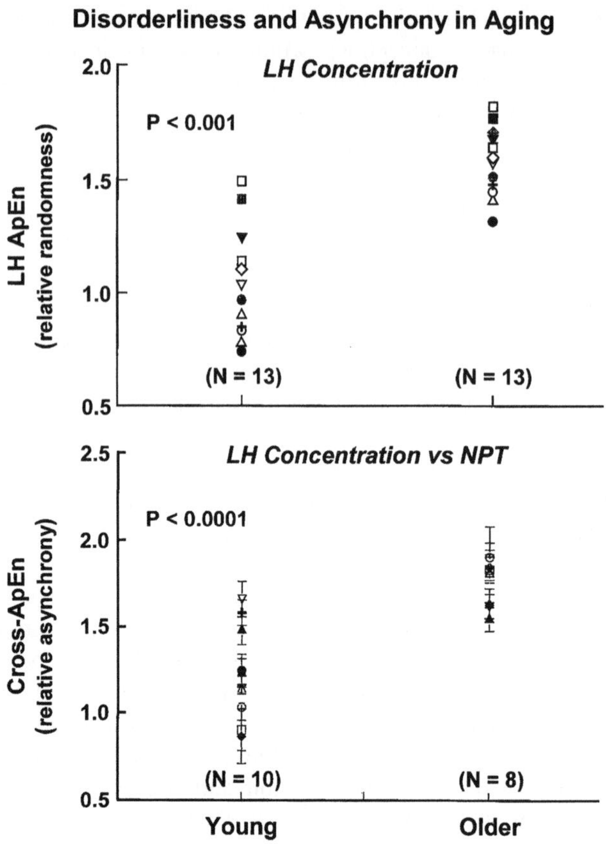

Fig. 10.9 (continued) ***Panel A***. Concept of approximate entropy (ApEn) as a measure of relative randomness. High ApEn designates irregularly irregular (unpredictable or disorderly) patterns (*top*). Low ApEn denotes more orderly or reproducible patterns (*bottom*). ***Panel B***. *Top*: High ApEn signifies less orderly LH patterns in older than young men. *Bottom*: High cross-ApEn of LH-NPT in aging individuals defines loss of synchrony of neurohormone and autonomic outflow. NPT = nocturnal penile tumescence

one controlling the GnRH pulse generator and the other autonomic nervous-system activity (Veldhuis et al., 2000). Aging also erodes synchrony between LH and FSH, LH and prolactin, and LH and electroencephalographic variations in men.

Comparable data have not been obtained in women.

10.5 Summary

Endocrine systems pulses generate at intervals ranging from several min (PTH, ADH, insulin) to several hours (LH and progesterone secretion in the luteal phase of the menstrual cycle). The size of secretory bursts varies over a scale of three orders of magnitude (GH). Analysis of pulsatility confers insights into primary pulse-generating mechanisms. Pulsatility is a physiological mechanism to augment hormone concentrations rapidly and to convey distinct signaling information to selected target tissues. GnRH pulses illustrate these points, inasmuch as GnRH secretory bursts reflect multiunit neuronal firing in the hypothalamus, amplify LH secretion during the preovulatory LH surge, and sustain normal reproductive function. Continuous GnRH delivery downregulates LH secretion, and thereby depletes gonadal sex steroids. Downregulation by this means is used to treat sex hormone-dependent diseases, such as precocious puberty, prostatic cancer and endometriosis. Pulsatile administration of GH is more lipolytic than continuous administration, but the latter induces hepatic IGF-I production and the LDL receptor more effectively. Oxytocin pulses are equivalent or superior to constant oxytocin infusion in stimulating uterine contractions during parturition. Insulin pulses can repress hepatic gluconeogenesis more effectively than continuous insulin delivery. Glucagon pulses may promote greater hepatic glucose output than continuous glucagon stimulation. The clinical implications of high-frequency pulses of PTH, ADH and steroid hormones (estradiol, testosterone, cortisol, aldosterone, progesterone) are uncertain. However, once-daily PTH administration is more osteotrophic than continuous administration. Current analytical models estimate both secretion and elimination rates from hormone-concentration profiles (deconvolution analysis), and reconstruct interlinking dose-response functions noninvasively (ensemble analysis). The approximate-entropy statistic (ApEn) quantifies insidious failure of regulatory connections in aging and premorbid states. Multidisciplinary efforts will be needed to advance these themes further.

Acknowledgments We thank Kay Nevinger for sedulous support of manuscript preparation; Ashley Bryant for excellent data analysis and graphics; the Mayo Immunochemical Laboratory for assay assistance; and the Mayo research nursing staff for implementing the protocol. Supported in part via the Center for Translational Research awarded to the Mayo Clinic and Foundation from the National Center for Research Resources (Rockville, MD), and R01 NIA AG029362, R21 DK072095, DK 063609, RR019991 from the National Institutes of Health (Bethesda, MD).

References

Biermasz NR, Pereira AM, Frolich M, Romijn JA, Veldhuis JD, Roelfsema F 2004. Octreotide represses secretory-burst mass and nonpulsatile secretion but does not restore event frequency or orderly GH secretion in acromegaly. Am J Physiol Endocrinol Metab 286: E25–E30

Bowers CY, Granda R, Mohan S, Kuipers J, Baylink D, Veldhuis JD 2004. Sustained elevation of pulsatile growth hormone (GH) secretion and insulin-like growth factor I (IGF-I), IGF-binding protein-3 (IGFBP-3), and IGFBP-5 concentrations during 30-day continuous subcutaneous infusion of GH-releasing peptide-2 in older men and women. J Clin Endocrinol Metab 89:2290–2300

Chase DJ, Schanbacher B, Lunstra DD 1988. Effects of pulsatile and continuous luteinizing hormone (LH) infusions on testosterone responses to LH in rams actively immunized against gonadotropin-releasing hormone. Endocrinology 123:816–826

Clarke IJ, Cummins JT 1982. The temporal relationship between gonadotropin-releasing hormone (GnRH) and luteinizing hormone (LH) secretion in ovariectomized ewes. Endocrinol 111:1737–1739

Conn PM, Crowley Jr. WF 1991. Gonadotropin-releasing hormone and its analogs. New Engl J Med 324:93–103

Engler D, Pham T, Fullerton MJ, Ooi G, Funder JW, Clarke IJ 1989. Studies of the secretion of corticotropin-releasing factor and arginine vasopressin into the hypophysial-portal circulation of the conscious sheep. I. Effect of an audiovisual stimulus and insulin-induced hypoglycemia. Neuroendocrinol 49:367–381

Evans WS, Sollenberger MJ, Booth Jr. RA, Rogol AD, Urban RJ, Carlsen EC, Johnson ML, Veldhuis JD 1992. Contemporary aspects of discrete peak detection algorithms. II. The paradigm of the luteinizing hormone pulse signal in women. Endocr Rev 13:81–104

Farhy LS, Veldhuis JD 2004. Putative GH pulse renewal: periventricular somatostatinergic control of an arcuate-nuclear somatostatin and GH-releasing hormone oscillator. Am J Physiol 286:R1030–R1042

Farhy LS, Bowers CY, Veldhuis JD 2007. Model-projected mechanistic bases for sex differences in growth-hormone regulation in humans. Am J Physiol Regul Integr Comp Physiol 292: R1577–R1593

Foresta C, Bordon P, Rossato M, Mioni R, Veldhuis JD 1997. Specific linkages among luteinizing hormone, follicle stimulating hormone, and testosterone release in the peripheral blood and human spermatic vein: evidence for both positive (feed-forward) and negative (feedback) within-axis regulation. J Clin Endocrinol Metab 82:3040–3046

Gibson-Berry KL, Chase DJ 1990. Continuous and pulsatile infusions of luteinizing hormone have identical effects on steroidogenic capacity and sensitivity of Leydig cell in rats passively immunized against gonadotropin-releasing hormone. Endocrinol 126:3107–3115

Giustina A, Veldhuis JD 1998. Pathophysiology of the neuroregulation of growth hormone secretion in experimental animals and the human. Endocr Rev 19:717–797

Goodman WG, Veldhuis JD, Belin TR, Van Herle A, Juppner H, Salusky IB 1998. Calcium-sensing by parathyroid glands in secondary hyperparathyroidism. J Clin Endocrinol Metab 83:2765–2772

Hartman ML, Pincus SM, Johnson ML, Matthews DH, Faunt LM, Vance ML, Thorner MO, Veldhuis JD 1994. Enhanced basal and disorderly growth hormone secretion distinguish acromegalic from normal pulsatile growth hormone release. J Clin Invest 94:1277–1288

Hwa V, Little B, Adiyaman P, Kofoed EM, Pratt KL, Ocal G, Berberoglu M, Rosenfeld RG 2005. Severe growth hormone insensitivity resulting from total absence of signal transducer and activator of transcription 5b. J Clin Endocrinol Metab 90:4260–4266

Jaffe CA, Turgeon DK, Lown K, DeMott-Friberg R, Watkins PB 2002. Growth hormone secretion pattern is an independent regulator of growth hormone actions in humans. Am J Physiol Endocrinol Metab 283:E1008–E1015

Keenan DM, Veldhuis JD 2003a. Cortisol feedback state governs adrenocorticotropin secretory-burst shape, frequency and mass in a dual-waveform construct: time-of-day dependent regulation. Am J Physiol 285:R950–R961

Keenan DM, Veldhuis JD 2003b. Mathematical modeling of receptor-mediated interlinked systems. Encyclopedia of Hormones. Academic, San Diego, CA, pp. 286–294

Keenan DM, Veldhuis JD, Yang R 1998. Joint recovery of pulsatile and basal hormone secretion by stochastic nonlinear random-effects analysis. Am J Physiol 275:R1939–R1949

Keenan DM, Licinio J, Veldhuis JD 2001. A feedback-controlled ensemble model of the stress-responsive hypothalamo-pituitary-adrenal axis. Proc Natl Acad Sci USA 98:4028–4033

Keenan DM, Alexander SL, Irvine CHG, Clarke IJ, Canny BJ, Scott CJ, Tilbrook AJ, Turner AI, Veldhuis JD 2004. Reconstruction of *in vivo* time-evolving neuroendocrine dose-response properties unveils admixed deterministic and stochastic elements in interglandular signaling. Proc Natl Acad Sci USA 101:6740–6745

Keenan DM, Chattopadhyay S, Veldhuis JD 2005. Composite model of time-varying appearance and disappearance of neurohormone pulse signals in blood. J Theor Biol 236:242–255

Keenan DM, Takahashi PY, Liu PY, Roebuck PD, Nehra AX, Iranmanesh A, Veldhuis JD 2006. An ensemble model of the male gonadal axis: illustrative application in aging men. Endocrinol 147:2817–2828
Knobil E 1990. The neuroendocrine control of the menstrual cycle. Rec Prog Horm Res 36:53–88
Leung KC, Doyle N, Ballesteros M, Sjogren K, Watts CK, Low TH, Leong GM, Ross RJ, Ho KK 2003. Estrogen inhibits GH signaling by suppressing GH-induced JAK2 phosphorylation, an effect mediated by SOCS-2. Proc Natl Acad Sci USA 100:1016–1021
Meier JJ, Veldhuis JD, Butler PC 2005. Pulsatile insulin secretion dictates systemic insulin delivery by regulating hepatic insulin extraction in humans. Diabetes 54:1649–1656
Moos F, Fontanaud P, Mekaouche M, Brown D 2004. Oxytocin neurones are recruited into coordinated fluctuations of firing before bursting in the rat. Neuroscience 125:391–410
Morley P 2005. Delivery of parathyroid hormone for the treatment of osteoporosis. Expert Opin Drug Deliv 2:993–1002
Pincus SM 2000. Irregularity and asynchrony in biologic network signals. Methods Enzymol 321:149–182
Pincus SM, Mulligan T, Iranmanesh A, Gheorghiu S, Godschalk M, Veldhuis JD 1996. Older males secrete luteinizing hormone and testosterone more irregularly, and jointly more asynchronously, than younger males. Proc Natl Acad Sci USA 93:14100–14105
Pincus SM, Hartman ML, Roelfsema F, Thorner MO, Veldhuis JD 1999. Hormone pulsatility discrimination via coarse and short time sampling. Am J Physiol 277:E948–E957
Pincus SM, Veldhuis JD, Rogol AD 2000. Longitudinal changes in growth hormone secretory process irregularity assessed transpubertally in healthy boys. Am J Physiol Endocrinol Metab 279:E417–E424
Porksen N, Munn S, Steers J, Veldhuis JD, Butler P 1996. Effects of somatostatin on pulsatile insulin secretion; selective inhibition of insulin burst mass. Am J Physiol 270: E1043–E1049
Porksen N, Nyholm B, Veldhuis JD, Butler PC, Schmitz O 1997. In humans at least seventy-five percent of overall insulin secretion arises from punctuated high-frequency insulin secretory bursts. Am J Physiol 273:E908–E914
Porksen N, Hollingdal M, Juhl CB, Butler P, Veldhuis JD, Schmitz O 2002. Pulsatile insulin secretion: detection, regulation and role in diabetes. Diabetes 51:S245–S254
Redekopp C, Irvine CH, Donald RA, Livesey JH, Sadler W, Nicholls MG, Alexander SL, Evans MJ 1986. Spontaneous and stimulated adrenocorticotropin and vasopressin pulsatile secretion in the pituitary venous effluent of the horse. Endocrinol 118:1410–1416
Ritzel RA, Veldhuis JD, Butler PC 2003. Glucose stimulates pulsatile insulin secretion from human pancreatic islets by increasing secretory burst mass: dose-response relationships. J Clin Endocrinol Metab 88:742–747
Roelfsema F, Pincus SM, Veldhuis JD 1998. Patients with Cushing's disease secrete adrenocorticotropin and cortisol jointly more asynchronously than healthy subjects. J Clin Endocrinol Metab 83:688–692
Roelfsema F, Biermasz NR, Veldman RG, Veldhuis JD, Frolich M, Stokvis-Brantsma WH, Wit J-M 2000. Growth hormone (GH) secretion in patients with an inactivating defect of the GH-releasing hormone (GHRH) receptor is pulsatile: evidence for a role for non-GHRH inputs into the generation of GH pulses. J Clin Endocrinol Metab 86:2459–2464
Salamalekis E, Vitoratos N, Kassanos D, Loghis C, Panayotopoulos N, Sykiotis C 2000. A randomized trial of pulsatile vs continuous oxytocin infusion for labor induction. Clin Exp Obstet Gynecol 27:21–23
Schaefer F, Daschner M, Veldhuis JD, Oh J, Qadri F, Scharer K 1994. In vivo alterations in the gonadotropin releasing hormone (GnRH) pulse generator and the secretion and clearance of luteinizing hormone in the uremic castrate rat. Neuroendocrinol 59:285–296
Schmitz O, Porksen N, Nyholm B, Skjaerback C, Butler PC, Veldhuis JD, Pincus SM 1997. Disorderly and nonstationary insulin secretion in relatives of patients with NIDDM. Am J Physiol 35:E218–E226

Shapiro BH, MacLeod JN, Pampori NA, Morrissey JJ, Lapenson DP, Waxman DJ 1989. Signaling elements in the ultradian rhythm of circulating growth hormone regulating expression of sex-dependent forms of hepatic cytochrome P450. Endocrinol 125:2935–2944

Shuto Y, Shibasaki T, Otagiri A, Kuriyama H, Ohata H, Tamura H, Kamegai J, Sugihara H, Oikawa S, Wakabayashi I 2002. Hypothalamic growth hormone secretagogue receptor regulates growth hormone secretion, feeding, and adiposity. J Clin Invest 109:1429–1436

Siragy HM, Vieweg WVR, Pincus SM, Veldhuis JD 1995. Increased disorderliness and amplified basal and pulsatile aldosterone secretion in patients with primary aldosteronism. J Clin Endocrinol Metab 80:28–33

Urban RJ, Evans WS, Rogol AD, Kaiser DL, Johnson ML, Veldhuis JD 1988. Contemporary aspects of discrete peak detection algorithms. I. The paradigm of the luteinizing hormone pulse signal in men. Endocr Rev 9:3–37

Vance ML, Kaiser DL, Evans WS, Furlanetto R, Vale WW, Rivier J, Thorner MO 1985. Pulsatile growth hormone secretion in normal man during a continuous 24-hour infusion of human growth hormone releasing factor (1–40). J Clin Invest 75:1584–1590

Vance ML, Kaiser DL, Martha Jr. PM, Furlanetto R, Rivier J, Vale WW, Thorner MO 1989. Lack of in vivo somatotroph densensitization or depletion after 14 days of continuous growth hormone (GH)-releasing hormone administration in normal men and a GH-deficient boy. J Clin Endocrinol Metab 68:22–28

Veldhuis JD, Iranmanesh A, Lizarralde G, Johnson ML 1989a. Amplitude modulation of a burst-like mode of cortisol secretion subserves the circadian glucocorticoid rhythm in man. Am J Physiol 257:E6–E14

Veldhuis JD, O'Dea LS, Johnson ML 1989b. The nature of the gonadotropin-releasing hormone stimulus-luteinizing hormone secretory response of human gonadotrophs in vivo. J Clin Endocrinol Metab 68:661–670

Veldhuis JD, Iranmanesh A, Johnson ML, Lizarralde G 1990. Twenty-four hour rhythms in plasma concentrations of adenohypophyseal hormones are generated by distinct amplitude and/or frequency modulation of underlying pituitary secretory bursts. J Clin Endocrinol Metab 71:1616–1623

Veldhuis JD, Iranmanesh A, Godschalk M, Mulligan T 2000. Older men manifest multifold synchrony disruption of reproductive neurohormone outflow. J Clin Endocrinol Metab 85:1477–1486

Veldhuis JD, Pincus SM, Mitamura R, Yano K, Suzuki N, Ito Y, Makita Y, Okuno A 2001a. Developmentally delimited emergence of more orderly luteinizing hormone and testosterone secretion in late prepuberty in boys. J Clin Endocrinol Metab 86:80–89

Veldhuis JD, Anderson SM, Shah N, Bray M, Vick T, Gentili A, Mulligan T, Johnson ML, Weltman A, Evans WS, Iranmanesh A 2001b. Neurophysiological regulation and target-tissue impact of the pulsatile mode of growth hormone secretion in the human. Growth Horm IGF Res 11:S25–S37

Veldhuis JD, Straume M, Iranmanesh A, Mulligan T, Jaffe CA, Barkan A, Johnson ML, Pincus SM 2001c. Secretory process regularity monitors neuroendocrine feedback and feedforward signaling strength in humans. Am J Physiol 280:R721–R729

Veldhuis JD, Evans WS, Bowers CY 2002. Impact of estradiol supplementation on dual peptidyl drive of growth-hormone secretion in postmenopausal women. J Clin Endocrinol Metab 87:859–866

Veldhuis JD, Iranmanesh A, Keenan DM 2004. An ensemble perspective of aging-related hypoandrogenemia in men. In: Winters SJ (ed.), Male Hypogonadism: Basic, Clinical, and Theoretical Principles. Humana, Totowa, NJ, pp. 261–284

Veldhuis JD, Roemmich JN, Richmond EJ, Bowers CY 2006. Somatotropic and gonadotropic axes linkages in infancy, childhood, and the puberty-adult transition. Endocr Rev 27:101–140

Chapter 11
Ultradian Rhythms as the Dynamic Signature of Life

F.E. Yates[1,2*] and L.B. Yates[3,4]

Abstract In this chapter we first define ultradian rhythms and offer a brief tutorial on spectroscopic methods to extract them from data, including some cautionary comments regarding false positive identifications. We then contrast features of ultradian rhythms with those of circadian rhythms and conjecture that the former are the basic rhythmic signature of life. To justify that conjecture we introduce a physical heuristic – *homeodynamics* – that presents quantized, cyclic action (the product of energy × time) as an essential thermodynamic characteristic of open systems that persist over times that are long compared to their internal process times. From that perspective we argue that the start-up of terrestrial life began with energetics (metabolism) as dynamics manifested by ultradian rhythms, only later acquiring informational molecular memories and constraints as relatively static codes.

Keywords and phrases Action cycles and the thermodynamic marginal stability of open systems, biospectroscopy, contrast between ultradian and circadian rhythms, dynamics and the origin of life, homeostasis and homeodynamics

11.1 Introduction

Studies of biological rhythms have been dominated by attention to the approximately 24h, circadian rhythm because of its ubiquity, large amplitude and easily understood significance: it synchronizes the business of living organisms to the

[1]Ralph and Marjorie Crump Professor of Medical Engineering, University of California, Los Angeles (Emeritus)

[2]Scientific Advisor, The John Douglas French Alzheimer's Foundation, Los Angeles, California

[3]Division of Aging, Brigham and Women's Hospital, Boston, Massachusetts

[4]Harvard Medical School, Boston, Massachusetts

*Correspondence to: The John Douglas French Alzheimer's Foundation, 11620 Wilshire Boulevard, Suite 270, Los Angeles, CA 90025, USA. Email: gyates@jdfaf.org

day/night geophysical cycle. Twenty-four hour 'clock' genes have been found and their networks identified for many classes of organisms. In contrast, we here propose that the seemingly lesser ultradian rhythms are in fact the primary rhythmic signature of the living state, and that the circadian rhythm is an evolutionary 'afterthought'. To justify this conjecture, we offer an update on the historic concept of homeostasis to provide basis for considering a thermodynamic foundation for ultradian rhythms. We discuss several distinctive features of ultradian rhythms and provide a brief tutorial on their spectroscopic analysis. Finally, we propose an answer to the question: Why are ultradian cycles the dynamic signature of life?

11.2 From Homeostasis to Homeodynamics

In 1926 Walter Cannon introduced the term *homeostasis* and expanded the concept in a now classic article in Physiological Reviews (Cannon, 1929). He was interested in the 'wisdom of the body', marked by its ability to recover from perturbations imposed by either internal events or a vicissitudinous environment. He offered many examples and also cited prior works by Bernard, Pflüger, Fredericq and Richet, all of whom were attracted by the stability of living organisms. Contrary to common misinterpretations, however, the notion of homeostasis did not require constancy (*homeo* = same; *stasis* = condition, not static constancy). In fact, Cannon allowed for oscillations as one form of homeostatic condition.

Cannon's concept of homeostasis may be considered an early theory of an organism, although it was not cast in mathematical, engineering, or physical terms. Nevertheless, it can be represented as an elementary, negative feedback construct with set points. (See Riggs, 1963; Yates, 1996, 2005 for details.) More recently, taking advantage of progress in systems sciences and expanding the notion of homeokinetics as a science for complex systems (Iberall, 1977; Iberall and Soodak, 1987; Soodak and Iberall, 1978), Yates (Yates, 1982, 1994, 1996) has updated Cannon's homeostasis into a *homeodynamic* heuristic. Based on an extension of physics, homeodynamics provides a foundation for both temporal and spatial organization, with an emphasis on the centrality of action as a coordinating principle. It rationalizes the ubiquity of biological rhythms, as will be discussed later.

11.3 Various Perspectives on Ultradian Rhythms

Many different views could be taken of the character and 'meaning' of ultradian rhythms. Metaphorically, we suggest that they might play any of the following biological roles:

1. They are dynamic analogs of architectural spandrels (Gould and Lewontin, 1979). That is, rather than existing for some adaptive purpose, they are accidental

and nonessential byproducts of essential dynamic actions occurring elsewhere in the organism.
2. They are supporting players in an operatic ensemble, each with a choreographed turn to sing.
3. They train and maintain the preparedness of the system to respond at high frequencies.
4. They are alerting signals to other parts of the system.
5. They are merely parasitic oscillations indicating instabilities.
6. They are significant homeodynamic signatures of a marginal stability that keeps the system able to adapt quickly.

11.4 Defining Ultradian Rhythms by Spectroscopic Characterization

If a time history of observations on a biological process of interest is queried for any recurring content, some criteria for detection are needed in advance. A rhythm may be any orderly recurrence of an event or a value of a variable, regardless of shape. However, in the broadest view, the concept of recurrence covers more possibilities than does the narrower spectral property of periodicity or rhythm (Zak et al., 1997). In this discussion we focus on the special, but common problem of measuring the interval between orderly recurrences of similar events and assessing their regularity or randomness. If ultradian rhythms are sought, then the intervals should have an average separation length of less than 12 h, on the assumption that a 12 h periodicity may be an harmonic of a circadian rhythm and not the sign of an independent oscillation. When present (and they are ubiquitous), circadian rhythms usually account for a very large fraction of the variance spectrum if the record is 24 h or longer. Ultradian rhythms, in contrast, are usually of much less power (squared product of amplitude), and they may be intermittent. Such is never the case for a circadian rhythm.

11.5 Spectroscopy: The Classical Approach to Characterizing Periodic Behaviors

A model for a periodic function has four characterizing parameters: mean value, amplitude, period (or its reciprocal, frequency), and phase. The classic approach to identifying periodic functions is the Fourier series, implemented as the Fast Fourier Transform (FFT), which supplies a linear series of harmonic sine and cosine functions that, when summed, can fit (in principle) any periodicity in a time series data set, regardless of the shape of the periodic element.

In practice, however, the FFT result is very shape-dependent, and the method is inappropriate to analyze spike trains, square waves, or data with any recurrent shape of sharp corners or vertical elements. To fit a sharp spike, for example, an infinite series of sines and cosines with their rounded shapeliness is required. For practical reasons one must therefore truncate the series with a limited number of terms, perhaps only two, e.g., the fundamental (lowest frequency) and second harmonic.

Unlike the FFT, the Fast Orthogonal Search (FOS) method (Korenberg, 1988) generates a series of non-harmonic sinusoidal functions which need not be commensurate, nor integral multiples of the fundamental frequency corresponding to the record length. This algorithm is robust against missing or unequally spaced data, but some cut-off also may be needed to limit the number of features modeled to fit periodic data. The Approximate Entropy (ApEn) method (Pincus, 1991) is an alternative approach to spectroscopic analysis in that it provides a measure of 'regularity' (not necessarily periodicity) that may lurk in a time series.

All of these methods benefit from long records, if the data are continuous (e.g., blood pressures), or many events, if the data are discrete (e.g., neural spike trains).

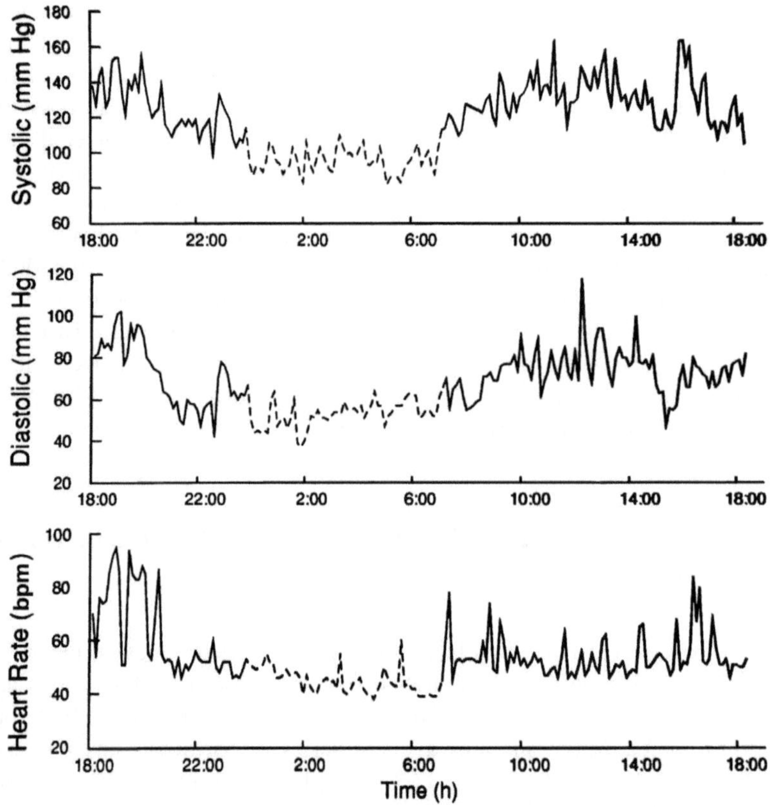

Fig. 11.1 Twenty-four hour ambulatory monitoring record of systolic and diastolic blood pressures and heart rate from a 57 year old, healthy man (Yates and Benton, 1991). Measurements were made every 7.5 min while the subject pursued his routine activities (solid lines = data while awake; dashed lines = data during sleep)

11.5.1 Examples of Spectroscopic Analysis of Ultradian Rhythms

Figure 11.1 displays raw data records of blood pressures and heart rate obtained continuously over 24 h in a healthy man. A circadian cycle in systolic pressure is easily detected by eye, but ultradian rhythms (if any) are less obvious. Figure 11.2

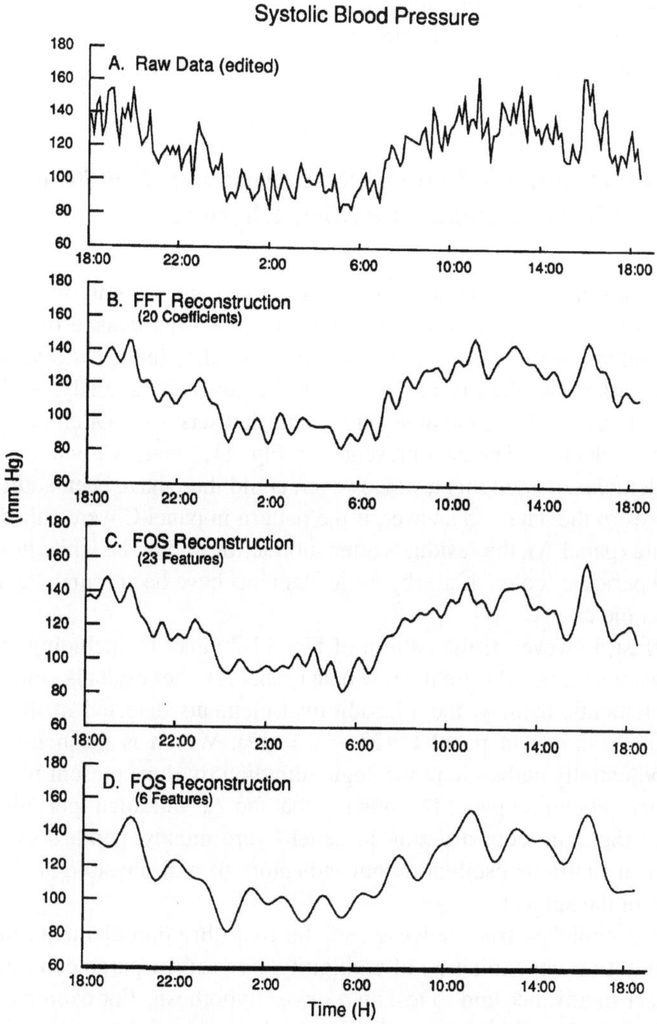

Fig. 11.2 Analysis and reconstruction of systolic blood pressure data in Fig. 11.1 (panel **A**) using the Fast Fourier Transform (FFT) with 20 coefficients (panel **B**) and the Fast Orthogonal Search (FOS) method (panels **C** and **D**). In panel **C**, 23 sinusoidal features were selected when the FOS cut-off criterion was >2% reduction of total mean square error. In panel **D** the cut-off criterion was >5% reduction. The reconstruction in panel **D** captured 50% of the total variance of the raw data record using six sinusoidal features, including a circadian period and five ultradian periods

demonstrates the results of 'searching' the systolic pressure record for ultradian rhythms by FFT (panel B) and by FOS (panels C and D, using two different cut-off settings). Seemingly periodic patterns in the ultradian range appear in the outputs of both methods. Panel D shows a reconstruction of the data using the fewest sinusoidal features, which include the circadian rhythm plus five ultradian periods. Although the reconstruction appears to be a reasonably good approximation of the raw data (panel A), the question remains: How many of the five ultradian periodicities were physiologically real (i.e., arising from actual biochemical, physiological or even behavioral activities)? It would be a fool's errand to look for oscillators in the data if the ultradian periods extracted mathematically existed only mathematically, not physically.

11.6 Determining When the 'Spectroscope' Has Found All Mathematically Justified Rhythms

Rhythms in a time history create an ordered set of data, but only the natural order is meaningful. One way to assess the mathematical significance of any reported rhythmic element is to 'shuffle' the sequence of the data in the history, breaking up the natural order, and then re-run the spectral analyzer. The analyzer should then fail to detect any rhythmic elements in the shuffled sets. (See Odell et al., 1975 for methodologic details.) The data presented in Fig. 11.2 panel C, which included 23 periodic elements, passed this test. However, could there have been additional periodic features in the data? To answer, if the pattern in panel C were subtracted from the raw data (panel A), the residuals after subtraction should be white noise with no remaining periodic 'color' if all rhythmic elements have been found. As noted, that was indeed the case.

In contrast, however, if the pattern of Fig. 11.2 panel D, including six periodic features, were subtracted from the raw data (panel A), the residuals would still have periodic elements, namely, the 17 additional elements detected in the 23-feature reconstruction shown in panel C (23 − 6 = 17). Which is the better pattern to describe potentially authentic physiologic ultradian rhythms present in the original data? We would choose panel D, knowing that the 22 'ultradian' periodic elements comprising the data reconstruction in panel C are mostly, perhaps even entirely, mathematical fictitious oscillators, not indicators of real physiological oscillatory generators in the subject.

This also could be true, however, of the five ultradian elements of panel D. One solution to escape this trap of ambiguity is to set the parameters of the spectral analyzer in advance and to test an *a priori* hypothesis. For example, if it were suspected that the ubiquitous, 90 min, basic rest-activity cycle (BRAC) of Kleitman (Kleitman, 1982) were present in the systolic blood pressure record, the spectroscope could be set to find a 24 h circadian cycle and one ultradian cycle with period of 90 min, suppressing all other periodic features. If the 90 min period were detected, relevant statistical tests (including the shuffling test) could be applied to estimate its validity. If statistically valid, the hypothesis would be supported. (There are, of course, other issues concerning Type 1 and Type 2 errors; in this case there is no issue about a false negative, although there could be a false positive.)

11.7 Spurious Rhythms: Aliasing, Beat Frequencies and Chaotic Dynamics

When a periodic process is sampled discretely by a periodic process of a different frequency, unreal and illusory frequencies may appear. Such aliasing by a camera shutter is a familiar sight in western movies showing a covered wagon traveling forward while the spokes of the wheels appear to be rotating backward. Aliasing is an illusion created by sampling, and aliasing frequencies are not present in the physical situation generating the data.

Beat frequencies, on the other hand, are illusions from summing and are created by linear superposition of two periodic processes that have different frequencies (periods). As an example, consider the simulation of two superimposed pure sine waves, each with mean value zero, amplitude one, and phase zero. One sine wave has a 24 h period, while the period of the other is 21.33 h. Sampling occurs every 4 h and is continued for 43 days (1,032 h). Spectroscopic analysis (not shown here) will reveal three significant frequencies in the combined record: 24 h and 21.33 h periods, as expected, as well as an additional frequency with a period of approximately 200 h (8 days). This gratuitous period is the beat frequency, real in the spectrum, but not indicating the presence of a physical generator.

Chaotic dynamics do not produce periodicities, but under some conditions they can appear to be near-periodic. For example, it is well known that one trajectory to chaos leads through a succession of period doublings, and that 'period three implies chaos'. However, the motions on chaotic attractors are not strictly periodic. (See Glass and Mackey, 1988 for details.)

11.8 Non-Spectroscopic Methods to Extract Patterns from Time History Data

High-order polynomials can also be used to model a time series. Polynomial fits of data have the potential to draw a curve that passes through every point of a discrete time series, which shows the versatility of polynomial fitting, but yields little information about the presence of ultradian rhythms, if the latter is the intended analysis. A defect of a polynomial basis function for fitting is that it cannot be used to extend the model reliably beyond the last data point, lest it extrapolate to infinity. Other methods of time series analysis include autocorrelations and periodograms, for describing periodicities, and autoregression and correlograms to detect features other than periodic. Box Jenkins autoregressive- integrated-moving-average (ARIMA) models are useful for forecasting, but such models are not needed to extract ultradian patterns. More advanced techniques to explore temporal organization in living systems, including fractal time, are available (Yates, 1992).

11.9 Intermittency of an Ultradian Rhythm

Figure 11.3 displays records of blood flow and cortisol secretion rate from the left adrenal gland of an unanesthetized, resting dog. Even without using a 'spectroscope' the eye can appreciate the presence of an ultradian rhythm with a period of 2 min in the blood flow (10 peaks in 20 min, from minute 10–30). The pattern in secretion rate is more difficult to analyze visually, although in 15 min of the record (from minute 15–30) there were seven 'bumps', suggesting a period of 2 min, as seen in the corresponding blood flow. (This impression was confirmed by formal cross-spectral analysis, not shown here.)

Figure 11.3 also shows an equally interesting feature in the flat portion of the blood flow record from minute 0–10, when the oscillation began. Thus the ultradian

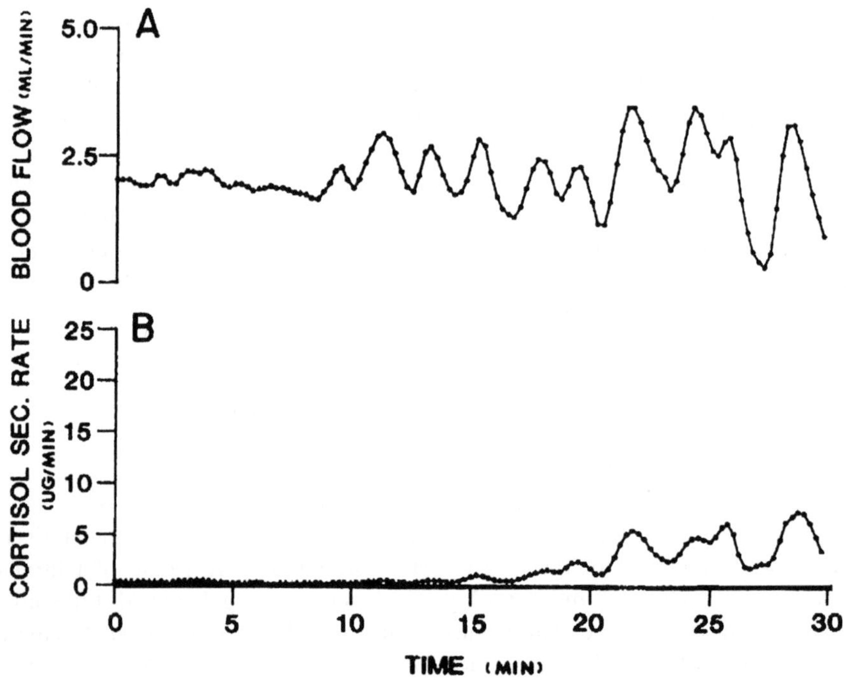

Fig. 11.3 Simultaneous time histories of blood flow and cortisol secretion rate from the left adrenal gland of an unanesthetized, resting dog (Benton and Yates, 1990). Measurements were made every 15 s for 30 min

rhythm, although physiologically real and of good amplitude, was intermittent. Similar intermittency is seen in secretion rate, although the flat, non-oscillatory segment lasted 15 min.

Differences between the blood flow and secretion rate records are informative, because secretion rate was calculated from blood flow and the concentration of cortisol in the blood leaving the adrenal gland (not shown). Thus while the blood flow and secretion rate records must be correlated, they clearly are not alike. Their differences suggest that some other activity, beyond washout by blood flow, was affecting the availability of cortisol for secretion.

11.10 Summary: Features of Ultradian Rhythms and Methods to Analyze Them

- There are many mathematical methods to extract rhythmic features from a time history record, but some of the extracted features may be spurious. The mathematical presence of a periodic feature is not irrefutable evidence for the existence of a real oscillator with that period in the system that generated the time series data. Caution is needed in interpreting mathematically-extracted periodicities.
- Ultradian rhythms may be intermittent – now you see them, now you don't. As a result, they may have arbitrary phase, in contrast to circadian rhythms, which have strongly defined phase (as anyone who has suffered jet lag after traveling quickly across multiple time zones can attest).
- The power of an ultradian rhythm may vary from beat to beat.
- Generators of real ultradian rhythms may have noisy periods, i.e., they can 'wobble'.

11.11 Rationalizing Ultradian Rhythms as Essential Dynamics for Stability

The preceding discussion deals with some aspects of detecting and describing ultradian rhythms. In this section we offer our view concerning their functional significance. Our interest here is in the dynamic contribution of ultradian rhythms to the operational stability of an organism. Of course, a rhythm can be an informative signal – the frequency-modulated hormonal regulation of ovulation is a stunning example (Knobil, 1981). However, there can be no functionality without gradients, potentials, and energy to pay thermodynamic costs. We believe that the perspective of homeodynamics (Yates, 1994) illuminates the functional significance of ultradian rhythms through the physical concept of action (energy × time).

11.11.1 Why Cycles? The Quantization of Action

The basic form of temporal organization in physics and in biology is recurrence (i.e., cycles). That much is definitional, but is there a deeper basis for the ubiquity of cycles in nature? We believe that they are necessary expressions of the thermodynamics of open systems. Although the classical thermodynamics of isolated systems is a completed science, there are vigorous, continuing disputes about the proper thermodynamic formulations for open systems. The heuristic of homeodynamics offers a physical framework for the bookkeeping of energetic and entropic changes and transformations in open systems and their environments. Within the limits of this chapter we cannot supply the detailed reasoning justifying this claim, but we do list a few of the assumptions and principles on which the homeodynamic perspective rests.

1. Open systems with an environment obey the First and Second Laws of thermodynamics, as would be expected in general, although they do so in a special way.
2. In all natural processes in open systems the requirements of the First Law are satisfied instantaneously as the process runs. However, the bookkeeping on the Second Law may have delayed closure, with the reckoning not finished until an internal cycle has been completed. Of course, for balance the bookkeeping must take account of both internal and external (environmental) changes.
3. The product of energy × time (i.e., action) is the fundamental quantized variable in dynamics (the science of motion and change) at all levels of organization, starting at the quantum level with Planck's constant (which has the physical dimensions of action).
4. A thermodynamic cycle expresses a characteristic packet of energy at each completed period, yielding a scaled unit of action per cycle at every level of organization in a complex, hierarchical system.
5. Ultradian rhythms express 'biological time' that is the functional coordination by a living system of its internal events.
6. Ultradian rhythms express the action fields that are the signature of life. They are the coordinators.

11.12 Conjecture: Ultradian Rhythms at the Startup of Life

Structural biological 'order' is not physically remarkable. By physical entropic units a rock has more order than a mammal. Living systems are physically remarkable instead for their *functional order* (Carieri, 1984). Although the details of the transition from geochemistry to biochemistry 4 billion years ago remains unknown, we suggest that the startup was dynamic (i.e., energetic; motion and change), not informational ('on', 'off'; zero or one). Energy is needed for geochemistry to become biochemistry, prior to the evolution of informational codes and their elaborate, energetically costly consequences.

The requirements for any chemical process *arranged to persist* are, first, that there be some physical boundary for constraint or confinement, although also open to chemical fluxes. Second, there must be energetic cycles so that initial conditions are recovered when a process has been completed. Homeodynamics emphasizes the importance of energetic packets that quantize physical action and achieve persistence through cycles. These cycles would need to be ultradian in periodicity to meet rapidly changing reaction conditions and to achieve stability of organisms.

We suggest that the primordial sequence for the origin of life, as crudely viewed from a homeodynamic perspective, might have been: geochemistry → physical constraints to sustain local concentrations of reactants → action packet thermodynamic ultradian cycles → synthesis of molecules that function as informational constraints on dynamics → simple organism → circadian rhythm adaptations→ parasitism and symbiosis.

If this 'metabolism first' scenario intersects the truth, then the display of the varieties and beauties of ultradian rhythms that are elegantly illustrated in this book more than justifies our considering it a benchmark. (The continuing, vigorous debate between the 'replicator first' and the 'metabolism first' thought collectives has recently been well-reported (Shapiro, 2007)).

In summary, we propose that ultradian rhythms serve as the dynamic signature and coordinators of life. Metabolism is a self-orchestrated and choreographed chemical 'dance'. Ultradian rhythms in an organism cue other partners in that dance and the next steps appear.

References

Benton LA, Yates FE (1990). Ultradian adrenocortical and circulatory oscillations in conscious dogs. Am J Physiol 258:R578–590

Cannon WB (1929). Organization for physiological homeostasis. Physiol Rev 9:399–431

Carieri G (1984). Order and Disorder in Matter. Benjamin/Cummings, Menlo Park

Glass L, Mackey MC (1988). From Clocks to Chaos: The Rhythms of Life. Princeton University Press, Princeton, NJ

Gould SJ, Lewontin RC (1979). The spandrels of San Marco and the Panglossian paradigm: a critique of the adaptationist programme. Proc R Soc Lond B Biol Sci 205:581–598

Iberall AS (1977). A field and circuit thermodynamics for integrative physiology. I. Introduction to the general notions. Am J Physiol 233:R171–180

Iberall AS, Soodak H (1987). Thermodynamics and complex systems. In: Yates FE (ed) Self-Organizing Systems: The Emergence of Order. Plenum, New York, p 499–519

Kleitman N (1982). Basic rest-activity cycle – 22 years later. Sleep 5:311–317

Knobil E (1981). Patterns of hypophysiotropic signals and gonadotropin secretion in the rhesus monkey. Biol Reprod 24:44–49

Korenberg MJ (1988). Identifying nonlinear difference equation and functional expansion representations: the fast orthogonal algorithm. Ann Biomed Eng 16:123–142

Odell RH, Jr., Smith SW, Yates FE (1975). A permutation test for periodicities in short, noisy time series. Ann Biomed Eng 3:160–180

Pincus SM (1991). Approximate entropy as a measure of system complexity. Proc Natl Acad Sci USA 88:2297–2301

Riggs DS (1963). The Mathematical Approach to Physiological Problems. MIT, Boston, MA

Shapiro R (2007). A simpler origin for LIFE. Scientific American 296:46–53
Soodak H, Iberall AS (1978). Homeokinetics: a physical science for complex systems. Science 201:579–582
Yates FE (1982). The 10th J. A. F. Stevenson memorial lecture. Outline of a physical theory of physiological systems. Can J Physiol Pharmacol 60:217–248
Yates FE (1992). Fractal applications in biology: scaling time in biochemical networks. Methods Enzymol 210:636–675
Yates FE (1994). Order and complexity in dynamical systems: homeodynamics as a generalized mechanics for biology. Math Comput Model 19:49–74
Yates FE (1996). Homeostasis. In: Birren J (ed) Encyclopedia of Gerontology. Academic, New York, pp 679–686
Yates FE (2005). Homeostasis. In: Schulz R (ed) The Encyclopedia of Aging. Springer, New York
Yates FE, Benton LA (1991). Characteristics of ultradian and circadian rhythms of selected cardiovascular variables. Diagnostic and therapeutic implications. Ann NY Acad Sci 618:38–56
Zak M, Zbilut JP, Meyers RE (1997). From Instabilities to Intelligence. Springer, New York

Chapter 12
The Mammalian Circadian Timekeeping System

U. Schibler

Abstract In most mammalian species, physiological processes undergo daily oscillations that are controlled by the circadian timekeeping system. This system consists of a master pacemaker located in brain's suprachiasmatic nucleus (SCN) and peripheral slave oscillators in virtually all body cells. The SCN, whose phase is entrained by daily light-dark cycles, imposes overt rhythms in behaviour and physiology by a variety of neuronal, humoral, and physical outputs. While some of these SCN outputs have direct consequences for circadian behaviour, others serve as inputs to synchronize the countless circadian oscillators in peripheral cell types. Daily feeding-fasting cycles are the major Zeitgebers (timing cues) for the synchronization of oscillators in many peripheral organs. Circadian gene expression and physiology have been particularly well studied in the liver. In this organ, local circadian clocks play an important role in the coordination of food processing and xenobiotic detoxification. Although all investigated mammals contain a light-entrainable master clock in the SCN, some species display ultradian locomotor and feeding rhythm. For example, the common vole *Microtus arvalis* forages in bouts of 150 min throughout the day. In this ultradian rodent circadian clock and clock-controlled genes are expressed rhythmically in the SCN, but at constant intermediate levels in peripheral tissues.

Keywords *Mus musculus*, *Microtus arvalis*, circadian clocks, suprachiasmatic nucleus (SCN), peripheral oscillators

All biological processes are timed, either by hourglass clocks or oscillators. Hour glass clocks determine the duration of processes that occur only once during the life time of a molecule, a cell, or an individual organism. Such timers determine the average life time of macromolecules and the length of gametogenesis, embryonic development, cell differentiation, reaching sexual maturity, ageing, lifespan, and evolution. In contrast, biological oscillators control regularly reoccurring

Department of Molecular Biology, University of Geneva, 30, Quai Ernest Ansermet, CH-1211 Geneva-4, Switzerland
Email: ueli.schibler@molbio.unige.ch

events within cells or organisms, such as metabolic rhythms, heartbeat frequency, rest-activity cycles, female oestrus cycles, and seasonal rhythms. Somewhat arbitrarily, oscillators are subdivided into circadian oscillators with period lengths of approximately 24 h, ultradian oscillators with period lengths from milliseconds to a few hours, and infradian oscillators with period lengths of days to years. Circadian oscillators are found in virtually all light sensitive organisms from cyanobacteria to humans. They have been discovered as early as in 1729 by the French astronomer Jean-Jacques Dertous de Mairan, who noticed that the mimosa plants closed their leaves in a daily cycle even if not exposed to the photoperiod (De Mairan, 1729). As indicated by their name – circa dies means about a day – circadian clocks can measure a 24-h day only approximately. Hence, they have to be synchronized every day. The deviation of circadian time from geophysical time depends on the species, but in wild-type organisms always remains within the capacity of phase entrainment (flexibility) of the circadian clock of the respective species. For example, humans have a limited daily phase-shifting capacity (1–3 h, depending on lighting conditions (Minors et al., 1991; Khalsa et al., 2003), and a sudden large phase alteration such as the one caused by a transatlantic flight results in jet lag (Waterhouse et al., 2007). In contrast, some unicellular organisms have remarkably flexible clocks that can adapt many hours, sometimes even corresponding to a complete phase inversion (Edmunds and Laval-Martin, 1984; Roenneberg and Taylor, 1994). The phase shifting behaviour of circadian clocks is generally represented by phase response (PRC) and phase transition curves (PTC) (Glass and Winfree, 1984). In phase response curves the magnitudes of phase delays (negative phase shifts) and phase advances (positive phase shifts) resulting from phase-shifting stimuli (e.g. light pulses) delivered along the circadian day (X-axis) are plotted on the Y-axis. In phase transition curves, the new phase (Y-axis) is plotted against the old phase (X-axis). Circadian oscillators with a strong phase shifting capacity yield type-zero phase transition curves, because the slope (Y/X) of the curve is around zero. Such oscillators are reset to the same new phase irrespective of when phase shifting stimuli have been delivered. The circadian locomotor activity of laboratory rodents displays a type-one phase transition curve, because the deviations of the new phase from the old phase are rather small, and the average slope is thus approaching a value of one (Comas et al., 2006). This chapter focuses on mammals, in which nearly every body cell harbours its own circadian oscillator. These clocks can display a type-one or a type-zero phase shifting behaviour, depending on their hierarchical status (Nagoshi et al., 2004; and references therein). This chapter presents a brief overview over the molecular makeup and the cellular architecture of the mammalian circadian timing system.

12.1 Molecular Oscillators: What Is the Underlying Principle?

Vertebrate and dipterian circadian oscillators share many components and are thought to rely on negative feed-back loops in clock gene expression (Reppert and Weaver, 2001; Yu and Hardin, 2006). According to the currently held model for mammalian oscillators (Fig. 12.1), the helix-loop-helix transcription factors BMAL1, CLOCK, and NPAS2 activate the genes encoding two Period 1 (PER1, PER2) and two cryptochrome (CRY1, CRY2) isoforms. These four proteins form heterotypic multisubunit complexes that attenuate the transactivation potential of BMAL1-CLOCK or BMAL1-NPAS2 heterodimers and thereby repress the transcription of their own genes. As a consequence, CRY and PER protein concentrations decrease to levels at which they can no longer repress their genes, and a new cycle of *Cry* and *Per* transcription can ensue. The transcription of Per/Cry genes is antiphasic to that of Bmal1 transcription. This is accomplished by the antagonistic action of members belonging to the ROR and REV-ERB families of orphan nuclear receptors that bind to two RORE cis-acting elements within the Bmal1 promoter. Bmal1 transcription is stimulated by ROR activators and inhibited by REV-ERB repressors (Preitner et al., 2002; Sato et al., 2004). Since the cyclic expression of

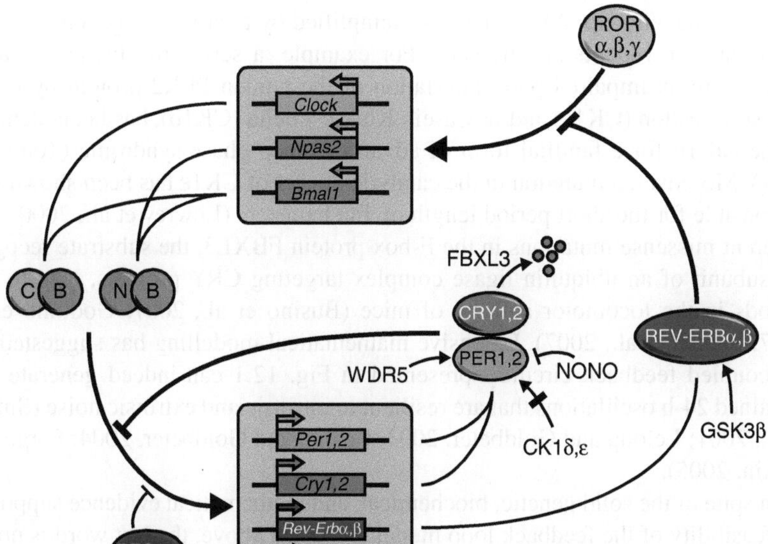

Fig. 12.1 Molecular model of the circadian oscillator in mammals. The transcription of *Per* and *Cry* genes is activated by heterodimers between BMAL1 (B) and either of the two related proteins CLOCK (C) or NPAS2 (N). The polycomb protein EZH2 interacts with these heterodimers and thereby facilitates their action. The accumulation and activity of PER and CRY proteins are also influenced by phosphorylation by protein kinases (CK1δ,ε), by ubiquitination via a complex containing the F-box protein FBXL3 (specific for CRYs), by the histone methyl transferase binding protein WDR5, and by NONO, an RNA and DNA binding protein. DEC1 and DEC2 (D1,2) compete with BMAL1-CLOCK/NPAS2 heterodimers for E-box binding and thereby reduce E-box-mediated transactivation. A accessory feedback loop, employing the nuclear orphan receptors RORα, RORβ, and RORγ (RORα,β,γ) as activators, and REV-ERBα and REV-ERBβ (REV-ERBα,β) as repressors, regulates the circadian transcription of *Bmal1*. See text for further explanations (Reproduced from Schibler, 2007)

Rev-erb genes is regulated by the transcriptional activators and repressors that also control Per and Cry genes, the resulting cyclic accumulation of REV-ERBα/β results in the circadian repression of Bmal1 transcription. The coupling of Bmal1 and Per/Cry expression by this accessory feedback loop, albeit not essential for rhythm generation, renders the clockwork circuitry more accurate and increases its resilience towards perturbations by phase shifting stimuli (Preitner et al., 2002). Additional transcriptional regulatory proteins, including the RNA/DNA-binding protein NONO (Brown et al., 2005), the histone methyl transferase-binding protein WDR5 (Brown et al., 2005), the polycomb group member EZH2 (Etchegaray et al., 2006), and the two orange domain helix-loop-helix proteins DEC1 and DEC2 (Honma et al., 2002) have been reported to be implicated in circadian rhythm generation.

Several additional interactions may further stabilize mammalian circadian oscillators. In fact, posttranslational modifications of clock proteins by a variety of kinases, phosphatases, and ubiquitin ligases have been shown to play important functions in rhythm generation and period length determination (see Fig. 12.1) (Gallego and Virshup, 2007). This is exemplified by a variety of genetic observations made in rodents and humans. For example, a serine to glycine mutation, resulting in an impaired phosphorylation of the human PER2 protein by Casein Kinase 1 epsilon (CK1ε) and/or Casein Kinase 1 delta (CK1δ), has been identified as the culprit for a familial form of advanced sleep phase syndrome (Toh et al., 2001). Moreover, a mutation in the catalytic pocket of CK1ε has been shown to be responsible for the short period length of Tau hamsters (Lowrey et al., 2000). Two different missense mutations in the F-box protein FBXL3, the substrate recognizing subunit of an ubiquitin ligase complex targeting CRY proteins, lead to long periods in the locomotor activity of mice (Busino et al., 2007; Godinho et al., 2007; Siepka et al., 2007). Extensive mathematical modelling has suggested that the coupled feedback circuitry presented in Fig. 12.1 can indeed generate self-sustained 24-h oscillations that are resilient to intrinsic and extrinsic noise (Smolen et al., 2001; Leloup and Goldbeter, 2003; Leloup and Goldbeter, 2004; Forger and Peskin, 2005).

In spite of the solid genetic, biochemical, and mathematical evidence supporting the feasibility of the feedback loop model presented above, the last word is not yet out. Thus, we cannot rigorously exclude that the oscillations of clock gene expression are manifestations of a more fundamental mechanism, perhaps based on posttranslational interactions yet to be identified. Breathtaking studies on the cyanobacterial circadian oscillator have recently revealed that a protein phosphorylation/dephosphorylation oscillator keeps ticking in the absence of transcription and translation (Tomita et al., 2005). This oscillator can be reconstituted *in vitro* with just three purified proteins – KaiA, KaiB, and Kai C – and adenosine triphosphates (ATP) (Nakajima et al., 2005; Kageyama et al., 2006). It involves the phosphorylation and dephosphorylation of a serine and a threonine within KaiC, which can act both as an autokinase and an auto-phosphatase. The phosphorylation and dephosphorylation operation modes of KaiC are mediated by its interactions with KaiA and KaiB. Remarkably, the 24-h phosphorylation-dephosphorylation cycles persist for many days in the test tube and are temperature compensated. Mathematical models accounting for the rhythmic KaiC phosphorylation have recently been presented (Miyoshi et al., 2007; Rust et al., 2007; van Zon et al., 2007).

12.2 The Mammalian System: A Hierarchical Network of Oscillators

Surgical lesion and transplantation experiments have unambiguously identified the suprachiasmatic nucleus in the ventral hypothalamus as the mammalian master pacemaker (Ralph et al., 1990; Silver et al., 1996). In the rat, each of the two SCNs is composed of only about 10,000 neurons. Dissociated SCN neurons display robust cycles in circadian firing frequency and gene expression, suggesting that the rhythm generating principle operates in a cell-autonomous fashion (Welsh et al., 1995; Jin et al., 1999; Liu et al., 2007). However, the period lengths recorded for individual neurons vary enormously, and these cellular oscillators become thus desynchronized within a few hours in cultures of dissociated SCN neurons. However, strong phase coherence between SCN neurons is observed in organotypic brain slice cultures, in which SCN neurons remain coupled via synaptic and/or diffusible signals (Kriegsfeld and Silver, 2006). This coupling also renders SCN oscillators more resilient to genetic lesions. For example, SCN neurons homozygous for a Cry1 null allele are mostly arrhythmic in dissociated cultures but still rhythmic in organotypic slice cultures (Liu et al., 2007).

The SCN gets phase adjusted daily by the photoperiod (Fig. 12.2). Light signals perceived by classical photoreceptors in the outer retina layer and melanopsin-containing ganglion cells in the inner retina layer are transmitted to SCN neurons via the retino-hypothalamic tract. The synaptic release of glutamate and pituitary adenylate cyclase-activating polypeptide (PACAP) leads to an influx of Ca^{++} in postsynaptic SCN neurons, and this elicits the activation of a variety of protein kinases (Kriegsfeld and Silver, 2006). In turn this triggers the activation of the immediate early transcription factor CREB (cAMP responsive element-binding protein). CREB then binds to cyclic AMP response elements (CREs) in promoter and enhancer regions of the Per1 and Per2 genes and thereby provokes an increase in PER1 and PER2 accumulation (Gillette and Mitchell, 2002; Jakubcakova et al., 2007). Light-induced Per1/2 expression during dawn and dusk advance and delay, respectively, the phase of circadian PER accumulation, and this keeps circadian rhythms tuned to the photoperiod (Albrecht et al., 2001). The phase-entrained SCN then communicates with downstream brain regions and peripheral organs through various neuronal and humoral output pathways. For example, it activates corticotrophin releasing factor (CRF) producing neurons in the hypothalamus, resulting in the diurnal release of ACTH by corticotrophs of the pituitary gland and glucocorticoids by the adrenal cortex (Vrang et al., 1995; Kalsbeek et al., 1996). The pathways by which the SCN governs daily rest-activity cycles are less well

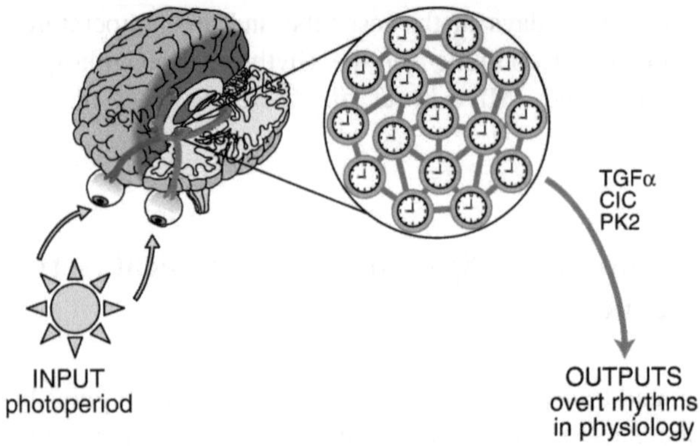

Fig. 12.2 Inputs into and outputs from the SCN. The SCN is composed of neurons that contain cell-autonomous, self-sustained oscillators. Phase coherence is maintained through the coupling of cellular oscillators (depicted by red lines between the clocks). These clocks are synchronized daily by photic inputs from the retina via the retinohypothalamic tract. The SCN master pacemaker conveys circadian information to other brain regions and to the periphery, resulting in overt cycles in physiology and behaviour. The three chemokines tumour growth factor alpha (TGFα), cardiotrophin-like cytokine (CIC), and prokineticin 2 (PK2) have been proposed to regulate circadian rest-activity rhythms. See text for further explanations

understood, but tumour growth factor alpha (TGFα), prokineticin 2 (PK2), and cardiotrophin-like cytokine (CIC), three peptides secreted by the SCN, may participate in this endeavour (Kramer et al., 2001; Cheng et al., 2002; Kraves and Weitz, 2006).

Surprisingly, self-sustained and cell-autonomous circadian oscillators exist not only in the SCN but in virtually all tissues, and even in cultured fibroblasts (Gachon et al., 2004). In contrast to the oscillators of SCN neurons, those of peripheral cell types are not coupled, and hence they rapidly loose phase coherence in SCN-lesioned animals (Yoo et al., 2004; Guo et al., 2006). Therefore, they must be periodically synchronized by the SCN master pacemaker. Cultured fibroblast can be transiently synchronized by a bewildering variety of signals, encompassing activators of protein kinases (MAPK, PKA, PKC), glucocorticoid and retinoic acid nuclear receptors, calcium ionophors, and glucose (for review, see Gachon et al., 2004). In animals daily feeding-fasting cycles are the dominant Zeitgebers for peripheral clocks in many organs, including liver, kidney, pancreas, and heart (Damiola et al., 2000; Le Minh et al., 2001; Stokkan et al., 2001). However, the molecular signalling pathways by which this is accomplished remain unknown. The identification of immediate early genes that interpret cyclic systemic signals in peripheral organs may be a valuable strategy in this endeavour. The expression of such genes is expected to remain rhythmic even in the absence of functional peripheral oscillators. Indeed, a transgenic mouse model in which hepatocyte clocks can be switched on and off has provided a list of systemically driven genes, and some of these are attractive candidates for players involved in physiologically relevant synchronization pathways. These include the genes specifying the core clock component mPER2 and the liver hormone fibroblast growth factor 21 (FGF21) (Kornmann et al., 2007; Kornmann et al., 2008). Moreover, FGF19, a hormone related to FGF21 that is secreted by intestinal cells, has also been reported to undergo diurnal oscillations in the blood plasma (Lundasen et al., 2006). Conceivably, daily feeding-fasting cycles drive the cyclic secretion of FGF19 and FGF21, resulting in the rhythmic activation of mitogen activated protein kinases (MAPK). In turn MAPK may phosphorylate the transcription factor CREB and thereby stimulate its transactivation capacity. As mentioned above, the cyclic activity of CREB has already been shown to play a pivotal role in the synchronization of SCN neurons.

The conditional liver clock mouse model, in conjunction with a novel proteomics approach allowing the identification of diurnally active transcription factors, has unveiled another, rather unexpected signalling pathway. Every day, heat shock transcription factor 1 (HSF1) is activated during a few hours, supposedly by body temperature and feeding cycles (Reinke et al., 2008). During the resting phase HSF1 forms an inert cytoplasmic complex with the HSP90 and some co-chaperones. The levels of this transcription factor are not subject to temporal variations, but at the onset of the activity period HSF1 is released from its inert complex, migrates to the nucleus, and activates its target genes. It is likely that some of the HSF1 controlled genes, perhaps even heat shock protein genes, participate in the fine tuning of circadian oscillator function, since the locomotor activity of Hsf1 knockout mice

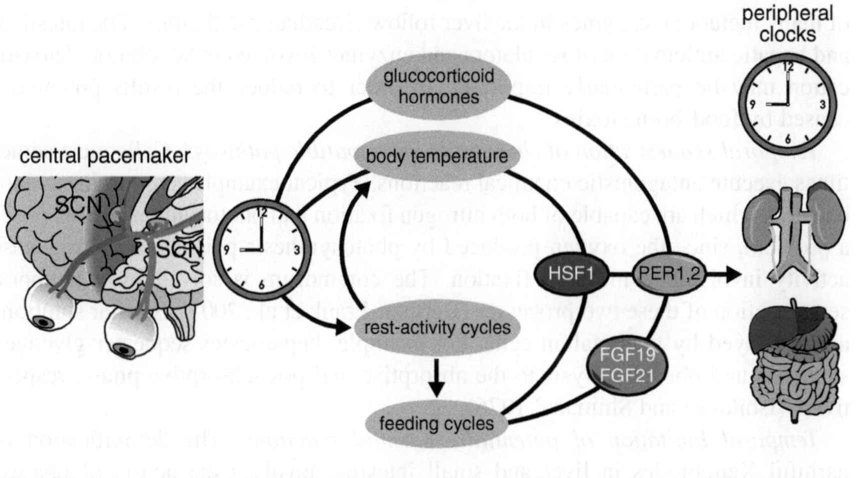

Fig. 12.3 Synchronization of peripheral oscillators. The SCN synchronizes circadian slave oscillators in the periphery through a variety of chemical and physical signals. Heat shock transcription factor 1 (HSF1) and two members of the fibroblast growth factor 19 family (FGF19 and FGF21) may modulate Per1 and Per2 gene expression in peripheral cell types and thereby contribute to the phase entrainment to peripheral clocks. See text for further explanations

display a significantly longer period than wild-type mice (Reinke et al., 2008). The daily activation of HSF1 may also participate in the phase entrainment of peripheral oscillators, given that temperature cycles are strong Zeitgebers for circadian clocks of cultured fibroblasts (Brown et al., 2002; C. Saini and U.S., 2007). A speculative model of signalling routes involved in the synchronization of peripheral clocks is displayed in Fig. 12.3.

12.3 Outputs: Circadian Metabolism

Genome-wide transcriptome profiling studies suggest that many metabolic pathways in liver and other peripheral organs are subject to daily fluctuations (Kornmann et al., 2001; Akhtar et al., 2002; Panda et al., 2002; Storch et al., 2002; Ueda et al., 2002; Walker and Hogenesch, 2005). Likewise, in yeast grown under aerobic conditions many metabolic activities are controlled by an ultradian respiratory clock (see Tu and McKnight, 2006; Lloyd and Murray, 2007; and other chapters of this book). The temporal fine tuning of metabolism may serve three purposes acting in concert with each other: Optimization of metabolic efficiency by anticipation, sequestration of chemically conflicting pathways to different time windows, and limiting the expression of potentially harmful enzymes to time spans during which they are needed.

Anticipation: It would make sense to prepare the food processing machinery before nutrients are absorbed. Indeed, the secretion of HCl into the stomach (Bumm et al., 1987), the production and secretion of digestive enzymes by the pancreas (Keller and Layer, 2002; Zabielski, 2004; Woods et al., 2006), and the expression of many metabolic enzymes in the liver follow circadian oscillations. The intestinal and hepatic anticipation of regulators and enzymes involved in xenobiotic detoxification may be particularly important, in order to reduce the insults potentially caused by food-borne toxins.

Temporal sequestration of chemically incompatible pathways: Cells must sometimes execute antagonistic chemical reactions. Typical examples are marine cyanobacteria, which are capable of both nitrogen fixation and photosynthesis. This poses a problem, since the oxygen produced by photosynthesis poisons the nitrogenase activity involved in nitrogen fixation. The conundrum is solved by a temporal sequestration of these two processes (Berman-Frank et al., 2001). Similar solutions are employed by mammalian cells. For example, hepatocytes sequester glycogen synthesis and phosphorolysis to the absorptive and post-absorptive phase, respectively (Ishikawa and Shimazu, 1976).

Temporal limitation of potentially harmful reactions: The detoxification of harmful Xenobiotics in liver and small intestine involves the action of phase I enzymes (cytochrome p450 enzymes), phase II, enzymes (conjugating enzymes), and phase III transporters. (e.g. P-glycoprotein, multidrug resistance proteins, organic anion transporting polypeptide 2) (Xu et al., 2005). Cytochrome p450 enzymes are heme-containing monooxygenases that hydroxylate Xenobiotics.

The resulting hydroxyl groups are esterified by phase II conjugating enzymes, and these modifications render Xenobiotics water soluble and thereby excretable. In the absence of substrate, cytochrome p450 monooxygenases can produce genotoxic reactive oxygen species from molecular oxygen (Zangar et al., 2004). It may thus be advantageous to limit the expression of such enzymes to the time when they are needed. Likewise, the synthesis of bile acids and their secretion into the gut are limited to the absorptive phase (Balabaud et al., 1975; Noshiro et al., 2007), perhaps because the constant presence of these detergents would be corrosive to the intestinal epithelium.

Circadian regulation of metabolic pathways works in close cooperation with inducible control mechanisms. For example, in liver several components involved in cholesterol/bile acid homeostasis, such as INSIG2, HMG CoA reductase, HMG Co A synthase, squalene monooxygenase, sterol 14 alpha-demethylase, isopentenyl-diphosphate δ isomerase and cholesterol 7α hydroxylase (Cyp7a1), are expressed in a diurnal manner. The cyclic expression of some of these enzymes depends on functional hepatocyte clocks (Gwendal Le Martelot and U.S., 2008). However, most of the genes associated with cholesterol synthesis are target genes of Sterol Regulatory Element Binding Protein (SREBP), a transcription factor sensing sterol concentrations in membranes of the endoplasmic reticulum (ER) (Brown and Goldstein, 1999; McPherson and Gauthier, 2004). The precursor of SREBP, a two transmembrane domain protein, forms a complex with SREBP cleavage activating protein (SCAP), which itself is tethered to either of the two ER resident proteins INSIG1 and INSIG2. Upon depletion of cholesterol in ER membranes, SCAP loses its affinity to INSIG1/2, the SREBP-SCAP complex gets translocated to the Golgi vesicles, and SREBP gets cleaved through the action of two proteinases associated with Golgi membranes. The N-terminal moiety of SREBP, a basic helix-loop-helix transcription factor, gets released into the cytosol, migrates to the nucleus, and activates genes involved in cholesterol synthesis and transport. Once cholesterol is replenished in the ER, newly synthesized SREBP remains anchored to ER membranes in the inert complex described above. The cyclic processing of SREBP may thus not be directly controlled by hepatocyte circadian oscillators, but by the daily variation of cholesterol levels. In turn, these fluctuations may be the consequence of circadian bile acid synthesis (which consumes massive amounts of cholesterol during the absorptive phase) and the reflux of bile acids from the gastrointestinal tract. The nuclear receptors FXR and LXR play important functions in this endeavour (Houten et al., 2006; Kalaany and Mangelsdorf, 2006). FXR is directly activated by bile acids, and this elicits an inhibitory mechanism that attenuates the transactivation potential of LXR for Cyp7a1 transcription.

An intricate interplay between circadian and inducible regulation is also operative in xenobiotic detoxification. Constitutive androstane receptor (CAR), a nuclear orphan receptor serving as a sensor of many Xenobiotics, is expressed in a circadian fashion (Handschin and Meyer, 2003). In the absence of CAR activating Xenobiotics (e.g. barbiturates), CAR is sequestered in an inert cytosolic complex with chaperones and co-chaperones. In the presence of activators, CAR dissociates from the chaperones, moves to the nucleus, and activates cytochrome p450 enzymes that metabolize the xenobiotic activators. As CAR is more abundant during the absorptive phase, CAR-inducing compounds elicit a much more powerful stimulation of cytochrome p450 enzyme encoding genes during the absorptive as compared to the postabsorptive phase (Gachon et al., 2006).

12.4 Interaction Between Circadian and Ultradian Rhythms

Mice, hamsters, and rats whose circadian organization has been disrupted through gene knockout or SCN lesion experiments still display rhythmic locomotor activity. However, depending on the genetic or surgical insult, the period length of behaviour is now between 4 and 8 h (Redlin and Mrosovsky, 1999; van der Horst et al., 1999; Bunger et al., 2000; Bae et al., 2001; Zheng et al., 2001). These ultradian rhythms are not readily observed in actograms and periodograms of intact mice, rats or hamsters, suggesting that they are normally masked by circadian rhythms in these species. The 4–8-h ultradian rhythms could not yet be associated with molecular rhythms, and the underlying mechanisms thus remain obscure (but see below).

Ultradian behavioural rhythms are more overtly manifested in the common vole (*Microtus arvalis*) (Gerkema and van der Leest, 1991; Gerkema et al., 1993). Both in the field and in the laboratory this hind gut fermenting rodent forages at regular bouts of about 150 min throughout the 24-h day (Fig. 12.4A, top panel). The brain region driving this ultradian behaviour is not yet identified unambiguously, but lesion experiments suggest that the arcuate nucleus and the retrochiasmatic region appear to be implicated (Gerkema et al., 1990). Voles possess a perfectly operational and light-entrainable SCN (Gerkema et al., 1990; Gerkema et al., 1994; van der Veen et al., 2006), but the circadian signals delivered by this circadian master pacemaker must be masked by dominant ultradian commands from other brain centres. Interestingly, if voles are kept in cages equipped with running wheels, their locomotor activity displays a strong circadian component in addition to residual ultradian bouts (Fig. 12.4A, bottom panel) (van der Veen et al., 2006). Hence, similar to what has been observed in mice, hamsters, and rats, circadian and ultradian rhythms appear to interact with each other in the common voles. However, while circadian rhythms dominate ultradian rhythms in mice, hamsters and rats, the opposite is true in voles.

As outlined earlier in this chapter, feeding-fasting cycles are potent Zeitgebers for the synchronization of peripheral clocks. So what is the impact of ultradian feeding behaviour on circadian transcription in peripheral organs of ultradian and circadian rodents (see van der Veen et al., 2006)? In voles kept under normal housing conditions, all examined core clock- and clock-controlled genes are expressed at constant intermediate levels in liver and kidney (Fig. 12.4B, left panel). Yet, voles subjected to a circadian feeding regimen exhibit robustly circadian gene expression in liver (Fig. 12.4B, right panel) and kidney (not shown). This indicates that the

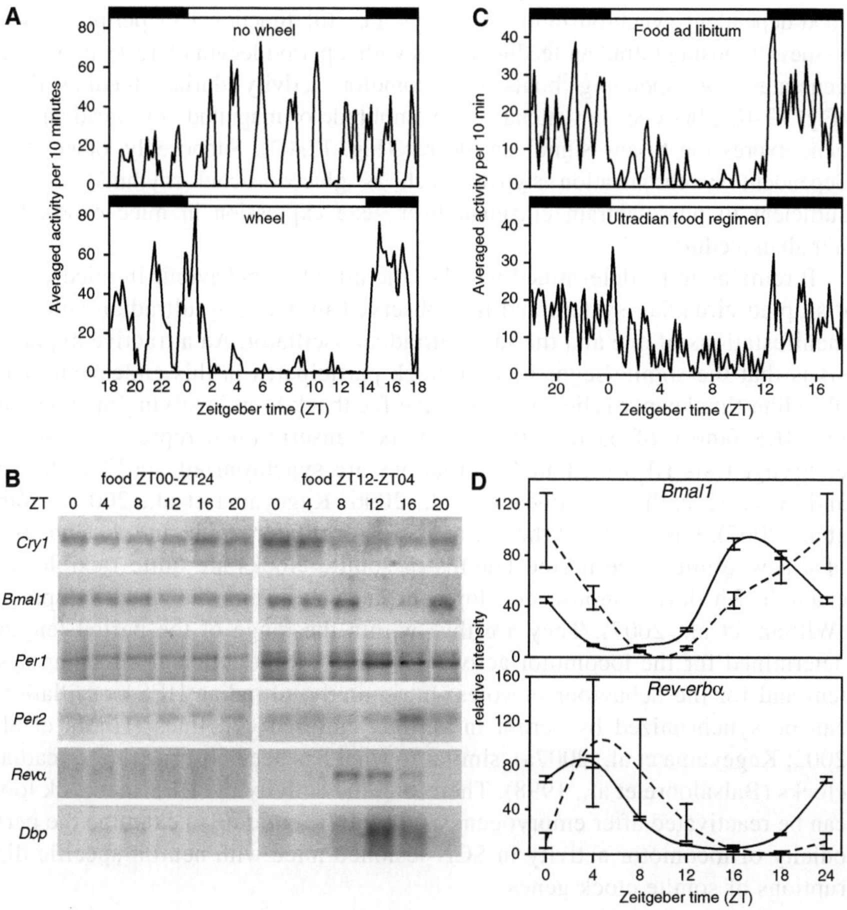

Fig. 12.4 Circadian and ultradian oscillations in mice. **A**. Five-day averaged activity pattern of a vole housed without (top panel) and with (bottom panel) a running wheel. Environmental light conditions are shown at the top of the panels. **B**. Temporal accumulation of transcripts encoded by clock and clock-controlled genes in the liver of voles subjected to different feeding conditions (indicated on top of the panels). Voles were housed in cages without running wheels. **C**. Spontaneous locomotor activity, as measured by infrared beam breaks, was recorded during 10 days for animals fed at 150 min intervals (N = 11) and animals fed ad libitum (N = 7). The activities of all animals (of the respective groups) monitored during 10 consecutive days were compiled in diagrams showing the average activity of 10-min bins during a day. **D**. Mice, kept in cages without a running wheel and exposed to a 12 h light/12 h dark regimen, were fed during 11 days with meals delivered every 150 min by a computer-driven feeding machine. On the 11th day the temporal accumulation of Bmal1 mRNA and Rev-erbα mRNA was recorded using Taqman® real-time RT-PCR. Control animals were housed under identical conditions, but had unlimited access to food. Note that ultradian feeding advanced the phase of circadian liver gene expression by approximately 4 h, but had little influence on the amplitude (Reproduced from van der Veen et al., 2006)

food-dependent synchronization pathways are still functional in peripheral vole tissues. Imposing ultradian feeding cycles with a period length of 150 min on mice generates corresponding bouts in locomotor activity during feeding times (Fig. 12.4C), but does not attenuate the amplitude or magnitude of circadian liver gene expression to any significant degree (Fig. 12.4D). Supposedly, other SCN-dependent synchronization signals, such as glucocorticoid rhythms, are still sufficient to phase-entrain circadian liver gene expression in mice exposed to ultradian feeding.

It remains to be determined whether the ultradian behaviour in mice with a disrupted circadian system and that observed in naturally ultradian voles are manifestations of one and the same ultradian oscillator. An attractive hypothesis is that the somite/segmentation clock participates in this endeavour. This ultradian timekeeper relies on a negative feedback loop involving members of the HES family of basic helix-loop-helix transcriptional repressors. During embryogenesis HES mediated oscillations are synchronized via FGF, Notch, and WNT signalling (Dequeant et al., 2006; Kageyama et al., 2007b; Wahl et al., 2007). Each wave of the segmentation oscillator results in the deposition of a new somite. The period lengths of somite clocks are quite flexible, for example 2 h during mouse development and 5 h during human development (William et al., 2007). They are thus within the range of the period lengths determined for the locomotor activity of mice with a deficient circadian system and for the behaviour of voles. Intriguingly, ultradian HES1 oscillations can be synchronized by serum in various cultured cell lines (Hirata et al., 2002; Kageyama et al., 2007a), similar to what has been observed for circadian clocks (Balsalobre et al., 1998). Therefore, the activity the HES feedback loop can be reactivated after embryogenesis. It will be enticing to examine the periodicity of locomotor activity in SCN-lesioned mice with neuron-specific disruptions of somite clock genes.

12.5 Conclusions and Perspectives

During the past decade we have witnessed an impressive boost of molecular circadian rhythm research. Many essential clock genes have been identified and studied by employing biochemical and genetic approaches (for a review, see Schibler, 2007). Owing to advanced genomics approaches, circadian transcriptomes containing hundreds of rhythmically expressed genes could be assembled for several tissues (Walker and Hogenesch, 2005). These studies underscore the importance of circadian clocks in the coordination of metabolism and other physiological processes. The discovery of self-sustained circadian oscillators in peripheral tissues and cultured cells has rendered these clocks amenable to biochemical dissection, both with regard to the molecular core clockwork circuitry and input and output pathways.

In spite of all these remarkable accomplishments, many important issues remain to be addressed. Thus, the currently publicized molecular feedback loop models still suffer from many inconsistencies (Lakin-Thomas, 2006). We can thus not rigorously exclude that the observed waves in clock gene expression are manifestations of more fundamental oscillators, perhaps relying on posttranslational mechanisms. Indeed, in both cyanobacteria and filamentous fungi, protein phosphorylation/dephosphorylation cycles appear to play pivotal roles in circadian rhythm generation (Nakajima et al., 2005; Tomita et al., 2005; Schafmeier et al., 2006). The hierarchical architecture of the mammalian circadian timing system raises the question of how the master pacemaker in the SCN establishes phase coherence in the periphery. The investigation of the involved signalling pathways is not a trivial undertaking, since many functionally redundant signalling routes appear to coexist. As a consequence, monitoring the steady-state phase of circadian gene expression in animals deficient for only one relevant signalling pathway will probably not reveal decisive information. Measuring phase shifting kinetics would be a more sensitive approach (Le Minh et al., 2001; Kornmann et al., 2008), but this requires novel whole body imaging technologies that permit the recording of peripheral gene expression in real time and in life animals. Last but not least, the selective advantage of possessing circadian clocks remains subject to speculation, at least for mammalian species. Clearly, it would be foolish to argue that the complex circadian clockwork circuitry has been phylogenetically conserved for hundreds of millions of years just to provide a research subject for chronobiologists. However, compelling evidence associating functional clocks with increased fitness has not yet been accomplished in mammals, and although several morbidities have been attributed to the disruption of clock genes, this does not necessarily identify disturbed rhythms as the culprits. Genes frequently get their names according to the research field of the scientists who had discovered them. For example, *Clock* stands for *Circadian Output Cycles Kaput*, since this gene has been identified by chronobiologists in a forward genetic screen for mice with altered wheel running activity (King et al., 1997, and references therein). A few years later, *Clock* mutant mice were shown to be susceptible to “metabolic syndrome” (MetS) (Turek et al., 2005; Scott et al., 2007), a disorder combining insulin resistance, hyperinsulinemia, hyperglycemia, hypertension, and hypercholesteremia that is frequently observed in overweight people. Had *Clock* been first discovered by a diabetologist in a screen for quantitative trait loci (QTLs) associated with metabolic syndrome (Dandona et al., 2005), it may have been dubbed *MetS-X*. Indeed, all of the health disorders reported to be correlated to clock gene mutations may, in fact, have little to do with circadian rhythms. What is needed is the demonstration that increased fitness is the result of time-dependent gene functions. This has been beautifully accomplished for the cyanobacterium *Synechococcus elongatus* and the green plant *Arabidopsis thaliana* (Ouyang et al., 1998; Woelfle et al., 2004; Dodd et al., 2005), by showing that organisms with a clock that resonates with environmental light-dark cycles possess a selective advantage over organisms with a non-resonating oscillator. Such experiments should now be possible with hamsters and mice harbouring circadian period length mutations.

Acknowledgements Research in the author's laboratory was supported by the Swiss National Foundation (through an individual research grant and the National Center of Competence in Research Program *Frontiers in Genetics*), the State of Geneva, the Louis Jeantet Foundation of Medicine, the Bonizzi-Theler Stiftuung, and the 6th European Framework Project EUCLOCK.

References

Akhtar, R.A., Reddy, A.B., Maywood, E.S., Clayton, J.D., King, V.M., Smith, A.G., Gant, T.W., Hastings, M.H., and Kyriacou, C.P. 2002. Circadian cycling of the mouse liver transcriptome, as revealed by cDNA microarray, is driven by the suprachiasmatic nucleus. *Curr Biol* **12**(7): 540–550.

Albrecht, U., Zheng, B., Larkin, D., Sun, Z.S., and Lee, C.C. 2001. MPer1 and mper2 are essential for normal resetting of the circadian clock. *J Biol Rhythms* **16**(2): 100–104.

Bae, K., Jin, X., Maywood, E.S., Hastings, M.H., Reppert, S.M., and Weaver, D.R. 2001. Differential functions of mPer1, mPer2, and mPer3 in the SCN circadian clock. *Neuron* **30**(2): 525–536.

Balabaud, C., Noel, M., Beraud, C., and Dangoumau, J. 1975. Circadian rhythm of bile secretion in the rat. *Experientia* **31**(11): 1299–1301.

Balsalobre, A., Damiola, F., and Schibler, U. 1998. A serum shock induces circadian gene expression in mammalian tissue culture cells. *Cell* **93**(6): 929–937.

Berman-Frank, I., Lundgren, P., Chen, Y.B., Kupper, H., Kolber, Z., Bergman, B., and Falkowski, P. 2001. Segregation of nitrogen fixation and oxygenic photosynthesis in the marine cyanobacterium Trichodesmium. *Science New York, NY)* **294**(5546): 1534–1537.

Brown, M.S. and Goldstein, J.L. 1999. A proteolytic pathway that controls the cholesterol content of membranes, cells, and blood. *Proc Natl Acad Sci USA* **96**(20): 11041–11048.

Brown, S.A., Zumbrunn, G., Fleury-Olela, F., Preitner, N., and Schibler, U. 2002. Rhythms of mammalian body temperature can sustain peripheral circadian clocks. *Curr Biol* **12**(18): 1574–1583.

Brown, S.A., Ripperger, J., Kadener, S., Fleury-Olela, F., Vilbois, F., Rosbash, M., and Schibler, U. 2005. PERIOD1-associated proteins modulate the negative limb of the mammalian circadian oscillator. *Science* **308**(5722): 693–696.

Bumm, R., Blum, A.L., Bauerfeind, P., and Emde, C. 1987. Lessons from prolonged gastric pH monitoring. *Aliment Pharmacol Ther* **1**(Suppl 1): 518S–526S.

Bunger, M.K., Wilsbacher, L.D., Moran, S.M., Clendenin, C., Radcliffe, L.A., Hogenesch, J.B., Simon, M.C., Takahashi, J.S., and Bradfield, C.A. 2000. Mop3 Is an Essential Component of the Master Circadian Pacemaker in Mammals. *Cell* **103**(7): 1009–1017.

Busino, L., Bassermann, F., Maiolica, A., Lee, C., Nolan, P.M., Godinho, S.I., Draetta, G.F., and Pagano, M. 2007. SCFFbxl3 controls the oscillation of the circadian clock by directing the degradation of cryptochrome proteins. *Science (New York, NY)* **316**(5826): 900–904.

Cheng, M.Y., Bullock, C.M., Li, C., Lee, A.G., Bermak, J.C., Belluzzi, J., Weaver, D.R., Leslie, F.M., and Zhou, Q.Y. 2002. Prokineticin 2 transmits the behavioural circadian rhythm of the suprachiasmatic nucleus. *Nature* **417**(6887): 405–410.

Comas, M., Beersma, D.G., Spoelstra, K., and Daan, S. 2006. Phase and period responses of the circadian system of mice (Mus musculus) to light stimuli of different duration. *J Biol Rhythms* **21**(5): 362–372.

Damiola, F., Le Minh, N., Preitner, N., Kornmann, B., Fleury-Olela, F., and Schibler, U. 2000. Restricted feeding uncouples circadian oscillators in peripheral tissues from the central pacemaker in the suprachiasmatic nucleus. *Genes Dev* **14**(23): 2950–2961.

Dandona, P., Aljada, A., Chaudhuri, A., Mohanty, P., and Garg, R. 2005. Metabolic syndrome: a comprehensive perspective based on interactions between obesity, diabetes, and inflammation. *Circulation* **111**(11): 1448–1454.

De Mairan, J. 1729. Observation botanique. *Histoire de l'Academie Royale des Sciences*: 35–36.

Dequeant, M.L., Glynn, E., Gaudenz, K., Wahl, M., Chen, J., Mushegian, A., and Pourquie, O. 2006. A complex oscillating network of signaling genes underlies the mouse segmentation clock. *Science (New York, NY)* **314**(5805): 1595–1598.

Dodd, A.N., Salathia, N., Hall, A., Kevei, E., Toth, R., Nagy, F., Hibberd, J.M., Millar, A.J., and Webb, A.A. 2005. Plant circadian clocks increase photosynthesis, growth, survival, and competitive advantage. *Science (New York, NY)* **309**(5734): 630–633.

Edmunds, L.N. Jr. and Laval-Martin, D.L. 1984. Cell division cycles and circadian oscillators in Euglena. *Chronobiol Int* **1**(1): 1–9.

Etchegaray, J.P., Yang, X., DeBruyne, J.P., Peters, A.H., Weaver, D.R., Jenuwein, T., and Reppert, S.M. 2006. The polycomb group protein EZH2 is required for mammalian circadian clock function. *J Biol Chem* **281**(30): 21209–21215.

Forger, D.B. and Peskin, C.S. 2005. Stochastic simulation of the mammalian circadian clock. *Proc Natl Acad Sci USA* **102**(2): 321–324.

Gachon, F., Nagoshi, E., Brown, S.A., Ripperger, J., and Schibler, U. 2004. The mammalian circadian timing system: from gene expression to physiology. *Chromosoma* **113**(3): 103–112.

Gachon, F., Fleury Olela, F., Schaad, O., Descombes, P., and Schibler, U. 2006. The circadian PAR-domain basic leucine zipper transcription factors DBP, TEF, and HLF modulate basal and inducible xenobiotic detoxification. *Cell Metab* **4**(1):25–36.

Gallego, M. and Virshup, D.M. 2007. Post-translational modifications regulate the ticking of the circadian clock. *Nat Rev* **8**(2): 139–148.

Gerkema, M.P. and van der Leest, F. 1991. Ongoing ultradian activity rhythms in the common vole, Microtus arvalis, during deprivations of food, water and rest. *J Comp Physiol* **168**(5): 591–597.

Gerkema, M.P., Groos, G.A., and Daan, S. 1990. Differential elimination of circadian and ultradian rhythmicity by hypothalamic lesions in the common vole, Microtus arvalis. *J Biol Rhythms* **5**(2): 81–95.

Gerkema, M.P., Daan, S., Wilbrink, M., Hop, M.W., and van der Leest, F. 1993. Phase control of ultradian feeding rhythms in the common vole (Microtus arvalis): the roles of light and the circadian system. *J Biol Rhythms* **8**(2): 151–171.

Gerkema, M.P., Van der Zee, E.A., and Feitsma, L.E. 1994. Expression of circadian rhythmicity correlates with the number of arginine-vasopressin-immunoreactive cells in the suprachiasmatic nucleus of common voles, Microtus arvalis. *Brain Res* **639**(1): 93–101.

Gillette, M.U. and Mitchell, J.W. 2002. Signaling in the suprachiasmatic nucleus: selectively responsive and integrative. *Cell Tissue Res* **309**(1): 99–107.

Glass, L. and Winfree, A.T. 1984. Discontinuities in phase-resetting experiments. *Am J Physiol* **246**(2 Pt 2): R251–258.

Godinho, S.I., Maywood, E.S., Shaw, L., Tucci, V., Barnard, A.R., Busino, L., Pagano, M., Kendall, R., Quwailid, M.M., Romero, M.R., O'Neill, J., Chesham, J.E., Brooker, D., Lalanne, Z., Hastings, M.H., and Nolan, P.M. 2007. The after-hours mutant reveals a role for Fbxl3 in determining mammalian circadian period. *Science (New York, NY)* **316**(5826): 897–900.

Guo, H., Brewer, J.M., Lehman, M.N., and Bittman, E.L. 2006. Suprachiasmatic regulation of circadian rhythms of gene expression in hamster peripheral organs: effects of transplanting the pacemaker. *J Neurosci* **26**(24): 6406–6412.

Handschin, C. and Meyer, U.A. 2003. Induction of drug metabolism: the role of nuclear receptors. *Pharmacol Rev* **55**(4): 649–673.

Hirata, H., Yoshiura, S., Ohtsuka, T., Bessho, Y., Harada, T., Yoshikawa, K., and Kageyama, R. 2002. Oscillatory expression of the bHLH factor Hes1 regulated by a negative feedback loop. *Science (New York, NY)* **298**(5594): 840–843.

Honma, S., Kawamoto, T., Takagi, Y., Fujimoto, K., Sato, F., Noshiro, M., Kato, Y., and Honma, K. 2002. Dec1 and Dec2 are regulators of the mammalian molecular clock. *Nature* **419**(6909): 841–844.

Houten, S.M., Watanabe, M., and Auwerx, J. 2006. Endocrine functions of bile acids. *The EMBO J* **25**(7): 1419–1425.

Ishikawa, K. and Shimazu, T. 1976. Daily rhythms of glycogen synthetase and phosphorylase activities in rat liver: influence of food and light. *Life Sci* **19**(12): 1873–1878.

Jakubcakova, V., Oster, H., Tamanini, F., Cadenas, C., Leitges, M., van der Horst, G.T., and Eichele, G. 2007. Light entrainment of the mammalian circadian clock by a PRKCA-dependent posttranslational mechanism. *Neuron* **54**(5): 831–843.

Jin, X., Shearman, L.P., Weaver, D.R., Zylka, M.J., de Vries, G.J., and Reppert, S.M. 1999. A molecular mechanism regulating rhythmic output from the suprachiasmatic circadian clock. *Cell* **96**(1): 57–68.

Kageyama, H., Nishiwaki, T., Nakajima, M., Iwasaki, H., Oyama, T., and Kondo, T. 2006. Cyanobacterial circadian pacemaker: Kai protein complex dynamics in the KaiC phosphorylation cycle in vitro. *Mol Cell* **23**(2): 161–171.

Kageyama, R., Masamizu, Y., and Niwa, Y. 2007a. Oscillator mechanism of Notch pathway in the segmentation clock. *Dev Dyn* **236**(6): 1403–1409.

Kageyama, R., Ohtsuka, T., and Kobayashi, T. 2007b. The Hes gene family: repressors and oscillators that orchestrate embryogenesis. *Development (Cambridge, England)* **134**(7): 1243–1251.

Kalaany, N.Y. and Mangelsdorf, D.J. 2006. LXRS and FXR: the yin and yang of cholesterol and fat metabolism. *Annu Rev Physiol* **68**: 159–191.

Kalsbeek, A., van Heerikhuize, J.J., Wortel, J., and Buijs, R.M. 1996. A diurnal rhythm of stimulatory input to the hypothalamo-pituitary-adrenal system as revealed by timed intrahypothalamic administration of the vasopressin V1 antagonist. *J Neurosci* **16**(17): 5555–5565.

Keller, J. and Layer, P. 2002. Circadian pancreatic enzyme pattern and relationship between secretory and motor activity in fasting humans. *J Appl Physiol* **93**(2): 592–600.

Khalsa, S.B., Jewett, M.E., Cajochen, C., and Czeisler, C.A. 2003. A phase response curve to single bright light pulses in human subjects. *J Physiol* **549**(Pt 3): 945–952.

King, D.P., Zhao, Y., Sangoram, A.M., Wilsbacher, L.D., Tanaka, M., Antoch, M.P., Steeves, T.D., Vitaterna, M.H., Kornhauser, J.M., Lowrey, P.L., Turek, F.W., and Takahashi, J.S. 1997. Positional cloning of the mouse circadian clock gene. *Cell* **89**(4): 641–653.

Kornmann, B., Preitner, N., Rifat, D., Fleury-Olela, F., and Schibler, U. 2001. Analysis of circadian liver gene expression by ADDER, a highly sensitive method for the display of differentially expressed mRNAs. *Nucleic Acids Res* **29**(11): E51–51.

Kornmann, B., Schaad, O., Bujard, H., Takahashi, J.S., and Schibler, U. 2007. System-driven and oscillator-dependent circadian transcription in mice with a conditionally active liver clock. *PLoS Biol* **5**(2): e34.

Kornmann, B., Schaad, O., Reinke, H., Saini, C., and Schibler, U. 2007. Regulation of circadian gene expression in liver by systemic signals and hepatocyte oscillators. *Cold Spring Harb Symp Quant Biol.* **7**: 319–330.

Kramer, A., Yang, F.C., Snodgrass, P., Li, X., Scammell, T.E., Davis, F.C., and Weitz, C.J. 2001. Regulation of daily locomotor activity and sleep by hypothalamic EGF receptor signaling. *Science (New York, NY)* **294**(5551): 2511–2515.

Kraves, S. and Weitz, C.J. 2006. A role for cardiotrophin-like cytokine in the circadian control of mammalian locomotor activity. *Nat Neurosci* **9**(2): 212–219.

Kriegsfeld, L.J. and Silver, R. 2006. The regulation of neuroendocrine function: timing is everything. *Horm Behav* **49**(5): 557–574.

Lakin-Thomas, P.L. 2006. Transcriptional feedback oscillators: maybe, maybe not. *J Biol Rhythms* **21**(2): 83–92.

Leloup, J.C. and Goldbeter, A. 2003. Toward a detailed computational model for the mammalian circadian clock. *Proc Natl Acad Sci USA* **100**(12): 7051–7056.

Leloup, J.C. and Goldbeter, A. 2004. Modeling the mammalian circadian clock: sensitivity analysis and multiplicity of oscillatory mechanisms. *J Theor Biol* **230**(4): 541–562.

Le Minh, N., Damiola, F., Tronche, F., Schutz, G., and Schibler, U. 2001. Glucocorticoid hormones inhibit food-induced phase-shifting of peripheral circadian oscillators. *EMBO J* **20**(24): 7128–7136.

Liu, A.C., Welsh, D.K., Ko, C.H., Tran, H.G., Zhang, E.E., Priest, A.A., Buhr, E.D., Singer, O., Meeker, K., Verma, I.M., Doyle, F.J., 3rd, Takahashi, J.S., and Kay, S.A. 2007. Intercellular coupling confers robustness against mutations in the SCN circadian clock network. *Cell* **129**(3): 605–616.

Lloyd, D. and Murray, D.B. 2007. Redox rhythmicity: clocks at the core of temporal coherence. *Bioessays* **29**(5): 465–473.

Lowrey, P.L., Shimomura, K., Antoch, M.P., Yamazaki, S., Zemenides, P.D., Ralph, M.R., Menaker, M., and Takahashi, J.S. 2000. Positional syntenic cloning and functional characterization of the mammalian circadian mutation tau. *Science (New York, NY)* **288**(5465): 483–492.

Lundasen, T., Galman, C., Angelin, B., and Rudling, M. 2006. Circulating intestinal fibroblast growth factor 19 has a pronounced diurnal variation and modulates hepatic bile acid synthesis in man. *J Intern Med* **260**(6): 530–536.

McPherson, R. and Gauthier, A. 2004. Molecular regulation of SREBP function: the Insig-SCAP connection and isoform-specific modulation of lipid synthesis. *Biochem Cell Biol* **82**(1): 201–211.

Minors, D.S., Waterhouse, J.M., and Wirz-Justice, A. 1991. A human phase-response curve to light. *Neurosci Lett* **133**(1): 36–40.

Miyoshi, F., Nakayama, Y., Kaizu, K., Iwasaki, H., and Tomita, M. 2007. A mathematical model for the Kai-protein-based chemical oscillator and clock gene expression rhythms in cyanobacteria. *J Biol Rhythms* **22**(1): 69–80.

Nagoshi, E., Saini, C., Bauer, C., Laroche, T., Naef, F., and Schibler, U. 2004. Circadian gene expression in individual fibroblasts: cell-autonomous and self-sustained oscillators pass time to daughter cells. *Cell* **119**(5): 693–705.

Nakajima, M., Imai, K., Ito, H., Nishiwaki, T., Murayama, Y., Iwasaki, H., Oyama, T., and Kondo, T. 2005. Reconstitution of circadian oscillation of cyanobacterial KaiC phosphorylation in vitro. *Science (New York, NY)* **308**(5720): 414–415.

Noshiro, M., Usui, E., Kawamoto, T., Kubo, H., Fujimoto, K., Furukawa, M., Honma, S., Makishima, M., Honma, K., and Kato, Y. 2007. Multiple mechanisms regulate circadian expression of the gene for cholesterol 7alpha-hydroxylase (Cyp7a), a key enzyme in hepatic bile acid biosynthesis. *J Biol Rhythms* **22**(4): 299–311.

Ouyang, Y., Andersson, C.R., Kondo, T., Golden, S.S., and Johnson, C.H. 1998. Resonating circadian clocks enhance fitness in cyanobacteria. *Proc Natl Acad Sci USA* **95**(15): 8660–8664.

Panda, S., Antoch, M.P., Miller, B.H., Su, A.I., Schook, A.B., Straume, M., Schultz, P.G., Kay, S.A., Takahashi, J.S., and Hogenesch, J.B. 2002. Coordinated transcription of key pathways in the mouse by the circadian clock. *Cell* **109**(3): 307–320.

Preitner, N., Damiola, F., Luis Lopez, M., Zakany, J., Duboule, D., Albrecht, U., and Schibler, U. 2002. The Orphan Nuclear Receptor REV-ERBalpha Controls Circadian Transcription within the Positive Limb of the Mammalian Circadian Oscillator. *Cell* **110**(2): 251–260.

Ralph, M.R., Foster, R.G., Davis, F.C., and Menaker, M. 1990. Transplanted suprachiasmatic nucleus determines circadian period. *Science (New York, NY)* **247**(4945): 975–978.

Redlin, U. and Mrosovsky, N. 1999. Masking by light in hamsters with SCN lesions. *J Comp Physiol* **184**(4): 439–448.

Reinke, H., Saini., C., Fleury-Olela, F., Dibner, C., Benjamin I.J., and Schibler, U. 2008. Differential Display of DNA-Binding Proteins Reveals Heat Shock Factor 1 as a Circadian Transcription Factor. *Genes Dev* **22**(3): 331–345.

Reppert, S.M. and Weaver, D.R. 2001. Molecular analysis of mammalian circadian rhythms. *Ann Rev Physiol* **63**: 647–676.

Roenneberg, T. and Taylor, W. 1994. Light-induced phase responses in Gonyaulax are drastically altered by creatine. *J Biol Rhythms* **9**(1): 1–12.

Rust, M.J., Markson, J.S., Lane, W.S., Fisher, D.S., and O'Shea, E.K. 2007. Ordered phosphorylation governs oscillation of a three-protein circadian clock. *Science (New York, NY)* **318**(5851): 809–812.

Sato, T.K., Panda, S., Miraglia, L.J., Reyes, T.M., Rudic, R.D., McNamara, P., Naik, K.A., FitzGerald, G.A., Kay, S.A., and Hogenesch, J.B. 2004. A functional genomics strategy reveals Rora as a component of the mammalian circadian clock. *Neuron* **43**(4): 527–537.

Schafmeier, T., Kaldi, K., Diernfellner, A., Mohr, C., and Brunner, M. 2006. Phosphorylation-dependent maturation of Neurospora circadian clock protein from a nuclear repressor toward a cytoplasmic activator. *Genes Dev* **20**(3): 297–306.

Schibler, U. 2007. The daily timing of gene expression and physiology in mammals. *Dialogues Clin Neurosci* **9**(3): 257–272.

Scott, E.M., Carter, A.M., and Grant, P.J. 2008. Association between polymorphisms in the Clock gene, obesity and the metabolic syndrome in man. *Int J Obes (Lond)* **32**(4): 658–662. Epub 2007 Dec 11.

Siepka, S.M., Yoo, S.H., Park, J., Song, W., Kumar, V., Hu, Y., Lee, C., and Takahashi, J.S. 2007. Circadian mutant Overtime reveals F-box protein FBXL3 regulation of cryptochrome and period gene expression. *Cell* **129**(5): 1011–1023.

Silver, R., LeSauter, J., Tresco, P.A., and Lehman, M.N. 1996. A diffusible coupling signal from the transplanted suprachiasmatic nucleus controlling circadian locomotor rhythms. *Nature* **382**(6594): 810–813.

Smolen, P., Baxter, D.A., and Byrne, J.H. 2001. Modeling circadian oscillations with interlocking positive and negative feedback loops. *J Neurosci* **21**(17): 6644–6656.

Stokkan, K.A., Yamazaki, S., Tei, H., Sakaki, Y., and Menaker, M. 2001. Entrainment of the Circadian Clock in the Liver by Feeding. *Science (New York, NY)* **291**(5503): 490–493.

Storch, K.F., Lipan, O., Leykin, I., Viswanathan, N., Davis, F.C., Wong, W.H., and Weitz, C.J. 2002. Extensive and divergent circadian gene expression in liver and heart. *Nature* **417**(6884): 78–83.

Toh, K.L., Jones, C.R., He, Y., Eide, E.J., Hinz, W.A., Virshup, D.M., Ptacek, L.J., and Fu, Y.H. 2001. An hPer2 phosphorylation site mutation in familial advanced sleep phase syndrome. *Science (New York, NY)* **291**(5506): 1040–1043.

Tomita, J., Nakajima, M., Kondo, T., and Iwasaki, H. 2005. No transcription-translation feedback in circadian rhythm of KaiC phosphorylation. *Science (New York, NY)* **307**(5707): 251–254.

Tu, B.P. and McKnight, S.L. 2006. Metabolic cycles as an underlying basis of biological oscillations. *Nat Rev* **7**(9): 696–701.

Turek, F.W., Joshu, C., Kohsaka, A., Lin, E., Ivanova, G., McDearmon, E., Laposky, A., Losee-Olson, S., Easton, A., Jensen, D.R., Eckel, R.H., Takahashi, J.S., and Bass, J. 2005. Obesity and metabolic syndrome in circadian Clock mutant mice. *Science (New York, NY)* **308**(5724): 1043–1045.

Ueda, H.R., Chen, W., Adachi, A., Wakamatsu, H., Hayashi, S., Takasugi, T., Nagano, M., Nakahama, K., Suzuki, Y., Sugano, S., Iino, M., Shigeyoshi, Y., and Hashimoto, S. 2002. A transcription factor response element for gene expression during circadian night. *Nature* **418**(6897): 534–539.

van der Horst, G.T., Muijtjens, M., Kobayashi, K., Takano, R., Kanno, S., Takao, M., de Wit, J., Verkerk, A., Eker, A.P., van Leenen, D., Buijs, R., Bootsma, D., Hoeijmakers, J.H., and Yasui, A. 1999. Mammalian Cry1 and Cry2 are essential for maintenance of circadian rhythms. *Nature* **398**(6728): 627–630.

van der Veen, D.R., Minh, N.L., Gos, P., Arneric, M., Gerkema, M.P., and Schibler, U. 2006. Impact of behavior on central and peripheral circadian clocks in the common vole Microtus arvalis, a mammal with ultradian rhythms. *Proc Natl Acad Sci USA* **103**(9): 3393–3398.

van Zon, J.S., Lubensky, D.K., Altena, P.R., and ten Wolde, P.R. 2007. An allosteric model of circadian KaiC phosphorylation. *Proc Natl Acad Sci USA* **104**(18): 7420–7425.

Vrang, N., Larsen, P.J., and Mikkelsen, J.D. 1995. Direct projection from the suprachiasmatic nucleus to hypophysiotrophic corticotropin-releasing factor immunoreactive cells in the paraventricular nucleus of the hypothalamus demonstrated by means of Phaseolus vulgaris-leucoagglutinin tract tracing. *Brain Res* **684**(1): 61–69.

Wahl, M.B., Deng, C., Lewandoski, M., and Pourquie, O. 2007. FGF signaling acts upstream of the NOTCH and WNT signaling pathways to control segmentation clock oscillations in mouse somitogenesis. *Development (Cambridge, England)* **134**(22): 4033–4041.

Walker, J.R. and Hogenesch, J.B. 2005. RNA profiling in circadian biology. *Methods Enzymol* **393**: 366–376.

Waterhouse, J., Reilly, T., Atkinson, G., and Edwards, B. 2007. Jet lag: trends and coping strategies. *Lancet* **369**(9567): 1117–1129.

Welsh, D.K., Logothetis, D.E., Meister, M., and Reppert, S.M. 1995. Individual neurons dissociated from rat suprachiasmatic nucleus express independently phased circadian firing rhythms. *Neuron* **14**(4): 697–706.

William, D.A., Saitta, B., Gibson, J.D., Traas, J., Markov, V., Gonzalez, D.M., Sewell, W., Anderson, D.M., Pratt, S.C., Rappaport, E.F., and Kusumi, K. 2007. Identification of oscillatory genes in somitogenesis from functional genomic analysis of a human mesenchymal stem cell model. *Dev Biol* **305**(1): 172–186.

Woelfle, M.A., Ouyang, Y., Phanvijhitsiri, K., and Johnson, C.H. 2004. The adaptive value of circadian clocks: an experimental assessment in cyanobacteria. *Curr Biol* **14**(16): 1481–1486.

Woods, S.C., Lutz, T.A., Geary, N., and Langhans, W. 2006. Pancreatic signals controlling food intake; insulin, glucagon and amylin. *Philos Trans R Soc London* **361**(1471): 1219–1235.

Xu, C., Li, C.Y., and Kong, A.N. 2005. Induction of phase I, II and III drug metabolism/transport by xenobiotics. *Arch Pharmacol Res* **28**(3): 249–268.

Yoo, S.H., Yamazaki, S., Lowrey, P.L., Shimomura, K., Ko, C.H., Buhr, E.D., Siepka, S.M., Hong, H.K., Oh, W.J., Yoo, O.J., Menaker, M., and Takahashi, J.S. 2004. PERIOD2::LUCIFERASE real-time reporting of circadian dynamics reveals persistent circadian oscillations in mouse peripheral tissues. *Proc Natl Acad Sci USA* **101**(15): 5339–5346.

Yu, W. and Hardin, P.E. 2006. Circadian oscillators of Drosophila and mammals. *J Cell Sci* **119** (Pt 23): 4793–4795.

Zabielski, R. 2004. Reefs in experimental gastroenterology – cyclic activities of the gastrointestinal tract. *J Physiol Pharmacol* **55**(Suppl 2): 19–32.

Zangar, R.C., Davydov, D.R., and Verma, S. 2004. Mechanisms that regulate production of reactive oxygen species by cytochrome P450. *Toxicol Appl Pharmacol* **199**(3): 316–331.

Zheng, B., Albrecht, U., Kaasik, K., Sage, M., Lu, W., Vaishnav, S., Li, Q., Sun, Z.S., Eichele, G., Bradley, A., and Lee, C.C. 2001. Nonredundant roles of the mPer1 and mPer2 genes in the mammalian circadian clock. *Cell* **105**(5): 683–694.

Part IV
Ultradian and Circadian Rhythms in Human Experience

Chapter 13
Ultradian Cognitive Performance Rhythms During Sleep Deprivation

C.M. LaJambe[1] and F.M. Brown[2]

Abstract Ultradian rhythms that modulate daytime human behavior and cognitive performance exist. However, their subtleness may make them susceptible to masking effects from heightened arousal, attention, and motivation. Experimental designs using sleep attenuation and total sleep-deprivation appear to unmask certain ultradian rhythms. This chapter reviews the few studies designed to evaluate rhythms in waking EEG and task performances during sleep deprivation. Some EEG studies demonstrate an approximate 90-min rhythm in arousal. No confirmation has been reached for an underlying common basic rest-activity cycle (BRAC) that ties the phases of the nightly 90-min NREM-REM sleep cyclicity to those of any ensuing similar-period wake time alertness and performance rhythms. Studies have also found a slower 4-h rhythm. Other arousal and performance rhythms shorter than 90 min and longer than 4 h have been suggested, but are not as well substantiated. The 90-min and 4-h rhythms seem to dominate the morning and early afternoon circadian rhythm portion during wakefulness, but are attenuated later by the rising circadian rhythm, or perhaps other slower ultradian cycles. Ultradian rhythm expression appears related to a greater susceptibility for reduced arousal and sleepiness. Impairment of cognitive-behavioral performance occurs at times of low circadian influences and increased ultradian fluctuations. These effects, in combination with any amount of sleep deprivation, raise concerns about naturally occurring potentially widespread performance decrements. Concern stems from the widespread increase in chronic partial-sleep loss in cultures deliberately abridging adequate nightly sleep in favor of attempting to maintain a 24/7 mode of existence.

[1] Pennsylvania Transportation Institute, College of Engineering, 201 Transportation Research Building, The Pennsylvania State University, University Park, PA 16802, USA
Email: cml149@psu.edu

[2] Human Performance Rhythms Laboratory, Department of Psychology, 354 Moore Building, The Pennsylvania State University, University Park, PA 16802, USA
Email: f3b@psu.edu

13.1 Introduction

The large number of research investigations during the 1950s and 1960s saw successful demonstration of a basic endogenous circadian (near 24-h) rhythmicity in nearly all physiological systems in all warm-blooded animals studied in past decades. This was followed later by confirmation of its genetic origins in genes *Per*, *Tim*, *Cloc* and *Cry*, and their variants in a variety of animals. On the heels of the well-demonstrated circadian cyclicity was the suggestion by Nathaniel Kleitman (1961) of a sub-harmonic wake-time rhythm of the circadian cycle, an approximately 90-min cycle. This rhythm was thought to include basic survival activities of physiology and behavior; he thought it was derived from the NREM-REM cyclic architecture of sleep, discovered by Eugene Azerinsky and Nathaniel Kleitman (1953) and characterized by William Dement (1955) as the cyclical deep sleep-dreaming pattern in nocturnal sleep known today. This 90-min cycle was labeled the basic rest-activity cycle or BRAC. The relationship between waking- and sleep-state activities culminating in a single-system BRAC has not reached the well-substantiated level of the circadian rhythm. Nonetheless, research generated by its attempted falsification has demonstrated a myriad of cognitive and behavioral rhythmic manifestations of varying periods both below and above 90 min. Certain rhythms of arousal and performance appear to relate to the BRAC. Regardless, these rhythms wax and wane with environmental and endogenous physiological changes. This chapter will discuss the effects of extended wakefulness on rhythms of arousal and performance.

13.2 Daytime Ultradian Alertness Rhythms in EEG

Using the reasonable assumption that basic daytime alertness fluctuations influence levels of performance, we reviewed such studies to examine emergence of ultradian rhythms that would precede changes in performance levels. An early EEG rhythms study (Nakagawa, 1980) tested for presence of the daytime BRAC rhythm. Subjects were placed under quiet relatively supine restfulness conditions in minimized natural lighting, sound, and time information. Following an acclimatization night of sleeping in the laboratory, a night of EEG recording was followed by 10–12 h of additional daytime recording. Although subjects were instructed not to fall asleep, the point of the study was to examine the frequency and occurrence of spontaneous napping and any sleep onset REM periods that might occur. While extension into daytime of the nightly approximate 90-min REM-NREM was not found, about 90 min post-awakening a peak in sleep occurred, followed 4 h later by a second peak of group-calculated episodes of napping. Nakagawa concluded no connection of the daytime napping with the prior night's approximate 90-min NREM-REM sleep architecture and no support for the BRAC rhythm. Instead, the author suggested that a 3–4 h rhythm might be related to the diurnal work-rest rhythm.

In another test of Kleitman's 90–100-min BRAC rhythm, Lavie and Shearson (1981) used a unique 15-min awake and 5-min attempted-sleep procedure to test for any underlying rhythm longer than 40 min. Three different sleep conditions preceded the common testing routine. The first (baseline) condition consisted of the nine subjects entering the laboratory in the morning following a night of regular sleep at home. Beginning mid-morning (1000–1155 h) and continuing for 12 h, thirty-six 20-min sessions each collected 5-min of EEG. Following each 15-min awake period, the subjects were asked to lie on a bed during the 5 min of EEG recording in a sound-attenuated bedroom, and were told to try to fall asleep. In the second condition (considered REM deprivation), six of the original nine subjects were brought into the lab on an evening following a usual day, but with no napping. Sleep onset was begun at 2300 h. Upon onset of each nightly REM stage, subjects were awakened for 10 min before resuming sleep. After awakening at 0700 h followed by breakfast, as in the baseline condition the thirty-six 15–5 EEG recording sessions began about 0840 h and continued throughout the day. The third experimental condition was total sleep deprivation, in which four of the original subjects participated. They stayed all night in the laboratory under *ad libitum* conditions and under the scrutiny of the experimenters to remain awake throughout.

EEG monitoring began under the 15–5 schedule earlier in the morning at 0700 h for two subjects and at 0840 h for the remaining two. Results from the 36 sessions of 5-min EEG recording showed that the Stage 1 daytime sleep followed a statistically significant 100-min (14.4 cycles/day) rhythmicity. Stage 2 sleep fluctuated at 3.6 cycle/day during baseline while it showed peaks at 7:30, 9:30–10:00, and 14:00–15:00 following total sleep deprivation. In contrast, Stage 1 BRAC-like rhythmicity occurred only during the baseline day, but neither in the REM-deprivation nor the total sleep-deprivation conditions. The authors concluded this 90-min Stage 1 to be in support of Kleitman's prediction by '...demonstrating rhythmic variations in subjects' ability to fall asleep throughout the day' (p. 171). However, it does not attempt to substantiate Kleitman's main BRAC idea that the phase relationship of daily 90-min rhythmicity and the NREM-REM cycle are identical. Lavie and Shearson (1981) suggested that Stage 1 diurnal rhythmicity is a 'sleepability gate' where during peak times, if conditions so warranted, a person could fall asleep easily for napping.

A subsequent study by Lavie and Zomer (1984), using an upgrade of the same general experimental design as in the Lavie and Shearson study (1981), demonstrated a phase synchronizing between the sleep NREM-REM and awake 'sleepability' cycles. This study awakened subjects either at the beginning or at the end of the REM phase after about four completed cycles. A computed spectral-analysis demonstrated a 7.2 cycle/day (3.3 h) rhythm present for both morning and evening time series. Furthermore, when individual morning data were synchronized with the onset of the last nocturnal REM period, the morning sleepability cyclicity showed 14.4 cycles/day, in synchrony with the NREM-REM cycle. The authors concluded that sleepability, a tendency to be drowsy during the day, is regulated by two ultradian rhythms as well as the circadian rhythm: one approximates the 90-min NREM-REM period, and a slower period ranges between 3.5 to 4 h. These findings are similar to Nakagawa's (1980) finding of a prominent 4-h interval that occurred between two morning EEG peaks.

Tsuji and Kabayashi (1988) performed a principal-components analysis on a power spectral analysis of EEG along with MSLT ratings on quietly sitting subjects across 8.5 h of daytime wakefulness (between 1100 and 1930 h). The MSLT was used in repeating 20-min test periods, similar but not identical in fashion to the 15–5 min sleep testing in the earlier cited Lavie and co-workers studies. The results affirmed the presence of a 100-min cyclicity in alpha and beta activity, the two being negatively correlated. Demonstrated as well was a longer 3–8 h (and particularly a 5 h) rhythm in theta and fast alpha activity, again negatively correlated. The authors concluded that ultradian arousal consists of a short (100 min BRAC) rhythm representing rest/activity and a long rhythm representing wakefulness/ drowsiness. Furthermore, they noted a mid-afternoon (around 1600 h) 'breaking point' in the arousal state, after which time the ultradian rhythms of the arousal state were less prominent. The authors suggested that the shorter rhythms appear to be masked by the longer ones. In other words, EEG ultradian rhythms of rest/activity rhythm and levels of consciousness associated with the alertness/drowsiness rhythm intersect. The former rhythms are attenuated with increasing alertness until the homeostatic pressure of nighttime sleepiness takes over just before bedtime.

Results from other EEG-based studies of alertness fluctuations found short BRAC-like rhythms and longer 3- to 6-h rhythms. Okawa et al. (1984) reported vigilance fluctuations of 60–110 min that were most evident during the first half of the day. Chapotot et al. (2000) studied EEG-measured brain-area involvements during a repetitive 70-s visual-fixation task. Every 10 min from 0900 to 1800 h EEG measurements were taken as subjects fixed their gaze on a projected visual stimulus spot on a wall 2 m in front of them. A 30-s warning preceded each stimulus presentation. If drowsiness was apparent and the subject did not respond, the warning was repeated. EEG results indicated high frequency (22.5 + Hz) frontal activity that alternated with both alpha and 1–3 Hz band activities. The high frequencies peaked in late morning and decreased in mid-afternoon while the lower frequencies peaked in mid-afternoon. A rise occurred in the lower EEG frequencies around time of the 'post-lunch dip,' which usually attends increased drowsiness. Furthermore, frontocentral high frequencies and parietal alpha activity showed a 3–4 h ultradian rhythm. The authors concluded that the decrease of higher EEG frequencies (over 22.5 Hz) and increase of parietal alpha activity reflects central cortical deactivation associated with reduced alertness. They further stated that the slow ultradian rhythm represented an endogenous fluctuating of arousal underlying the oscillations in vigilance level.

Conclusion reached from these EEG studies suggest that the 60–120 min BRAC-like rhythm manifests itself as part of the daily morning awakening stages, even under conditions of relatively adequate nightly sleep, as a mechanism for returning to sleep early in the day. Thereafter the circadian pressure of the two-process system exerts itself and the short early daytime ultradian sleep/rest cycle dissipates until re-emerging the next day following awakening. Also, as suggested from the Lavie and Zomer study (1984), any lack of synchrony between the sleep NREM-REM cycle and daily BRAC cycle is probably the result of daily re-setting of the circadian portion of the two-process system upon morning awakening. Thus, with adequate nightly sleep no expectation ought to occur for a continuous short ultradian rhythm of 'sleepability' throughout the day. It is dissipating as body temperature begins approaching its maximum level mid- to later afternoon. As for the 3-h or longer episodic, if not rhythmic, EEG events, Tsuji and Kobayashi (1988) suggest occurrences are a result of possible environment events. They listed as possible events the intervals between meals, while Chapotot et al. (2000) suggested the possibility of the noon meal and 'post-lunch dip,' events that must be closely investigated and ruled out.

13.3 Ultradian Alertness and Cognitive Performance Rhythm Interactions

The connection between performance and EEG-based arousal rhythms was examined in a study by Hayashi et al. (1994). Thirty-six 15-min test sessions were conducted from 0830 to 1730h. Data were collected on EEG, visual-detection reaction time, and task performance measures of auditory vigilance, addition, and logical reasoning, in addition to self-reported ratings of sleepiness. Results showed a 12–14 cycle/day (BRAC range) rhythm and six cycle/day (4-h rhythm) for EEG components. Subjective sleepiness and performance tasks primarily fluctuated at 12 cycles/day. However, no definitive conclusions could be drawn about the relationship between ultradian rhythms of performance and EEG-related alertness.

In essence, the research discussed so far shows that ultradian fluctuations in alertness/drowsiness involve multiple components that can change in their expression over time. On one hand, ultradian rhythms such as clear 1.5-h cycles that are present during the first part of the day give way with increased sleep pressure later in the day to slower ultradian oscillations of 3 h or longer that then mask them (Tsuji and Kobyashi, 1988; Chapotot et al., 2000). Perhaps, as Lavie has suggested, 1.5-h cycles are present but hidden or 'masked' by slower rhythms or interact with them, or the faster cycles are non-stationary and have shifted over time. In any case, their relationship with performance remains unclear. Lavie has further suggested that a number of 'masking' effects may impede the detection of ultradian performance rhythms during experimentation. For instance, simple tasks appear not to be sensitive to ultradian fluctuations, while knowledge of results can increase motivation and arousal, which then suppress ultradian rhythms. Manseau and Broughton (1984) studied subjects from 0930 to 1700 h in partial isolation during a monotonous 'unperformed' task while they visually fixated on a projected dot. Results demonstrated that 16 cycle/day (1.5 h) and eight cycle/day (3 h) rhythms emerged for 4–20 Hz frontal EEG activity. The 16 cycle/day rhythm was associated with fluctuations in alertness/sleepiness.

A systematic EEG study combined masking effects of motivation and subjective sleepiness and fatigue on ultradian fluctuations in arousal (Hayashi et al., 1998).

Subjects were presented with uninteresting landscape video scenes that increased their subjective sleepiness and fatigue. They also were presented with interesting animation scenes that increased their arousal levels. Results indicated that alpha activity for landscape scenes produced 22 cycles/day while animation scenes produced 11 cycles/day. Beta activity was eight cycles per day for landscape and two cycles/day for animation videos. The authors concluded that ultradian sleepiness cycles were unmasked by low levels of arousal, which they called low motivation, while viewing uninteresting landscape scenes. Conversely, high arousal, or motivation with the absence of sleepiness or fatigue apparently masked ultradian arousal cycles and resulted in fewer cycles per day. Furthermore, they concluded that extended wakefulness, which can lead to reduced arousal or motivation and precipitate elevated levels of sleepiness, might unmask ultradian rhythms. Lavie (1989) found 1.5-h rhythms (16/day) in actigraphy measures during a second night of total sleep deprivation, rhythms that are typically not detectable under normal free-living conditions. These results suggest that ultradian rhythms in cognitive performance may emerge during periods of extended wakefulness. Additional evidence for this conclusion is discussed next.

13.4 Sleep Deprivation: Effects on Cognitive Performance Rhythms

13.4.1 Sleep Deprivation, Circadian Rhythms, and Cognitive Performance

Effects of sleep deprivation on cognitive performance are evident with as little as 24 h of continuous wakefulness. After one night of sleep deprivation, subjects missed significantly more targets on a monotonous vigilance task (Gillberg and Åkerstedt, 1998) and a two- and three-letter cancellation task (Casagrande et al., 1997). Significant decrements on a complex problem-solving task also were found (Linde and Bergstrom, 1992), suggesting impairment in encoding (selective attention) and inability to monitor mental 'computations.' In addition, both slowed reaction times and increased errors of commission on a letter recognition task have been demonstrated after a night of sleep deprivation (Wimmer et al., 1992). Cognitive performance decrements after the first 24 h of sleep deprivation generally range from 30–40%, after 48 h 55–70%, and less precipitous declines thereafter up to 72 h (Angus and Heslegrave, 1985; Heslegrave and Angus, 1985; Pigeau et al., 1995). In addition, the influences of circadian rhythms on performance decrements are most prominent during the 2nd and 3rd days of sleep deprivation (Mukulincer et al., 1989).

Alteration in circadian performance rhythms usually occur following sleep deprivation, in addition to the characteristic monotonic decline in performance. These changes include a significant peak-to-trough amplitude increase as well as a phase shift in the rhythms. Table 13.1 gives examples of these changes for specific performance tasks. In general, simple monotonous repetitive types of tasks are more sensitive to sleep deprivation than are novel and more complex types of tasks (Dinges and Kribbs, 1991). Likewise, the dependent measure of speed is more sensitive to sleep deprivation than is accuracy (although the reason for this remains unclear). For example, decrements were confined to speed on both logical reasoning and serial reaction time tasks while accuracy did not change during the testing sessions (Heslegrave and Angus, 1985). In addition, number of hits for target letters (number correct) for three different variations of a letter cancellation task (search and memory) decreased monotonically over 72 h of sleep deprivation (Babkoff et al., 1988). Interestingly, the false positive responses (error rate) were not monotonic. Instead, they showed a 24-h rhythmicity across the 72 h, suggesting that error rate is more vulnerable to circadian effects than to sleep deprivation. Number of attempted items on the logical reasoning task was reduced by 65% after more than 48 h of sleep deprivation (Haslam, 1982). Another study also found a significant decrease in percent correct for actively worded sentences as compared with passively worded sentences over two continuous workdays (Ryman et al., 1985). Logical-reasoning task sensitivity to moderate sleep deprivation was greatest for total number of items attempted, followed by number correct, and least sensitive for accuracy score (Blagrove et al., 1995).

Table 13.1 Sleep derivation and circadian effects on physiological and behavioral variables (Babkoff et al., 1988, 1989, 1991a, b, 1992). Sleep deprivation × circadian effects (column 4) refer to circadian changes with progressive sleep deprivation

Measure	Sleep deprivation effect	Circadian effect	Sleep deprivation × circadian effect
Core body temperature	Reduced mean body temperature	Peak: 1800–2000 Trough: 0400–0600	Increased variability in peak/trough timing; no peak-to-trough amplitude changes
Self-reported alertness	Linear increase in sleepiness (decrease in alertness)	Bimodal peak in alertness: Peak: 1000; 1800–2000 Trough: 0200–0600	Greater circadian rhythm peak-to-trough amplitude on days 2 and 3 v. day 1
Vigilance (search and memory)	Linear decrease in speed, with accuracy being more sensitive to task variations in memory load (2, 4, 6 letter versions) than speed	*Speed (% trials attempted)* Peak: 1600–2200 Trough: 0200–0600 *Accuracy (% correct)* Peak: 1600–2000 Trough: 0200–0600	Increasing peak-to-trough amplitude changes in accuracy but not speed; sleep deprivation delayed performance peaks and troughs by 2–4 h
Logical reasoning	Linear decrease in speed; accuracy reported as insensitive to sleep deprivation	*Speed (% trials attempted)* Peak: 1600–2200 Trough: 0200–0600 *Accuracy (% correct)* Peak: 1600–2000 Trough: 0200–0600	Increased peak-to-trough amplitude in speed (accuracy not evaluated) on days 2 and 3 v. day 1

As a circadian marker of the underlying endogenous rhythm, core body temperature (CBT) appears to share a unique circadian relationship with cognitive performance. That is, cognitive performance rhythms generally parallel the CBT rhythm in peak and trough times, although they are not caused by it directly. Rather, like CBT, cognitive performance rhythms are influenced by sleep-debt status and the endogenous circadian clock (Dijk et al., 1992; Johnson et al., 1992). Also according to the latter study, during a normal waking day, the performance rhythm of some cognitive tasks (particularly those involving short-term memory and mathematical calculations) does not appear to parallel CBT. Cognitive performance rhythms on many types of cognitive tasks eventually align themselves with the CBT rhythm as hours of sleep deprivation increase (Babkoff et al., 1988; Johnson et al., 1992; Monk et al., 1997). For instance, Babkoff et al. (1988) found across 3 days of sleep deprivation an increasing phase alignment in the results of the two cognitive performance tasks, logical reasoning and digit symbol substitution, with CBT. In essence, phase of the CBT can serve as a time-of-day predictor of cognitive performance rhythms as sleep deprivation increases (see Table 13.1).

To some extent, the changes in subjective alertness during sleep deprivation also reflect cognitive performance changes. During 72 h of total-sleep deprivation, subjective alertness, as measured by the Stanford Sleepiness Scale (SSS), declined along with a general performance score (total number of responses) on a Search and Memory letter cancellation task. The correlation between subjective alertness and accuracy scores, however, was less substantial (Mikulincer et al., 1989). As sleep deprivation was extended, circadian rhythm effects on speed (for a short-term prose memory task and a mathematical calculation task) aligned with both subjective alertness and CBT rhythms (Johnson et al., 1992). Mikulincer et al. (1989) also found that the highest correlation between general- and accuracy performance and subjective alertness occurred on the 2nd day of sleep deprivation (see Table 13.1). This was accompanied by a significant increase in the amplitude of the circadian performance and subjective alertness rhythms. These results suggest that subjective alertness measures, when taken during extended sleep deprivation, provide a general predictor of the timing of performance peaks and troughs. Subjective alertness shows the most prominent peak between 1800 and 2000 h. However, it may be biphasic during a normal waking day, with peaks occurring at about 1000 h and again at 2000 h, and troughs at about 1400 h and again at 0400 h (Richardson et al., 1982).

13.4.2 Sleep Deprivation Effects on Ultradian Performance Rhythms

Few research studies are available on ultradian rhythm expression during extended wakefulness (beyond the normal 16 h or so waking period). At least in part, this kind of work is limited by the logistical difficulties and practical hardships assumed by subjects who often are unable to tolerate frequent testing in monotonous situations under increasing sleep pressure. Testing of BRAC and other fast ultradian rhythms is particularly difficult because of the need to apply a nearly continuous testing regimen. However, some work done in the area is presented below.

In an early study using sleep deprivation, Kripke (1972) described 10–20 cycles/day 'ultradian oscillations' in the EEG records of six young male subjects restricted to isolation chambers that involved partial perceptual and sleep deprivation. Using spectral analysis of awake behavioral and both asleep and awake EEG data his results tended to support earlier assumptions of a round-the-clock ultradian rhythm similar to the REM-NREM cycle found in sleep. No conclusion was drawn as to its origin, other than it was similar to a basic physiological rhythm observed throughout the 24 h. However, the subjects all fell asleep repeatedly throughout the 24-h deprivation period. This made it difficult to conclude that the ultradian oscillations were functionally present during expected intervals of daytime wakefulness.

In a complex study of both EEG and cognitive performance under sleep deprivation conditions, Kripke et al. (1985) measured the concordance of the phase relationship of both asleep and awake ultradian rhythms to the alleged 90-min REM-NREM BRAC rhythm. Sixty predominantly male university students were divided into six groups of 10 each, of which three groups each were assigned to one of two 42-h experiments. In the first experiment, one group performed continuously, while a second group was allowed a 6-h nap after performing from 18 to 24 h, and a third group was allowed a 1-h nap after each 6 h of performance. The continuously repeating 10-min cycle of performance tasks consisted of 3 min each of a target tracking task, a visual pattern memory task for assessing nondominant hemisphere functioning, and an arithmetic calculation task for assessing dominant hemisphere functioning. Each was accompanied by a simultaneously presented Wilkinson Auditory Vigilance (WAV) task. In the second experiment under similar conditions, using the remaining three groups but without using the WAV task, the three tasks ran 2.5 min each, plus addition of the self-rating Stanford Sleepiness Scale and a measure of daydreaming to complete each 10-min performance session. EEG, EOG, and EMG were recorded, along with assessment per 10-min interval of alpha and amount of sleep.

Kripke and colleagues used variance-spectrum analysis on the normalized data designed specifically to detect any 1 cycle/100 min or 14.4 cycles/day rhythmic peak. No further filtering was used which might introduce spurious peaks. Results of the performance tasks, as well as most of the physical maintenance tasks (eating, drinking, bathroom, etc.), daydreaming, and brain lateralization assessments indicated only suggestions of rhythms, but no significant ultradian peaking at 14.4 cycles/day. In comparing phase consistency after awakening for the two experiments, only the addition task was significant at 14.4 cycles/day in both experiments, as it was for a 9.6 cycles/day rhythm. In contrast, several variables were significant for 14.4 cycles/day, but only for one or the other experiments. Of the nightly REM-NREM cycle phases, these were consistent at 9.6 and 14.4 cycles/day in both experiments. However, no consistency was found in the performance tasks of the two experiments with the prior night's REM-NREM cyclicity. Only drinking was considered as phase-locked to the night RE-NREM cyclicity. Nonetheless, consistency did exist for tracking and visual pattern memory performance with the 9.6 cycle/day EEG NREM-REM cyclicity.

The authors concluded that the types of tasks and the conditions under which they were presented probably were not the most effective means for displaying the ultradian rhythms. In fact, Lavie (1989) has pointed out that in Kripke's study ultradian rhythms may have suffered from masking: Subject's were given immediate knowledge of results about their performance, which may have elevated motivation and arousal. As for the lack of REM-NREM phase synchronization with the awake performance rhythms, the suggestion was made of an awakening re-setting of the underlying ultradian oscillator. This would be in line with the observations of the early chronobiologist, F.A. Brown, Jr. (1959), who, based upon his many animals and plant studies years earlier, suggested a general and consistent 'autophasing' of the circadian rhythm on a daily basis to prominent time-givers (*Zeitgebers*), such as sunlight upon awakening. Thus, if daily autophasing is substantiated, the 'cart may be before the horse' by seeking a phase synchronization of daily activity with the prior night's REM-NREM ultradian rhythms. Instead, the hypothesis could test for the nightly NREM-REM ultradian cyclicity as a direct function of an indwelling, but not overtly expressive, daytime 14.4 cycle/day rhythmicity that is reset daily, and not the other way around as had been concluded.

More extended sleep deprivation studies subsequently were conducted at the Walter Reed Army Institute of Research. In an explicitly designed psychophysical study to measure signal detection accuracy under sleep deprivation conditions, Orr et al. (1974) found that a 90-min rhythm emerged but only toward the end the 48-h of sleep deprivation. Subjects were trained to a task criterion several days prior to data collection under sleep deprivation conditions. The task was to detect programmed deflections from the null needle positions on one of three unlit dials, and then immediately toggle as a detection response the corresponding dial switch. A button could be pressed to illuminate for 0.1 s the dials and observe their needle positions. Data were collected for needle-movement detection hits, button presses for dial illuminatings, and heart rate. The task was nearly continuous, except for a 30-min break allowed within a 3-h test session. Subjects were told that they could terminate the experiment when they felt they could no longer do the task. Only 2 of 11 male subjects completed 44 h of sleep deprivation, while all completed at least 21 h. Results of interest to this topic were the hits, and illumination-button presses.

Analysis of their data used a special complex-demodulation program set for detecting a 90-min period. Regardless of the length of time subjects remained under the sleep-deprivation condition, their results were similar. A very modest emergence of a 90-min rhythm for movement detection occurred during the first half to two-thirds of the 48-h sleep-deprivation recordings, followed by a clear increase in their amplitude during the latter half of the study. The illumination-button presses followed a similar course. In particular, subjects who performed longer than 24 h had dampened response rhythms during the first 24 h. Subsequently, within a few hours before self-terminating the experiment, they showed a substantial increase in the amplitude of the 90-min cycle, which peaked just prior to termination. The authors interpreted the increased amplitude in the ultradian oscillations as reflecting a cumulative ‘stress reaction’ produced by the increasing sleep deprivation. We would propose as an alternate interpretation: Ultradian rhythms were unmasked in the later part of the experiment. This came about as subjects experienced significant increases in sleepiness and fatigue coupled with decreasing motivation just prior to its ending. As noted in the last section (see Table 13.1), circadian rhythm peak-to-trough amplitudes can increase significantly as sleep loss progresses. In the case of ultradian rhythms, typically much lower in amplitude than circadian rhythms, progressive sleep loss would increase their relative amplitude and ease of detection.

In a follow-up article explaining the use of complex demodulation in the analysis of their previous study Orr et al. (1976) re-affirmed their conclusion: When endogenous rhythms exist, increasing fatigue from sleep deprivation produces greater declines in arousal. This results in cognitive behaviors becoming more susceptible to their underlying rhythmic influences. Conversely, they further state that when performance levels are relatively stable, as with adequate sleep, performance oscillations would not be expected. In summary, a 90-min ultradian performance rhythm would emerge at the point of some threshold level of fatigue with a concomitant decrease in arousal. What the level is of that emergent rhythmic threshold was not speculated. Again, like other studies, this one suggests that emergence of a 1–2 h ultradian performance rhythm comes only under non-ordinary conditions. In this case, it appears to be under sleep deprivation when pressure by the homeostatic process of sleepiness is increasing and cognitive arousal drops below some unsustainable stable level.

A more recent approximately 35-h sleep-attenuation study by Hayashi et al. (1992) reported detection of several different ultradian rhythms. Variables measured included eyes-open EEG recordings, self-rated sleepiness, and performance measures of logic, alphanumeric character detection, and four-choice position tasks. One session of 20 h of continuous work (CW1) was first measured, followed by half the subject napping for 3 h while the other half rested without bed rest. Thereafter, another 12 h of continuous work was measured (CW2). Using power spectra analysis on CW1 and CW2 data separately, they confirmed for the logical reasoning task the presence of 12 cycles/day during CW1 and 10 cycles/day for CW2. For the detection task, they detected 12 cycles/day and 14 cycles/day, for CW1 and CW2, respectively. Sleepiness showed an eight cycle/day (3 h) rhythm during both testing periods. The EEG alpha activity showed a 16 cycles/day BRAC-like rhythm for CW1, and a somewhat slower 14 cycles/day rhythm for alpha 2-band activity. The authors noted that the 3-h rhythm in sleepiness was not consistent with previous findings of a 90-min sleepiness cycle. On the other hand, Lavie (1989) has pointed out that slower cyclic components such as those seen in the later half of a normal waking day tend to mask the faster oscillations. An amplification of slower components resulting from sleep deprivation indeed could mask the BRAC-like rhythm. The source of these slower rhythms is apparently as of yet unknown.

Performance data collected over an extended period of sleep deprivation often contain ultradian components, in addition to dominant circadian and monotonic components. The work done at Walter Reed Army Institute of Research by Babkoff et al. (Table 13.1) often detected 12-h ultradian frequencies and faster components as well following analysis by complex demodulation. Frequently, the ultradian characteristics of data collected during sleep-deprivation protocols are either ignored or smoothed. Yet they may contain important information about systematic changes in performance during sleep deprivation, and can account for additional variance in a model of sleep loss as well.

During a 36-h sleep-deprivation study, we collected performance data from 15 university students in the Human Performance Rhythms Laboratory at Penn State University. Subjects were healthy, nonsmoking 18–25 years old, with a habitual bedtime of 2400 ± 1 h and rise time of 0800 ± 1 h. All sleep deprived subjects, and a comparable group of 13 control subjects, were 'I-types.' These are intermediate chronotypes who were neither morning- or evening-preferring, as classified by their scores on both the Basic Language Morningness Scale (Brown, 1993) and Morningness-Eveningness Questionnaire (Horne and Östberg, 1976). Subjects reported to the lab at 1730 h and the sleep-deprivation subjects remained there until 2230 h the next evening. Those subjects remained awake at all times under the watchful eyes of shifts of laboratory assistants. They were allowed only food and beverages served in the laboratory containing no caffeine and low sugar content. When not testing, they were allowed to watch videos, play cards and board games, and socialize. They completed a battery of tests including the Psychomotor Vigilance Task (PVT) (Dinges and Powell, 1985) in computerized form (Thorne et al., 1985) every 2 h starting at 1900 h for 14 test sessions. The first session was a practice session not used in subsequent analyses. The PVT is a simple reaction-time task. Subjects are instructed to press the response button when a running time display randomly appears. Task duration is 10 min. The test reportedly shows no learning effects with repeated use. All control subjects completed the same testing regimen, except for the night and early morning testing sessions that occurred while they were at home sleeping.

PVT Data from six subjects in the sleep deprivation group are shown in Fig. 13.1. Notable changes from the baseline level of mean reaction time were not seen until after 0500 h, more than 5 h after their average typical bedtime of 2400 h. Subject 2 showed an 8-h rhythm with large performance degradations over the next 16 h. Subjects 31 and 84 were somewhat less variable in performance and exhibited a 4-h fluctuation in performance. Notably, the 4-h rhythm of subject 31 was phase delayed by 2 h relative to that of subject 84. Subject 7 showed some performance impairment, although no clear ultradian rhythm was apparent. Subjects 1 and 33 were quite resistant to the effects of sleep deprivation and showed little change from baseline performance. The individual variability in expression of ultradian rhythms is quite clear although these subjects had very similar bedtimes and rise times, and were all Intermediate chronotypes. It is also interesting to note that around the time of their circadian peak, the ultradian rhythms were attenuated. For reasons unknown, data from other subjects (not shown on this graph) exhibited less organized rhythmicity in their performance variability, with the exception of their circadian rhythms. Nevertheless, some subjects inevitably showed systematic oscillations that can be characterized as ultradian rhythms and warrant further study. Careful consideration of individual subjects' variations in the phase and period of their rhythms is crucial. This is because averaging across subjects to summarize a group

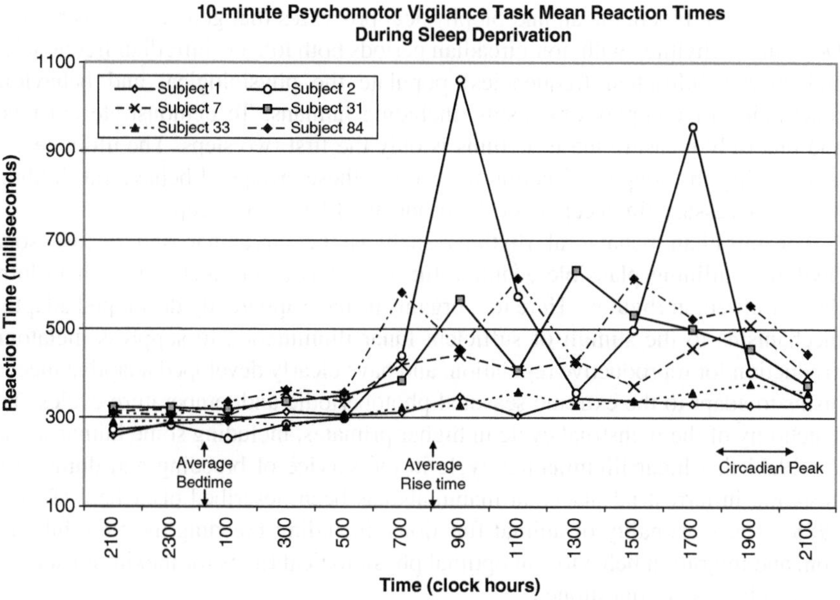

Fig. 13.1 Mean reaction times from a 10-min PsychomotorVigilance Task (unpublished data, Penn State University)

mean might cancel otherwise significant rhythmic oscillations. This perennial problem can lead to masking of both ultradian (Lavie, 1989) and circadian (Babkoff et al., 1991b) rhythms when individual data are pooled and presented as a group statistic. A trait like Morningness-Eveningness, typically associated with circadian rhythms, also may affect ultradian rhythm expression, especially during the early part of the day when their characteristic chronotype differences are most evident. Therefore, careful evaluation of individual data series prior to group averaging is essential when studying ultradian rhythms.

Does sleep deprivation enhance or suppress ultradian rhythms? Studies so far in this area do not lead to definitive conclusions. What they do suggest is that sleep deprivation may amplify both underlying fast BRAC-like rhythms and slower oscillations. The slower oscillations may at times mask the faster oscillations, as suggested by Lavie (1989), although no mechanism for this masking is definitive. Then in addition, the underlying mechanisms that promote ultradian cyclicity and increase in activity following sleep deprivation have yet to be delineated.

13.5 Significance of Ultradian Performance Rhythms

'To have life is to have rhythmicity' (Brown, 1982, p. 1). The last quarter of a century has solidly affirmed this simple observation. Although this statement originally was made for rhythmic behaviors, those behaviors only mirror the underlying physiological, biochemical, and biophysical processes that generate the behaviors. Domains of rhythms with non-circadian periods both longer (infradian frequencies) and shorter (ultradian frequencies) populate the physiological and behavioral functioning of complex organisms, including humans. To demonstrate existence and precisely measure these rhythms is only the first two steps. The ultimate step is to specify the adaptive function for each of these groups of behavioral rhythms, be they necessary timekeepers or the products of basic timekeepers.

For infradian behavioral rhythms such as the circalunar and circaseasonal rhythms, outlining plausible adaptive functions has been much easier than doing that for ultradian rhythms. Here too, organisms have apparently developed adaptive mechanisms to the stimuli of sufficient lunar illumination to suppress melatonin production for reproductive regulation, and have clearly developed adaptive mechanisms to adapt to the extreme seasonal photoperiodic and temperature cycles. The synchrony of the menstrual cycle in higher primates, including some humans, with the 29.53-day lunar illumination cycle in the service of breeding and duration of gestation in terrestrial placental mammals has been described elsewhere (Brown, 1988). The seasonality of animal functions, including breeding, nesting, hibernation, and migration behaviors at optimal physiological times for maximum survival adaptability, is not questioned.

Extensive research has affirmed the genetically based origins of the circadian rhythm, possibly the primary rhythm because of its direct link to the universal near-24 light-dark cycling of planet Earth. The assumption underlying that hypothesis, of course, is that all life on Earth evolved under a similar light-dark patterning with the concomitant daily rhythms of temperature and humidity that would affect biochemical reactions. In this context, ultradian rhythms might assume a support role to the dominant circadian rhythm. Even more intriguing is the suggestion that coupled ultradian rhythms are the source of circadian rhythms, proposed earlier through work on drosophila (Dowse and Ringo, 1987). Interest in this area has lead to some recent mathematical modeling such as that of Paetkau et al. (2006). However, as of yet those rhythms have no discovered geophysical correlate, as does the circadian period (Dowse and Ringo, 1992).

Ultradian rhythms of physiological functioning, from ultra-second brain waves, ultra-minute heartbeat and breathing, to ultradian cycles in eating, digestion, and hormone excretion have been well-documented (Stupfel and Pavely, 1990; Lloyd, 1992). They are basic for the survival of all warm-blooded organisms. In contrast, those ultradian rhythms that emerge in cognitive functions and behaviors, collectively called ‘performance rhythms,’ have been considered as secondary, if that, although they, too, are important to survival. There is clear evidence for the presence of ultradian oscillations in alertness, sleepiness, and cognitive performance. However, their role in general functioning remains unknown and therefore unrecognized in models of alertness and performance. Lavie (1989) has postulated that ultradian rhythms remain neglected for two reasons: First of all, compared to circadian rhythms, they are of relatively low amplitude, relatively unstable, and often require special statistical and mathematical techniques for detection. Second, their origins until now have been linked primarily to Kleitman's BRAC model, which may be inadequate as an explanation for the presence ultradian rhythms in general.

As with circadian and infradian rhythms, ultradian rhythms may contribute as well to biological survivability. Lloyd and Stupfel (1991) have pointed out that many physiological ultradian processes are temporally adjusted to the needs of individual species. They also suggest that behavioral ultradian rhythms such as rest-activity periods allow vertebrates to optimize vigilance during activity, thereby avoiding possible dangers related to predator-prey situations. They suggest that this flexibility occurs because ultradian rhythms appear to be endogenously driven rather than heavily influenced by environmental factors (such as the circadian system with the 24-h light-dark cycle), and are therefore less stringently regulated. Rather ironically, they point out that this variability in ultradian rhythms makes them more difficult to study.

Broughton (1985) has proposed certain important characteristics of ultradian rhythms that lead to increased survivability of a species:

(a) Ultradian frequencies in both animals and humans frequently occur in 2:1 integer ratios related to the circadian system (at 12, 6, 3, and 1.5 periodicities), and perhaps in the 3:1 ratio as well. A consistent ratio of integer periods results in constant phase relationships between sub-oscillator systems at the period of the slowest oscillator, resulting in optimal consistency and efficiency in temporal organization of the organism's functions. This ultimately results in greater energy efficiency and increased survivability of the organism.
(b) Species survival requires having the capability to maintain constant phase relationships between endogenous sub-oscillators and important natural periodicities in the geophysical environment such as environmental zeitgebers or imposed time cues. The above-mentioned sub-multiple ultradian frequencies are prominent in nature as they are most adaptive and therefore have been retained during biological evolution.

The scientific history of a 90–100 min BRAC emerging in diurnal awake activity is usually attributed to Nathaniel Kleitman (1961). Thereafter this rhythm was hypothesized as paralleling the 90–100 min NREM-REM cycle sleep architecture discovered by Dement and Azerinsky in the 1960s. The assumption was that this daytime rhythm appearance was caused by a fundamental rhythm governing both awake and sleep activities. The more recent attempts to affirm the common origin of the two approximately 90–100 min cycles have met with mixed results. Nonetheless, the existence of ultradian cognitive/behavioral rhythms is apparent. What is not well substantiated are the conditions under which such rhythms emerge and whether they can be considered as emerging under normal or only under unusual or pathological conditions such as sleep deprivation. This question remains basic in this chapter's discussions and will remain unanswered pending more extensive work in the area of sleep deprivation and its effect on ultradian performance rhythms.

13.6 Ultradian Rhythms: Implications for Health and Safety

Clearly, we have seen that circadian performance rhythms in memory, perception, and logical reasoning systematically wax and wane with situations involving levels of sleep deprivation. No clear explanations are apparent as to the specific survival role of the 75 min to 12 h reported ultradian oscillations that emerge during sleep loss. The enigma of the existence of the BRAC remains until research affirms the connection of the nightly NREM-REM cyclicity to phase-related, if not phase-locked, daytime performance. Although a 90-min performance rhythm is fairly well substantiated, nevertheless it is still equivocal. The same can be said for the 4-h rhythm. Less information is available in support of ultradian performance rhythms below 90 min and above 4 h. Nonetheless, one area where understanding implications of ultradian rhythms for health seem evident is that for morning behaviors and judgments. When night sleep is both adequate and deprived, what consistently emerges is a short span around 90 min post-awakening when arousal is most likely to decrease. Lavie's concept of the 'sleep gate' and greater 'sleepability' at this time suggests that vulnerability may be quite strong for lapses in awareness, motivation, and good cognitive processing.

Vulnerability to sleepiness during the period just following morning awaking may be further exacerbated by 'chronic partial-sleep deprivation.' Chronic partial-sleep deprivation is a condition under which adequate restful sleeping is truncated by artificial awakening demands. These include being awakened by some sort of alarm or external arousal that shortens sleep duration before spontaneously awakening at the end of a REM cycle and feeling refreshed. Chronic partial-sleep deprivation is a malady affecting a significant proportion of the human population of advanced cultures. Within them societal demands are assuming a daily 24-h urgency. Unfortunately, the biological requirements of adequate sleep are being relegated to a nuisance inconvenience. Consequently, the duration of primary nightly sleep has been shrinking noticeably over the last two generations. On-the-job, in-transit, and driving sleepiness, lapses in conscious judgment and falling asleep have increased, with fatal consequences in some cases. If the conclusions hold that several ultradian performance rhythms emerge as sleepiness increases, then chronic partial-sleep loss, like total sleep deprivation, has serious implication for performance.

Earlier chronobiological studies indicating a peak in morning commute vehicular accidents may be the combined result of driving during a relative low circadian point on the descending limb of a 90-min ultradian sleepiness oscillation. Considering the best estimates by the US National Sleep Foundation, upwards of 40% of the US population is chronically partially sleep deprived. Because of that, the amplitude of the ultradian oscillation can be amplified especially during the early portion of day activity, and then attenuate as the circadian oscillation continues climbing toward its peak in the latter half of the day. Several of the studies that we have cited demonstrate early-day emergence of several ultradian rhythms, with the near 90-min cycle predominating, followed by later-day attenuation. These studies lead us to conclude that while inadequate nightly sleep impairs performance, its combination with negative circadian-ultradian interactions results in a dangerous rise in performance impairment peaking from 1 to 2 h after awakening.

Less dramatic episodic cognitive-behavioral events occur daily. They can be considered as predictably repetitive phenomena, like daily TV programming. However, that does not make them a rhythmic phenomenon with oscillatory characteristics. These can include the experience of the post-lunch dip in alertness and the later afternoon dip in physical vigor. Their degree is dependent upon antecedent conditions such as amount of sleep the night before, amount of physical activity, and type of food and drink consumed before their occurrence. However, these have not yet been studied to connect them directly as events accompanying a normal night of restful sleep or as a consequence of lack of a good night's sleep. Clearly, much more research is necessary to separate endogenous ultradian cognitive-behavioral rhythms from predictably episodic exogenous phenomena caused by the random vicissitudes of ambient conditions.

Existence of ultradian cognitive-behavioral rhythms under sleep-deprivation conditions is not questioned. However, undetermined is just how little or much sleep deprivation must occur before rhythmic changes are evident. In addition, it is not clear whether sleep deprivation differentially affects ultradian cognitive-behavioral performance rhythms of particular periodicities, and whether these rhythms are amplified or attenuated. We have suggested that ultradian rhythms during arousal and performance interact with circadian rhythms and accumulating sleep deprivation to create potential performance vulnerabilities. This point requires further investigation. Clearly, it takes on an urgency, especially under the increasing socially driven chronic partial-sleep deprivation of our 24/7 culture. These question implore continued research.

References

Angus, R. G., and Heslegrave, R. J. (1985). Effects of sleep loss on sustained cognitive performance during a command and control simulation. *Behav Res Meth Instrum Comp*, *17*, 55–67.

Aserinsky, E., and Kleitman, N. (1953). Regularly occurring periods of eye motility, and concomitant phenomena, during sleep. *Science*, *118*, 273–274.

Babkoff, H., Mikulincer, M., Caspy, T., Kempinski, D., and Sing, H. C. (1988). The topology of performance curves during 72 hours of sleep loss: A memory and search task. *Q J Exp Psychol A*, *40A* (4), 737–756.

Babkoff, H., Mikulincer, M., Caspy, T., Carasso, R. L., and Sing, H. C. (1989). The implications of sleep loss for circadian performance accuracy. *Work and Stress*, *3* (1), 3–14.

Babkoff, H., Caspy, T., and Mikulincer, M. (1991a). Subjective sleepiness ratings: The effects of sleep deprivation, circadian rhythmicity and cognitive performance. *Sleep*, *14* (6), 534–539.

Babkoff, H., Caspy, T., Mikulincer, M., and Sing, H. (1991b). Monotonic and rhythmic influences: A challenge for sleep deprivation research. *Psychol Bull*, *109* (3), 411–428.

Babkoff, H., Mikulincer, M., Caspy, T., and Sing, H. C. (1992). Selected problems of analysis and interpretation of the effects of sleep deprivation on temperature and performance rhythms. In D. Friedman and G. E. Bruder (Eds.), Psychophysiology and experimental psychopathology: A tribute to Samuel Sutton. Ann NY Acad Sci, 658. New York: New York Academy of Sciences, pp. 93–110.

Blagrove, M., Alexander, C., and Horne, J. A. (1995). The effects of chronic sleep reduction on the performance of cognitive tasks sensitive to sleep deprivation. *Appl Cogn Psychol*, *9*, 21–40.
Broughton, R. J. (1985). Three central issues concerning ultradian rhythms. *Exp Brain Res*, *12* [Suppl.], 217–231.
Brown, F.A. Jr. (1959). Living clock. Science, 130 (3388), 1535–1544.
Brown, F. M. (1982). Rhythmicity as an emerging variable for psychology. In F. M. Brown and R. C. Graeber (Eds.), *Rhythmic aspects of behavior*. Hillsdale, NJ: Lawrence Erlbaum, pp. 3–38.
Brown, F. M. (1988). Common 30-day multiple in gestation time of terrestrial placentals. *Chronobiol Int*, *5* (3), 195–210.
Brown, F. M. (1993). Psychometric equivalence of an improved Basic Language Morningness (BALM) scale using industrial population within comparisons. *Ergonomics*, *36* (1–3), 191–197.
Casagrande, M., Violani, C., Curcio, G., and Bertini, M. (1997). Assessing vigilance through a brief pencil and paper letter cancellation task (LCT): Effects of one night of sleep deprivation and the time of day. *Ergonomics*, *40* (6), 613–630.
Chapotot, F., Jouny, C., Muzet, A., Buguet, A., and Brandenberger, G. (2000). High frequency waking EEG: Reflection of a slow ultradian rhythm in daytime arousal. *Neuroreport*, *11* (10), 2223–2227.
Dement, W. (1955). Dream recall and eye movements during sleep in schizophrenics and normals. *J Nerv Ment Dis*, *122* (3), 263–269.
Dijk, D. J., Duffy, J. F., and Czeisler, C. A. (1992). Circadian and sleep/wake dependent aspects of subjective alertness and cognitive performance. *J Sleep Res*, *1* (2), 112–117.
Dinges, D. F., and Kribbs, N. B. (1991). Performing while sleep: Effects of experimentally induced sleepiness. In T. H. Monk (Ed.), *Sleep, sleepiness, and performance*. Chichester, UK: Wiley, pp. 97–128.
Dinges, D. F., and Powell, J. W. (1985). Microcomputer analyses of performance on a portable, simple visual RT task during sustained operations. *Behav Res Meth Instrum Comp*, *17*, 652–655.
Dowse, H. B., and Ringo, J. M. (1987). Further evidence that the circadian clock in Drosophila is a population of coupled ultradian oscillators. *J Biol Rhythms*, *2*, 65–76.
Dowse, H. B., and Ringo, J. M. (1992). Do ultradian oscillators underlie the circadian clock in Drosophila? In D. Lloyd and E. L. Rossi (Eds.), *Ultradian rhythms in life processes*. London: Springer-Verlag, pp. 105–117.
Gillberg, M., and Åkerstedt, T. (1998). Sleep loss and performance: No 'safe' duration of a monotonous task. *Physiol Behav*, *64* (5), 599–604.
Haslam, D. R. (1982). Sleep loss, recovery sleep, and military performance. *Ergonomics*, *25*, 163–178.
Hayashi, M., Arakawa, Y., and Hori, T. (1992). Ultradian rhythms in task-performance and EEG activities during prolonged continuous work. *Jpn J Psychiatr Neurol*, *46* (1), 238–239.
Hayashi, M., Tsukamoto, H., and Hori, T. (1994). Ultradian rhythms in task-performance and EEG activities. *Japn J Psychiat Neurol*, *48* (2), 498–499.
Hayashi, M., Minami, S., and Hori, T. (1998). Masking effect of motivation on ultradian rhythm. *Percept Mot Skills*, *86* (1), 127–136.
Heslegrave, R. J., and Angus, R. G. (1985). The effects of task duration and work-session location on performance degradation induced by sleep loss and sustained cognitive performance. *Behav Res Meth Instru Comp*, *17* (6), 592–603.
Horne, J. A., and Östberg, O. (1976). A self-assessment questionnaire to determine morningness-eveningness in human circadian rhythms. *Chronobiol Int*, *4*, 97–110.
Johnson, M. P., Duffy, J. F., Dijk, D. J., Ronda, J. M., Dyal, C. M., and Czeisler, C. A. (1992). Short-term memory, alertness and performance: a reappraisal of their relationship to body temperature. *J Sleep Res*, *1* (1), 24–29.
Kleitman, N. (1961). The nature of dreaming. In G. E. W. Wolstenhome and M. O'Connor (Eds.), *The nature of sleep*. London: Churchill, pp. 349–364.

Kripke, D. F. (1972). Ultradian biologic rhythm associated with perceptual deprivation and REM sleep. *Psychosom Med, 34* (3), 221–234.

Kripke, D. F., Mullaney, D. J., and Fleck, P. A. (1985). Ultradian rhythms during sustained performance. *Exp Brain Res, 12* [Suppl.], 201–216.

Lavie, P. (1989) Ultradian rhythms in arousal–the problem of masking. *Chronobiol Int, 6* (1), 21–28.

Lavie, P., and Scherson, A. (1981). Ultrashort sleep-waking schedule. I. Evidence of ultradian rhythmicity in 'sleepability'. *Electroencephalogr Clin Neurophysiol, 52* (2), 163–174.

Lavie, P., and Zomer, J. (1984). Ultrashort sleep-waking schedule. II. Relationship between ultradian rhythms in sleepability and the REM-Non-REM cycles and effects of the circadian phase. *Electroencephalogr Clin Neurophysiol, 57*, 35–42.

Linde, L., and Bergstrom, M. (1992). The effect of one night without sleep on problem-solving and immediate recall. *Psychol Res, 54* (2), 127–136.

Lloyd, D. (1992). Intracellular timekeeping: Epigenetic oscillations reveal the functions of an ultradian clock. In D. Lloyd and E. L. Rossi (Eds.), *Ultradian rhythms in life processes* London: Springer, pp. 5–22.

Lloyd, D., and Stupfel, M. (1991). The occurrence and functions of ultradian rhythms. *Biol Rev Camb Philos Soc, 6*, 275–299.

Manseau, C., and Broughton, R. J. (1984). Biaterally synchronous ultradian EEG rhythms in awake adult humans. *Psychophysiology, 2*, 265–273.

Mikulincer, M., Babkoff, H., Caspy, T., and Sing, H. C. (1989). The effects of 72 hours of sleep loss on psychological variables. *Br J Psychol, 80*, 145–162.

Monk, T. H., Buysse, D. J., Reynolds C. F. III, Berga, S. L., Jarrett, D. B., Begley, A. E., and Kupfer, D. J. (1997). Circadian rhythms in human performance and mood under constant conditions. *J Sleep Res, 6*, 9–18.

Nakagawa, Y. (1980). Continuous observation of EEG patterns at night and in daytime of normal subjects under restrained conditions. I. Quiescent state when lying down. *Electroencephalogr Clin Neurophysiol, 49*, 524–537.

Okawa, M., Matoušek, M., and Petersén, I. (1984). Spontaneous vigilance fluctuations in the daytime. *Psychophysiology, 21*, 207–211.

Orr, W. C., Hoffman, H. J., and Hegge, F. W. (1974). Ultradian rhythms in extended performance. *Aerosp Med, 45* (9), 995–1000.

Orr, W. C., Hoffman, H. J., and Hegge, F. W. (1976).The assessment of time-dependent changes in human performance. *Chronobiologia, 3* (4), 293–305.

Paetkau, V., Edwards, R., and Illner, R. (2006). A model for generating circadian rhythm by coupling ultradian oscillators. *Theor Biol Med Model, 3*, 12.

Pigeau, R., Naitoh, P., Buguet, A., McCann, C., Baranski, J., Taylor, M., Thompson, M., and Mack, I. (1995). Modafinil, d-amphetamine and placebo during 64 hours of sustained mental work. I. Effects on mood, fatigue, cognitive performance and body temperature. *J Sleep Res, 4*, 212–228.

Richardson, G. S., Carskadon, M. A., Orav, E. J., and Dement, W. C. (1982). Circadian variation of sleep tendency in elderly and young adult subjects. *Sleep, 5* [Suppl. 2], S82–94.

Ryman, D. H., Naitoh, P., and Englund, C. E. (1985). Decrements in logical reasoning performance under conditions of sleep loss and physical exercise: The factor of sentence complexity. *Percept Motor Skills, 61*, 1179–1188.

Stupfel, M., and Pavely, A. (1990). Ultradian, circahoral and circadian structures in endothermic vertebrates and humans. *Comp Biochem Physiol, 96A* (1), 1–11.

Thorne, D. R., Genser, S. G., Sing, H. C., and Hegge, F. W. (1985). The Walter Reed performance assessment battery. *Neurobehav Toxicol Teratol, 7*, 415–418.

Tsuji, Y., and Kobayashi, T. (1988). Short and long ultradian EEG components in daytime arousal. *Electroencephalogr Clin Neurophysiol, 70* (2), 110–117.

Wimmer, F., Hoffman, R. F., Bonato, R. A., and Moffit, A. R. (1992). The effects of sleep deprivation on divergent thinking and attention processes. *J Sleep Res, 1*, 223–230.

Chapter 14
High Frequency EEG and Its Relationship to Cognitive Function

H.C. Sing

Abstract Objective assessment of the sleep-wake states in clinical and laboratory settings is determined from the electroencephalogram (EEG) as recordings of brain electrical signals. The EEG record is visually scored according to the awake state and 4 stages of varying sleep depth, as well as a "rapid eye movement" (REM) or the dream stage. Drowsiness or the state between awake and sleep is not considered in this standard EEG scoring. Awareness of drowsiness is critical for individuals such as commercial truck drivers, pilots, railroad engineers and nuclear plant operators who are charged with maintaining high alertness in the performance of work that impacts public safety. Prevention of drowsiness progression to the sleep state once detected in these individuals is essential to avoid potential catastrophic events. The omission to include characterization of a drowsiness state was the impetus for a systematic investigation of high frequency EEG as potential marker for this purpose.

Recordings of EEG from study volunteers while performing various cognitive tasks and the sleep latency test were collected at the high sampling rate of 1000 Hz instead of the conventional 256 Hz. Discrete Fourier Transform of the time domain EEG to the frequency domain yielded 500 spectral components. From these components, a 1-15 Hz low frequency band (LFEEG) and a 201-500 Hz high frequency band (HFEEG) were extracted for comparison of relative energy proportion during each testing session.

In the sleep deprived condition, cognitive performance of the volunteers falters and concomitantly, the LFEEG energy proportion increases relative to the HFEEG. Tracking the EEG on a second-by-second basis in synchrony with the performance output provides a means for determining incipient drowsiness when responses to task stimuli are either prolonged or missing. Individual variability is markedly expressed in the examples given here of 2 volunteers: one resilient and one not resilient to the effects of sleep deprivation.

An Index of Alertness/Drowsiness and Cognitive Capacity has been developed based on quantitative comparison of the HFEEG and LFEEG spectral components and is currently in the patent application phase. Integration of this Index in an

Walter Reed Army Institute of Research, Division of Psychiatry and Neuroscience,
Silver Spring, MD 20910 (USA)

ambulatory EEG monitoring system will not only provide real time assessment of alertness and cognitive state but also permit emission of an audio or other warning signals if drowsiness is detected. This has application in the military as well as in the public safety arena for monitoring personnel in critical occupations requiring constant alertness over 24 hrs or longer.

Keywords EEG, high frequency EEG, spectral components, energy proportion, cognitive capacity, vigilance, alertness, drowsiness, sleep deprivation

14.1 Introduction

Sleep deprivation is inevitable in the military environment where the battlefield situation often involves 24 h or more of continuous operation. Maintaining a high level of alertness and cognitive performance under demands of constant readiness around the clock is not only individually difficult, but also impossible to assess without a means of direct monitoring. The Army has long been concerned about the potential for catastrophic outcomes as consequences of sleep deprivation which result in poor judgment and performance on the part of military personnel whose decision making ability is impaired. This concern has been realized in the immediate past with the capture of soldiers in Iraq who took a wrong turn on a road and found themselves in unknown enemy territory. It was later revealed that the soldiers on this expedition had been without sleep for greater than 24 h, were cognitively impaired and unaware that they had misread the map (Boot, 2003).

Consequential incidents due to sleep deprivation and sleep restriction pertain not only to the military, but also to the public sector in areas of transportation, nuclear facilities, emergency support, and health care providers among the more immediate concerns. Many incidents or near accidents occurring in the public arena are not publicized especially those involving pilot fatigue. Congress is currently investigating why these incidents are not reported to the public and cites the case on March 4, 2004, where both pilot and co-pilot flew three sequential "red eyes" between Denver and Baltimore with only 1 h in between flights. During the last 45 min of the third flight as it was approaching Denver, both pilot and co-pilot were sound asleep and missed all calls from the air traffic controller while the plane was traveling at 590 mph instead of less than 290 mph. Fortunately, the pilot did suddenly awake to hear the air traffic controller's frantic calls and was able to follow his instructions resulting in a safe landing (FoxNews.com, 2007).

Even more alarming are the results from a 1992 survey of tractor trailer truckers which found that 19% of the truckers admitted to having fallen asleep at the wheel in the previous month (Braver et al., 1992). A report from the Center for National Truck Statistics (1994) included the disturbing statistic that annually over 5,000 fatalities and 110,000 injuries resulting from motor vehicle accidents involve commercial trucks in the United States. Knipling (1994) estimated that the percentage of vehicle crashes in which fatigue was a factor could be as high as 56%. Although the Department of Transportation (DOT) regulates work hours permitted for truck drivers, pilots, airport controllers and railroad engineers, there is no routine checking for status of alertness (hence being well rested) just prior to start of duty or during duty hours.

A more recent example documented security guards in a Pennsylvania nuclear plant were regularly asleep on the job for as long as an hour or more. This public exposure resulted in loss of employment and dismissal of the security company providing the staff, but offered no remedy as to how this could be prevented in the future (Weinberger, 2007). The impetus to develop an ambulatory monitoring system for real-time assessment of alertness level in individual military personnel and provide warning when indication of drowsiness and potentially sleep is imminent, has been an on-going objective for the U.S., Army. The intent was to target individuals showing high probability towards increasing drowsiness and low probability of maintaining cognitive function, in order to initiate countermeasures such as sleep hygiene or pharmacological agents in a timely manner. The Army's Warfighter Physiological Status Monitor is a project that aims to integrate multiple channels of vital physiological measures such as body temperature, blood pressure, heart rate (EKG), respiration, and neurophysiological measure as brain activity (electroencephalography [EEG]), in a single ambulatory device. Thus far, the physiological measures have been undergoing field testing successfully, but EEG measures have not yet been incorporated (USAIREM). There are current ambulatory devices available commercially, but none so far has been able to satisfy all of the Army's requirements especially in the area of drowsiness detection. These devices usually employ EEG as the measure for determining the alertness level on the basis of being in the awake or asleep state with acceptable accuracy, but fail in the gray area classification of drowsiness.

In our laboratory at the Walter Reed Army Institute of Research (WRAIR), we have developed an EEG-based methodology for estimating alertness/drowsiness and cognitive function capacity. The different states of alertness, cognitive capacity, drowsiness, and sleep are quantitatively derived from the EEG and corroborated with the performance and sleep latency tests administered in the study. The methodology and its applications will be described in this chapter.

14.2 EEG Defined

Brain electrical activity, commonly referred to as EEG, is the manifestation of neuronal communication which may be discerned and recorded at the surface of the scalp by electrode sensors and subsequently displayed, measured, and analyzed. Clinically, the EEG is used for detection of brain pathology such as tumors, epileptic seizures, and behavioral abnormalities such as narcolepsy and attention deficit hyperactive disorder (ADHD). The brain signals, collectively referred to as an electroencephalograph, are analyzed for their constituent frequencies (rhythmic oscillations) and/or selective characteristic wave shapes to detect deviations from normal. In the sleep research laboratory, the EEG is used not only for determination of sleep/wake states and for quantification of sleep amount during nighttime sleep but also to track sleepiness level during the course of sleep deprivation studies of normal, healthy individuals. Our focus here will be the investigation of EEG with respect to characteristics of alertness, drowsiness and cognitive capacity.

Polysomnography is the methodology for defining the awake and sleep states from observation of EEG signals over an extended time period. As its name implies, other physiological measures are recorded synchronously with the EEG to aid in differentiating the awake from the sleep state as well as marking the different stages of sleep. Multiple electrode sensors are attached to the scalp, face, and body of the individual under study to record both the neurophysiological (EEG) and basic physiological measures such as electrooculogram (EOG) for recording eye movements; submental electromyogram (EMG) from the chin for detecting muscle movement; and electrocardiogram (EKG) for heart rate. Although the EEG is the main determinant of sleep characteristics, the EOG and EMG aid in defining Rapid Eye Movement (REM) sleep more commonly known as the dream stage in which it is conjectured that memory consolidation occurs. REM is thus distinguished from non-Rapid Eye Movement (NREM) sleep which defines all other sleep stages. Rolling eye movements observed in the EOG are characteristic during REM simultaneously with muscle atonia as noted in the EMG. During night time sleep, the REM state alternates with non-REM sleep in ultradian cycles of approximately 90 min and increases in length as non-REM length decreases in the progression towards the end of the sleep period. The EKG provides continuous monitoring of heart rate not only to assure normal functioning, but also to confirm the deeper sleep stages when the reduced rate of heart beats indicates slowing of body functions.

Although the frequency realm of EEG is in cycles per second or Hertz (Hz) and several orders of magnitude higher than that of ultradian frequencies (i.e., cycles/24 h), the same fundamental principles of rhythmic behavior apply. The EEG signal as visually observed in its entirety is a combination of all the frequencies selected for recording in the acquisition process. Overall circadian rhythmicity is observed in the oscillation of the frequencies and depending on the frequency band, the cyclic variation mimics the circadian or is out of phase by 180°. That is, the band of low frequencies peaks in the hours of sleep while the band of high frequencies peaks during the waking active period.

14.3 EEG Scoring System

Sleep researchers have devised a scoring system, considered to be the "gold standard" for evaluating sleep depth according to specific frequencies and patterns of EEG waveforms as established by Rechtschaffen and Kales (1968). The system consists of six levels in which sleep is scored within a 20 or 30 s standard epoch as: Wake; Stage 1; Stage 2; Stage 3; Stage 4; and REM. These stages are illustrated in Fig. 14.1. By conventional Rechtschaffen and Kales practice, the EEG during the awake state consists mainly of frequencies between 12–50 Hz and is known as beta frequency although it is sometimes subcategorized as gamma in the 30–40 Hz range. Stage 1 is considered light sleep and the EEG is defined by a mix of predominantly alpha (7–14 Hz) and some theta (5–7 Hz) frequencies. It is not difficult to be awakened at this stage. Stage 2 is deeper sleep dominated by theta, along with some alpha. This stage is characterized by intrusions of specific wave patterns described as K-complexes and spindles because they resemble these descriptions and appears to be the threshold to actual sleep whereas Stage 1 is more the transitional state between awake and sleep. Stages 3 and 4 are marked by delta (1–4 Hz) frequencies with Stage 3 showing all of these frequencies while Stage 4 have both greater percentage and higher amplitude of 1 and 2 Hz frequencies. Stage 4 represents the deepest sleep stage where frequency of neuronal communication is lowest and judging from the high amplitude of these lowest frequencies, indicative that the sum of

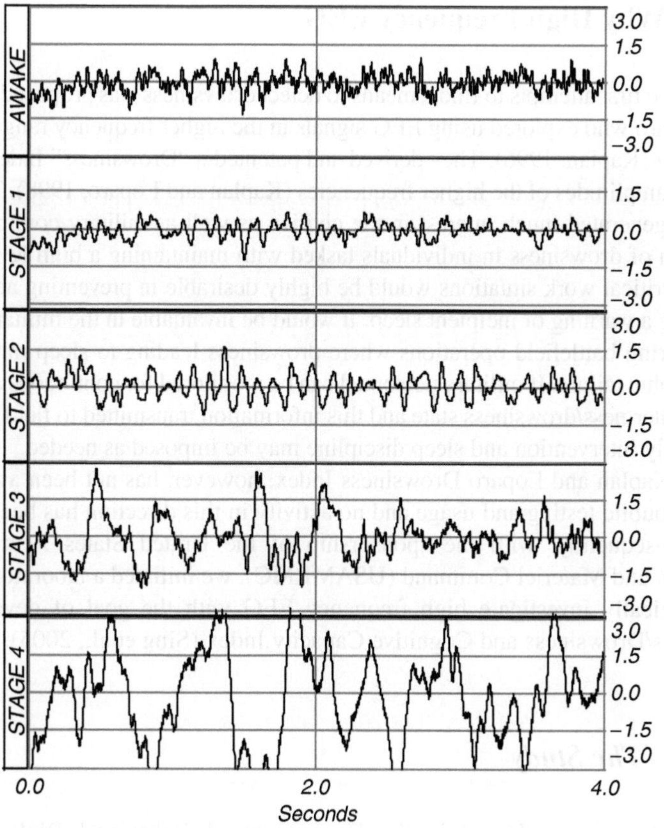

Fig. 14.1 Sleep stages according to Rechtschaffen and Kales visual scoring scheme for EEG

active brain function is essentially minimal throughout the brain, i.e., the brain has essentially "shut down". Arousal from this sleep stage is extremely difficult.

Sleep scorers mark latency to sleep with the appearance of K-complexes or sleep spindles seen in Stage 1 or 2. Latency of 5 min or less is considered pathological in the clinical setting under normal conditions, but in the sleep research laboratory, this is quite often the case in sleep restricted or sleep deprived individuals with no existing pathology.

It is to be noted that EEG records are usually visually scanned and manually scored – a long, tedious process with emphasis of the process on either the sleep or awake state and little or no attention directed to the between state of drowsiness. There has been little change in manual EEG sleep stage scoring for over 35 years, until recently where attempts have been made to automate the procedure with some measure of accuracy by following the Rechtschaffen and Kales guidelines as well as the more recent American Association of Sleep Medicine's *The AASM Manual for the Scoring of Sleep and Associated Events* (Anderer et al., 2005).

14.4 Why High Frequency EEG

One of the first attempts to find a means to detect drowsiness was proposed by Kaplan and Loparo who explored using EEG signals in the higher frequency range from 100 to 400 Hz (Kaplan, 1996). They derived and patented a "Drowsiness" Index based on spectral amplitudes of the higher frequencies (Kaplan and Loparo, 1998). This development generated much interest in the civilian as well as military community. The detection of drowsiness in individuals tasked with maintaining a high level of alertness in critical work situations would be highly desirable in preventing accidents by signaling a warning of incipient sleep. It would be invaluable in the military environment during battlefield operations where drowsiness leading to sleep may result in catastrophic failure. If military personnel can be monitored for continuous assessment of their alertness/drowsiness state and this information transmitted to field commanders, timely intervention and sleep discipline may be imposed as needed.

The Kaplan and Loparo Drowsiness Index, however, has not been available for general public testing and usage and no activity in this direction has been discernible. Consequently, with the sponsorship of the United States Army Medical Research and Materiel Command (USAMRMC), we initiated a laboratory study to systematically investigate high frequency EEG with the goal of developing an Alertness/Drowsiness and Cognitive Capacity Index (Sing et al., 2005).

14.4.1 The Study

The study was conducted in the Department of Behavioral Biology's Sleep Research Suite at WRAIR under a protocol approved by the Institute's Internal

Review Board (IRB) and the USAMRMC Human Subjects Research Review Board (HSRRB).

14.4.1.1 Study Subjects

Thirteen military and civilian subjects of both genders (6 m, 7 f; mean age 30.5 ± 6.0 years) participated in this study after medical examination and clinical laboratory tests confirmed their physical and mental well being as healthy and normal. All subjects were briefed on the study's purpose and procedures and provided informed consent prior to their enrollment. The study was conducted over 4 days and included a 40 h continuous wakefulness period. Because of instrument limitation only one subject was studied in each run. The subjects were paid for their participation at the conclusion of each study run

14.4.1.2 Study Schedule

The subjects resided in the Sleep Suite for the entire 4 days of the study and followed a set schedule described in Table 14.1. The cognitive tasks administered in this experiment are listed in Table 14.2, and discussed below under *Study Measures*.

14.4.1.3 Study Instrumentation

The neurophysiological and physiological signals described above were recorded for each subject during cognitive testing sessions and overnight sleep. The Biopac MP150 (Biopac Systems, Inc., Santa Barbara, CA), recorded these signals on its analog channels. The sampling rate for the signal acquisition was set at 1,000 Hz for EEG and EKG; at 500 Hz for EOG and EMG. This meant that for each 1 s epoch of EEG and EKG, 1,000 data points described the signal wave form – an unusually high resolution for EEG and EKG. A unique feature of this particular instrument allowed stimulus input and subject response from testing computers to be stamped as markers on its digital channels. Consequently, for computerized tasks which generate sequential stimuli on the monitor in front of the subjects, the exact response times can be determined from the markers on the digital channels. However, the most valuable aspect of the instrument was the simultaneous time synchrony of all analog and digital channels, such that the properties of the EEG signals may be exactly known prior to stimulus presentation, during the response time and post response for the entire test session. A SONY laptop computer (VAIO NV170P) was interfaced with the MP150 and controlled the initiation and termination of each test session. These two points were also stamped as markers on a separate digital channel. The raw data acquired and digitized by the MP150 were transmitted to a high capacity hard drive for storage and subsequent analysis. All test sessions were videotaped to provide visual assessment of the subject's alertness/drowsiness level during performance.

Table 14.1 Study schedule of events

Time	Task
Day 1	
0800	Arrive at laboratory. Provide urine sample 1. Vital signs taken. Sensors attached
1000–1200	PVT, CVPT, PAB, MSLT instruction and practice. Vital signs
1200–1250	Vital signs. Lunch. Electrodes and actigraph check
1300–1350	PVT, CVPT
1400–1450	Vital signs. PAB, PAB-PVT, MSLT
1500–1550	PVT, CVPT
1600–1650	Vital signs. PAB, PAB-PVT, MSLT
1700–1750	PVT, CVPT
1800–1850	Vital signs. Dinner. Electrodes and actigraph check
1900–1950	PVT, CVPT
2000–2050	Vital signs. PAB, PAB-PVT, MSLT
2100–2150	PVT, CVPT
2200–2250	Vital signs. PAB, PAB-PVT, MSLT
2300–2350	PVT, CVPT
0000–0800	Lights out. 8-h sleep period
Day 2	
0800–0850	Awaken. Electrodes and actigraph check. Vital signs. Breakfast
0900–0950	PVT, CVPT
1000–1050	Vital signs. PAB, PAB-PVT, MSLT
1100–1150	PVT, CVPT
1200–1250	Vital signs Lunch. Electrodes check
1300–1350	PVT, CVPT
1400–1450	Vital signs. PAB, PAB-PVT, MSLT
1500–1550	PVT, CVPT
1600–1650	Vital signs. PAB, PAB-PVT
1700–1750	PVT, CVPT
1800–1850	Vital signs. Dinner. Electrodes and actigraph check
1900–1950	PVT, CVPT
2000–2050	Vital signs. PAB, PAB-PVT, MSLT
2100–2150	PVT, CVPT
2200–2250	Vital signs. PAB, PAB-PVT
2300–2350	PVT, CVPT
Day 3	
0000–0050	Urine sample 2. Vital signs. PAB, PAB-PVT, MSLT
0100–0150	PVT, CVPT
0200–0250	Vital signs. PAB, PAB-PVT
0300–0350	PVT, CVPT
0400–0450	Vital signs. PAB, PAB-PVT
0500–0550	PVT, CVPT
0600–0650	Vital signs. PAB, PAB-PVT, MSLT
0700–0750	PVT, CVPT
0800–0850	Vital signs. Electrodes and sctigraph check. Breakfast
0900–0950	PVT, CVPT
1000–1050	Vital signs. PAB, PAB-PVT, MSLT
1100–1150	PVT, CVPT
1200–1250	Vital signs. Lunch. Electrodes and actigraph check
1300–1350	PVT, CVPT
1400–1450	Vital signs. PAB, PAB-PVT, MSLT
1500–1550	PVT, CVPT
1600–1650	Vital signs. PAB, PAB-PVT

(continued)

Table 14.1 (continued)

Time	Task
1700–1750	PVT, CVPT
1800–1850	Vital signs. Dinner. Electrodes and actigraph check
1900–1950	PVT, CVPT
2000–2050	Vital signs. PAB, PAB-PVT, MSLT
2100–2150	PVT, CVPT
2200–2250	Vital signs. PAB, PAB-PVT
2300–2350	PVT, CVPT
Day 4	
0000–1000	Lights out. 10-h sleep period
1000–1050	Awaken. Urine sample 3. Electrodes and actigraph check. Vital signs. Breakfast
1100–1150	PVT, CVPT
1200–1250	Lunch. Electrodes check
1300–1350	PVT, CVPT
1400–1450	Vital signs. PAB, PAB-PVT, MSLT
1500–1550	PVT, CVPT
1600–1650	Vital Signs. PAB, PAB-PVT, MSLT
1700–1750	PVT, CVPT
1800–1850	Vital signs. Remove sensors. Dinner. Debrief
1900–2000	Optional nap. Release from study. Depart laboratory

PAB – Performance Assessment Battery; **CVPT** – Choice Visual Perception Task; **PVT** – Psychomotor Vigilance Task (variants described in text above); **MSLT** – Multiple Sleep Latency Test

Table 14.2 Cognitive tasks administered during scheduled test sessions

Task name	# Of test sessions	EEG Total C3 & C4 analyses
CVPT	19	38
MSLT	12	24
PAB	20	40
PALM PVTB	30	60
PALMPVTC	30	60
PCPVT	20	40
PVT192	30	60
Sleep (overnight)	2	4
Total – per subject	163	326

CVPT – Choice Visual Perception Test; **MSLT** – Multiple Sleep Latency Test; **PAB** – WRAIR Performance Assessment Battery; **PALM PVTB** – PDA-sized Psychomotor Vigilance Task-B; **PALM PVTC** – PDA-sized Psychomotor Vigilance Task-C

14.4.1.4 Study Measures

Besides the neurophysiological and physiological measures mentioned above, other physiological measures including tympanic temperature, pulse, and blood pressure were recorded at 2-h intervals to ascertain that the subject's vital signs remained normal.

Cognitive performance tasks administered to the subjects during the study included: (1) Walter Reed Performance Assessment Battery (PAB) (Thorne, 1985); (2) The Choice Visual Perception Task (CVPT) (Russo et al., 2004; Russo, 2005), and (3) Psychomotor Vigilance Task (PVT) (Dinges and Powell, 1985).

The assessment of sleepiness during the 40 h continuous waking period was determined eight times by the Multiple Sleep Latency Test (MSLT) (Carskadon et al., 1986).

Descriptions of these tasks are given here:

1. *The Performance Assessment Battery*: *a* computerized battery consisting of several tasks administered every 2 h during the 40 h awake period and lasting 8–12 min, depending on the response times of the subject to the stimuli presented. The tasks comprising the PAB were always presented in the same order during each of 19 testing sessions and are listed below.

   (a) *The Profile of Mood Scale (POMS)*: a subjective assessment checklist of 65 adjectives such as anger, fatigue, vigor, measuring current mood states based on a scale of 1 (not at all) to 5 (extremely).
   (b) *Serial Add/Subtract*: a mental arithmetic test in which two randomly selected digits followed by either a plus or minus sign, and then a prompt signal are sequentially presented. The subject must perform the indicated operation. If the result is a positive value, then the least significant digit must be entered. If the result is a negative value, then 10 must be added to this value and the result entered.
   (c) *Two-Choice Serial Reaction Time*: a modified Wilkinson Reaction Time test in which two 1/2-in. squares are displayed on the monitor. A red dot appears on one of the squares and the subject must press the corresponding keypad key, either 1 or 2. It was designed to determine if the 'learning curve' would be steeper and stabilize more quickly.
   (d) *Manikin*: a drawn human figure is presented holding an object. The subject determines which hand is holding the object from the perspective of the figure. The human figures may appear by front or back view and/or upright or inverted. The task was designed to investigate egocentric perspective transformations (Ratcliff, 1979).
   (e) *Ten-Choice Reaction Time*: a task that serves as control for the Serial Add/ Subtract test. The speed of response on the same keypad as used in the Serial Add/Subtract test is measured, but does not include the cognitive effort involved in the arithmetical operations portion.
   (f) *Running Memory*: the digits 1, 2, or 3 were presented randomly at 1 s intervals and the subject was required to temporarily memorize the current one but remember and enter the preceding one (See Babkoff et al., 1988, for discussion on memory tasks).

   The Serial Add/Subtract task will be the only task discussed at length here. Effects of other tasks in the battery have been published elsewhere (Babkoff et al.).

2. *Choice Visual Perception Task*: a visual perception task originally designed as a secondary task for pilots training on the Air Refueling Part Task Trainer (ARPTT) simulator to assess central and peripheral visual field awareness along the horizontal meridian in a high cognitive load, multitask environment. Attention was divided between performance of the task and active flight. The visual perception task display is a semicircular-shaped small-caliber tube in which light emission diodes (LEDs) are embedded at 15° intervals spanning from 75° left to 75° right of the center of the tube. The device was installed onto the instrument panel of an U.S. Air Force C-141 simulator cockpit linked to the ARPTT just below the simulated horizon and against the instrument panel's black background at a distance of 18 in. from the pilot's head to the center of the tube. The display presented 150 sequential single stimulus of a lit LED or dual-light stimuli just below eye level with 15 stimulus types presented 10 times each. Each stimulus was 0.25 s in duration and inter-stimuli intervals ranged randomly from 3 to 15 s. The presentation of two simultaneous stimuli provided the choice possibility of observing one, both, or none of the stimuli. For our study, this task was modified for laboratory testing and mounted on a table against a black wall background as shown in Fig. 14.2, and administered as a primary task instead of a divided attention task. The test required approximately 20–26 min to complete.

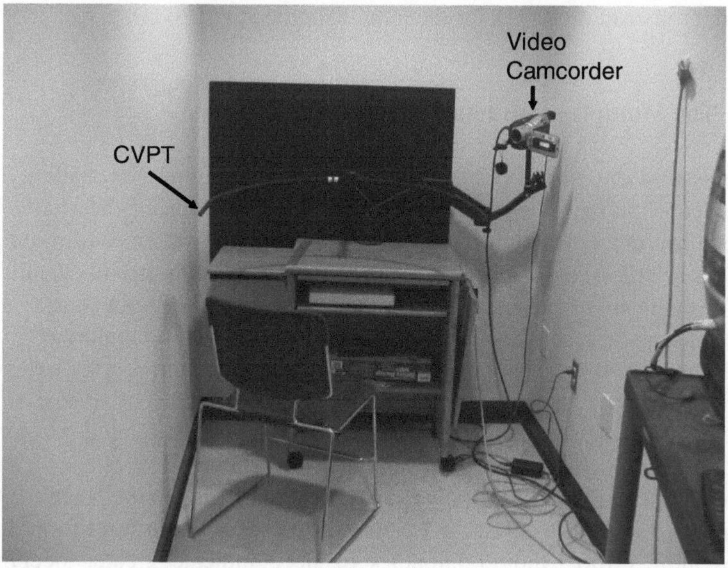

Fig. 14.2 Set-up of the choice visual perception task. The subjects focus on the white spots at the center of the arc and determine when and where LEDs (i.e., the stimuli) are lit by peripheral vision

3. *Psychomotor Vigilance Task*: the original Psychomotor Vigilance Task (PVT-192) is considered the "gold standard" by which other vigilance and reaction time tests are compared. Its operational simplicity and output of a straightforward response time provide immediate feedback of cognitive awareness. The test is administered on a hand-held device of approximately the size of a brick with two response buttons. A four digit numeric LED display is located at the top center of the device. After the test is initiated, the numeric LED display initiates a running count of elapsed time in milliseconds. The subject is instructed to press the preferred response button as soon as the digital display appears. This halts the counting and provides feedback in the form of numerical display of the reaction time. During scheduled sessions the test is performed for 10 min providing approximately 100 trials. The inter-stimulus interval is varied randomly from 1 to 5 s in 1 s increments.

A computerized version of the PVT, the PC-PVT, replicates the manual PVT using the monitor and keyboard of the Personal Computer (PC) to provide visual (scrolling numeric digits) and tactile (buttons) components. The advantage of the PC-PVT, like the other PC-based PAB measurements, is the registering of a strobe light indicating stimulus/response events that are recorded on a digital channel of the MP150.

In addition, handheld PDA versions of the PVT task, the Palm PVTB and Palm PVTC were developed and tested in this study (see Table 14.2). Because of their compactness and portability, the potential of utilizing these test instruments in field operations for rapid assessment of alertness would be advantageous. The Palm PVTs were designed to run at half the time length of the standard PVTs, i.e., 5 min for the former instead of 10 min for the latter (Thorne et al., 2005).

14.4.1.5 The Multiple Sleep Latency Test (MSLT)

Objective assessment of the current level of subject sleepiness is made by the Multiple Sleep Latency Test administered several times during the study. The subject reclines on a comfortable bed in a darkened, sound-attenuated room and is instructed to fall asleep if possible. At the beginning of the study, 20 min are allowed for the test, but as sleep deprivation cumulates resulting in shorter sleep latencies, subsequent tests are restricted to 15 min to prevent premature recovery sleep and ensure a more accurate measure of actual sleepiness over extended wakefulness. Post-hoc scoring of sleep latency to the first 30 s of stage 1 sleep, and to the first 30 s of stage 2, are recorded off-line from the digitized polysomnograms, using the Rechtschaffen and Kales scoring scheme. Both Stages 1 and 2 are counted as sleep in the quantitative summary, the distinction between the two stages being that arousal from Stage 1 is not difficult whereas arousal from Stage 2 requires more effort. Shorter latencies to sleep indicate greater sleepiness (Carskadon et al., 1986) and sleep latencies of less than 5 min duration are considered pathological.

14.5 Analysis Procedures

14.5.1 EEG

EEG recorded during test sessions are delineated from the raw EEG data files and analyzed on a second to second basis, that is to say, the unit epoch of analysis is 1 s and not 20 or 30 s as in conventional sleep stage scoring. Each second of the digitized EEG, is converted from the time domain as recorded to the frequency domain by the mathematical procedure of the Discrete Fourier Transform detailed in Smith (1997). The result is a spectrum of frequencies and their respective magnitudes or energy, i.e., a spectral density mapping of 500 frequency components. This is the consequence of sampling the digitized EEG signals at 1,000 Hz, which according to the Nyquist limit theorem (Smith, 1997), allows only half that number of frequencies as output. We speak of the original EEG signals, i.e., waveforms, shown in Fig. 14.1, as being in the time domain and the spectrum of frequencies shown in Fig. 14.3, as being in the frequency domain.

The 500 frequency components are summed to obtain a Total Amplitude for each second, a value that will be referred to whenever validity of EEG for a particular second is questioned. This will be discussed in the Results. The amplitude/energy values of the frequencies obtained are combined through simple addition into the following bands: 1–15 Hz; 16–50 Hz; 51–100 Hz; 101–200 Hz; 201–300 Hz; 301–400 Hz; 401–500 Hz.

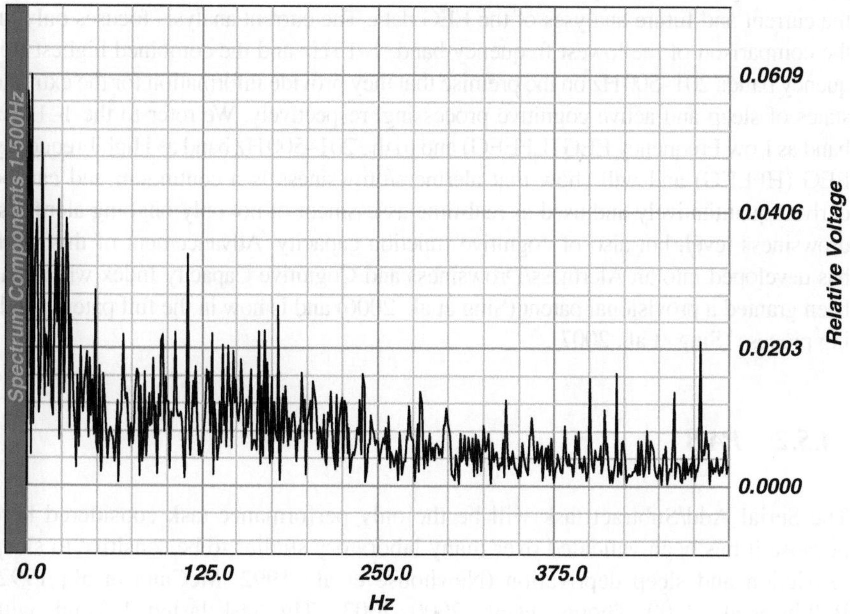

Fig. 14.3 Spectrum of frequency components derived from the Discrete Fourier Transform of the EEG signals sampled at 1,000 Hz. The frequencies from 1 to 500 Hz are represented on the abscissa and their respective magnitudes are given as relative voltage on the ordinate

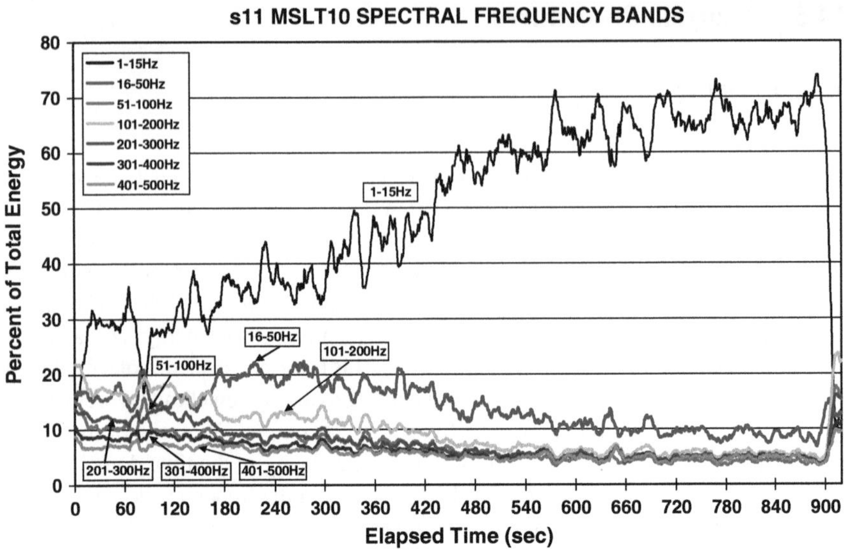

Fig. 14.4 The partition of the frequency components into bands and their changing amplitudes on a second by second basis recorded during a Multiple Sleep Latency Test (MSLT)

An example of the partition of these frequency bands is shown in Fig. 14.4. These particular partitions may seem arbitrary, but have an underlying rationale for the current and future analyses of the EEG data. The current analysis focuses only on the comparison of the lowest frequency band, 1–15 Hz and the combined highest frequency bands 201–500 Hz on the premise that they provide information for the extreme states of sleep and active cognitive processing, respectively. We refer to the 1–15 Hz band as Low Frequency EEG (LFEEG) and to the 201–500 Hz band as High Frequency EEG (HFEEG) and will show that alertness/drowsiness is a continuum and can be derived quantitatively and used as real-time assessment of not only ongoing alertness/drowsiness level, but also of cognitive function capacity. Advancement of this work has developed into an Alertness/Drowsiness and Cognitive Capacity Index which has been granted a provisional patent (Sing et al., 2006) and is now in the full patent pending process (Sing et al., 2007).

14.5.2 PAB

The Serial Add/Subtract task will be the only performance task considered here because it has been validated over many laboratory studies to be sensitive to sleep restriction and sleep deprivation (Newhouse et al., 1992; McCann et al., 1992; Balkin et al., 2000; Thomas et al., 2000, 2003). This task lasted 2–3 min, with stimulus/response markers recorded in synchrony with the EEG. Test parameters of accuracy and mean response time are calculated for each session and will be compared with the EEG to show changes between the rested and sleep deprived conditions for those sessions discussed.

14.5.3 CVPT

The CVPT provides a wealth of data output due to its complexity and varied and multiple stimuli types. Among the measures obtained from this task are: (a) accuracy in responding to the 150 stimuli, (b) number of incorrect responses, (c) lapses defined as non-response ≥ 3 s, (d) mean, and (e) median response times. These measures may in fact be also calculated as subsets for each of the 11 stimuli types. We will, however, discuss only the overall performance in the test session in comparison with its corresponding EEG.

14.5.4 PVT

The original PVT task was developed and validated on the PVT-192, a brick-sized handheld stand-alone device. The data results were stored internally in the device and downloaded to a computer after completion of testing session for further analysis. Dr. Thorne of our laboratory has developed a computerized version of the PVT, the PC-PVT. Following after the same principle of transmitting stimulus/response to the Biopac MP150 as in the Serial Add/Subtract task, PC-PVT performance can now also be correlated with EEG to track changes in alertness/drowsiness or cognitive function changes over sleep deprivation.

14.5.5 MSLT

The EEG data obtained in the MSLT sessions were not scored by the standard Rechtschaffen and Kales criteria. Instead, the results are given as comparisons between the LFEEG and HFEEG bands. Because the MSLT can clearly demonstrate the progression from awake to sleep in a sleep deprived individual, it should also be readily apparent that the concomitant EEG will objectively show the changes this process encompasses, in terms of both frequency changes and energy distribution between these bands. It would thus provide the basis for a quantitative index defining the different states in the awake to sleep continuum.

14.6 Results

Selected results from the study are presented. Due to random selection, the subject population of studies involving sleep deprivation usually consists of both resilient and non-resilient individuals. The individual variability in performance as a consequence of extended wakefulness is manifested by the changes in the response distributions of tests taken over the course of the study (Thorne, 2007). Cognitive performance test results and their corresponding EEG analysis for both a resilient and a non-resilient subject in the rested and sleep deprived conditions will be given.

14.6.1 MSLT

One of our interests in the inclusion of the MSLT in the study aside from estimating sleepiness level, was to capture boundaries of transition between the awake and sleep states in order to define a state of drowsiness. The conventional method for recording EEG is to sample the brain signals at 256 Hz, then exclude frequencies above 128 Hz by filtering, followed by down sampling to attain a final number of 64 sample points in each second. All these steps are necessary to avoid "aliasing" of frequencies and distortion of wave forms in the frequencies of interest. Since the Rechtschaffen and Kales sleep stage scoring system requires visual scoring, any distortion or artifactual enhancement of the wave forms compromises the score results.

The EEG in our study is analyzed at 1,000 sample points per second which provides not only an almost 16-fold higher resolution, but also the means to establish an association between cognitive activity and higher frequencies *vs.* association of drowsiness/sleep characteristics at the lower frequency end. The between state of drowsiness is not defined in the standard sleep stage scoring in polysomnography. This omission may be due to the low resolution of 64 sample points per second and the limitation of the visual/manual scoring technique which evaluates EEG epochs of 30 s as the unit size in determining sleep stages. Subtle changes in the EEG waveforms may not be easily detected when scanning visually across this relatively long (as compared to 1 s) period of time. The EEG data from our study are analyzed using 1 s epochs so that gradual transitions from the HFEEG of the awake state to the LFEEG of the asleep state are observed aiding in detection of the between state of drowsiness. Graphs best illustrate the difference between the HFEEG and LFEEG and provide the rationale for using comparisons between them to show the shift from awake to sleep state, i.e., from the actively cognitive processing state to drowsiness to sleep. This is supported by the EEG waveforms showing the transition from higher frequency (illustrated in Fig. 14.5a) during an MSLT when well rested to lower frequency (illustrated in Fig. 14.5b) during an MSLT after continuous sleep deprivation. Recall that EEG waveforms are represented in the time domain and their transform to the frequency domain allow synthesis of the HFEEG and of the LFEEG bands.

Results from two MSLTs of a subject taken in the rested and in the sleep deprived conditions compare the differences of the calculated HFEEG and LFEEG bands as shown in Fig. 14.6a and b, respectively. The abscissa displays the elapsed time in seconds from initiation of the test to termination (20 min in length) and the ordinate displays the relative proportion of the total energy for the HFEEG and

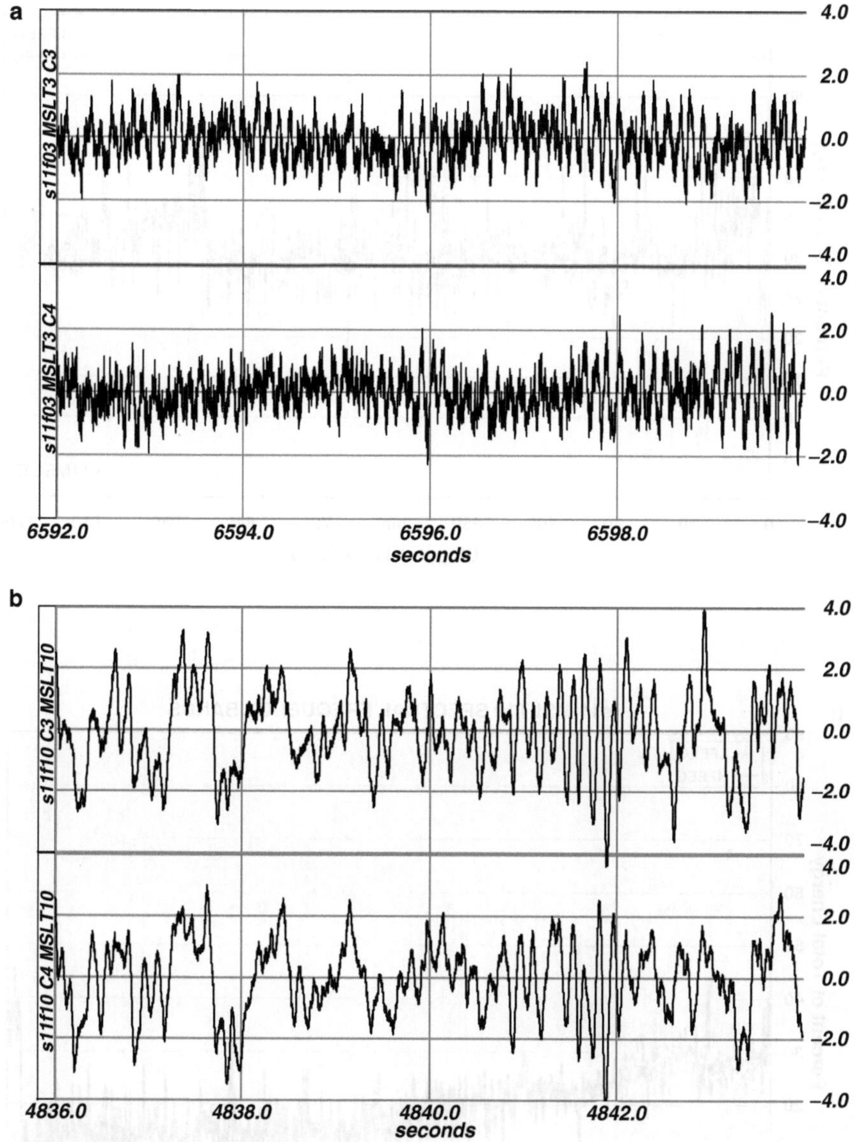

Fig. 14.5a and b Eight second snapshots of the EEG during MSLT in the rested condition for the resilient subject (**a**) and in the sleep deprived condition (**b**). Note the higher frequency range in the rested EEG waveform in contrast with the lowest frequency range for the sleep deprived condition

LFEEG in each second. These plots of the direct output from second by second calculation of the HFEEG and LFEEG bands do not show constancy, but rather continuous oscillations of the frequency band values. This makes it difficult to

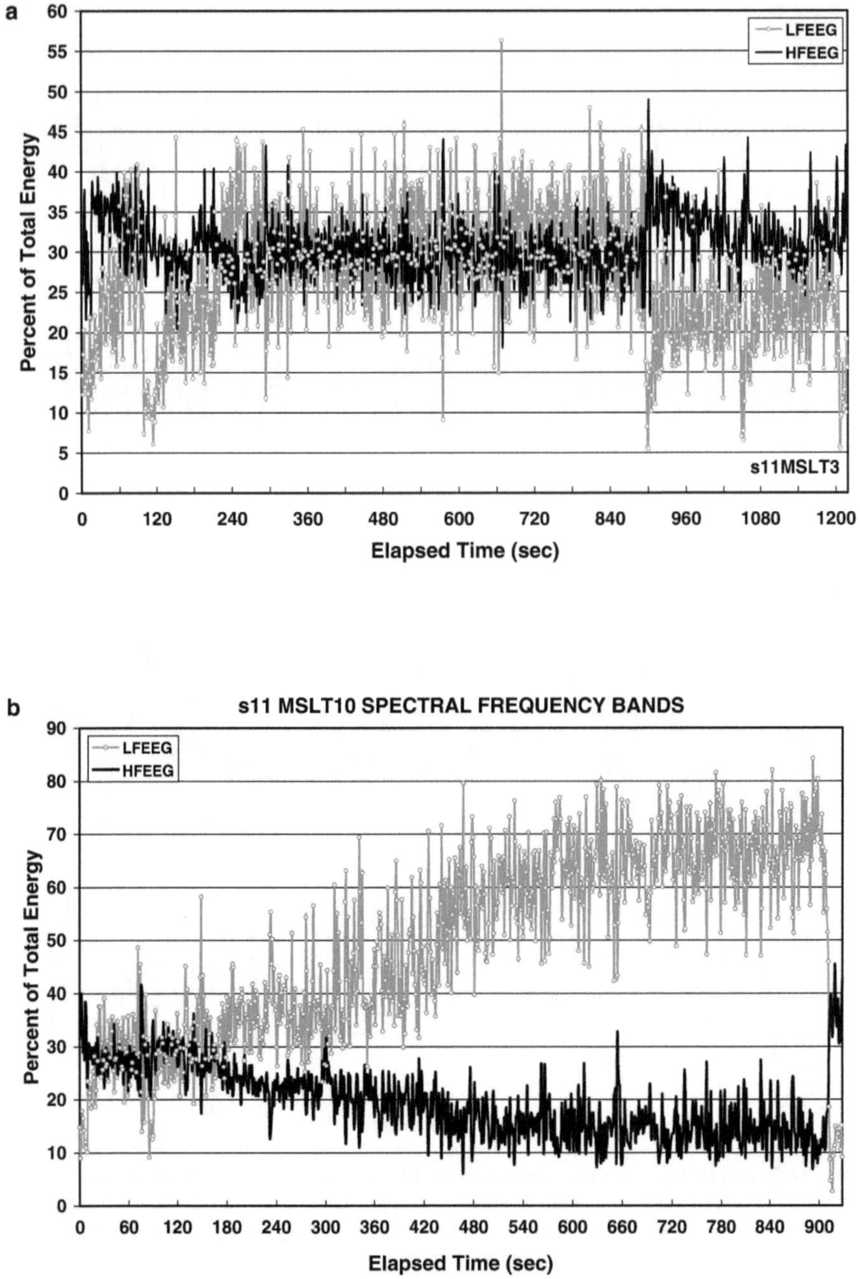

Fig. 14.6a and b Results of Fourier transform of the EEG recorded in the MSLTs of the resilient subject during the rested condition (**a**) and during the sleep deprived condition (**b**). Shown here are the direct second by second output of the calculated HFEEG band (201–500 Hz) and the LFEEG band (1–15 Hz). Note the rapid amplitude oscillations from 1 s to the next, makes detection of trends difficult as the test progresses. The solution is shown in Fig. 14.7a and b

visualize and assess the trend of each frequency band instantaneously. To resolve this difficulty, a moving average function was applied to each frequency band output to "smooth" the values. The moving average function distributes the energy of the *n* number of seconds chosen in each moving average step by initially averaging the first *n* values of the original frequency output to obtain the first value of a new output series. Each subsequent moving averaging step excludes the first value of the previously calculated *n* values of the original frequency output and then averages the next *n* values of the original frequency output to obtain the sequence of the new output series. This process results in the decrease of *n* values in the new output series, but improves visualization of the nature of the output and at the same time not change the inherent energy content of the output. The moving average function shown in our results here uses $n = 10$ (i.e., 10 s). This is a pragmatic selection, but is practical and addresses the need for assessing drowsiness with sufficient warning before sleep prevails. However, any reasonable *n-point* may be used depending on the degree of smoothing desired and what the objective is. Figures 14.7a and 14.7b illustrate the result of the moving average smoothing.

In the rested condition, the HFEEG and LFEEG bands share approximately equal energy, that is, they contribute in equal amounts to the total energy of that second as shown in Fig. 14.7a. The energy distribution of the EEG in the sleep deprived MSLT, however, is quite different. Although both HFEEG and LFEEG are equal in energy at the start of the test when the subject is awake, the LFEEG gradually increases in proportionate energy until it dominates the energy distribution at the termination of the MSLT as shown in Fig. 14.7b. The corresponding differences in their respective EEG waveforms have been shown in Fig. 14.5a and b respectively. The EEG of the rested condition consisted mainly of high frequency waveforms while that of the sleep deprived is mainly delta waveforms of 1–4 Hz, the polysomnographic criterion of Stage 4, the deepest sleep stage.

The sleep deprived subject is unable to maintain wakefulness and the progression to deep sleep is not a long process here. Along the way, the transitions from awake through the different sleep stages are tracked by the increasing proportion of the low frequency band in sequential seconds.

The above MSLT results were from a resilient subject whose cognitive performance tests showed little deficit even during sleep deprivation. The following MSLT results are from a subject best described as non-resilient to sleep deprivation. The MSLT for this subject during the height of sleep deprivation shows immediate transition to the sleep state within 1 min of the initiation of the test, well within the pathological criterion. This is indicated by the LFEEG steadily increasing in proportion of energy while the HFEEG is simultaneously decreasing as the test progresses until the LFEEG dominates at approximately 70–80% of total energy. Figures 14.8a and b compare the rested and sleep deprived MSLT results, respectively, of this non-resilient subject.

HFEEG and LFEEG bands derived from the frequency domain provide efficient, accurate and systematic assessment of the sleep/wake process at 1 s epoch or greater time length resolution. This methodology does not supplant visual/manual sleep stage

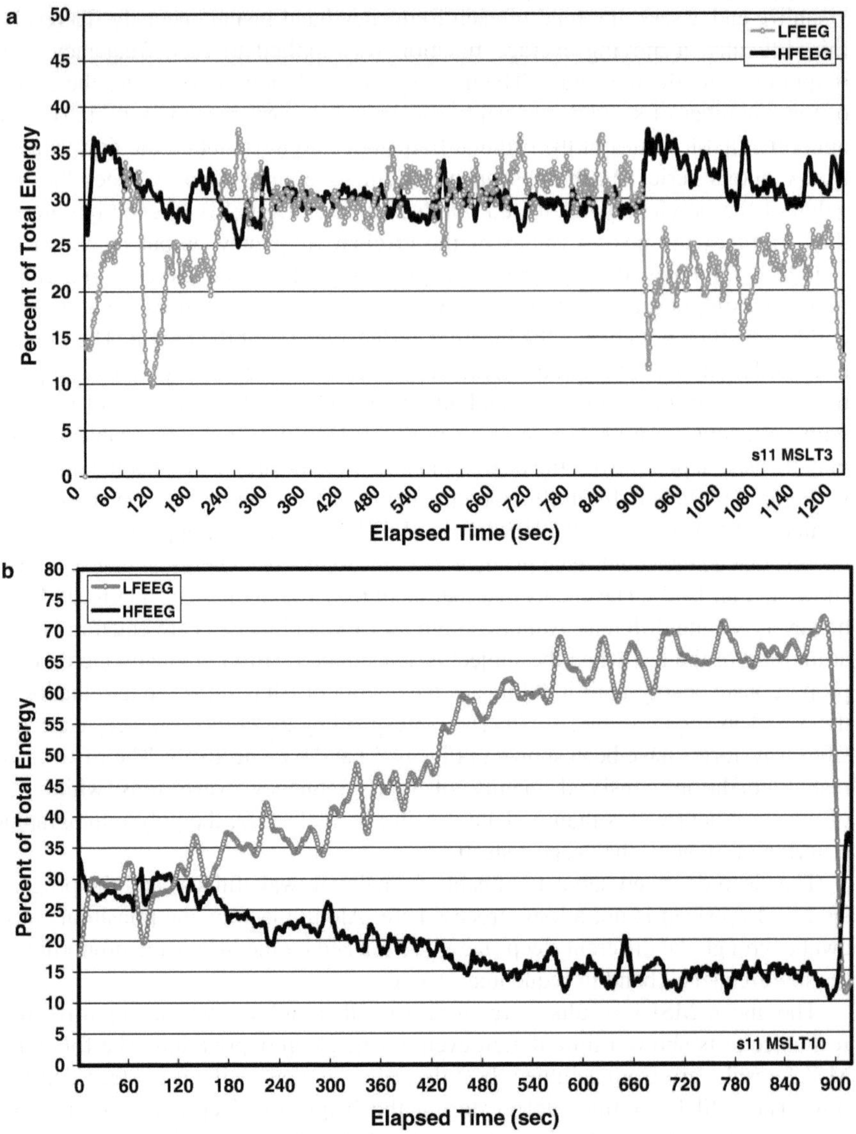

Fig. 14.7a and b A 10-point moving average function applied to the calculated outputs of the HFEEG and LFEEG bands (shown in Fig. 14.6a and b) results in the "smoothing" of these bands and allows improved visualization of trend detection. This is shown in (**a**) for the rested condition and in (**b**) for the sleep deprived condition of the resilient subject

scoring of time domain EEG if characteristic waveforms are sought and required for specific identification of sleep traits. Even this pursuit may be helped by examination of the second by second EEG sampled at 1,000 Hz in which the EEG can change from LFEEG in 1 s to HFEEG in the next second as shown in Fig. 14.9.

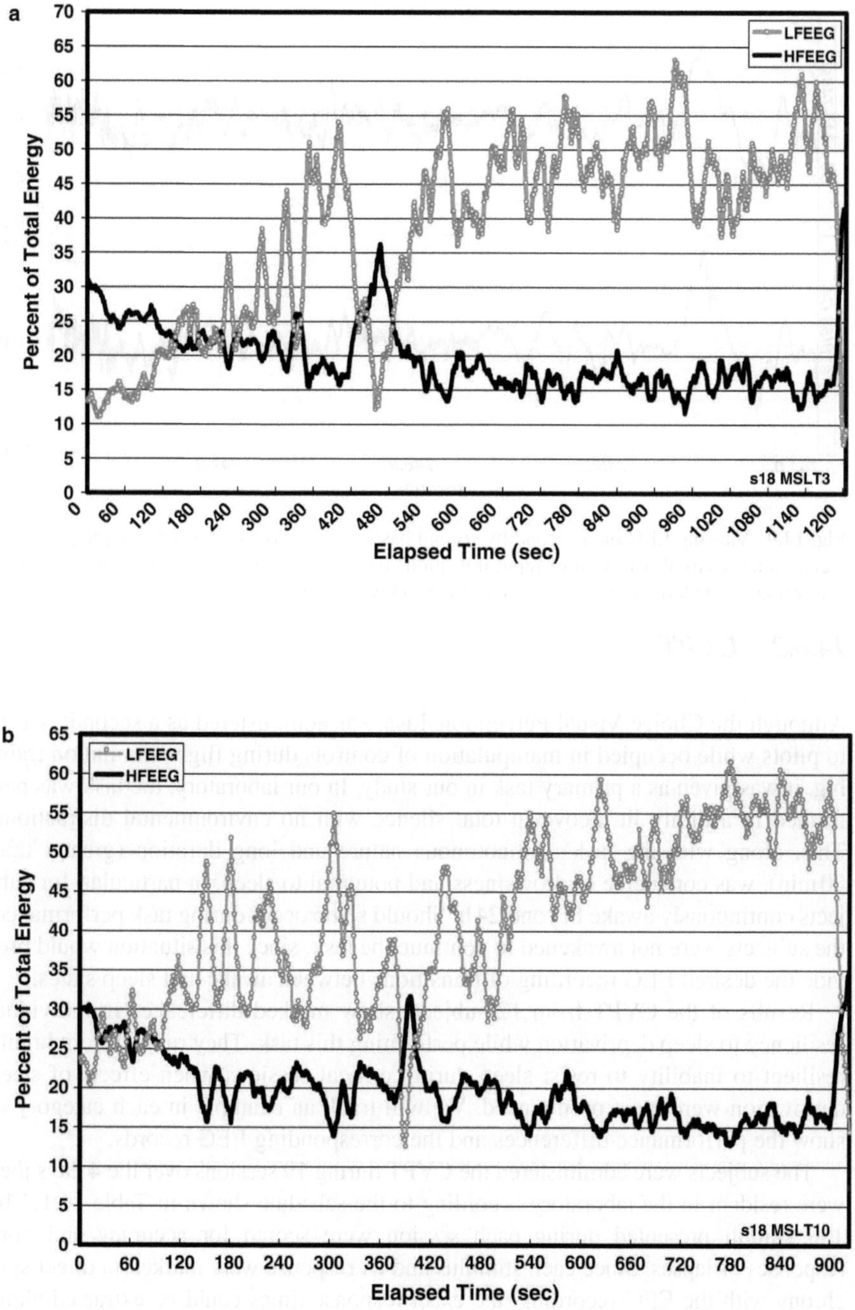

Fig. 14.8a and b The same smoothing process applied to the HFEEG and LFEEG derived from the MSLTs of the non-resilient subject in the rested condition (**a**) and in the sleep deprived condition (**b**). This subject's progression to the sleep state in both conditions began within 2 min of the MSLT

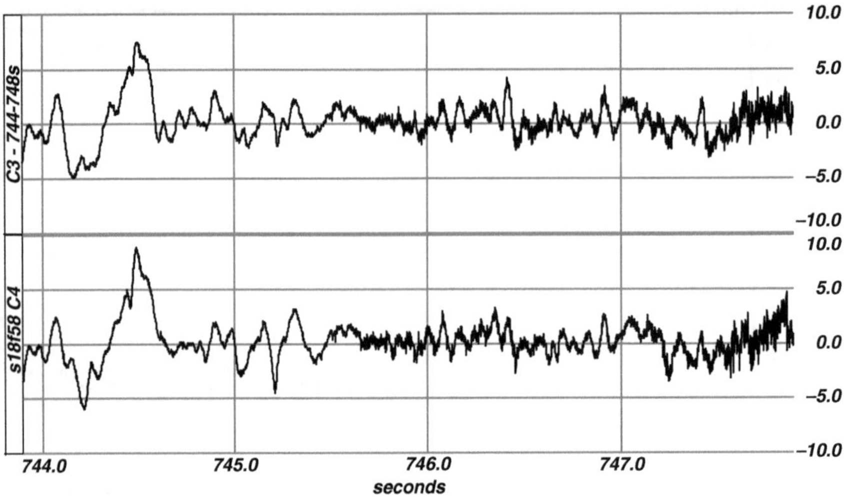

Fig. 14.9 Viewing EEG on a second by second basis at a resolution of 1,000 sample points per second allows visual detection of rapid transitions between low and high frequencies as well as discerning the underlying fine structure of the EEG waveforms

14.6.2 CVPT

Although the Choice Visual Perception Task was administered as a secondary task to pilots while occupied in manipulation of controls during flight simulation training, it was given as a primary task in our study. In our laboratory, the task was performed in a dimly lit alcove in total silence with no environmental distractions. This, along with the task's monotonous nature and long duration (greater than 20 min), was conducive to drowsiness and potential to sleep, in particular, for subjects continuously awake beyond 24 h. Should sleep occur during task performance, the subjects were not awakened to continue the task, since this situation would provide the desired EEG recording of transitions between awake and sleep states.

Results of the CVPT from 12 subjects show marked differences in individual resiliency to sleep deprivation while performing this task. They ranged from highly resilient to inability to resist sleep during critical sessions when effects of sleep deprivation were most pronounced. We will track an example in each category to show the performance differences and the corresponding EEG records.

The subjects were administered the CVPT during 19 sessions over the 4 days they were resident in the laboratory according to the schedule shown in Table 14.1. The 150 stimuli presented during each session were scored for accuracy and non-responses or lapses. Since each stimulus and its response were marked in direct synchrony with the EEG recording, the exact response times could be extracted along with the concomitant EEG. This will be shown in later figures. The changes in these scores and response times will be seen to relate closely with cumulative sleep loss. Tables 14.3a and b compare the overall performance of the two subjects. Most noteworthy is the number of lapses commited by the non-resilient subject compared with

Table 14.3a CVPT performance for sleep deprivation resilient subject

Subject 11

Test #	Time	Mean response time (ms)	Median response time (ms)	# Lapses	Accuracy (%)
1	11:00	1,448	1,350	1	98.7
2	13:30	1,201	1,180	0	100
3	15:30	1,283	1,300	0	99.3
4	19:30	1,120	1,095	0	100
5	23:30	1,116	1,050	1	97.3
6	9:30	1,082	1,040	1	98
7	13:30	1,131	1,095	1	98
8	17:30	1,107	1,070	0	99.3
9	21:30	1,035	1,025	0	100
10	1:30	1,134	1,045	4	95.3
11	5:30	1,276	1,135	6	95.3
12	9:30	1,317	1,160	10	90
13	13:30	1,191	1,140	2	97.3
14	17:30	1,206	1,140	2	98
15	21:30	1,283	1,260	1	98.7
16	23:30	1,350	1,275	3	98
17	11:30	1,282	1,240	0	98
18	15:30	1,259	1,250	0	98.7
19	17:30	1,246	1,230	1	98.7
	AVERAGE	*1,215.2*	*1,162.1*	*1.7*	*97.8*
	STDEV	*105.68*	*98.48*	*2.56*	*2.31*
	STERR	*24.25*	*22.59*	*0.59*	*0.53*

Highlighted test numbers are the sessions most affected by sleep deprivation

Table 14.3b CVPT performance for sleep deprivation non-resilient subject

Subject 18

Test #	Time	Mean response time (ms)	Median response time (m)	# Lapses	Accuracy (%)
1	11:00	1,948	1,895	4	94.7
2	13:30	1,537	1,420	7	91.3
3	15:30	1,521	1,435	1	96.7
4	19:30	1,570	1,440	7	91.3
5	23:30	1,272	1,195	1	95.3
6	9:30	1,530	1,270	9	91.3
7	13:30	1,417	1,245	13	85.3
8	17:30	1,322	1,260	3	96.0
9	21:30	1,134	1,070	0	99.3
10	1:30	1,253	1,190	0	98.0
11	5:30	2,741	2,980	125	13.3
12	9:30	2,864	2,980	140	6.7
13	13:30	2,856	2,980	138	7.3
14	17:30	2,727	2,980	121	19.3
15	21:30	1,652	1,510	12	90.7
16	23:30	1,361	1,235	3	96.7
17	11:30	1,352	1,285	2	94.0
18	15:30	1,002	940	0	96.7
19	17:30	1,104	1,075	0	98.0
	AVERAGE	*1,692.8*	*1,651.8*	*30.8*	*76.9*
	STDEV	*623.68*	*732.35*	*53.42*	*34.88*
	STERR	*143.08*	*168.01*	*12.26*	*8.00*

Highlighted test numbers are the sessions most affected by sleep deprivation the resilient subject. Babkoff et al. (1985), have noted the effects of sleep deprivation with respect to response lapses and response accuracy in cognitive testing. We will look in detail at a rested performance versus a sleep deprived one for each subject.

The spectral output of the energy proportions represented by the LFEEG and HFEEG bands show that in the rested state, the HFEEG band is always greater than the LFEEG for the resilient subject and approximately equal for the non resilient subject as shown in Fig. 14.10a and b.

The spectral values for the sleep deprived state, however, reveal marked differences between the subjects. Quantitatively, the level of the HFEEG band for the resilient subject is consistently greater than that of the LFEEG bands as shown in Fig. 11a. The spectral output of the non resilient subject (Fig. 14.11b) shows the level of the LFEEG band gradually increasing as the task proceeds until it is 80% of the total energy.

The EEG signals recorded during this task in the rested condition as shown in Fig. 14.12a and b for the resilient and non resilient subject, respectively, show that the HFEEG band is prominent in both subjects and that the response time to the stimulus presented was similar for each. In the sleep deprived condition, the EEG signals of the resilient subject continue to indicate active cognitive functioning as represented by the presence of high frequency waveforms (Fig. 14.13a) and response to stimulus albeit requiring increased response time. On the other hand, Fig. 14.13b, reveals the EEG signals of the non resilient subject in the deepest stage of sleep showing 1 and 2 Hz waveforms with high amplitudes and no response (i.e., lapse) to the stimulus presented.

The EEG signals captured synchronously while subjects performed the CVPT task corroborate not only the alertness/drowsiness states, but also the capacity for cognitive performance. Quantitatively, it signifies the greater proportion of energy vested in the HFEEG relative to the LFEEG. When HFEEG is prevalent, the subject is actively performing the task and conversely, when LFEEG energy predominates, the subject is no longer able to perceive and act cognitively but is essentially in a drowsy state tending towards sleep if not already asleep.

14.6.3 PAB – Serial Add/Subtract Task

The CVPT is a task that requires vigilance and attention in the detection of randomly lit LEDs embedded in a semi-circular arc. The Serial Add/Subtract task involves not only attention, but also engages multiple brain regions in arithmetic processing, selection of correct numeric key and its location in response as set forth by Thomas et al. (2000, 2003). Tasks which are more cognitively challenging and shorter in duration appear to provide greater resilience to effects of sleep deprivation, even for those individuals who are susceptible and have little resistance against sleep induction during the CVPT. The results of the Serial Add/Subtract test, occupying the first 2–3 min of the PAB task, appears to illustrate this premise. The results of the resilient subject are shown in Fig. 14.14a and b and of the non resilient subject in Fig. 14.15a and b, during rested condition and sleep deprived conditions, respectively. For the resilient subject, the HFEEG proportion of total

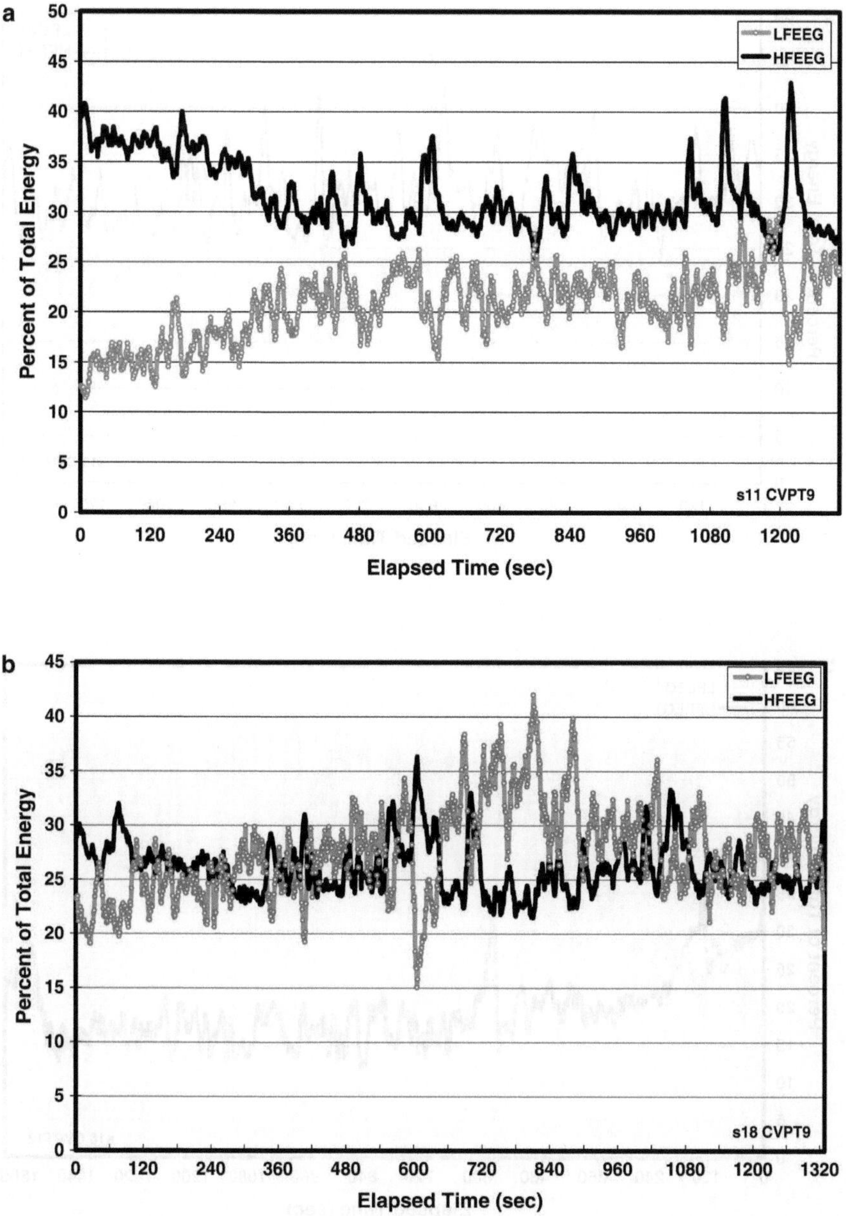

Fig. 14.10a and b The HFEEG and LFEEG bands from the CVPT test during the rested condition for the resilient subject (**a**) and the non-resilient subject (**b**). The HFEEG of the resilient subject is consistently greater than the LFEEG indicating maintenance of alertness and cognitive function capacity to perform from beginning to end of test session with 100% accuracy. The non-resilient subject showed signs of flagging alertness midway through the test session, indicated by the greater LFEEG proportion than HFEEG, but still retained sufficient cognitive capacity to perform at 99% as shown in Table 14.2

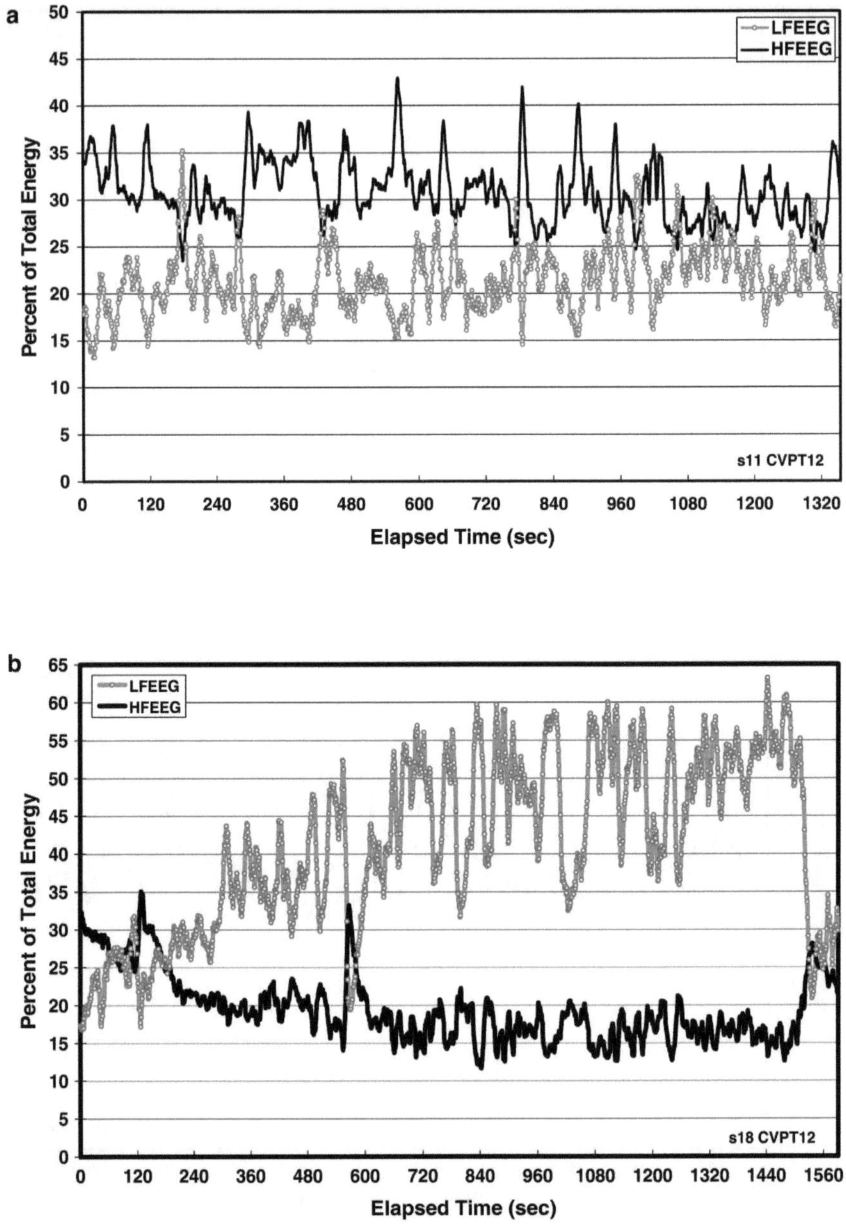

Fig. 14.11a and b The HFEEG and LFEEG bands from the CVPT test during the sleep deprivation condition for the resilient subject (**a**) and the non-resilient subject (**b**). The resilient subject is still capable of maintaining sufficient cognitive function capacity to perform at 90% accuracy (see Test 12, Table 14.3a) indicated by the HFEEG consistently greater than the LFEEG. The non-resilient subject is not able to maintain wakefulness and his LFEEG increases resulting in sleep. This is borne out by his performance in missing 140 of the 150 stimuli resulting in an accuracy of only 6.7% (see Test 12, Table 14.3b)

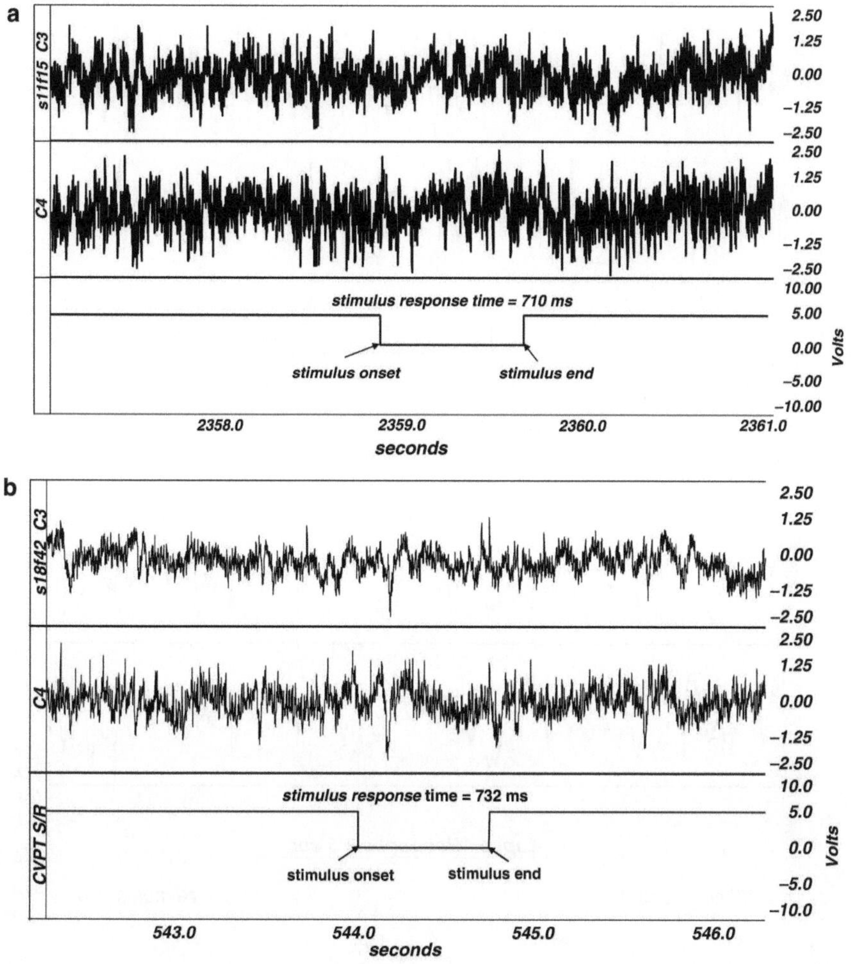

Fig. 14.12a and b Four second EEG snapshots during the rested condition of the resilient subject (**a**) and non-resilient subject (**b**) showing an example of a stimulus and response recorded on the digital channel in synchrony with the EEG. Both subjects show EEG waveforms in the high frequency range and approximately the same response time to the stimulus

energy is consistently greater than that of the LFEEG during the entire Serial Add/Subtract test in both conditions, indicating that the subject was actively performing. The only difference in performance between the two conditions is subtle as shown in the slightly longer time (7 s) required to complete the test when sleep deprived. For the not so resilient subject, performance on the test was inconsistent when both rested and sleep deprived. The results show that the HFEEG proportion at the beginning of the test in both cases, is greater than the LFEEG, but for only a minute or so, then in general trends lower than the LFEEG with an occasional instant of

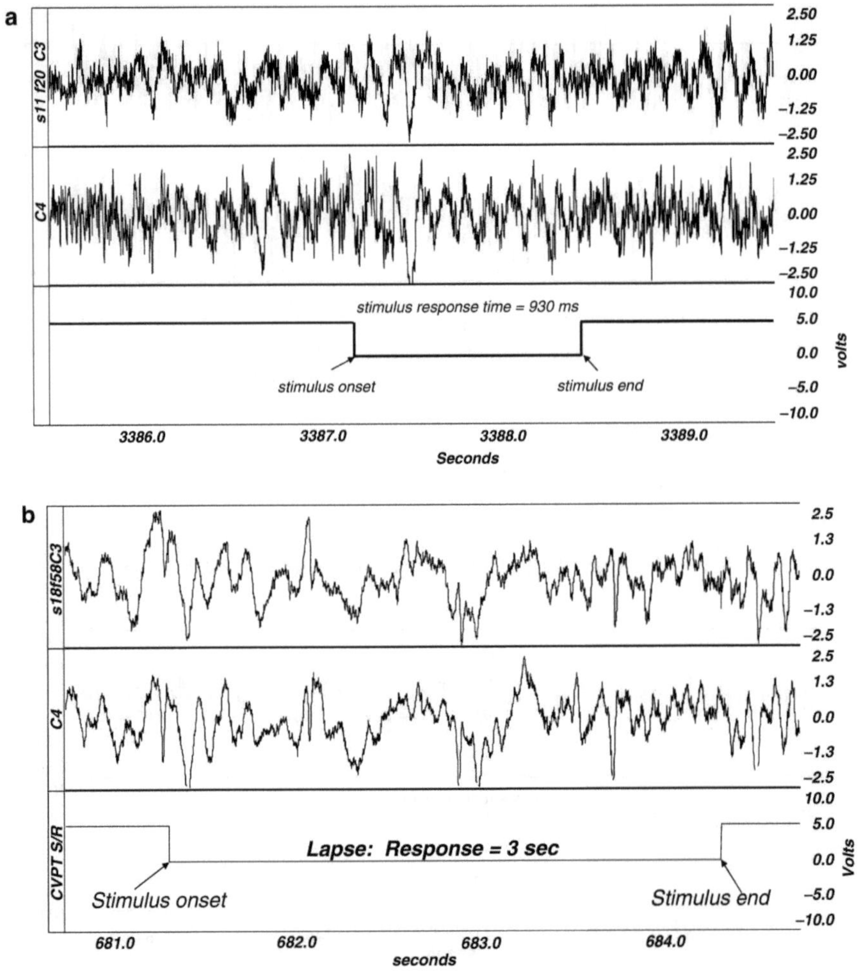

Fig. 14.13a and b Four second EEG snapshots during the sleep deprived condition of the resilient subject (**a**) and non-resilient subject (**b**) showing an example of a stimulus and response recorded on the digital channel in synchrony with the EEG. Both subjects show EEG waveforms in the lowest frequency range. For the resilient subject there is evidence of the presence of high frequencies seen on waveform peaks and cognitive capacity to perform with response time of less than 1 s. For the non-resilient subject, there was no resistance to the effects of sleep deprivation and sleep dominated the test session resulting in response lapses for almost all the stimuli presented The EEG waveforms in the delta range (1–4 Hz) confirm the sleep state of this subject

reversing to slightly greater magnitude than the LFEEG. During the last minute of the test, the LFEEG proportion of total energy is dominant, indicating that the subject is clearly in a drowsy state, but still attempting to respond to the test stimuli. The difference in performance between these two subjects is apparent in the test

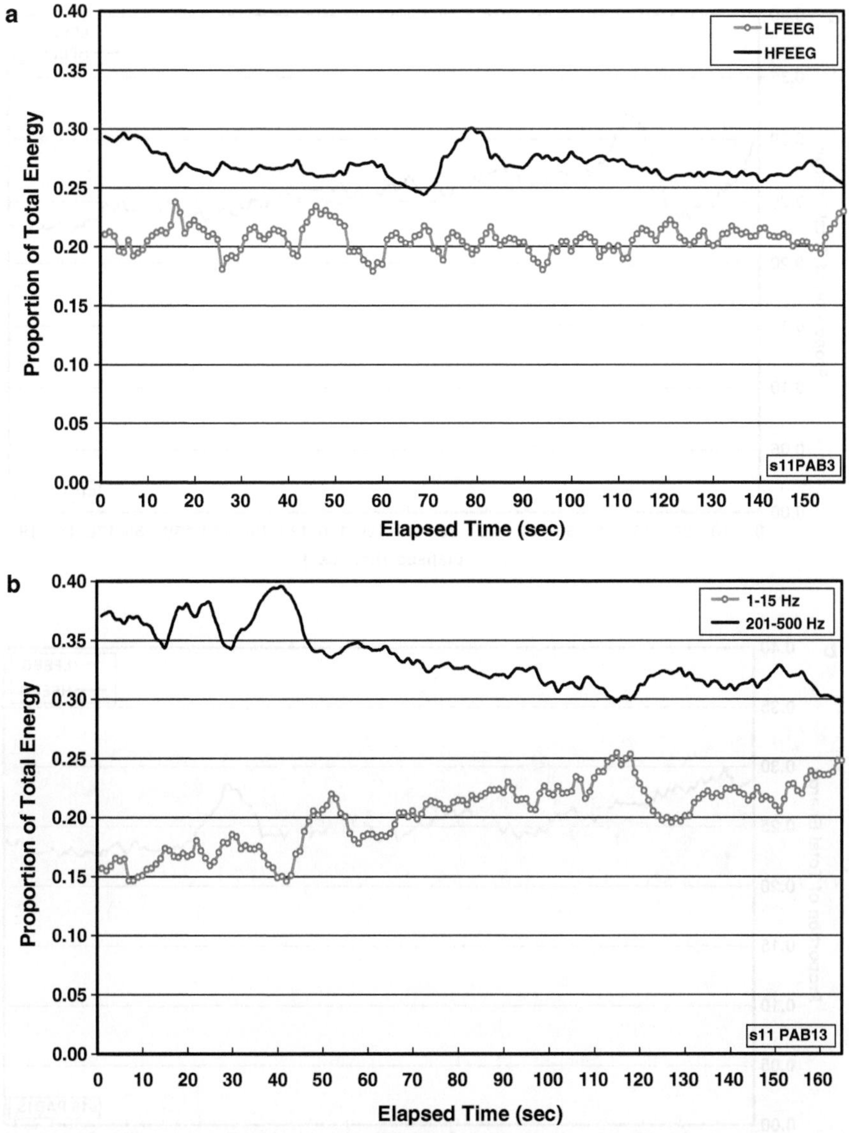

Fig. 14.14a and b The HFEEG and LFEEG bands from the Serial Add/Subtract test during the rested condition (**a**) and sleep deprived condition (**b**) for the resilient subject. In both conditions, the HFEEG is consistently greater than the LFEEG indicating again maintenance of alertness and cognitive function capacity to perform this 2–3 min test. The only difference between these two conditions for this subject is the slightly longer time to complete the test in the sleep deprived condition

durations. The resilient subject completed the Serial Add/Subtract test in approximately 2 1/2 min under both conditions, while the non-resilient subject required 40 s longer in responding to the same number of stimuli.

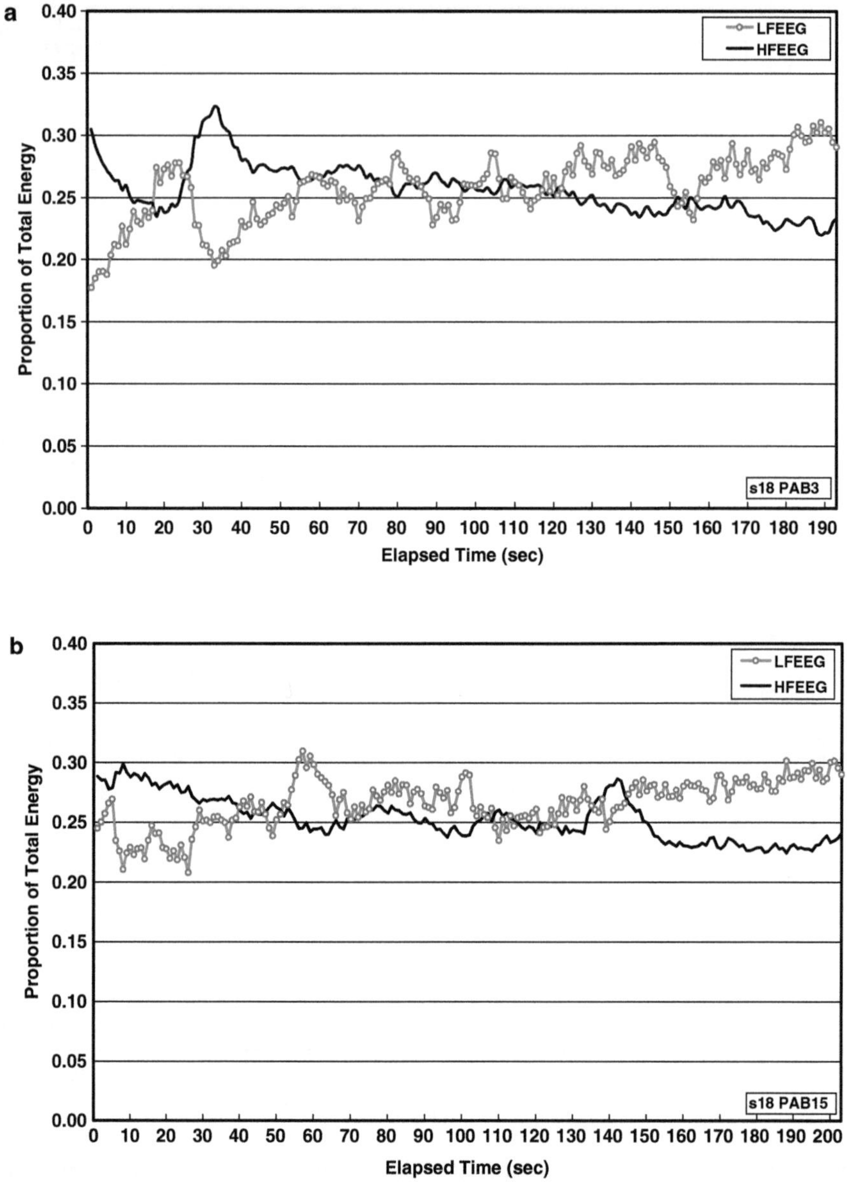

Fig. 14.15a and b The HFEEG and LFEEG bands from the Serial Add/Subtract test during the rested condition (**a**) and sleep deprived condition (**b**) for the non-resilient subject. In both conditions, the HFEEG and LFEEG bands are at approximately the same proportion of energy but distributed differently over the test interval. The HFEEG of the rested condition is greater than the LFEEG for the initial 2 min, then trends downward and is lower than the LFEEG for the remainder of the test. In the sleep deprived condition, the HFEEG is greater than the LFEEG for only 40 s and thereafter declines below the LFEEG. Again, completion of the test in the sleep deprived condition required slightly longer time than in the rested condition

14.6.4 PC – PVT

The computerized version of the PVT provides stimulus and response markers to the MP150 in the same manner as the CVPT and the Serial Add/Subtract task to allow measurement of exact response times as well as assessment of the concomitant EEG and other physiological signals during the stimulus and response intervals. This is one of the main differences with the original handheld device, the PVT-192, which records only a response time to internal memory storage. Another difference between the two PVT tasks is in the administration of the tasks. The subject is not constrained to a set sitting position in taking the PVT-192, although encouraged to do so. Consequently, shifting posture and/or changing the level at which the device is held can occur during periods of drowsiness in order to maintain wakefulness. The PC-PVT version of the task requires attentive focus on the fixed screen of the monitor and manual sensing of the proper response keys, i.e., without looking down at the keyboard.

Since the PC-PVT task is more comparable to the CVPT and Serial Add/Subtract task in its execution, results of this task rather than the PVT-192 will be presented here. Figures 14.16a and b show the rested and sleep deprived results respectively, for the resilient subject. In the rested condition, the proportion of total energy for the HFEEG band is greater than the LFEEG during the entire task indicating that the subject is awake, alert and with sufficient cognitive capacity to perform the task. On the other hand, this subject's performance while sleep deprived shows obvious decline in cognitive capacity in which both HFEEG and LFEEG are at approximately the same proportion of total energy or alternating between the two in showing increase or decrease relative to the other. This is brought out in terms of response speed expressed over the entire session. Although average response times were 321.8 and 337 ms respectively for rested and sleep deprived conditions, the span of response times over the 100 stimuli were quite different. For the rested condition response times ranged from 243 to 768 ms, while in the sleep deprived condition, the range extended from 231 to 2,932 ms, corroborating the relationship of the HFEEG and LFEEG bands in each condition.

The results of the non-resilient subject were surprisingly similar to the resilient subject for this task and are shown in Fig. 14.17a and b, for the rested and sleep deprived conditions respectively.

The HFEEG band was distinctly greater than the LFEEG in the rested condition, although showing a declining trend when both bands were equivalent for a very brief moment before the HFEEG resumed a consistent level to completion of task. On the other hand, in the sleep deprived condition, the difference between these two frequency bands narrows markedly to about equal proportions of energy with the HFEEG usually being slightly greater. The unusual spike at the 2.5 min point in the task is due to an "electrode pull" where the sensor wiring is temporarily disturbed (i.e., by head movement) and introduces a sudden high frequency signal masking the EEG. The mean response times for this subject during rested and sleep deprived conditions are respectively, 442.1 and 643.2 ms with the range of these

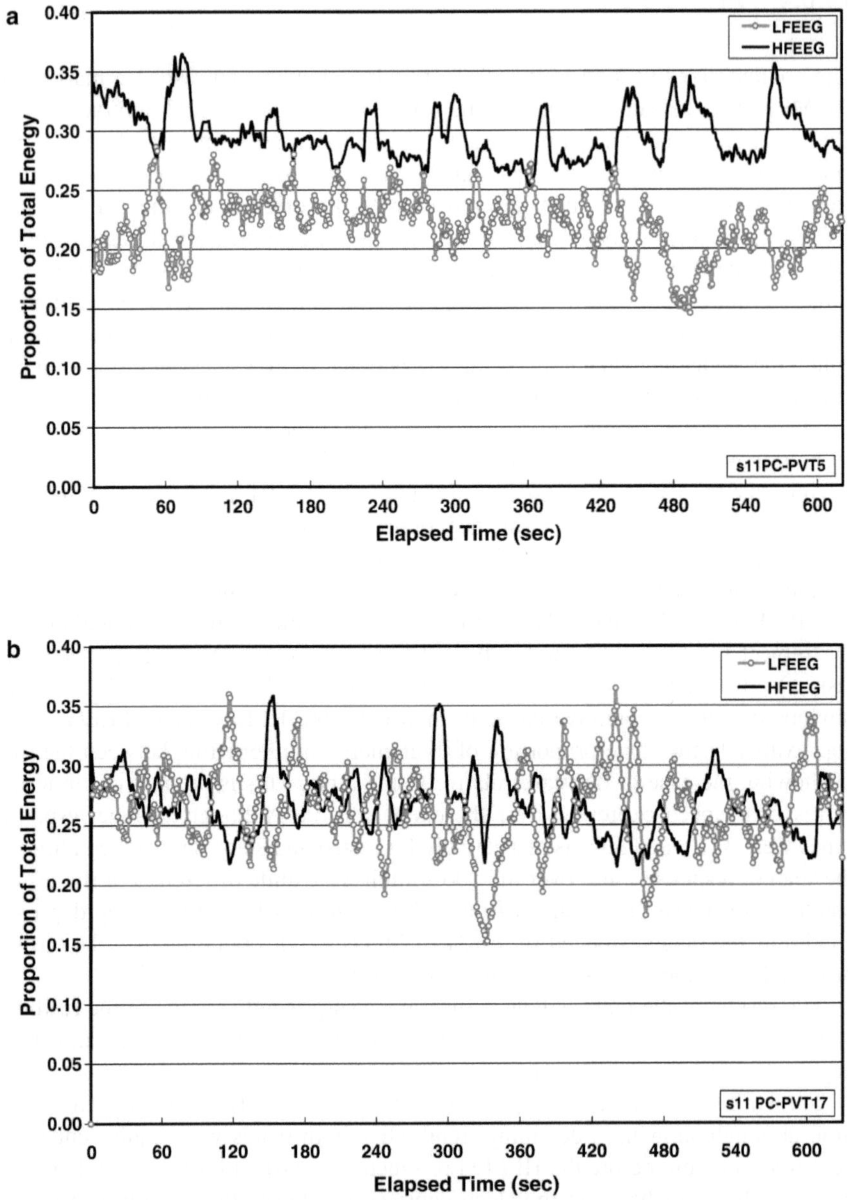

Fig. 14.16a and b The HFEEG and LFEEG bands from the PC-PVT test during the rested condition (**a**) and sleep deprived condition (**b**) for the resilient subject. Similar to the results of the Serial Add/Subtract test in the rested condition, the HFEEG is greater than the LFEEG during the entire test period confirming that subject is alert and performing. The sleep deprived condition presented a different result with both HFEEG and LFEEG at approximately the same level, but alternating with peak oscillations indicating a struggle to maintain cognitive capacity to perform simultaneously with a drive to drowsiness and sleep

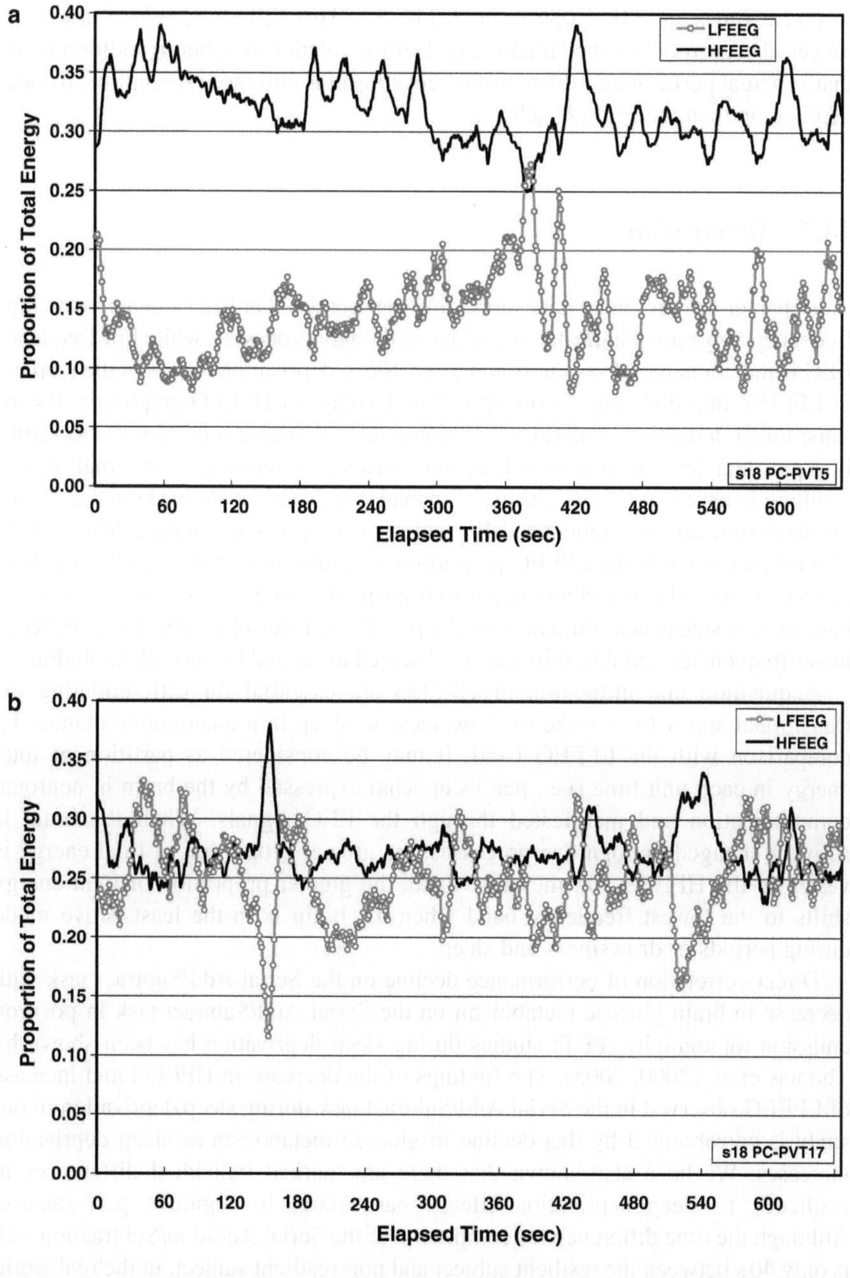

Fig. 14.17a and b The HFEEG and LFEEG bands from the PC-PVT test during the rested condition (**a**) and sleep deprived condition (**b**) for the non-resilient subject. In the rested condition, the HFEEG is consistently greater than the more depressed LFEEG energy proportion, confirming that the subject is awake and performing the entire 10 min of the test. In the sleep deprived condition, the HFEEG is slightly greater than the LFEEG with only a few instances when this is reversed with the LFEEG greater

response times as 238–1,056 ms and 248–3,883 ms. These response times and ranges also corroborate the HFEEG and LFEEG relationship, but in addition reveal that in actual performance the non-resilient subject is still more susceptible to sleep deprivation than the resilient subject.

14.7 Discussion

This chapter has presented data supporting the premise that EEG signals in the high frequency range are manifestations of active thought processes, while low frequency EEG is the domain of drowsiness and sleep. Sleep deprivation results in the increase of LFEEG amplitude and a corresponding decrease in HFEEG amplitude. By the same token, a decline in alertness, or conversely, an increase in drowsiness results in equivalent loss of cognitive function capacity to respond to external stimuli resulting in apparent lapses. The experimental data from our study show that during the sleep state all high frequency EEG activity is suppressed to a basal level of 0.20 of total energy while the LFEEG proportion of energy is at maximum level of 0.80 of total energy. The underlying hypothesis posits that the process of declining alertness to drowsiness and ultimately to sleep is the transfer of energy from higher to lower frequencies and this shift can be observed at second by second resolution.

Acquisition and utilization of HFEEG are essential for differentiating the transitional states from wake to drowsiness to sleep in a quantitative manner by comparison with the LFEEG band. It may be considered as partition of total energy in each unit time (i.e., per 1 s epochs) expressed by the brain in neuronal communication and manifested through the EEG signals. When the brain is actively engaged in cognitive processes, the greater proportion of total energy is vested in the HFEEG. On the other hand, the greater proportion of total energy shifts to the lowest frequency band when the brain is in the least active mode during periods of drowsiness and sleep.

Direct correlation of performance decline on the Serial Add/Subtract task with decrease in brain glucose metabolism on the Serial Add/Subtract task in positron emission tomography (PET) studies during sleep deprivation has been shown by Thomas et al. (2000, 2003). The findings of the decrease in HFEEG and increase of LFEEG observed in the Serial Add/Subtract task during sleep deprivation in our study is corroborated by this decline in glucose metabolism as sleep deprivation increases. We have also shown that there are marked individual differences in resiliency to sleep deprivation effects, particularly in cognitive performance. Although the time difference for completion of the Serial Addition/Subtraction task is only 40 s between the resilient subject and non-resilient subject, in the real world where decisions must be made on a moment's notice, this could be catastrophic. A span of 40 s is an unacceptable long pause when a critical life or death decision must be made in an instant. Other factors separating the performance of the resilient subject from the non-resilient subject may well include sensitivity to circadian desynchrony during sustained sleep deprivation (Babkoff, 1989).

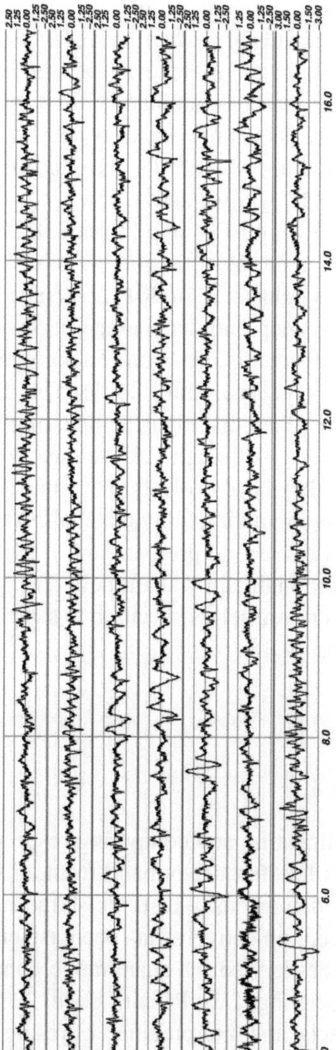

Fig. 14.18 A 72 s EEG snapshot of the non-resilient subject from the sleep deprived MSLT to illustrate the Cyclic Alternating Pattern (CAP) of high amplitude low frequency EEG alternating with low amplitude high frequency EEG

Among other advantages of EEG signals sampled at 1,000 Hz is the visual clarity of the EEG traces presented on a second by second scale. At the LFEEG end of the scale, frequencies of 1 to about 20 Hz can be deduced by just counting the number of waveform peaks in a second. Hence, it is immediately apparent that stage 3 sleep is prominent when the peaks number from 3 to 7, or stage 4 sleep is dominant when the number of peaks is only one or two. Although peaks cannot be enumerated for exact determination of HFEEG, the high density of peaks quite obviously point to the awake and cognitive active state.

There are other features in viewing EEG on a 1 s scale. In the polysomnographic interpretations of sleep EEG, a phenomenon known as Cyclic Alternating Pattern (CAP) is observed for scoring epochs of 20 or 30 s. It is considered a periodic EEG activity occurring in non-REM sleep and "is characterized by sequences of transient electrocortical events that are distinct from background EEG activity and recur at up to 1 min intervals" (Terzano et al., 2001). This EEG activity may be indicative of sleep instability or sleep disturbance appearing spontaneously, but it can also occur in sleep pathologies. Although CAP sequences were originally considered as arousal phenomena, they have evolved to encompass both the process of sleep maintenance and sleep fragmentation (Terzano et al., 2001). This pattern may also be observed in the second by second presentation of EEG signals in some of our subjects in futile attempts to maintain wakefulness during cognitive testing in the sleep deprivation phase, but nevertheless succumb to sleep at least for brief time periods. This is captured in Fig. 14.18.

14.8 Conclusion

The novel methodology of delineating and computing respective energy contribution of a high frequency and a low frequency band from an EEG spectrum width of 500 Hz to provide a quantitative means to distinguish the awake state and transition to drowsiness, then to sleep, objectively, has been established here. An alertness/drowsiness index has been derived by quantitative comparison of high and low frequency bands (Patent pending, U.S. Patent Office, 2007). The simplicity of computation, removal of artifacts, and interpretation of results lead to development of algorithms which may be incorporated in ambulatory monitoring systems for on-line acquisition and analysis of EEG signals for assessment of sleep, wake, or incipient drowsiness and a measure of cognitive function capacity. As a public safety measure, application of such a monitoring system would provide on-going assessment of personnel performance in sensitive positions described in the Introduction and allow immediate implementation of countermeasures to drowsiness or sleepiness.

Where the use of sensors necessitated in EEG acquisition is impractical, it has been proposed by Papadelis et al. (2007) that other surrogate measures may be used, such as oculometrics (e.g. eye blinks) from the EOG. These would also be objective measures but would require corroboration with the more direct EEG parameters. There are in fact systems currently exploring these possibilities. For example, Dr. William C. Torch (2007) has developed a system, the Eye-Com® (EYE-COM Corporation, 75 Pringle Way, Suite 906, Reno, NV 89502), that not only detects eye blinks, but also measures pupil diameter, percent of eyelid closure and saccadic velocity. Each of these measures is sensitive to sleep deprivation effects and their quantitative correlation with the HFEEG and LFEEG will be corroborated by direct experimental procedures. The physical make-up of the EYE-COM is an eyeglass frame on which are embedded the eye tracking light source and the accompanying electronics. As an unobtrusive and non-obstructive device, its application is intended for pilots and military personnel in field operations as well as truck drivers.

The accelerated rate of development in the technical sector including miniaturization of electronic chips provides a glimpse to realization of advance instrumentation of monitoring, computing, assessing and remote transmission of neuro- and electro-physiologic states from individuals of interest to decision makers for timely intervention to assure public safety as well as to prevent catastrophic events.

Acknowledgements This work was supported by the Military Operational Medicine Research Program of the United States Army Medical Research and Materiel Command (USAMRMC), Ft. Detrick, MD and the Walter Reed Army Institute of Research (WRAIR), Silver Spring, MD. Funding for the study was provided through a USAMRMC Broad Agency Announcement (BAA) grant.

We thank our volunteers for their participation in the study and the WRAIR enlisted military and contract student staff for their technical assistance. Participation of the volunteers and the technical assistance of the WRAIR enlisted military and contract student staff in this study are acknowledged with gratitude. The collaboration of senior investigators who made this study possible include COL Michael Russo, COL (RET) Daniel Redmond, Dr. Mary Kautz, Dr. David Thorne, and Stanley Hall and is gratefully acknowledged. Thanks are due the study managers, Kim Warren, Dagny Johnson and Gregory Lounsberry. Software to automate analysis of the EEG data and provide an integrated data base were designed and implemented by Joshua Bailey. This greatly accelerated the analysis process and is acknowledged with gratitude.

Department of Defense Disclaimer The contents of this article have been reviewed by the Walter Reed Army Institute of Research and there is no objection to its publication. The opinions or assertions contained herein are the private views of the author and are not to be construed as official or as reflecting the position of the Department of the Army or of the United States Department of Defense. The study described in this chapter was approved by the Walter Reed Army Institute of Research Human Use Research Committee and the United States Medical Research and Materiel Command Human Subjects Research Review Board of the Army Surgeon General and was performed in accordance with the ethical standards laid down in the 1964 Declaration of Helsinki.

Citations of commercial organizations and trade names cited in this chapter do not constitute an official Department of the Army endorsement or approval of the products or services of these organizations.

References

Anderer P, Gruber G, Parapatics S, Woertz M, Miazhynskaia T, Klösch G, Saletu B, Zeitlhofer J, Barbanoj M, Danker-Hopfe H, Himanen S, Kemp B, Penzel T, Grözinger M, Kunz D, Rappelsberger P, Schlögl A, Dorffner G. An E-health solution for automatic sleep classification according to Rechtschaffen and Kales: validation study of the Somnolyzer 24x7 utilizing the Siesta database. *Neuropsychobiology*, 2005; 51(3):115–113.

Babkoff H, Genser SG, Sing HC, Thorne DR, Hegge FW. The effects of progressive sleep loss on the lexical decision task: response lapses and response accuracy. *Behavior Research Methods, Instruments & Computers*, 1985; 17(6):614–622.

Babkoff H, Mikulincer M, Caspy T, Kempinski D, Sing H. The topology of performance curves during 72 hours of sleep loss: a memory and search task. *The Quarterly Journal of Experimental Psychology*, 1988; 40A(4):737–756.

Babkoff H, Mikulincer M, Caspy T, Carasso RL, Sing H. The implications of sleep loss for circadian performance accuracy. *Work and Stress*, 1989; 3(1):3–14.

Balkin T, Thorne D, Sing H, Thomas M, Redmond D, Wesensten N, Williams J, Hall S, Belenky G. Effects of sleep schedules on commercial motor vehicle driver performance. FMCSA Report No. DOT-MC-00–133, May 2000.

Boot M. The new American way of war. *Foreign Affairs*, 2003; July/August.

Braver ER, Preusser CW, Preusser DF, Baum HM, Beilock R, Ulmer R. Long hours and fatigue: a survey of tractor-trailer drivers. *Journal of Public Health Policy*, 1992; 13(3):341–366.

Carskadon MA, Dement WC, Mitler MM, Roth T, Westbrook PR, Keenan S. Guidelines for the multiple sleep latency test (MSLT): a standard measure of sleepiness. *Sleep*, 1986; 9:519–524.

Center for National Truck Statistics. Truck and bus accident factbook 1994. Washington, DC: Federal Highway Administration Office of Motor Carriers, 1996 (Report no. UMTRI-96–40).

Dinges DF, Powell JW. Microcomputer analyses of performance on a portable, simple visual RT task during sustained operations. *Behavior Research Methods, Instruments, and Computers*, 1985; 17:652–655.

FOXNEWS.COM. Pilot, 1st Officer Slept While Approaching Denver, Lawmaker Says. October 31, 2007.

Kaplan R. Exploration of high frequency EEG signals, a doctoral dissertation. Case Western Reserve University, 1996.

Kaplan R, Loparo K. Patent No. 5,813,993; September 29, 1998; Alertness and Drowsiness Detection and Tracking System.

Knipling RR, Wang J. Crashes and fatalities related to driver drowsiness/fatigue: research note. Washington, DC: National Highway Traffic Safety Administration, 1994.

McCann UD, Penetar DL, Shaham Y, Thorne D, Gillin JC, Sing H, Thomas H, Belenky G. Sleep deprivation and impaired cognition: the role of catecholamines. *Biological Psychiatry*, 1992; 31(11):1082–1097.

Newhouse PA, Penetar DM, Fertig JB, Thorne DR, Sing HC, Thomas ML, Cochran JC, Belenky GL. Stimulant drug effects on performance and behavior after prolonged sleep deprivation: a comparison of amphetamine, nicotine, and deprenyl. *Military Psychology*, 1992; 4: 207–233.

Papadelis C, Chen Z, Kortidou-Papadeli, Bamidis PD, Chouvarda I, Bekiaris E, Maglaveras N. Monitoring sleepiness with on-board electrophysiological recordings for preventing sleep-deprived traffic accidents. *Clinical Neurophysiology*, 2007; 118:1906–1922.

Ratcliff, G. Spatial thought, mental rotation and the right cerebral hemisphere. *Neuropsychologia*, 1979; 17:49–54.

Rechtschaffen A, Kales A. A manual of standardized terminology, techniques and scoring system for sleep stages of human subjects. Washington, DC: Public Health Service, U.S. Government Printing Office, 1968 (reprinted 1971).

Russo MB. System and method for detecting visual alertness. U.S. patent #6849050, February 1, 2005.

Russo MB, Sing H, Santiago S, Kendall AP, Johnson D, Thorne D, Escolas S, Holland D, Hall S., Redmond D. Visual neglect: occurrence and patterns in pilots in a simulated overnight flight. *Aviation, Space, and Environmental Medicine*, 2004; 75(4–1):323–332.

Sing HC, Kautz MA, Thorne DR, Hall SW, Redmond DP, Johnson D, Warren K, Bailey J, Russo MB. High frequency EEG as potential measure of cognitive function capacity: a preliminary report. *Aviation, Space, and Environmental Medicine*, 2005; 76(7), Section II:C114–C135.

Sing HC, Russo MB, Kautz MA, Thorne DT. U.S. Patent Provisional Patent #182.0023P, 182.0023P2: Alertness/Drowsiness and Cognitive Function Capacity Index; 2006.

Sing HC, Russo MB, Kautz MA, Thorne DT, Hall SW, Bailey J. U.S. Patent pending: Alertness/ Drowsiness and Cognitive Function Capacity Index; 2007.

Smith SW. The scientist and engineer's guide to digital signal processing: discrete Fourier Transform. 1st Edition 1997. California: California Technical Publications, ISBN:0966017633.

Terzano MG, Parrino L, Sherieri A, Chervin R, Chokroverty S, Guilleminault, C, Hirschkowitz M, Mahowald M, Moldofsky H, Rosa A, Thomas R, Walters A. Atlas, rules, and recording techniques for the scoring of cyclic alternating pattern (CAP) in human sleep. *Sleep Medicine*, 2001; 2:537–553.

Thomas M, Sing H, Belenky G, Holcomb H, Mayberg H, Dannals R, Wagner H, Thorne D, Popp K, Rowland L, Welsh A, Balwinski S, Redmond D. Neural basis of alertness and cognitive performance impairments during sleepiness. I. Effects of 24 hours of sleep deprivation on waking human regional brain activity. *Journal of Sleep Research*, 2000; 9(4):335–352.

Thomas ML, Sing HC, Belenky G, Holcomb HH, et al. Neural basis of alertness and cognitive performance impairments during sleepiness. II. Effects of 48 and 72 h of sleep deprivation on waking human regional brain activity. *Thalamus and Related Systems*, 2003, August; 2(3):199–229.

Thorne DR. Problems in defining, identifying, and quantifying response lapses and distinguishing them from changes in distribution parameters. Manuscript submitted for publication.

Thorne DR, Genser SG, Sing HC, Hegge FW. The Walter Reed performance assessment battery. *Neurobehavioral Toxicology and Teratology*, 1985; 7:415–418.

Thorne DR, Johnson DE, Redmond DP, Sing HC, Belenky G, Shapiro JM. The Walter Reed palm-held psychomotor vigilance test. *Behavior Research Method*, 2005; 37(1):111–118.

Torch WC, Russo MB, Cardillo C, Thomas ML. Eye-Com® Biosensor, Communicator & Controller System for Assessing Cognitive State via Oculomotor Signals. In: Schmorrow DD, Nicholson DM, Drexler JM, Reeves LM. (Eds.). Foundations of Augmented Cognition. 4th Edition 2007. Falcon Books. California.

Chapter 15
Total Sleep Deprivation and Cognitive Performance: The Case for Multiple Sources of Variance

H. Babkoff[1], A. Goldstein[2], and G. Zukerman[3]

Abstract The present chapter reviews and summarizes some of the major findings of the studies of the impact of sleep deprivation on cognitive performance over the past half century. The studies discussed are limited to instances of total sleep loss and support the argument that total sleep loss may impact different cognitive functions differently. Some cognitive variables may be more sensitive to sleep loss than others and some cognitive variables may be more sensitive to the homeostatic pressures of accumulated wakefulness, while others may be more sensitive to the impact of the circadian cycle.

Three types of data were chosen for illustration: 1) those that are descriptive of the major effects of short- to-mid term (24-48 hrs) or long term (72 hrs and longer) sleep deprivation; 2) those that are descriptive of disassociations of the effects of sleep deprivation on cognitive performance; and 3) those that describe differential effects of treatments on different cognitive performance tasks. Evidence is presented to show that the effects of sleep deprivation on cognitive performance are not uniform, either with respect to task or with respect to the time since the last sleep epoch.

The propositions discussed in the present chapter have important theoretical and practical implications. If more than one neural mechanism is involved in performance decrement due to sleep deprivation and sleep restriction at any moment, then any proposed intervention that targets only one of the mechanisms cannot be fully successful in ameliorating the effects of sleep loss. Multiple sources of variance in causing performance decrement require multiple forms of treatment to eliminate or reduce decrement.

Keywords Sleep deprivation; Circadian rhythms; Cognitive performance;

[1]Department of Psychology, Bar-Ilan University, Ramat Gan 52900, Israel and Ashkelon Academic College, Ashkelon, 78211, Israel
Email: babkofh@mail.biu.ac.il

[2]Department of Psychology and Gonda Brain Research Center, Bar-Ilan University, Ramat Gan 52900, Israel Email: goldsa@mail.biu.ac.il

[3]Gonda Brain Research Center, Bar-Ilan University, Ramat Gan 52900, Israel
Email: zukgil@yahoo.com

D. Lloyd, E. Rossi (eds.), *Ultradian Rhythms from Molecules to Mind*,

15.1 Introduction

15.1.1 The Importance of Studies of the Effects of Sleep Loss

The negative effects of prolonged sleep deprivation on behavior and cognitive performance are of major concern to a technologically advanced modern society. To a great extent, the developments and demands of modern society have become dependent upon the ability of a fairly large number of professionals to continue to perform at peak level over the hours of a day following a night of sleep loss or of reduced sleep. Often the demands of a high tech-based industry and of health, rescue and defense services require the continued availability of highly trained personnel despite more than 48 h of sleep loss. One of many examples is that of modern air traffic piloting and control, both civilian and military, which has been cited as subject to problems of sleep loss. For example, pilot fatigue has been reported to be a significant problem in both civilian and military aviation (Caldwell, 2005). Official statistics indicate that pilot fatigue is involved in around 4–8% of aviation accidents. The problem of pilot fatigue becomes even more of a concern with modern round-the-clock flying schedules (Caldwell, 2005). Some components of the US military have, in fact, even authorized the use of specific pharmacological compounds as fatigue countermeasures (Caldwell and Caldwell, 2005).

Sleep loss and impairments related to the accompanying fatigue are also reported as common causes of errors among professionals working in health care (Owens, 2007). Fatigue combined with excessive workloads was implicated in medical mishaps in pediatric intensive care units (Montgomery, 2007). In a meta-analytic study of residents and non-physicians, Philibert (2005) reported that sleep loss of less than 30 h reduced physicians' overall cognitive and motor performance by one standard deviation and clinical performance by more than 1.5 standard deviations. Thus, Philibert (2005) concluded that the weekly hours and continued wakefulness permitted under the current US standards for residents may not completely guard against the impact of sleep loss on cognitive and clinical performance.

Although research on the impact of sleep deprivation on cognitive performance is more than a century old, there is still no clear answer, to date, as to how to either avoid sleep loss or of how to maintain or enhance cognitive abilities despite sleep loss. Consequently, there continues to be an interest in basic and applied research in this area.

In this chapter, we attempt to review and summarize some of the major findings of the studies of the impact of sleep deprivation on cognitive performance over the past half century but with emphasis on research within the last 2 decades. The studies discussed in this chapter are limited to instances of total sleep deprivation only, and have been chosen to help make the argument that total sleep deprivation may impact different cognitive functions differently. Some cognitive variables may be more sensitive to sleep loss than others. Some cognitive variables may be more sensitive to the homeostatic pressures of accumulated wakefulness, while others may be more sensitive to the impact of the circadian cycle. The impact of sleep loss on some cognitive tasks may be reduced by one type of intervention, but the same treatment may have no ameliorating effect at all on the performance of another cognitive task. These types of disassociations are discussed in this chapter and used as evidence to advance the argument that more than one neural mechanism is involved in the decrements in cognitive performance associated with sleep deprivation and that, consequently, for any intervention to be successful, it must address more than one causal factor.

We begin with a review of the effects of, what may be considered the most relevant manipulation in a sleep deprivation study, the extent of wakefulness or the amount of sleep deprivation.

15.1.2 Long Term Versus Short Term Sleep Deprivation Study Protocols

The effect of sleep deprivation and the circadian rhythm on cognitive performance has been the subject of research for over 100 years (Patrick and Gilbert, 1896). In general, sleep deprived individuals experience an increasing tendency to fall asleep, a decline in cognitive and motor ability, aggravated by lapses in performance, and a general decline in attention (Dinges and Kribbs, 1991; Koslowsky and Babkoff, 1992; Murray et al., 1958; Wilkinson, 1965, 1990). Long term sleep deprivation results in decrements in cognitive and psychomotor performance. Linear decreases in performance related to sleep deprivation are accompanied by circadian oscillations, with peaks around 1400–2000 and troughs between 0000 and 0600 (Babkoff et al., 1991a; Horne, 1988; Minors and Waterhouse, 1981). The amount of performance decrement that occurs during sleep deprivation depends on various stimulus and environmental variables (see, e.g., Babkoff et al., 1991a; Johnson, 1982). Different tasks may show differing sensitivities to sleep loss (Babkoff et al., 1991a; Dinges and Kribbs, 1991; Johnson, 1982). A number of recent studies have been designed to clarify the mechanisms responsible for performance deficits during sleep loss as well as to test interventions. Furthermore, although some researchers have noted that most if not all individuals are vulnerable to sleep deprivation (Frey et al., 2005), a number of recent studies have reported very large individual differences in vulnerability to sleep loss (Caldwell et al., 2005; Durmer and Dinges, 2005; Van Dongen et al., 2004, 2005), as well as proposing the possible underlying neural mechanisms that may serve as the basis for individual differences in sensitivity to sleep deprivation (Bell-McGinty et al., 2004; Caldwell et al., 2005; Chuah et al., 2006). Frey et al. (2004) also pointed out that individuals may differ in terms of their sensitivity to different tasks, such that some individuals are more sensitive to the effects of sleep loss on one type of cognitive task and less sensitive to sleep loss on another type of cognitive task, while others show opposite sensitivities to the same tasks when challenged by sleep loss.

In addition to fairly frequent reviews that consistently reported that total sleep deprivation (TSD) and measures of cognitive and psychomotor performance are negatively correlated, a number of meta-analyses, with more accurate summaries of the relationships between these variables, have also been reported over the past 2 decades. Koslowsky and Babkoff (1992) analyzed the results of 27 studies (from 1969–1990) with a total of 429 subjects that met the criteria for a quantitative analysis, and applied meta-analytic techniques to test several hypotheses. The correlations between decrements in performance and sleep deprivation were found to be highest for total sleep deprivation of greater than 45 h. Speed rather than accuracy measures of performance, and work-paced rather than self-paced tasks were most affected. The authors concluded that the findings were consistent with the "lapse hypothesis" that considers the lapse or an error of omission (defined behaviorally as a period of not responding that is longer than several standard deviations beyond the expected or mean response time) as the major dependent variable and posits the presence of microsleeps (defined as an episode of sleep lasting from 0.5 to 10 s) during long hours of sleep deprivation as the major factor in performance decline during long term sleep deprivation.

Is the overwhelming presence of lapses in long term sleep deprivation data evidence that the interference of microsleeps with task performance is the major decrement in cognitive processing associated with sleep loss? We believe that the answer may be positive for long term sleep deprivation, but, negative for short term sleep deprivation. First of all, the number of lapses in any performance data set increases as sleep deprivation increases (Van dongen et al., 2003, 2004). Second, the frequency of lapses in any performance data set is especially influenced by the circadian rhythm (Babkoff et al., 1985, 1991a; Frey et al., 2004; Gross et al., 2006). The number of lapses in any data set increases during the nadir of the circadian rhythm, and decreases after the nadir has passed. Performance lapses are probably one of the most robust effects of long term sleep deprivation, but are much less frequent in performance data during short term sleep deprivation, except at the nadir of the circadian rhythm (Babkoff et al., 1991a; Gross et al., 2006).

Third, we would add a caveat to the earlier meta-analysis by Koslowsky and Babkoff (1992). A fairly large number of the studies of long term sleep deprivation, i.e., of 72 h or longer used the within-subject design paradigm to study performance changes over time. The reasons for not using a control group of non sleep deprived subjects were numerous, but mainly because of the additional costs. Furthermore, the basic assumption was that the performance of the subjects during the first day could serve as their own baseline, so that within-subject comparisons were always available and used by a number of researchers as their control in the experimental design (see e.g., Babkoff et al., 1991a). The design was directed to compare changes in the dependent variables from the first to the second and third days, and these changes were used as evidence of the effects of sleep deprivation on the various tasks. This means that there were very few long term sleep deprivation studies that compared subjects' performance under conditions of no sleep deprivation with either their own or other subjects' performance under conditions of sleep deprivation, as is de rigueur in studies of short term sleep deprivation. Comparison groups were used in studies of drug interventions, but with all of the groups sleep deprived, one not treated by the intervention, the others, treated. Thus, the comparisons in the meta-analysis reported by Koslowsky and Babkoff (1992) were of performance deficits in short term versus long term sleep deprivation, and not of performance deficits under conditions of sleep deprivation versus non sleep deprivation. Their conclusion, therefore, that errors of omission were the main decrement resulting from sleep loss and that the decrement was seen mostly in work-paced rather than self-paced tasks, which perhaps, implies that response lapses is the major factor operating in sleep deprivation, should be understood as limited to a comparison of short term versus long term sleep deprivation only, not to a comparison of non sleep deprivation with sleep deprivation. In summary, one of the major factors differentiating the effects of short term from long term sleep deprivation is a very big increase in the presence of lapses in the data sets.

The meta-analysis by Koslowsky and Babkoff (1992) was followed several years later by Pilcher and Huffcutt (1996) who reported on a meta-analysis of 19 studies, that were selected based on whether they provided the following information: (1) means and standard deviations for sleep deprived and non sleep deprived groups; (2) they had to involve short-term sleep deprivation (<45 h), long-term sleep deprivation (>45 h) or partial sleep deprivation (<5 h in a 24 h period). The studies also had to use either a cognitive performance task, a motor performance task or a mood scale Their meta-analysis study showed that sleep deprivation affects performance, with sleep deprived subjects performing 1.37 standard deviations lower than non sleep deprived subjects indicating that a sleep deprived subject in the 50th percentile performs at the same level as a non sleep deprived subject in the 9th percentile. Short-term sleep deprivation had the greatest effect on complex and long tasks. Long-term sleep deprivation affected performance on short and simple tasks, while partial sleep deprivation was the most detrimental on either simple or long tasks. Partial sleep deprivation had a greater effect on performance than either short or long-term sleep deprivation; with partially sleep deprived subjects performing at a level two standard deviations lower than non sleep deprived subjects, compared with one standard deviation for short and long-term sleep deprived subjects.

A number of important shifts in research emphasis have also taken place over the years, accompanied by changes in the preferred experimental paradigms and protocols (Carrier and Monk, 2000). The earliest studies examined the effects of total sleep deprivation of a variety of durations, but did not always control for diurnal effects (Babkoff et al., 1991b). The use of experimental protocols that studied sleep deprivation and controlled for time of day effects by measuring performance frequently and at fixed times became widespread in the second half of the 20th century. Changes have also occurred in the experimental designs so that although during the latter part of the 20th century, emphasis was placed on studies of long term, 48–72 plus h of sleep deprivation by a large number of researchers, over the past decade emphasis seems to have shifted to include more studies of short term, i.e., less than 45 h of sleep deprivation.

The earlier research, in general, was more of a descriptive nature, with the emphasis on discovering and measuring the physiological, behavioral and environmental parameters that resulted in performance decrement and in decreased subjective arousal (Dinges and Kribbs, 1991). One of the reasons the emphasis was placed on long term sleep deprivation protocols was the fact that there was difficulty in finding substantial and consistent negative effects of sleep loss on performance with short term sleep deprivation as compared to the robust decrements often reported with long term sleep deprivation. In addition, a large number of studies were devoted to testing a variety of interventions to eliminate or reduce the effects of sleep loss on cognitive performance, from the "natural" (e.g., naps) to the pharmacological (Babkoff and Krueger, 1992). The long term sleep deprivation protocol was quite useful in providing a large data base of quantitative descriptive information in response to such applied research questions. It seems fairly clear that the use of long term sleep deprivation protocols led to the discovery of the presence of two distinct factors in sleep deprivation data, one, monotonic, the homeostatic component, associated with accumulated wakefulness, the other, cyclical, mainly associated with the circadian rhythm (Babkoff et al., 1991b).

15.1.2.1 Long Term Sleep Deprivation Studies

The following is a partial summary of the main findings of studies of the effects of long term sleep deprivation on psychomotor and cognitive performance (Babkoff et al., 1991a; Koslowsky and Babkoff, 1992):

1. The curves depicting changes in performance, sleepiness and mood during prolonged sleep deprivation (48–72 h) have both monotonic and rhythmic components. The components can be separated and the rhythmic components examined by Time Series Analyses (e.g., Cosinor, Complex Demodulation).
2. The monotonic component is mainly linear or exponential. The rhythmic component is mainly circadian, with some minor hemi-circadian cyclicity.
3. There is a dramatic change in most performance curves at the second circadian nadir of sleep loss, after approximately *44–46 h* of wakefulness (around 0400–0600), followed by dramatic *partial recovery* the third day (Babkoff et al., 1991b).
4. The circadian component becomes more pronounced as sleep deprivation progresses (Babkoff et al., 1991b).
5. With *prolonged* sleep deprivation, the *greater* the cognitive or psychomotor load, the *greater* the performance deficit due to sleep loss (Babkoff et al., 1988).
6. Time-related measures of performance, such as number of attempts, reaction times and response times, show greater deficit than do measures of accuracy.
7. Errors of omission (Lapses) increase to a greater extent during sleep deprivation than do errors of commission.
8. Performance becomes more erratic during prolonged sleep deprivation. Intra-individual variance increases and RT distributions develop "long tails" (Babkoff et al., 2002; Van dongen and Dinges, 2003).

Figures 15.1 and 15.2 are examples of the influence of both the homeostatic (monotonic) component and the cyclic component on performance of a logical rea-

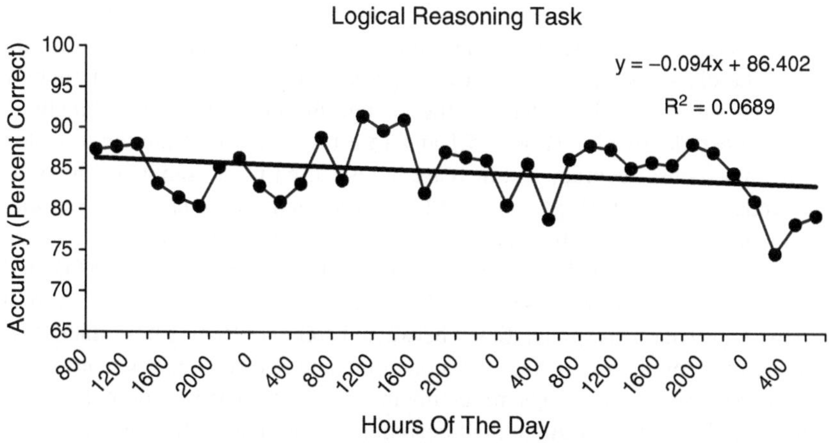

Fig. 15.1 Performance accuracy on the Logical Reasoning Task during 72 h of sleep deprivation is plotted on the ordinate as a function of time of day/night on the abscissa

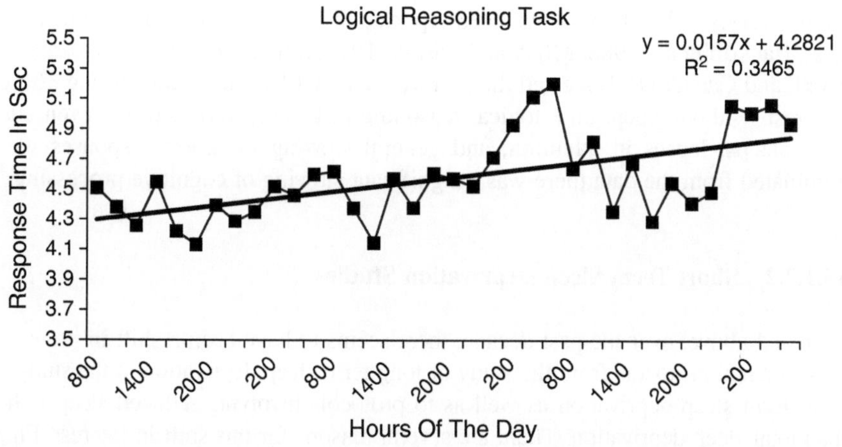

Fig. 15.2 Correct response time in the Logical Reasoning Task during 72 h of sleep deprivation is plotted on the ordinate as a function of time of day/night on the abscissa

soning task. The data are from an earlier study, one purpose of which was to examine the effects of long term sleep deprivation on cognitive performance when errors of omission (lapses) were removed from the data (Mikulincer et al., 1990). In that study, we tested eleven subjects on a logical reasoning task in a 72 h sleep deprivation experiment. The logical reasoning task had three levels of difficulty, in three blocks of trials each, presented in a random order during 36 sessions that were spaced 2 h apart. For all of the analyses, response lapses were removed form the data sets. Lapses were defined as any 10 s interval after stimulus presentation without a response. The 10 s period without a response was more than 5 standard deviations longer than the mean response time and 3.8 standard deviations longer than the longest mean response time among the eleven subjects. This illustration follows the classical presentation of average accuracy and response time data for the eleven subjects and emphasizes the change in performance as a function of hours of sleep deprivation (Figs. 15.1 and 15.2 respectively). Note the monotonic decline in response accuracy and increase in response time accompanied by prominent cyclic components (Figs. 15.1 and 15.2). The cyclic components are especially prominent in the response time data (Fig. 15.2) showing peaks (largest increases in response times) at the nadir of the circadian rhythm, at around 0400–0600 on the second and third mornings (48 and 72 h) after the first night without sleep and troughs, with shorter response times between 1200 and 1600. These data clearly illustrate the prominence of both the monotonic and rhythmic components of long term sleep deprivation data sets, reflecting the strong influences of both the homeostatic drive for sleep as wakefulness accumulates and that of the circadian sleep-wake cycle (Babkoff et al., 1991a; Van Dongen et al., 2003, 2004; Gross et al., 2006). These data thus reflect the influence of cumulative wakefulness and of the circadian rhythm in logical reasoning performance during long term sleep loss, even after lapses have been removed. Similar curves for a large variety of tasks having very similar shapes have been reported in a number of earlier publications (e.g., Babkoff et al., 1988, Monk and Carrier, 1997; Sing et al., 1985). For example, Monk and Carrier (1997) tested the performance of 18 young adults during 36 h of bed rest, without sleep, on a logical reasoning task. They reported that even after microsleeps, lapses in attention, and general slowing of motor responses were eliminated from the data there was a significant slowing of cognitive processing.

15.1.2.2 Short Term Sleep Deprivation Studies

As noted above, over the past decade, there seems to have been a shift in preferred experimental protocols from the study of long term sleep deprivation to the study of short term sleep deprivation as well as to protocols involving reduced sleep rather than total sleep deprivation. There are several reasons for this shift in interest. First, there has been a shift in research interests, directed away from descriptive research towards identifying the neural mechanisms that impact cognitive performance during sleep loss or sleep reduction. There are probably several reasons for this shift to short term sleep deprivation protocols. First, experimental protocols that test short term or moderate total sleep deprivation and partial sleep deprivation protocols may serve this research interest better, since as sleep deprivation accumulates, the relative weightings of the homeostatic and cyclical components seem to change (see, e.g., Babkoff et al., 1991a) such that the cyclical component may become the major source of variance and begin to overwhelm the data, whereas early into sleep deprivation, the two major factors may be more easily identified and studied, separately. Furthermore, sleep deprivation of less than 45 h and partial or reduced, rather than, total sleep deprivation, may also be more representative of many real life situations, especially situations that require the optimal performance of medical and security personnel.

However, there is a third, more important, reason for the shift in research interest to the short-to-moderate from long term sleep deprivation protocols, a shift in theoretical emphasis in the last 2 decades. Up until the 1990s, many researchers designed their studies and interpreted their results in terms of the Arousal theory of sleep deprivation. They argued that a reduction in sleep length affects performance by causing decreased arousal and slower cognitive processing (Babkoff et al., 1985; Harrison and Horne, 2000a; Kjellberg, 1977; Wilkinson, 1965; Williams et al., 1959; Wilkinson, 1990). Dinges and Kribbs (1991) had noted that the majority of studies of sleep-deprivation and performance reported global changes in performance ability, changes that may have been mediated as much by microsleeps and lapses in attention as by more subtle or profound alterations in information processing due to sleep deprivation. In fact, Wilkinson (1990) argued that sleep deprivation reduced the non-specific arousal level, but had no specific effect. According to this hypothesis, there were no intrinsic differences in performance decrement among different tasks, since the major source of variance impacted by sleep deprivation was basically the same for performance on all tasks, reduced arousal. The nature of the task itself was apparently of lesser importance. Moreover, according to the arousal hypothesis, the effect of sleep deprivation would be most evident while performing a monotonous task that measures basic skills, e.g., reaction time or vigilance.

The major alternative hypothesis, proposed by Horne and colleagues is the 'sleep-based neuro-psychological perspective', which posits that sleep loss affects different parts of the brain differentially (Harrison and Horne, 1998, 2000a, b). Harrison and Horne (2000a, b) have argued that sleep reduction primarily impacts the functions associated with pre-frontal cortex (PFC) activity. The prefrontal cortex is among the first brain regions to be affected by sleep deprivation (e.g., Drummond et al., 2000, 2001; Muzur et al., 2002; Thomas et al., 2000). Some recent work even indicates that prefrontal activation recovery may not be fully restored by recovery sleep. Wu et al. (2006) studied sleep deprivation and recovery sleep in normal subjects and reported that sleep deprivation resulted in a significant decrease in relative metabolism of the frontal cortex, thalamus, and striatum. They reported that recovery sleep had only a partial restorative effect on frontal lobe function and suggest that sleep may be especially important for maintenance of frontal lobe activity. According to the sleep based neuropsychological hypothesis, even with relatively mild to moderate sleep loss (24 h), tasks that are highly dependent on intact prefrontal cortical activity will show deficit. These researchers predict that deficit due to sleep deprivation mainly affects complex tasks that measure higher cortical related skills such as: language skills in communication, divergent thinking, creativity and cognitive flexibility. Cognitive and motor tasks demanding only basic skills, although requiring attention and arousal, should be relatively less affected after mild to moderate sleep loss.

The increased use of short term sleep deprivation protocols has been accompanied by the use of fMRI and PET measurements of regional and global changes in brain activity associated with sleep deprivation. Because of the recent emphasis given to the sleep based neuropsychological hypothesis many of the recent fMRI studies of sleep deprivation have tested brain activity during the performance of tasks involving some type of executive control and have focused on frontal brain regions. Performance on these tasks is affected by short term sleep deprivation so that significant findings may be obtained with less expenditure and less wear and tear on the subjects (see e.g., Drummond et al., 2000, 2001; Habeck et al., 2004; Lim et al., 2007; Stricker et al., 2006). The more recent studies that used short to moderate (<48 h of) sleep deprivation protocols can thus be said to have contributed much to our understanding of the involvement of regional brain regions in sleep loss.

15.1.3 A Multi-variate Approach to Understanding the Impact of Total Sleep Deprivation on Performance

In the following sections we present a broad discussion of the effects of total sleep deprivation on cognitive performance using representative data from long term and short term sleep deprivation protocols as well as discuss the contributions of the different protocols and their theoretical implications. Figures are shown to illustrate the main points. The main thrust of the present chapter is to present the point of view that the effects of sleep deprivation on cognitive performance are not uniform, either with respect to task or with respect to the time since the last sleep epoch. The argument will be developed to show that the impact of total loss of sleep on cognitive performance does not represent one source of variance but rather reflects multiple sources of variance. First, we believe that the relative weightings of the homeostatic and cyclic components in the data sets of the various physiological and behavioral dependent variables usually measured during sleep deprivation may vary over time and that the weightings may differ from one dependent variable to another. That is, that one task may be more sensitive to the homeostatic drive for sleep than to the diurnal or circadian rhythm, while another task may be more sensitive to the circadian rhythm than to the homeostatic drive for sleep. For example, the slope of performance decline for one task as a function of accumulating sleep loss may be steeper than for another task; while the amplitude of the oscillations in the performance data of the second task may be greater than that in the first task. Furthermore, since the two major factors that impact performance during sleep loss are the homeostatic drive for sleep resulting from accumulated wakefulness and the circadian rhythm, the relative weighting of the impact of one mechanism may be greater than that of the other at any given moment, i.e., the relative contributions of the two factors may change from early in sleep deprivation to later in sleep deprivation (see e.g., Babkoff et al., 1991a).

Second, we also believe, in accord with the sleep based neuro-psychological perspective of Harrison and Horne (1998) that different tasks that are based on intact activity of different parts of the brain are affected differently by sleep loss, depending on the impact of sleep loss on those regions of brain associated with performance of the tasks. The simplest expression of this second point might be, for example, that short term sleep loss may affect one task more than it affects another task. The simplest prediction is, then, that different psychomotor and cognitive tasks, associated with different brain mechanisms and/or brain regions may be affected differently by sleep deprivation. In addition, these propositions have important theoretical and practical implications as well. If more than one neural mechanism is involved in performance decrement due to sleep deprivation and sleep restriction at any moment, then any proposed intervention that targets only one of the mechanisms cannot be fully successful in ameliorating the effects of sleep loss. Multiple sources of variance in causing performance decrement require multiple forms of treatment to eliminate or reduce decrement. These points will be argued and illustrated throughout the chapter mainly by citing evidence from disassociations. The first argument, presented in the first section (A) relates to the relative weightings of the homeostatic and circadian components in two data sets measured during long term sleep deprivation. Evidence will be cited that separates the effects of the homeostatic from the circadian factors in long term sleep deprivation in two data sets, body temperature and ratings of subjective sleepiness, and shows that for the individual subject, only one factor is common to both dependent variables. The second argument will be presented in two sections, one section (B) cites evidence from a study of the effects of different countermeasures or treatments on two different performance tasks, both of which were affected by sleep loss, but that responded differently to the two treatments; while the last section (C) cites evidence from a study comparing the effects of short term sleep deprivation on different cognitive performance tasks.

15.1.3.1 Long Term Sleep Deprivation: Sleepiness and Body Temperature

Although long-term sleep deprivation results in a variety of behavioral, motivational and performance changes (Angus and Heselgrave, 1985; Babkoff et al., 1985, 1988, 1991a; Dinges and Kribbs, 1991; Johnson, 1982; Minors and Waterhouse, 1981; Mikulincer et al., 1989), the most obvious result is increased sleepiness (Babkoff et al., 1991b; Dinges and Kribbs, 1991; Dinges et al., 1984; Minors and Waterhouse, 1981). Sleep latency as measured by the multiple sleeplatency test (MSLT) reaches a minimum after approximately 24–30 h of sleep loss (Dinges et al., 1984). Subjective sleepiness ratings, however, continue to change dynamically during sleep deprivation and provide information on changes in sleepiness even during lengthy periods of sleep loss. Cognitive performance decrement also begins early and continues to increase during the entire sleep deprivation episode. A monotonic increase, reflecting the homeostatic effects of accumulated wakefulness, and oscillations, mainly circadian (Babkoff et al., 1991b, 1992; Dinges et al., 1984) characterizes subjective sleepiness ratings as well as psychomotor and cognitive performance data during sleep deprivation. The presence of rhythmic components becomes evident in the data after approximately 36 h of sleep loss even when subjects rest in bed, but do not sleep (Butesse, 1993).

The influence of both the homeostatic and cyclic components in physiological and performance data in long term sleep deprivation protocols will be discussed in this section and illustrated by several figures, taken from two unpublished studies that took place during the 1990s in two different laboratories, one in the USA, one in Israel.

There are two main reasons for using body temperature and subjective sleepiness as examples of the effects of long term sleep deprivation. The first reason is that body temperature, classically, has played such an important role as a predictor of sleep in normal, non sleep deprived, circumstances. For example, many years ago, poor sleepers were reported to have higher body temperature than good sleepers (Williams and Williams, 1966). Heating the body during exercise produces a relatively greater amount of slow wave sleep (SWS) in the EEG during the subsequent sleep episode (Horne, 1988). Passive body heating has been reported to delay the phase of the core body temperature rhythm, increase SWS and improve sleep quality in older female insomniacs (Dorsey et al., 1999). Experimentally induced lowering of rectal temperature was reported to increase stage 4-sleep (Sewitch et al., 1986). REM sleep may be especially sensitive to elevated body temperature particularly in menstruating women (Baker et al., 1999). Nevertheless, not all factors that affect body temperature impact sleep characteristics. For example, menstrual cycle phase and hormonal contraceptives may significantly influence body temperature without affecting sleep (Baker et al., 2001). In fact, it appears that it is the rhythmic component of body temperature that is most related to sleep characteristics. Although a low body temperature may not be a necessary prerequisite for sleep (Broughton, 1989), sleep duration seems to be dependent on body temperature at sleep onset (Akerstedt and Gillberg, 1981; Czeisler et al., 1980; Zully et al., 1981). Most awakenings by desynchronized subjects occur on the rising phase of the temperature rhythm (Czeisler et al., 1980). Furthermore, although body temperature retains an unchanged circadian rhythm, there is a small, but consistent overall reduction during sleep deprivation (Babkoff et al., 1991a; Dinges et al., 1987; Horne, 1988).

The second reason for illustrating the contribution to the information based on neural mechanisms associated with the effects of sleep deprivation by the long term sleep deprivation protocol with body temperature and subjective sleepiness data is the long standing hypothesis that especially during long term sleep deprivation, there is an inverse relationship between body temperature and subjective ratings of sleepiness. The possibility of such a relationship seems to be supported by Wright et al. (2002) who studied non sleep deprived subjects using a forced desynchrony experimental protocol, that allowed the assessment of the relationship between body temperature and performance, while controlling for circadian phase and hours awake. They found that working memory, subjective sleepiness and the slowest 10% of reaction times were improved when body temperature was elevated. They concluded that increased body temperature was associated with improved performance and alertness, and that body temperature modulates neurobehavioral function in humans. The results reported by Wright et al. (2002) for non sleep deprived subjects raises the question as to whether and what extent is subjective arousal or its inverse, subjective sleepiness, related to, or modulated by changes in body temperature during long term sleep deprivation. In fact, a number of studies of pharmaceuticals that counteract the effects of sleep deprivation seem to also point to a relationship between body temperature and subjective sleepiness. A large number of pharmaceuticals that reduce sleepiness and improve performance during prolonged sleep deprivation also tend to maintain or even increase body temperature, for example: amphetamine (Newhouse et al., 1992; Pigeau et al., 1995), methylphenidate (Babkoff et al., 1992), dexedrine (Caldwell et al., 1995), modafinil (McLellen et al., 2003; Pigeau et al., 1995), nicotine (Jessen et al., 2003) and caffeine (Beaumont et al., 2001; Wright et al., 1997a, b, c).

Over 30 years ago, Froberg (1977) illustrated a fairly clear inverse relationship between group averaged body temperature and group averaged subjective sleepiness ratings in sleep deprived subjects. The argument was made that when sleep is suspended, the influence of the endogenous circadian pacemaker (as measured by body temperature) predominates in determining the circadian oscillations in sleepiness, thus resulting in a parallelism between body temperature and subjective sleepiness (Monk, 1987, 1991). Such a relationship would be very important for modeling changes in sleepiness during periods of long term sleep loss. However, although assumed, the generality of the relationship between body temperature and subjective sleepiness during long term sleep deprivation has not been demonstrated. More important, however, is that, in accord with this hypothesis, body temperature should predict subjective sleepiness in the individual subject during sleep deprivation (Froberg, 1977; Monk, 1991). This latter hypothesis has also never been formally, directly tested.

The demonstration by Froberg (1977) that the group average sleepiness ratings during sleep deprivation is a mirror image of the average body temperature does not demonstrate the nature of the relationship of the two variables in the individual subject. First, both body temperature and subjective sleepiness change monotonically and rhythmically during sleep deprivation. Neither a comparison of group averages nor a correlation of the raw data alone can clarify which of the underlying components in the two data sets are related and which are not. Second, group data may obscure the form of the individual subjects' curves by smoothing or averaging out higher order oscillations (Babkoff et al., 1991a; Sing et al., 1985). Since changes in alertness may also have a hemicircadian component (Mitler et al., 1988), the second order diurnal oscillation may play a role in the relationship of body temperature and sleepiness, but be masked in the averaged data. Third, since the early part of long term sleep deprivation is identical with a nycthemeral schedule, the relation between subjective sleepiness and body temperature must change from one, which is not parallel, to one which is an inverse, mirror image, during the course of sleep deprivation. This implies some dynamic change in the relationship between body temperature and subjective sleepiness during sleep deprivation which may be difficult to analyze if only the raw data are correlated.

The analyses addressed two questions in the data base. The first question related to the generality of the purported mirror image relationship between group averaged body temperature and subjective sleepiness data (Froberg, 1977). This question is a restatement of the proposition noted above relating to the possibility that the homeostatic and cyclic factors, both prevalent in long term sleep deprivation data sets, may have different weightings for different dependent variables. The second question related to the extent and form of that relationship for the individual subject of body temperature and subjective sleepiness ratings during long-term sleep deprivation.

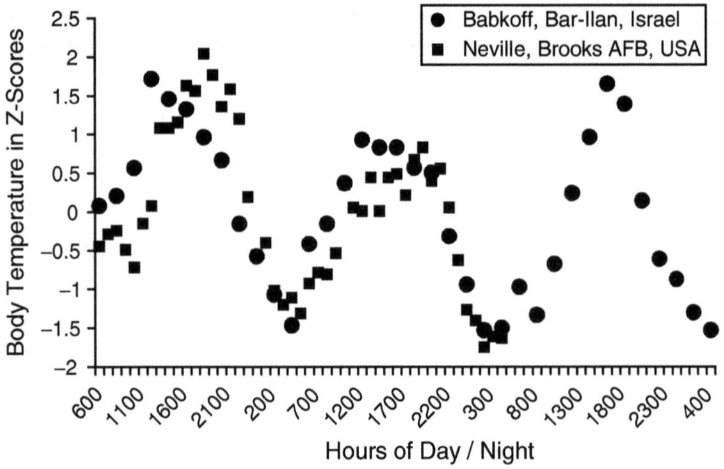

Fig. 15.3 Body temperature (in Z scores) is plotted on the ordinate as a function of time of day/night on the abscissa for two studies (see text for explanation)

Since ratings of subjective sleepiness and measurements of body temperature are a feature of most studies of long-term sleep deprivation, we analyzed data from two different, independent, studies to obtain a more general description of the changes in these variables. Changes in body temperature and subjective sleepiness during 46–72 h of sleep deprivation are described and the relationships between these variables are analyzed for individual subjects.

For this analysis we used data from two experiments that were conducted in the past. Changes in body temperature and ratings of subjective sleepiness during 46–72 h of sleep deprivation were measured in two independent experiments in two different locations: Bar-Ilan University, Ramat-Gan, Israel (1988) and Brooks Air Force Base, San Antonio, Texas, USA (1998) (French et al., 2006).

Average body temperature is shown as a function of time of day for the two studies in Fig. 15.3. The two data sets overlap extensively, especially from the morning of the second day, and their most salient feature is the rhythmic oscillation, indicating significant diurnal changes (Bar-Ilan: $F(11,88) = 15.94$, $p < 0.0001$; Brooks AFB: $F(23, 299) = 20.77$, $p < 0.0001$) whose peaks and troughs occur at around 19:00–20:00 and 03:00–06:00, respectively (Duncan Multiple Range Tests, $p < 0.05$). Comparison of data at the beginning of each study with data at the end of each study indicates a slight general reduction in temperature over time, 0.09°C over the 72 h of sleep deprivation in the Bar-Ilan study ($t(8) = 3.813$, $p < 0.002$) and 0.167°C over the 46 h of sleep deprivation in the Brooks AFB study ($F(1, 13) = 10.15$, $p < 0.0072$).

The small, but consistent drop in body temperature during long term sleep deprivation has been illustrated in a number of studies, beginning about 50 years ago. For example, Murray et al. (1958) recorded oral temperatures 10 times during each

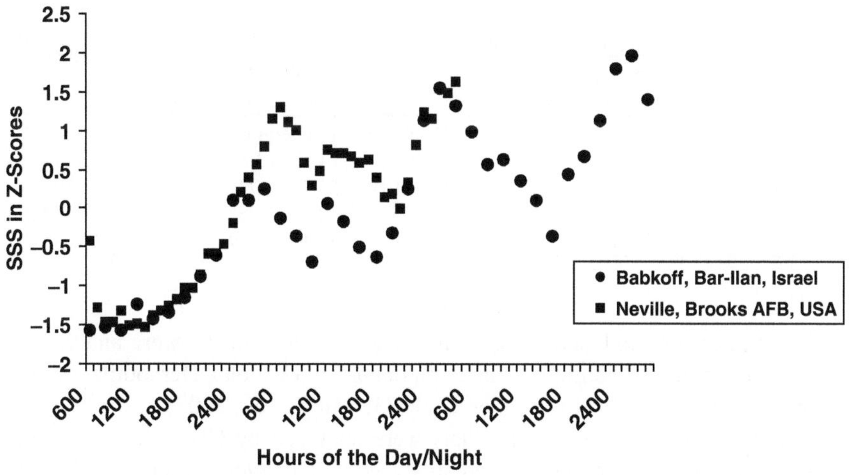

Fig. 15.4 Subjective sleepiness ratings on the Stanford Sleepiness Scale (SSS) (in Z scores) are plotted on the ordinate as a function of time of day/night on the abscissa for two studies (see text for explanation)

24-h period while subjecting participants to 98 h of sleep deprivation. While a circadian rhythm remained evident throughout the sleep deprivation study, temperature minimums and maximums were shifted progressively lower. The mechanism for the overall decrease in body temperature during sleep deprivation has not been positively identified (Kelly, 2007). Some researchers have claimed that this is due to the generally reduced activity; others have argued that it is due to a change in the sensitivity of thermoregulation (Savourey and Bittel, 1994). Increased sensitivity of the thermoregulatory system due to sleep loss was reported in a study of male subjects, who, when challenged by a cold test showed higher skin temperature after 27 h of sleep deprivation than non sleep deprived subjects (Savourey and Bittel, 1994). Similarly, Landis et al. (1998) suggested that sleep deprived individuals may have reduced ability to warm even at temperatures thought to be associated with thermal comfort.

Two-way ANOVA for repeated measures were performed on the sleepiness ratings to determine if they differed across studies or days. Subjective sleepiness ratings increase monotonically, but also show clear rhythmic oscillations, whose amplitude appears to increase over time (Fig. 15.4). There is a significant increase in sleepiness ratings from the first to the second day in the Brooks study and, in the case of the Bar-Ilan study to the second and third days (Bar-Ilan: $F(2,16) = 42.53$, $p < 0.0001$; Brooks AFB: $F(1,13) = 97.93$, $p < 0.0001$). Sleepiness ratings also changed significantly across the hours of the day (Bar-Ilan: $F(11,88) = 16.9$, $p < 0.0001$; Brooks AFB: $F(23,299) = 21.4$, $p < 0.0001$).

Table 15.1 Percent of explained variance of monotonic and rhythmic components analyzed by linear regression and complex demodulation (CD). Mean percent of explained variance (±SD)

	Bar-Ilan University Ramat-Gan, Israel		Brooks AFB, San Antonio, TX, USA	
Component	Temperature N = 9	SSS(H) N = 9	Temperature N = 14	SSS N = 14
Linear	13 (13.47)	48.29 (14.14)	14.77 (13.81)	60.58 (14.16)
Circadian	51.8 (13.85)	23.68 (19.19)	54.41 (9.99)	31.51 (9.37)
Hemicircadian	5.87 (4.44)	3.70 (3.75)	8.29 (3.06)	1.95 (1.92)
Total	70.67 (14.77)	75.68 (9.0)	77.47 (11.39)	94.04 (4.02)

The temperature and sleepiness data of each of the subjects were analyzed for linear,[1,2] circadian and hemicircadian components by Complex Demodulation (CD) (Babkoff et al., 1988, 1991a; Sing et al., 1985; Wang et al., 2007).

After linear regression, the residuals were analyzed by CD, using a circadian filter. The second order residuals were analyzed by a hemicircadian filter. Together the three components account for approximately 72–94% of the variance (Table 15.1), although each of the components differs in its contribution to the variance of body temperature and subjective sleepiness. The small overall decrease in body temperature is reflected by the 13–14.77% of the variance accounted for by the linear component in the two studies. In contrast, the linear component accounts for 48.29% of the variance of the SSS ratings in the Bar-Ilan study and for 60.58% of the variance in the Brooks AFB study, reflecting the large overall increase in subjective sleepiness. The circadian component accounts for 51.8–54.41% of the variance in body temperature in the two studies, but for only approximately 23.68–31.51% of the variance of SSS ratings in the two studies. The amount of variance accounted for by the hemicircadian component in body temperature is slightly larger (5.87–8.29%) than for subjective sleepiness (1.95–3.7%).

These results thus support the argument that the homeostatic and circadian factors do not distribute in the same way in the body temperature and subjective sleepiness data sets even in the same subject population subjected to the same conditions of sleep loss. The second question we asked of these data was: Do the two data sets share the same sources of variance? That is, can one predict the level of subjective sleepiness of a given subject during long term sleep deprivation from the changes in his/her body temperature? Correlations between subjective sleepiness and body

[1] Analysis of Circadian and Hemicircadian components by Complex Demodulation (CD) requires input of numbers divisible by 12 and 24. Since only 45 measurements were made in the Brooks AFB experiment, we added three dummy variables, one before the first measurement and two after the last measurement for the CD analysis. The additions were the averages of the two earlier and/or latter measurements. The data were tested by ANOVA without the additions and with the additions and found not to differ.

[2] Various alternatives to a linear slope were tested, e.g., log transformation of the dependent variables, and an exponential fit. The amount of variance accounted for by the alternative regression equations were within 3–4% of that explained by a linear equation. We, therefore, chose to use the best linear fit.

Table 15.2 Correlations between body temperature, and subjective sleepiness ratings. Raw data, linear, circadian and hemicircadian components. Mean r and 95% confidence intervals

	Bar-Ilan University, Ramat-Gan, Israel N = 9	Brooks AFB, San Antonio, TX, USA N = 14
Raw data (BT and SSS)	–0.42 (0.11)	–0.44 (0.136)
Linear	0.31	0.28
Circadian	–0.66[a] (0.15)	–0.57[a] (0.18)
Hemicircadian	–0.05 (0.34)	–0.48 (0.20)

[a]For df = 8; $r > 0.63$, $p < 0.05$. For df = 13; $r = 0.514$, $p < 0.05$.

temperature were performed for each subject separately, using his 36 data pairs in the Bar-Ilan study and 48 data pairs in the Brooks AFB study for the raw data, the circadian and hemicircadian components. This analysis yielded a single correlation per subject. The correlations were averaged for the subjects in each study and 95% confidence intervals assessed (Table 15.2). The correlation of the linear component was performed across subjects in each study since each subject provides only one measure of slope for body temperature and rating of sleepiness each (Table 15.2).

The correlations between the sleepiness ratings and body temperature were negative as expected, but were not significant. Neither the linear regression slope nor the hemicircadian components of body temperature were correlated with those of subjective sleepiness. Only the circadian components of body temperature and sleepiness were significant and negatively correlated ($-0.57 < r < -0.66$).

Despite the fact that experimental protocols, procedures and conditions were somewhat different in the two studies, the relation between body temperature and subjective sleepiness at the level of the individual participant were very similar. This finding reflects the robustness and generality of the circadian component of that relationship during long term sleep loss. The circadian components of body temperature and subjective sleepiness were significant ($-0.57 < r < 0.66$) and account for between 32–43.5% of the variance. However, one must recall that the circadian component accounts for 24–31% of the variance of subjective sleepiness, while it accounts for 51.8–54% of the variance of body temperature. Therefore, the predictive weighting of the circadian component would be around 17% of the raw data, maximally. Consequently, we conclude that the correlations are not large enough to support the hypothesis that body temperature can be used to predict subjective sleepiness, or vice versa. In addition, we conclude that although the linear, circadian and hemicircadian components may be significant sources of variance in body temperature and subjective sleepiness, only the circadian is common to both variables. Thus, the parallelism found between the averaged group body temperature and subjective sleepiness during long term sleep deprivation and forced desynchrony protocols (Froberg, 1977; Gillberg and Akerstedt, 1981; Monk, 1987, 1991; Johnson et al., 1992) is probably due mainly to the endogenous circadian rhythm that is common to both variables. Recent support for this hypothesis was provided by Krauchi et al. (2005) who reported that although accumulated sleep pressure increased subjective sleepiness and EEG slow wave activity it did not influence the thermoregulatory system as measured by core temperature. However the circadian rhythm of sleepiness was significantly correlated and phase locked with core body temperature. They conclude that the human thermoregulatory system seems to be independent of homeostatic pressure to sleep, but the circadian modulation of sleepiness is definitely associated with thermoregulatory changes.

The hemicircadian component accounted for the least amount of variance both in body temperature (5.87–8.29%) and subjective sleepiness (1.95–3.7%). This finding plus the absence of any significant correlation between the hemicircadian components of body temperature and subjective sleepiness may be of interest in light of the reported presence of a hemicircadian rhythm in fatigue-related traffic accidents (Mitler et al., 1988; Summala and Mikkala, 1994). Apparently, although there may be a significant hemicircadian component in sleepiness ratings by sleep deprived subjects, it is very small and not phase-locked to the hemicircadian component of their body temperature.

These conclusions are, thus, similar to the conclusions of other researchers regarding the relationship of body temperature and sleepiness during a normal 24 h schedule (Monk, 1987, 1991). Body temperature does not directly predict subjective sleepiness during long-term sleep deprivation, despite the appearance of parallelism between averaged group body temperature and sleepiness ratings (Figs. 15.3 and 15.4), probably because the two variables do not share all of the sources of variance to the same degree. The monotonic component of subjective sleepiness during sleep deprivation is very strong, while that of body temperature, although significant, is weaker. Both body temperature and subjective sleepiness have significant hemicircadian components, but they are not significantly correlated with each other. The circadian components of the two variables, on the other hand, are significantly correlated.

The suggestion of multiple sources of variance in body temperature and subjective sleepiness, some of which are shared and others, not, implies that there may be experimental manipulations that affect body temperature during sleep loss, but do not affect subjective sleepiness. This is especially true if the manipulation does not operate on the circadian mechanism, shared by the two variables. An example of a manipulation that resulted in a disassociation, i.e., the decoupling of body temperature from subjective sleepiness was reported 20 years ago (Dinges et al., 1987). Napping during the early part of a 54-h sleep deprivation protocol tended to prevent the decrease in body temperature during the subsequent sleep deprivation period, but did not affect sleepiness ratings (Dinges et al., 1987).

These data provide a good example of a disassociation, decoupling or non equivalence between the effects of sleep deprivation on different dependent variables that are likely due to differences between the sources of variance of the dependent variables. Such results may have important implications for attempts that have been made to model and predict changes in performance during moderate to long term sleep deprivation, since these findings imply that the relative weightings of the circadian and homeostatic factors in the effects of sleep deprivation may differ across the various dependent variables (see e.g., Gross et al., 2006). These illustrations also may serve as an introduction to the following sections that deal with the effects of sleep deprivation on the performance decrement on different tasks and of interventions or treatments meant to reduce it (see, e.g., Babkoff and Krueger, 1992; Babkoff et al., 2002).

15.1.3.2 Short Term Sleep Deprivation: Choice Reaction Time (CRT) Versus Working Memory – Effects of Sleep Deprivation, Bright Lights and Caffeine

Among the performance tasks most often found as part of the test battery both in long term and short term sleep deprivation studies are the choice reaction time (CRT) and one or another form of a verbal working memory task. (see, e.g., Babkoff et al., 1985, 1988, 1991, 2002; Dinges and Kribbs, 1991; Pilcher et al., 2007). These are tried and true tasks that generate very reliable and stable performance data that are affected both by short and long term sleep deprivation and usually respond to a variety of interventions. However, as we will try to develop below, a more detailed analysis should indicate that any conclusion of the equivalence of the neural mechanisms associated with the effects of sleep loss on performance or on the actions of various interventions on these two dependent variables might be incorrect.

The original purpose of the study that will be reanalyzed in this section (Babkoff et al., 2002) was to test two intervention treatments on psychomotor and cognitive performance during a night shift work schedule in individuals who had not slept during the day prior to the night shift work schedule. Two interventions were tested: (1) 1 h exposure to very bright lights mid-way in the night shift work schedule; and (2) the ingestion of 200 mg caffeine. The following section explains the background for that study.

Background to Bright Light-Caffeine Study

From the 1990s to the present, a large number of laboratory studies have shown that bright illumination (1,000–5,000 lx) can improve nocturnal cognitive performance, attenuate the circadian reduction in body temperature and enhance alertness in shift workers and sleep deprived subjects (Badia et al., 1991; Campbell and Dawson, 1990; Daurat et al., 2000; French et al., 1990, 1991; Hannon et al., 1992). Other researchers have reported that alertness and performance were improved when subjects were exposed to bright light throughout a 24-h period, although there was no effect on the circadian-related nocturnal drop in temperature (Daurat et al., 1993). Similarly, a large number of studies had shown that caffeine has a general psycho-stimulant effect in normal conditions as well as after restricted sleep and in irregular work schedules (Rosenthal et al., 1991). Caffeine reduces reaction time, enhances vigilance performance, increases self-rated alertness and improves mood (Bonnet et al., 1995; Jacobson and Edgley, 1987; Loke, 1988). The caffeine dose-response curve is non-monotonic (Lieberman et al., 1987). Although caffeine enhances performance at moderate levels (~200–300 mg), there is often a reversal at higher levels (~400–600 mg) (Bonnet et al., 1995; Jacobson and Edgley, 1987).

Although the site and mode of action of bright light and caffeine on melatonin may differ (Wright et al., 1997a, b) nevertheless, their overall effect is similar. Studies by Wright et al. (1997a, b) tested the effects of continuous exposure to a bright light and 200 mg caffeine (administered twice during the night), separately and together, on melatonin levels, body temperature, alertness and performance in sleep deprived subjects. Wright et al. (1997a, b) reported that either exposure to the bright light or the ingestion of caffeine suppressed melatonin, attenuated the circadian decrease in body temperature and enhanced alertness and performance. Caffeine together with exposure to the bright light suppressed melatonin, attenuated the decrease in body temperature, enhanced alertness and performance to a greater extent than either treatment alone.

For the purpose of the present section, we reanalyzed the performance data of the CRT and working memory tasks and include short explanations of the role of endogenous melatonin in the sleep-wake cycle and its suppression by bright light as well as the effects of caffeine on performance decrements due to sleep loss, so as to set the stage for the demonstration of a disassociation between the effectiveness of the bright light and caffeine treatments on choice reaction time and working memory.

Melatonin

Melatonin has been implicated in a number of physiological functions, such as, regulation of circadian rhythms, sleep and body temperature, sexual maturation, immune function, antioxidant mechanisms, regulation of mood, cardiovascular functions, etc. (see e.g., Arendt, 1995, 1998; Macchi and Bruce, 2004)). Among these various functions, there has been increased scientific interest in the soporific and sleep/wake rhythm regulating functions of melatonin, resulting in numerous studies undertaken in animals and in human subjects. The finding that melatonin is secreted primarily during the night (Lynch et al., 1975), the reports that suppression of melatonin production by procedures, such as the treatment with b-blockers were shown to correlate with insomnia (Brismar et al., 1988) and the many reports that implicated melatonin as a soporific (e.g., Campbell et al., 1999) has led to numerous studies of the role of melatonin in the sleep wakefulness circadian rhythm.

The hormone melatonin is secreted at night from the pineal gland, regulated by the suprachiasmatic nuclei (SCN), that serves as the circadian pacemaker, located in the hypothalamus that, in turn, receives its input on the presence or absence of daylight via the retino-hypothalamic tract among others (Cardinali and Pevet, 1998). The nocturnal increase in plasma melatonin concentration is accompanied by a decrease in core body temperature and an increase in sleep propensity (Wright et al., 1997a, b, c). The close relationship between the increase of plasma melatonin during the night, the timing of human sleep and the decrease in core body temperature have led many investigators to suggest that melatonin is involved in the physiological regulation of sleep. The onset of nighttime melatonin secretion occurs approximately 2 h before bed time and has been shown to correlate well with the onset of evening sleepiness (Tzischinsky et al., 1993; Zhdanova and Tucci, 2003; Zhdanova et al., 1996). The effect that melatonin has on sleep promotion is presumably due to its inhibitory effect on SCN activity (Liu et al., 1997; Hunt et al., 2001). Exposure of the SCN to melatonin causes inhibition of SCN metabolism and neuronal firing rate (Mason and Brooks, 1988). The firing rate of SCN neurons increases throughout the day and reaches its maximum at the evening, prior to the onset of melatonin production; this increase has been linked to the wakefulness promoting effect of the circadian pacemaker (Buysse et al., 1993; Long et al., 2005). As discussed above, it is thought that melatonin-induced inhibition of SCN underlies the 'opening of the sleep gates' by melatonin. Sleep is also associated with a decline of core body temperature. Though melatonin affects both sleep and thermoregulation, these two processes might be independent of each other (Zhdanova, 2005), or at least the monotonic component.

Melatonin onset has been used as a clinical tool to evaluate problems related to the onset or offset of sleep. The circadian rhythm of melatonin in saliva or plasma, or of the melatonin metabolite 6-sulphatoxymelatonin (aMT6S) in urine, is an accepted characteristic of suprachiasmatic nucleus (SCN) function, which has been identified as the endogenous oscillatory pacemaker. Pandi-Perumal et al. (2007) have argued that within this rhythmic profile, the onset of melatonin secretion under dim light conditions is the single most accurate marker for assessing the circadian pacemaker. The Dim light melatonin onset time has been identified as one of the most reliable markers of the phase of the circadian pacemaker and is currently being used as an effective metric for studies of sleep phase delay. Through the use of the dim light melatonin onset time marker, the phase advance and phase delays of the melatonin phase response curves can be differentiated, thus permitting a diagnosis of whether an individual is entrained to the 24-h light-dark cycle or is free running. Dim light melatonin onset time is therefore a very popular marker for identifying optimal application times for therapies such as bright light or exogenous melatonin treatment (Pandi-Perumal et al., 2007). Light induced melatonin suppression is believed to impact the circadian pacemaker directly (Kubota et al., 2002).

The proposed explanation for the improvement in alertness and performance when subjects are exposed to bright lights during the night is that the hormone melatonin, normally in abundance during the night, is suppressed by bright light (McIntyre et al., 1989). Exposure to bright light during the night in an industrial setting has also been reported to alter morning urinary melatonin levels (Budnick et al., 1995). The suppression of melatonin during exposure to bright light and its recovery after the light is extinguished is related to the intensity of the light (McIntyre et al., 1989). After exposure to a 3,000 lx light for 1 h, melatonin was reduced to ~30% of its level at peak (McIntyre et al., 1989). Since melatonin has been associated with an increase in fatigue (Lieberman et al., 1984), a reduction in melatonin may account for the alerting action of bright light.

Caffeine

In rested individuals caffeine is most effective with cognitive and psychomotor functions, associated with alertness and attention such as visual vigilance, choice reaction time, and fatigue (POMS), but less effective with cognitive functions that require complex information processing, such as memory or reasoning (Amendola et al., 1998; Battig et al., 1984; Lieberman et al., 1987). Several studies have found that complex tests of cognitive function are improved by caffeine in sleep-deprived volunteers (Beaumont et al., 2001; Penetar et al., 1993; Lieberman et al., 2002), but these effects may be secondary to the effects of caffeine on alertness and attention. For example, Lieberman et al. (2002) have suggested that when an individual is engaged in complex cognitive tasks during periods of sleep loss, the difficulty to maintain attention could result in performance decrement and suggest that such an interpretation may be consistent with the action of caffeine in the CNS. That may also explain why caffeine's effects on higher cognitive functions are only observed when individuals are sleep-deprived. It is widely agreed that caffeine's effects on the brain, when it is administered in doses commonly consumed by humans, are mediated by two classes of adenosine receptors, A1 and A2A, to which caffeine binds and blocks the action of agonists on them (for a review see Fredholm et al., 1999). These receptors, especially the A1 class, are closely associated with regulation of alertness. Mesopontine cholinergic neurons are associated with regulation of arousal level and are under tonic A1 receptor control (Basheer et al., 2000; Portas et al., 1997; Strecker et al., 2000). Adenosine A1 agonists, when delivered directly to the basal forebrain, inhibit neurons associated with maintaining wakefulness and A1 antagonists, including xanthines, increase the activity of these neurons (Basheer et al., 2000). It has been suggested that this region is a site where caffeine exerts its effects on arousal level by acting to block the inhibitory effects of endogenous adenosine (Rainnie et al., 1994). Caffeine's direct effect on neuronal systems associated with arousal may explain its relatively selective effects on behaviors associated with alertness, such as vigilance and fatigue, and the possible absence of direct effects on other neural systems such as those regulating learning, memory and perception.

Wyatt et al. (2004) studied the effectiveness of a high-frequency low-dose caffeine regimen in ameliorating performance decrement during extended wakefulness. Their experiment was designed to address the question whether the effectiveness of caffeine in antagonizing the effects of extended wakefulness was related to its interaction with the homeostatic or the circadian factors that modulate performance and tendency to fall asleep. Sixteen subjects participated in a double-blind, placebo-controlled, parallel-group design in a 29-day forced desynchrony paradigm in which the period of the sleep-wake cycle was scheduled to be 42.85 h. This design allowed for separate estimation of the sleep homeostatic, circadian, and caffeine contributions to changes in performance. The authors administered caffeine according to a schedule designed to increase the plasma concentration in parallel to the hypothesized rate of increase in the sleep homeostatic drive during wakefulness. The homeostatic drive for sleep during extended wakefulness has been hypothesized to follow the accumulation of adenosine (Benington et al., 1995; Porkka-Heiskanen et al., 2000; Radulovacki et al., 1982; Wyatt et al., 2004). Wyatt et al. (2004) explain the significant reduction of deficits seen on several measures of neurobehavioral functioning by caffeine as an attenuation of wake-dependent impairment. Repeated low-dose caffeine administration was effective in reducing decrement in the addition (ADD) and digit symbol substitution tasks (DSST), and the PVT. Caffeine also reduced/inhibited EEG-verified accidental sleep onsets during the scheduled wake periods. The authors argued that the observed reduction in accidental sleep onsets supports the conclusion that caffeine attenuates expression of the homeostatic sleep drive. They also suggested that since the plasma caffeine concentrations would be expected to affect only adenosine receptors and not other neuro-pharmacological mechanisms and because caffeine primarily affects the sleep-wake–dependent modulation of performance, their results support the hypothesis that adenosine mediates sleep-wake–dependent modulation of sleep propensity and associated variation in neurobehavioral functioning.

In summary, the evidence seems to favor the conclusion that the actions of bright lights and that of caffeine in restoring performance in sleep deprived individuals are very likely via different neural mechanisms. While exposure to bright lights very likely interacts with the actions of the circadian factor, the ingestion of caffeine very likely interacts with the homeostatic factor in sleep deprivation (see, e.g., Porkka-Heiskanen et al., 2000; Strecker et al., 2000).

The study by Babkoff et al. (2002)

In the study by Babkoff et al. (2002), six men and five women, aged 19–36 were tested four times over a 4 week period. The "shift-work schedule" consisted of eight testing sessions, the first of which began at 1730, and the last of which began at 0830 the following morning. Session-to-session onset time was 2 h, except for the period between the 2330 and 0230 sessions which included the 1 h rest period (from 0130 to 0230) as well.

During the 1 h rest period midway in the work-shift, subjects were exposed to one of four experimental conditions: (1) 1 h exposure to ~20–50 lx from 0130 to 0230 and placebo at 0140 (Dim Light-Placebo); (2) 1 h exposure to ~3,000 lx from 0130 to 0230 and placebo at 0140 (Bright Light-Placebo); (3) 1 h exposure to ~20–50 lx from 0130 to 0230 and caffeine at 0140 (Dim Light-Caffeine); (4) 1 h exposure to ~3,000 lx from 0130 to 0230 and Caffeine at 0230 (Bright Light-Caffeine). The experimental conditions were separated from each other by at least 1 week. During each session the participants performed four different computer tests for a total of ~90–95 min. Work-to-rest ratio was, therefore, approximately 0.75. All testing took place in dim light (less than 50 lx). The order of the four experimental conditions was randomized across groups of three participants each over the four test weeks. Although with regard to the bright/dim light conditions, the experiment was not double-blind, with regard to the caffeine/placebo condition, it was.

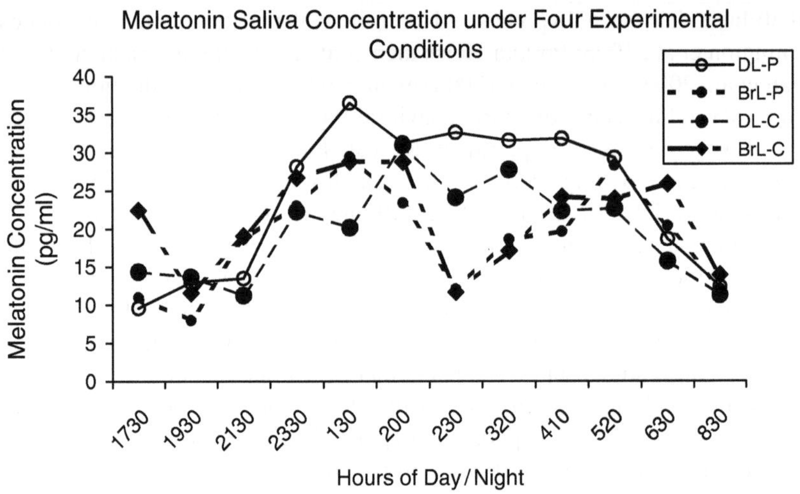

Fig. 15.5 Melatonin saliva concentration is plotted on the ordinate as a function of hour of the day/night on the abscissa. Data for the four different experimental conditions are plotted separately. See text for full explanation

The concentration of salivary melatonin as a function of time of day/night is shown in Fig. 15.5 (adapted from Babkoff et al., 2002). The overall saliva concentration of melatonin, measured from 0230 to 0410, differed significantly across the four conditions ($F(3, 36) = 9.384812$; $p < 0.00012$). Under the dim light-placebo condition the concentration of saliva melatonin began to increase between 2130 and 2330, reached a peak between 0130–0200 and remained at peak level until around 0510, when it began to decrease to the daytime level (Fig. 15.5, Dim Light-Placebo treatment (DL-P)). From 0230–0410, the mean saliva-melatonin concentration in the Dim Light-Placebo condition was significantly higher than in any other condition (30.845 pg/ml, sd = 8.95).

The shapes of the melatonin curves for the Bright Light-Placebo (BrL-P) and Bright Light-Caffeine (BrL-C) treatments are quite different. For the latter treatments, the concentration of salivary melatonin also rises from the evening level to a peak at around 0130, paralleling the DL-P condition, but at 0200, midway into the bright light treatment, both the the BrL-P and the BrL-C curves began to diverge sharply and to drop to their lowest levels, measured at 0230, at the time the bright light treatment was terminated. At 0230 saliva concentration under the BrL-P and BrL-C conditions, began to increase until ~0520, when all of the curves began to decrease in a parallel manner. Salivary melatonin concentration under the DL-C condition differed from the DL-P condition and showed a 21.3% decrease in melatonin concentration from 0230 until 0410 (24.26 pg/ml, sd = 8.93; $t(18) = 1.75$; $p < 0.048$). During the same period (from 0230 until 0410) the mean saliva melatonin concentration in the Bright Light-Placebo condition was reduced 42.4% (17.76 pg/ml, sd = 4.95; $t(18) = 4.3056$; $p < 0.0002$) and in the Bright Light-Caffeine condition, saliva melatonin concentration was reduced 47.4% (16.23 mg/ml, sd = 3.73; $t(18) = 5.4415$; $p < 0.0001$).

The bright light melatonin suppression effect is intensity and duration dependent (McIntyre et al., 1989). Recovery of plasma melatonin concentration begins as soon as subjects leave a bright light environment (McIntyre et al., 1989). Since the measurement of saliva melatonin concentration phase-lags that of plasma concentration, the salivary melatonin bright light suppression data of Babkoff et al. (2002) are plotted in Fig. 15.6 together with those published by McIntyre et al. (1989) for comparison with a phase advance of 25 min assuming the suppression in plasma preceded the detection of suppression in saliva by approximately that amount of time. Note that with a phase advance of 25 min the reduction in saliva melatonin concentration both under the bright light – placebo and the bright light – caffeine conditions (Babkoff et al., 2002) almost completely overlap the reduction in plasma melatonin concentration (McIntyre et al., 1989), especially during the suppression phase. Figure 15.6 stresses the equivalence of the dynamics of plasma and saliva melatonin concentration.

The nocturnal phase of the circadian rhythm apparently impacts both the CRT task and the working memory task in what appears to be a similar manner. Figure 15.7 shows reaction time on the CRT task and the number of correct responses (i.e., correct choice of letters in the test matrices on the working memory task). Both data sets have been transformed to standardized scores to facilitate comparison of the effects of a night work shift on individuals who had not slept during the previous day. Both curves show the classical impact of the circadian rhythm, and begin to increase around 2130. This rise indicates a slowing of speed on the CRT task and fewer hits on the working memory task. The two curves remain elevated until between 0430 and 0630 at which time both curves return to an earlier level of performance.

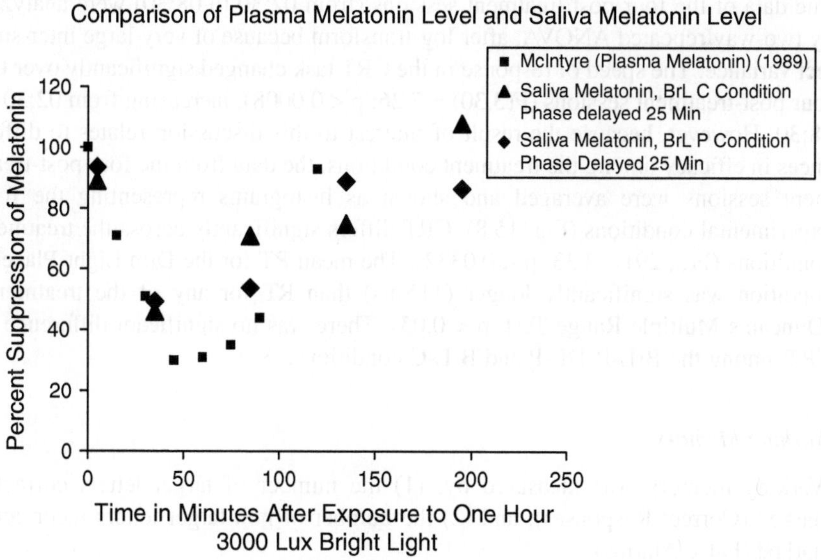

Fig. 15.6 Percent suppression of salivary/plasma melatonin concentration is plotted on the ordinate as a function of time (in minutes) after the onset of a 1-h duration 3,000 lx light. Compare the plasma melatonin levels with the saliva melatonin levels

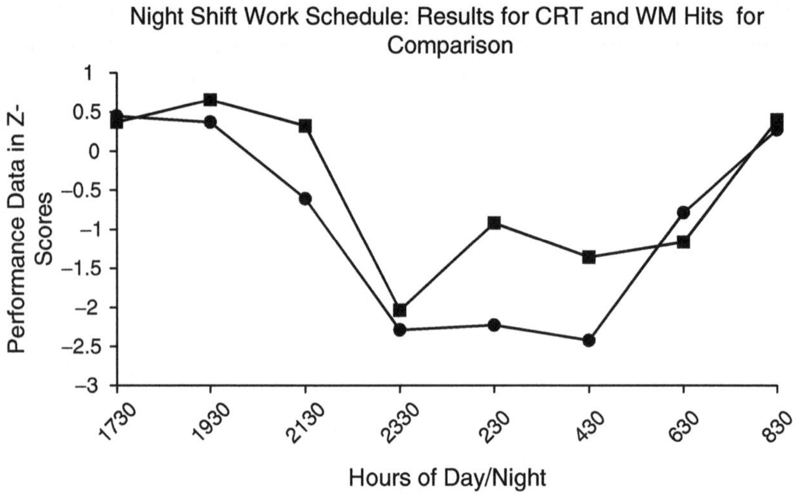

Fig. 15.7 Performance data (in Z scores) for choice reaction time (CRT) and working memory (WM) are plotted on the ordinate as a function of time of day/night on the abscissa when subjects had not slept the previous day before the beginning of the night work schedule

The Effectiveness of Two Interventions: Exposure to Bright Light and Ingestion of Caffeine

Choice Reaction Time

The data of the four post treatment sessions (from 02:30 to 08:30) were analyzed by two-way repeated ANOVA, after log-transform because of very large inter-subject variance. The speed of response in the CRT task changed significantly over the four post-treatment sessions ($F(3,30) = 7.26$; $p < 0.0008$), increasing from 02:30 to 08:30. However, because the result of interest to this discussion relates to differences in efficacy among the treatment conditions, the data from the four post-treatment sessions were averaged and shown as histograms representing the four experimental conditions (Fig. 15.8). CRT differs significantly across the treatment conditions ($F(3, 29) = 3.33$; $p < 0.0332$). The mean RT for the Dim Light-Placebo condition was significantly longer (115 ms) than RT for any of the treatments (Duncan's Multiple Range Test, $p < 0.05$). There was no significant difference in CRT among the BrL-P, DL-P and BrL-C conditions.

Working Memory

Working memory was measured by: (1) the number of target letters correctly marked (Correct Responses); and (2) the number of non-target letters incorrectly marked (False Alarms).

Correct Responses: The number of correct responses changed significantly over the last four sessions ($F(3, 30) = 3.28$; $p < 0.05$). The largest number of hits was

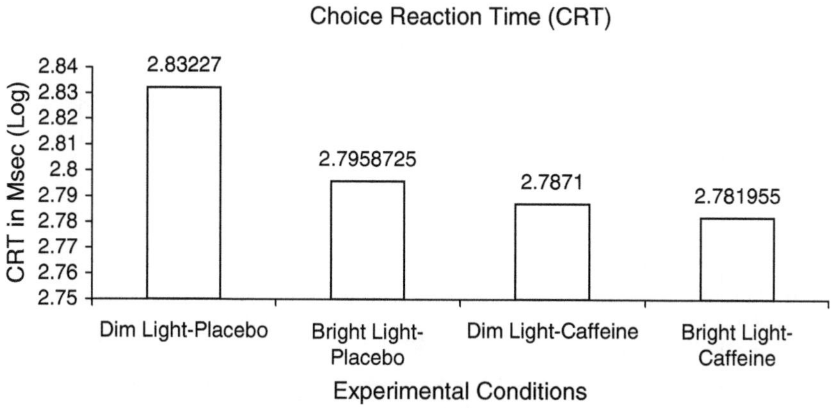

Fig. 15.8 Choice Reaction Time (CRT in $\log_{10}$) averaged over the last four post-treatment sessions is plotted on the ordinate for each of the four treatments. See text for full explanation

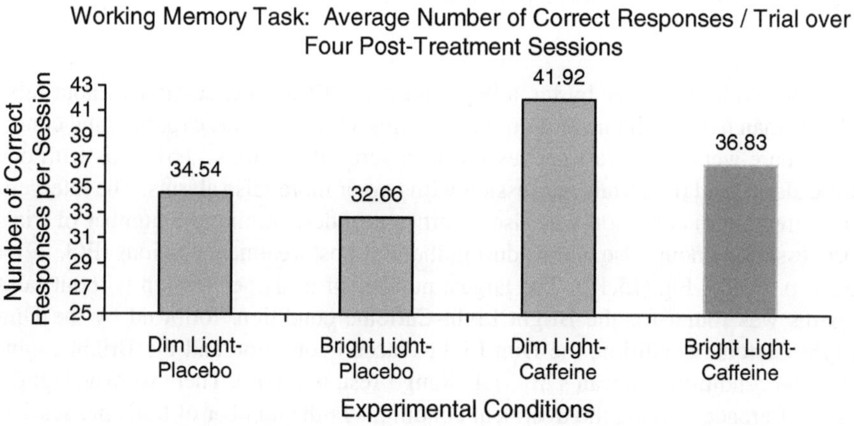

Fig. 15.9 The number of correct responses (hits) per trial averaged over the four post-treatment sessions is plotted for each of the four treatments. See text for full explanation

recorded for the Dim Light-Caffeine condition, which differed significantly from all the other treatments (Duncan's Multiple Range Test, $p < 0.05$) (Fig. 15.9).

False Alarms: Because many trials did not have any false alarms, the distribution of false alarms was exponential in shape, and parametric statistics were inappropriate to test the findings. The false alarm data were, therefore, grouped into three categories: (1) the number of trials per session with no false alarms; (2) the number of trials per sessions with one false alarm; and (3) the number of trials per session with two or more false alarms.

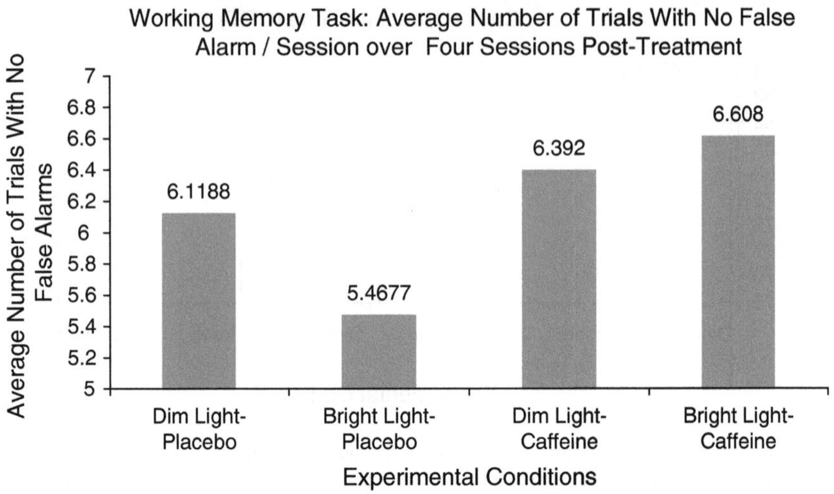

Fig. 15.10 The number of trials per session with no false alarms averaged over the four post-treatment sessions is plotted for each of the treatments. See text for full explanation

In general, there was, by far, a larger number of trials per session without false alarms than trials with one or more false alarms. Over all of the experimental conditions there were 6.41 trials per session with zero false alarms; 0.61 trials with one false alarm; and 0.52 trials per session with two or more false alarms. The intervention (treatment) condition was also significant in determining the number of trials per session without false alarms during the four post treatment sessions ($F(3, 29) = 3.35$; $p < 0.05$, Fig. 15.10). The largest number of trials per session without false alarms was found for the Bright Light-Caffeine condition, followed by the Dim Light-Caffeine condition, the Dim Light-Placebo condition and the Bright Light-Placebo condition (Duncan's Multiple Range Test, $p < 0.05$). There were no significant differences among the treatment conditions in the number of trials per session with one false alarm or two or more false alarms ($F(3, 29) = 1.51$, NS; $F(3, 29) = 1.29$, NS).

These findings indicate that the two tasks, CRT and working memory are affected differently by the two treatments, bright light and caffeine when subjects participate in a night shift work schedule without having slept during the previous day. By implication, a disassociation in the effectiveness of two different intervention treatments that are known to affect different brain mechanisms of decremented performance during sleep deprivation, means that the two tasks are associated with different brain mechanisms. We suggest that the overall variance in the CRT, which is a fairly simple visual discrimination and response selection task, with very little working memory requirement, and relatively less associated with the level of frontal cortical activity, is weighted both by the circadian component and by the monotonic homeostatic drive of accumulating wakefulness together. The circadian component, in turn, is strongly influenced by the concentration of plasma melatonin, which is suppressed by the bright light. The homeostatic component is antagonized by the caffeine. When either exposure to bright lights or the ingestion of caffeine interacts with its respective factor, it reduces the effect, either of the circadian or of the homeostatic factor, and as a result, CRT performance is enhanced or restored.

The working memory task, in contrast, is strongly associated with intact prefrontal cortex activity (see below for expanded discussion) and is most affected by the homeostatic component of accumulated wakefulness. Caffeine, which attenuates the expression of the homeostatic pressure for sleep, reduced the effects of sleep deprivation on performance in the working memory task, whereas, exposure to bright lights did not significantly reduce the effects of sleep deprivation on working memory.

15.1.3.3 A More Direct Test of the Sleep Based Neuro-Psychological Perspective Hypothesis: Comparison of Performance on Different Tasks

The final study to be presented in this chapter was part of a dissertation (Zuckerman, 2004), one of whose purposes was to test the sleep-based neuropsychological perspective more directly by comparing the performance of the same subjects on a number of cognitive and psychomotor tasks, when they were not sleep deprived and when they experienced short term sleep deprivation. The comparison included tasks strongly associated with intact prefrontal activation and those that are not strongly associated with intact prefrontal activation.

Rationale A fairly large number of studies over the past decade have provided evidence for the argument that sleep and sleep deprivation not only impact global brain functions but affect regional brain functions as well (Kajimura et al., 1999). The effects of sleep deprivation on brain function and associated cognitive performance were hypothesized to represent local as well as global brain processes. Specifically, since a large number of recent studies have targeted the prefrontal cortex as especially sensitive to sleep loss (e.g., Drummond et al., 2000, 2001; Muzur et al., 2002; Thomas et al., 2000) the hypothesis tested posited that cognitive tasks associated with intact prefrontal activation would be relatively more sensitive to the effects of short term sleep deprivation than tasks that are less so associated (see Babkoff et al., 2005 for fuller explanation).

Tasks The five tasks chosen for the purpose of comparing performance under conditions of short term sleep deprivation, were: (1) bimodality reaction time, i.e., simple RT to either tones or visual stimuli, with no discrimination component (BRT); (2) choice reaction time, compatible response to each of five spatially separated visual stimuli on computer monitor screen (CRT); (3) paired associates (task requires memory of 10 word lists; (4) memory for correct sequence of words appearing in the list; (5) working memory, i.e., the subject is instructed to mark the three letters (in a matrix of 100 letters) to which he/she was exposed for 30 s prior to the appearance of the test matrix. The first two tasks were considered tasks for which optimal performance may have little if any association, with intact prefrontal activation; the third task, paired associates activates areas in prefrontal cortex, but also activates many other areas of brain, while the last two were considered tasks for which optimal performance is strongly associated with intact prefrontal activation.

Most if not all cognitive performance tasks activate some part of the prefrontal cortex, especially if the task includes some type of sequencing and/or executive control. However, some cognitive tasks have a larger proportion, if not the majority of their activation nodes localized within prefrontal cortex, while other tasks have a more global and broader representation of activation within the brain. Among the tasks that have been identified as having large local representation of activity with the prefrontal cortex are working memory and memory for the order or the correct sequence of items in a list. Among the tasks that have local representation in prefrontal cortex, but are basically multi-nodal including many areas outside of prefrontal cortex, is the paired association memory for words.

Working memory Cognitive theorists have posited that higher functions such as language, planning, and problem-solving all rely on working memory, which acts to temporarily maintain and manipulate task-relevant information (Baddeley, 1986; Just and Carpenter, 1992; Shallice, 1988). Working memory enables us to hold in our 'mind's eye' the contents of our conscious awareness, even in the absence of sensory input, by maintaining an active representation of information for a brief period of time (Courtney et al., 1998; Curtis et al., 2000).

Reviews of the neurophysiological data from nonhuman primates suggest that prefrontal cortex is an important component of the neural substrate for working memory. In single-cell recording studies of dorsolateral prefrontal cortex, researchers have identified neurons that remain active during delay periods in tasks that require the subject to actively maintain an internal representation of a target stimulus (Fuster, 1989; Goldman-Rakic, 1987). In humans, reports of working memory impairment in patients with damage to the frontal cortex support the involvement and importance of this brain region in performance based on working memory (Shimamura, 1994; Stuss et al., 1994). Recently, neuroimaging data have also begun to contribute to our understanding of the neurobiology of human working memory. Studies using both positron emission tomography (PET) and functional magnetic resonance imaging (fMRI) have provided strong evidence of prefrontal cortex involvement in a wide variety of working memory tasks (Cohen et al., 1994; Courtney et al., 1998; Fiez et al., 1996; Jonides et al., 1993; McCarthy et al., 1994; Petrides et al., 1993; Swartz et al., 1995). Courtney et al. (1998) presented a comprehensive review of the neurophysiological and fMRI literature up to 1998 and concluded that the evidence from brain-imaging studies shows that prefrontal cortex has sustained activity during the delay period of visual working memory tasks, indicating that this cortex maintains on-line representations of stimuli after they are removed from view. They also posited that there are prefrontal areas brought into play during the monitoring and manipulation of information in working memory in addition to those engaged during the maintenance of this information.

Perlstein et al. (2001) used tasks that parametrically manipulated working memory load to compare fMRI prefrontal cortex activity in 17 schizophrenic patients and 16 normal controls. They reported that schizophrenic patients showed a deficit in physiological activation of the right dorsolateral prefrontal cortex in the context of normal task-dependent activity.

Patients with greater dorsolateral prefrontal cortex dysfunction performed more poorly. Perlstein et al. (2001) concluded that their results are consistent with the hypotheses that working memory dysfunction in patients with schizophrenia is caused by a disturbance of the dorsolateral prefrontal cortex and that this disturbance is selectively associated with cognitive deficit in more associative components of working memory.

Memory for sequences There is growing evidence that the memory for the sequence of items in a list is also strongly associated with intact activity in prefrontal cortical regions. Ninokura et al. (2003) reported that prefrontal cortical cells in primates are involved in maintaining temporal order information when that information is necessary for planning forthcoming motor behavior. In humans, memory for the spatial or temporal order of items or events has been associated with activity in ventrolmedial prefrontal cortex (Szatkowska et al., 2004). Amiez and Petrides (2007) recently reported the results of an fMRI study of regional brain activity while subjects were coding the precise order of a short sequence of abstract visual stimuli. The results demonstrated the involvement of the mid-dorsolateral prefrontal cortex. They concluded that the availability of such detailed serial-order information in working memory allows high-level cognitive planning and mental manipulation, functions that depend on the prefrontal cortex.

Memory of paired associates A network of prefrontal and posterior cortical structures (Warrington and Weiskrantz, 1982; Goldman-Rakic, 1988) which possesses distinct roles in the different components of episodic memory has been proposed by a number of researchers. Previous investigations delineated widespread brain systems in verbal episodic memory including the prefrontal cortex (e.g. Goldman-Rakic, 1987; Grasby et al., 1993; Fletcher, Dolan, et al., 1995, and Fletcher, Frith et al., 1995; Blaxton et al., 1996; Halsband et al., 1998) and the cingulate gyrus (e.g. Valenstein et al., 1987; Rudge and Warrington, 1991; Halsband et al., 1998). Krause et al. (1999) studied the functional anatomy of paired word associates memory tasks. They chose a paradigm in which the word pairs (highly imaginable as well as abstract) were semantically unrelated and presented visually or auditorily during either the encoding or retrieval phase in two groups of six normal subjects each. Regional cerebral blood flow (rCBF) was measured using PET with [15O] butanol during episodic encoding and retrieval. The authors reported that the precuneus activation occurred in visual or auditory presentation modalities and for both highly imaginable and abstract words. Their study therefore provides further evidence that the precuneus has a specific function in episodic memory retrieval as a multimodal association area. Thus it appears that associative memory is an example of a task that is associated with wide spread brain activation including some areas of prefrontal cortex.

A commonality of brain regions observed in a study by Hichwa (1995) suggests that the human brain may contain a distributed multinodal general memory system. Nodes on this network include the frontal, parietal, and temporal cortices, the thalamus, the anterior and posterior cingulate, the precuneus, and the cerebellum. There appears to be a commonality of components across tasks (e.g., retrieval, encoding) that is independent of content, as well as differentiation of some components that may be content-specific or tasks-specific. In addition, these results support a significant role for the cerebellum in cognitive functions such as memory (Hichwa, 1995). In summary, it appears that paired associates memory represents multi-nodal activation in a number of brain regions including some areas of the prefrontal cortex.

The study A number of recent studies of cognitive tasks for which there is neuroimaging evidence of activation of prefrontal cortex, reported not finding significant effects of short term sleep deprivation (see e.g., Harrison and Espelid, 2004). The authors (Harrison and Espelid, 2004) designed their study to test the effects of sleep deprivation in one session between 600–1800, so as to avoid the influence of the circadian cycle. Consequently their report reflects the results of one test period only. If the effect of sleep deprivation on the cognitive task that they used is not robust and inter-individual variance was relatively large, the lack of a positive significant finding should not be considered as contradictory evidence to the sleep-based neuropsychological hypothesis.

In contrast to some earlier studies (Harrison and Espelid, 2004), the present study was designed to allow multiple testing during a day following either sleep deprivation or normal nocturnal sleep. However, we were also apprehensive about the possible interaction of the circadian rhythm. The present study was, therefore, designed so that the first testing session began after the nadir of the circadian rhythm of the first day had passed in order to emphasize the impact of the homeostatic component of accumulated wakefulness, with minimal contribution by the circadian component. Similarly, all of the sessions were completed before the expected big decline in performance associated with the circadian rhythm of the second day. The six sessions took place at approximately 0830, 1030, 1245, 1530, 1830 and 2045. The purpose of the study was to compare the performance of the same subjects on cognitive tasks associated with intact prefrontal activation with their performance on tasks less or not at all associated with intact prefrontal activation under conditions of short term (24–40 h) sleep deprivation. A detailed description of the procedure and methods has been published elsewhere together with the electrophysiological results (Babkoff et al., 2005; Zuckerman et al., 2007).

The five tasks were administered during each of the six sessions on the day following normal nocturnal sleep and on the day following sleep deprivation. The same subjects participated in the non sleep deprived (NSD) and in the sleep deprived testing sessions. One week separated each of the two testing days. One test day followed normal nocturnal sleep; the other test day followed a night of sleep deprivation. Subjects spent the night prior to testing at home. Subjects were told that during the night they were sleep deprived they could read, watch TV or videos, but must refrain from physical exercises, smoking or ingesting any foods or beverages that contained caffeine or alcohol. In addition, they were to remain at home. Subjects were also instructed that on the sleep-deprivation night, they were to telephone the laboratory every 20 min beginning from 20:00 to 06:00 the next morning and to leave a short message verifying they were awake. Compliance with the stay-at-home instruction could be monitored by noting the phone number used to relay the message. Subjects were also told that if they failed to call even once during the sleep deprivation night, they would not be tested the following day. The eight subjects whose data were analyzed and are presented below complied fully with the instructions.

Results The results are shown in Table 15.3 and Figs. 15.11 and 15.12. Of the five tasks, sleep deprivation, as a main factor, only reduced performance significantly on the two tasks that are strongly associated with intact prefrontal activity, memory of sequence and working memory (F(1,7) = 12; p < 0.01 and F(1,7) = 8.2; p < .024, respectively). The diurnal rhythm did not affect performance of either of these tasks. In contrast, three tasks that are not generally associated with intact prefrontal

Table 15.3 Results of ANOVAs

Task	A. Sleep deprived vs non sleep deprived	B. Diurnal effect (session)	Interaction (A x B)
Bimodality RT	F(1,7) = 0.03; NS	F(5,35) = 5.095; p < 0.001	F(5,35) = 1.098; NS
Choice RT	F(1,7) = 1.007; NS	F(5,35) = 3.374; p < 0.014	F(5,35) = 0.45; NS
Paired associates	F91,7) = 0.168; NS	F(5,35) = 3.216; p < 0.017	F(5,35) = 1.286; NS
Memory for sequence	F(1,7) = 12; p < 0.01	F(5,35) = 0.68; NS	F(5,35) = 0.42; NS
Working memory $\log_{10}$ (letter check)	F(1,7) = 8.2; p < 0.024	F(5,35) = 2.243; NS	F(5,35) = 2.04; NS

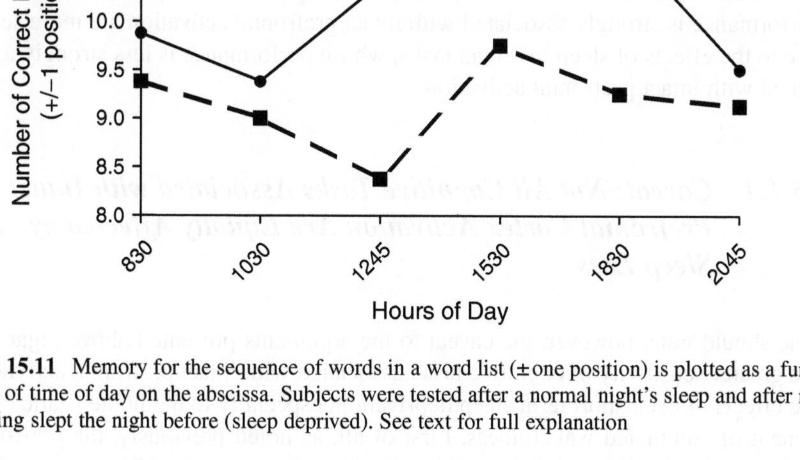

Fig. 15.11 Memory for the sequence of words in a word list (± one position) is plotted as a function of time of day on the abscissa. Subjects were tested after a normal night's sleep and after not having slept the night before (sleep deprived). See text for full explanation

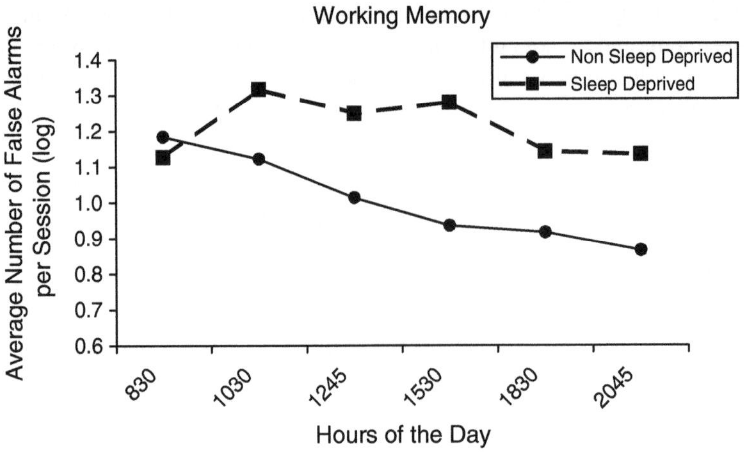

Fig. 15.12 Average number of false alarms/session (in $\log_{10}$) is plotted as a function of time of day on the abscissa. Subjects were tested after a normal night's sleep and after not having slept the night before (sleep deprived). See text for full explanation

activity, paired associates, bimodal RT and choice RT were not significantly impacted by sleep deprivation, but were significantly impacted by the diurnal rhythm ($F(5,35) = 3.216$; $p < 0.017$, $F(5, 35) = 5.095$; $p < 0.001$ and $F(5, 35) = 3.374$; $p < 0.014$, respectively). Figures 15.11 and 15.12 show the impact of sleep deprivation on performance of the memory of sequence task (number of correct sequences allowing for one positional error) and working memory task (number of incorrect responses or false alarms) respectively.

In conclusion, these results are another example of a disassociation between the effects of short term sleep deprivation on tasks that have been shown to be associated with activation in different parts of the brain. These data support the argument that under conditions of short term sleep deprivation and when there is minimal decrement due to the circadian rhythm, i.e., during the diurnal phase, tasks, whose performance is strongly associated with intact prefrontal activation are more sensitive to the effects of sleep loss than tasks, whose performance is less strongly associated with intact prefrontal activation.

15.1.4 Caveat: Not All Cognitive Tasks Associated with Intact Prefrontal Cortex Activation Are Equally Affected by Sleep Loss

One should note, however, the caveat to the arguments presented above regarding the greater sensitivity of cognitive tasks associated with intact prefrontal activity to the effects of even short term sleep deprivation, especially to the homeostatic component of cumulated wakefulness. First of all, as noted previously, the prefrontal cortex is not the only brain region activated when subjects perform the working memory task. Together with the overwhelming evidence that associates working memory with the prefrontal cortex, some studies have shown that working memory recruits other cortical regions as well, e.g., the medio-temporal networks (Honey et al., 2000); posterior parietal cortex, Broca's area, suggesting the possibility that working memory function is subserved by multiple brain regions. For example, Braver et al. (1997) used fMRI to probe prefrontal cortex activity during a sequential letter task in which memory load was varied in an incremental fashion. In all nine subjects, dorsolateral and left inferior regions of prefrontal cortex were identified that exhibited a linear relationship between activity and working memory load. The prefrontal areas were not the only brain regions activated during the performance of a working memory task. The same authors report a second experiment in which they used whole-brain imaging techniques on eight different subjects. They replicated the original findings and identified additional frontal regions and other more caudal brain regions, e.g., the bilateral posterior parietal cortex and the caudate nucleus of the basal ganglia that showed a linear relationship with load.

Recently, van Asselen et al. (2006) studied the areas of brain involved in spatial working memory by comparing the performance of patients with lesions in a number of different brain areas with normal controls and concluded that damage to the right posterior parietal and right dorsolateral prefrontal cortex impaired the ability to keep spatial information in working memory. Patients with damage to the right posterior parietal cortex, the right dorsolateral prefrontal cortex and the hippocampal formation bilaterally made more errors, suggesting to the authors the importance of these areas in maintaining spatial information in working memory over an extended time period. These and similar data seem to suggest the presence of a fairly widely distributed circuit that participates with the prefrontal cortex in processing working memory. Thus successful performance of a working memory task may not reflect an exclusive relation to the prefrontal cortex and the negative impact of sleep loss on working memory may also reflect its impact on other brain areas as well.

Second, not all cognitive tasks associated with intact prefrontal activation, are equally sensitive to short term sleep deprivation. Not all executive control mechanisms are equally sensitive to sleep deprivation. This makes it difficult to predict a priori which tasks that test executive control will be impacted by short term sleep loss. This may be due partially to the fact that apparently the inter-individual variance in sensitivity to sleep loss of cognitive abilities associated with executive processes after short term sleep deprivation is greater than after long term sleep deprivation (Chuah et al., 2006). For example, Jennings et al. (2003) reported on the effects of one night sleep deprivation on performance of a choice reaction time task that included compatible and incompatible responses in eight 90 min blocks of trials on the day following the night of sleep deprivation. The paradigm was designed to assess response inhibition, task shifting skill and task strategy, all components of supervisory attention. Sleep deprivation did not affect either response inhibition (i.e., responses strongly associated with the stimuli), or task shifting. But sleep deprivation did seem to affect the ability to optimize preparatory strategies. The authors concluded that sleep deprivation seems to selectively affect some, but not all components of prefrontal executive control processes. The suggestion has also been made that inter-individual differences in sensitivity to sleep loss may relate to differences in prefrontal activation during wakefulness (Chuah et al., 2006).

Harrison and Espelid (2004) reported that even after 34 h of sleep deprivation sleep deprivation did not negatively affect performance under conditions believed to facilitate inhibitory processes. However, under conditions believed to rely on memorial processes, sleep deprivation did affect performance. Distracter interference was also greater following sleep deprivation. Additional evidence for a selective rather than general negative effect of short term sleep deprivation on executive control mechanisms was offered by Heuer et al. (2005) who studied the effect of sleep deprivation on the random generation of key-presses, numbers and nouns, especially the suppression of prepotent responses and the selection of the next responses by applying a local heuristic. They reported that the suppression of prepotent responses with random key-presses was not affected by sleep loss, but when numbers and nouns were generated, suppression of prepotent responses was negatively affected by sleep deprivation. Huer et al. (2005) conclude that random generation involves multiple processes and sleep deprivation does not equally impact all of these processes.

15.1.4.1 Suggested Criteria for Sensitivity to Short Term Sleep Deprivation

In summary, it appears that not all tasks that seemingly fit the criterion of being associated with intact prefrontal activation are equally sensitive to the effects of short term sleep deprivation. In fact, the above review seems to indicate that the identification of a cognitive task as one whose performance is associated with intact activation of the prefrontal cortex is not a sufficient criterion in order to predict whether the task will be impacted by short term sleep deprivation or not. However, from the same review and from the data of the Zuckerman study discussed in detail above, there still seems enough evidence to claim that this criterion, although not sufficient by itself, may be considered a necessary criterion for identifying cognitive tasks that are sensitive to short term sleep loss. Since many cognitive tasks also activate areas of prefrontal cortex along with other brain areas, we suggest the following slight modification of this criterion: Although there may be other brain regions that are activated when performing the task, i.e., a multi-nodal distribution of activation, a task whose optimum performance is strongly associated with intact prefrontal cortex activation will be sensitive to the effects of short term sleep deprivation, if the major source of variance in regional brain activation is located in prefrontal regions so that when there are prefrontal lesions or the prefrontal activation is not at a normal level, the performance is significantly negatively impacted.

In addition to the neuroanatomical activation pattern, however, there may be an additional criterion that must be met for a cognitive task to display sensitivity to short term sleep deprivation. Harrison and Espelid (2004) suggested that although there is compelling evidence from functional neuroimaging studies during sleep and following sleep loss to support a focus of research attention on tasks associated with intact prefrontal cortex activation in studies of sleep deprivation, their own findings of only partial effects of sleep loss on inhibitory processes that are associated with prefrontal cortex and the results of other researchers, support an argument that the neuroanatomical association of a task with the prefrontal cortex alone is an insufficient explanation of the sensitivity of a cognitive task to the effects of sleep deprivation. Rather, one must also identify cognitive frameworks and conceptualize tasks in terms of underlying cognitive mechanisms to advance our understanding of the link between sleep and the prefrontal cortex. The problem then becomes how to identify the set of underlying cognitive mechanisms that share aspects of the processing necessary for optimal performance on those cognitive tasks that are sensitive to short term sleep deprivation. With regard to this latter, seemingly amorphous, criterion, there is very little experimental evidence available to aid in identifying the proposed mechanisms. However, if forced to speculate, we would suggest that any task whose cognitive processing, coding, decoding or executive control involves some type of timing or sequencing of components for optimal performance will be sensitive to the effects of short term sleep deprivation.

15.2 Conclusion

In conclusion, the arguments set forth in this chapter are based on an analysis (or reanalysis) of data that show disassociations reflecting the actions of more than one mechanism in the effects of sleep deprivation on cognitive performance. First, we presented evidence that the effects of sleep deprivation on cognitive performance are not uniform, either with respect to task or with respect to the time since the last sleep epoch. In the first section, we presented evidence to show that the relative weightings of the homeostatic and cyclic components in sleep deprivation data sets vary over time and that the weightings may differ from one dependent variable to another. This implies that one dependent variable may be more sensitive to the homeostatic drive for sleep than to the diurnal or circadian rhythm, while another dependent variable may be more sensitive to the circadian rhythm than to the homeostatic drive for sleep. Furthermore, different dependent variables may share a common source of variance with one factor, but not with any others. Specifically, we argued from an analysis of subjective sleepiness and body temperature data that the commonality, i.e., the inverse relationship between the averages of these variables often reported during long term sleep deprivation studies only involved the circadian component, but not the homeostatic or ultradian components.

Second, we reanalyzed data from a study of two types of intervention to ameliorate the effects of a night of sleep deprivation, one pharmacological, the other exposure to bright light. The argument, again by disassociation, was based on the difference between the effectiveness of the two treatments on two different tasks, one of which is strongly associated with intact prefrontal activation, working memory (WM) and one that is not, choice reaction time (CRT). Caffeine and exposure to bright lights were each alone and both together effective in reducing the effects of sleep loss on CRT performance; while only caffeine was effective in reducing the effects of sleep loss on WM. Since there is a large body of evidence that exposure to bright lights operates mainly on the circadian component of the sleep-wake cycle, while caffeine as an adenosine antagonist apparently operates mainly on the homeostatic drive for sleep, the analysis of the performance data can be understood to support the argument that the two tasks, that reflect the activation of different brain regions also differ in their sensitivity to the two major factors involved in sleep deprivation, homeostatic and circadian.

Third, we reanalyzed data from a recent dissertation that compared the performance of the same subjects on a number of cognitive and psychomotor tasks, when they were not sleep deprived and when they experienced short term sleep deprivation. The comparison included tasks strongly associated with intact prefrontal activation and those that are not strongly associated with intact prefrontal activation. Of the five tasks, sleep deprivation, as a main factor, only reduced performance significantly on the two tasks that are strongly associated with intact prefrontal activity, memory of sequence and working memory. We concluded that the results support the argument that under conditions of short term sleep deprivation and when there is minimal decrement due to the circadian rhythm, i.e., during the diurnal phase, tasks, whose performance is strongly associated with intact prefrontal activation are more sensitive to the effects of sleep loss than tasks, whose performance is less strongly associated with intact prefrontal activation. The argument again was based on a disassociation in the sensitivity of some of the cognitive tasks to short term sleep deprivation, when testing occurs during the diurnal phase of the circadian rhythm in the absence of its nadir. However, because of the mounting evidence from the literature that association of a cognitive task with intact prefrontal activation is not a sufficient predictor of the sensitivity of the task to short term sleep loss, we suggested some modification of that criterion and noted the arguments made by others (e.g., Harrison and Espelid, 2004) that an additional criterion should be added that relates to the underlying cognitive processing involved in optimal task performance as, perhaps, co-predictors of the sensitivity of a task to short term sleep loss.

The propositions discussed in the present chapter have important theoretical and practical implications as well. If more than one neural mechanism is involved in performance decrement due to sleep deprivation and sleep restriction at any moment, then any proposed intervention that targets only one of the mechanisms cannot be fully successful in ameliorating the effects of sleep loss. Multiple sources of variance in causing performance decrement require multiple forms of treatment to eliminate or reduce decrement.

Acknowledgements The authors would like to acknowledge and thank Drs. J. French and K. Neville for the use of data from a study of sleep deprivation conducted at Brooks Air Force Base, San Antonio Texas, USA (French et al., 2006).

References

Akerstedt, T. and Gillberg, M. (1981) The circadian variation of experimentally displaced sleep. *Sleep*, *4*, 159–169.

Amendola, C.A., Gabrieli, J.D.E. and Lieberman, H.R. (1998) Caffeine's effects on performance and mood are independent of age and gender. *Nutritional Neuroscience*, *1*, 269–280.

Amiez, C. and Petrides, M. (2007) Selective involvement of the mid-dorsolateral prefrontal cortex in the coding of the serial order of visual stimuli in working memory. *Publication of the National Academy of Sciences*, *104*, 13786–13791.

Angus, R.G. and Heselgrave R.J. (1985) Effects of sleep loss on sustained cognitive performance during a command and control simulation. *Behavior Research Methods, Instruments and Computers*, *17*, 55–67.

Arendt, J. (1995) *Melatonin and the Mammalian Pineal Gland.* Chapman & Hall, London (pp. 209).

Arendt, J. (1998) Melatonin and the pineal gland: Influence on mammalian seasonal and circadian physiology. *Review of Reproduction*, *3*, 13–22.

Babkoff, H. and Krueger, G. P. (1992) Use of stimulants to ameliorate the effect of sleep loss during sustained performance. *Military Psychology*, *4*, 191–205.

Babkoff, H., Genser, S.G., Sing, H.C., Thorne, D.R. and Hegge, F.W. (1985) The effects of progressive sleep loss on a lexical decision task: Response lapses and response accuracy. *Behavior Research Methods, Instruments and Computers*., 17: 614–622.

Babkoff, H., Mikulincer, M., Caspy, T., Kempinski, D. and Sing, H.C. (1988) The topology of performance curves during 72 hours of sleep loss: A memory and search task. *The Quarterly Journal of Experimental Psychology, Section A: Human Experimental Psychology*, *40*, 737–756.

Babkoff, H., Caspy, T., Mikulincer, M. and Sing, H.C. (1991a) Monotonic and rhythmic influences: A challenge for sleep deprivation research. *Psychological Bulletin*, *19*, 411–428.

Babkoff, H., Caspy, T. and Mikulincer, M. (1991b) Subjective sleepiness ratings: The effects of sleep deprivation, circadian rhythmicity and cognitive performance. *Sleep*, *14*, 534–539.

Babkoff, H., Kelly, T. L., Matteson, L.T., Gomez, S.A., Lopez, A., Hauser, S., Naitoh, P. and Assmus, J. (1992) Pemoline and methylphenidate: Interaction with mood, sleepiness and cognitive performance during 64 hours of sleep deprivation. *Military Psychology*, *4*, 235–265.

Babkoff, H., French, J., Whitmore, J. and Sutherlin, R. (2002) Single dose bright light and/or caffeine effect on nocturnal performance. *Aviation, Space and Environmental Medicine*, *73*, 341–350.

Babkoff, H., Zukerman, G., Fostick, L. and Ben-Artzi, E. (2005) The effect of the diurnal rhythm and twenty-four hours of sleep deprivation on dichotic temporal order judgment (TOJ). *Journal of Sleep Research*, *14*, 7–15.

Baddeley, A. D. (1986). Working Memory. Oxford: Oxford University Press.

Badia, P., Myers, B., Boecker, M., Culpepper, J. and Harsh, J. (1991) Bright light effects on body temperature, alertness, EEG and behavior. *Journal of Physiological Behavior*, *50*, 583–588.

Baker, F.C., Driver, H.S., Rogers, G.G., Paiker, J. and Mitchell, D. (1999) High nocturnal body temperatures and disturbed sleep in women with primary dysmenorrhea. *American Journal of Physiology*, *277*, E1013–E1021.

Baker, F., Waner, J., Vieira, E., Taylor, S., Driver, H. and Mitchell, D. (2001) Sleep and 24 hour body temperatures: A comparison in young men, naturally cycling women and women taking hormonal contraceptives. *Journal of Physiology*, *530*(Pt3), 565–574.

Basheer, R., Porkka-Heiskanen, T., Strecker, R.E., Thakkar, M.M. and McCarley, R.W. (2000) Adenosine as a biological signal mediating sleepiness following prolonged wakefulness. *Biological Signals Receptors*, *9*, 319–327.

Battig, K., Buzzi, R., Martin, J.R., and Feierabend, J.M. (1984) The effects of caffeine on psychological functions and mental performance. Experimentia, 40, 1218–1223.

Beaumont, M., Batejat, D., Pierard, C., Coste, O., Doireau, P., Van Beers, P., Chauffard, F., Chassard, D., Enseln, M., Denis, J.B. and Lagarde, D. (2001) Slow release caffeine and prolonged wakefulness: Effects on vigilance and cognitive performance. *Journal of Sleep Research, 10*, 265–276.

Bell-McGinty, S., Habeck, C., Hilton, H.J., Rakitin, B., Scarmeas, N., Zarahn, E., Flynn, J., DeLaPaz, R., Basner, R. and Stern, Y. (2004) Identification and differential vulnerability of a neural network in sleep deprivation. *Cerebral Cortex, 14*, 496–502.

Benington, J.H., Kodali, S.K., and Heller, H.C. (1995). Stimulation of A1 adenosine receptors mimics the electroencephalographic effects of sleep deprivation. Brain Research, 692, 79–85.

Blaxton, T.A., Bookheimer, S.Y., Zeffiro, T.A., Figlozzi, C.M., Gaillard, W.D., and Theodore, W.H. (1990) Functional mapping of human memory using PET: comparisons of conceptual and perceptual tasks. Canadian Journal of Experimental Psychology, 50, 42–56.

Bonnet, M.H., Gomez, S., Wirth, O. and Arand, D.L. (1995) The use of caffeine versus prophylactic naps in sustained performance. *Sleep, 18*, 97–104.

Braver, T.S., Cohen, J.D., Nystrom, L.E., Jonides, J. and Smith, E.E. (1997) A parametric study of prefrontal cortex involvement in human working memory. *Neuroimage, 5*, 49–62.

Brismar K., Hylander, B., Eliasson, K., Rössner, S., and Wetterberg, L. (1988) Melatonin secretion related to side-effects of beta-blockers from the central nervous system. Acta Medica Scandinavica, 223, 525–530.

Broughton, R.J. (1989) Chronobiological aspects and models of sleep and napping. In: D.F. Dinges and R.J. Broughton (Eds). *Sleep and Alertness: Chrononbiological, Behavioral, and Medical Aspects of Napping*. Raven, New York (pp. 71–98).

Budnick, L.D., Lerman, S.E. and Nicolich, M.J. (1995) An evaluation of scheduled bright light and darkness on rotating shiftworkers: Trial and limitations. *American Journal of Industrial Medicine, 27*, 771–782.

Buysse, D.J., Monk, T.H., Reynolds III, C.F., Mesiano, D., Houck, P.R. and Kupfer, D.J. (1993) Patterns of sleep episodes in young and elderly adults during a 36-hour constant routine. *Sleep, 16*, 632–637.

Caldwell, J.A. (2005) Fatigue in aviation. *Travel Medicine and Infectious Diseases, 3*(2), 85–96.

Caldwell, J.A. and Caldwell, J.L. (2005) Fatigue in military aviation: An overview of US military-approved pharmacological countermeasures. *Aviation, Space and Environmental Medicine, 76*(Suppl 7), C39–C51.

Caldwell, J.A., Caldwell, J.L., Crowley, J.S. and Jones, H.D. (1995) Sustaining helicopter pilot performance with dexedrine during periods of sleep deprivation. *Aviation, Space and Environmental Medicine, 66*, 930–937.

Caldwell, J.A., Mu, Q., Smith, J.K., Mishory, A., Caldwell, J.L., Peters, G., Brown, D.L. and George, M.S. (2005) Are individual differences in fatigue vulnerability related to baseline differences in cortical activation? *Behavioral Neuroscience, 119*, 694–707.

Campbell, S. and Dawson, D. (1990) Enhancement of nighttime alertness and performance with bright ambient light. *Journal of Physiological Behavior, 48*(2), 317–320.

Campbell, S.S., Murphy, P.J., van den Heuvel, C.J., Roberts, M.L. and Stauble, T.N. (1999) Etiology and treatment of intrinsic circadian rhythm sleep disorders. *Sleep Medicine Review, 3*, 179–200.

Cardinali, D.P., and Pévet, P. (1998) Basic aspects of melatonin action. Sleep Medicine Reviews, 2, 175–190.

Carrier, J. and Monk, T.H. (2000) Circadian rhythms of performance: New trends. *Chronobiology International, 17*, 719–732.

Chuah, Y.M.L., Venkatramen, V., Dinges, D.F. and Chee, M.W.L. (2006) The neural basis of inter-individual variability in inhibitory efficiency after sleep deprivation. *The Journal of Neuroscience, 26*, 7156–7162.

Cohen, J.D., Forman, S.D., Braver, T.S., Casey, B.J., Servan-Schreiber, D., and Noll, D.C. (1994). Activation of prefrontal cortex in a non-spatial working memory task with functional MRI. Human Brain Mapping, 1, 293–304.

Courtney, S.M., Petit, L., Haxby, J.V. and Untergleider, L.G. (1998) The role of prefrontal cortex in working memory: Examining the contents of consciousness. *Philosophical Transactions of the Royal Society London, B, 353*, 1819–1828.

Curtis, C.E., Zald, D.H. and Pardo, J.V. (2000) Organization of working memory within the human prefrontal cortex: A PET study of self-ordered object working memory. *Neuropsychologia, 38*, 1503–1510.

Czeisler, C.A., Weitzman, E.D., Moore-Ed, M.C., Zimerman, J.C. and Knauer, R.S. (1980) Human sleep: Its duration and organization depend on its circadian phase. *Science, 210*, 1264–1267.

Daurat, A., Aguirre, A., Foret, J., Gonnet,P., Keromes, A. and Benoit, O. (1993) Bright light affects alertness and performance rhythms during a 24-h constant routine. *Physiology and Behavior, 53*, 929–936.

Daurat, A., Foret, J., Benoit, O. and Mauco, G. (2000) Bright light during nighttime: Effects on the circadian regulation of alertness and performance. *Biological Signals Reception, 9*, 309–318.

Dinges, D.F. and Kribbs, M.B. (1991) Performing while sleepy: Effects of experimentally induced sleepiness. In T.H. Monk (Ed). *Sleep, Sleepiness and Performance*. Wiley, New York (pp. 97–128).

Dinges, D.F., Orne, M.T. and Orne, E.C. (1984) Sleepiness during sleep deprivation: The effects of performance demands and circadian phase. *Sleep Research, 13*, 189.

Dinges, D.F, Orne, M.T., Whitehouse, W.G. and Orne, E.C. (1987) Temporal placement of a nap for alertness: Contributions of circadian phase and prior wakefulness. *Sleep, 10*, 313–329.

Dorsey, C.M., Teicher, M.H., Cohen-Zion, M., Stefanovic, L., Satlin, A., Tartarini, W., Harper, D. and Lukas, S.E. (1999) Core body temperature in sleep of older female insomniacs before and after passive body heating. *Sleep, 22*, 891–898.

Drummond, S.P.A., Brown, G.G., Gillin, J.C., Stricker JL, et al. (2000) Altered brain response to verbal learning following sleep deprivation. *Nature, 403*, 655–657.

Drummond, S.P.A., Gillin, J.C. and Brown, G.G. (2001) Increased cerebral response during a divided attention task following sleep deprivation. *Journal of Sleep Research, 10*, 85–92.

Durmer, J. and Dinges, D.F. (2005) Neurocognitive consequences of sleep deprivation. *Seminar in Neurology, 25*, 117–129.

Fiez, J.A., Tallal, P., Raichle, M.E., Miezin, F.M., Katz, W.F., and Petersen, S.E. (1995) PET studies of auditory and phonological processing: effects of stimulus characteristics and task demands. Journal of Cognitive Neuroscience, 7, 357–375.

Fletcher, P.C., Dolan, R.J., and Frith, C.D. (1995) The functional anatomy of memory. Experientia, 51, 1197-207.

Fletcher, P.C., Frith, C.D., Grasby, P.M., Shallice, T., Frackowiak, R.S., and Dolan, R.J. (1995) Brain systems for encoding and retrieval of auditory-verbal memory. An in vivo study in humans. Brain, 118, 401–416.

Fredholm, B.B., Battig, K., Holmen, J., Nehlig, A., Zvartau, E.E. (1999) Actions of caffeine in the brain with special reference to factors that contribute to its widespread use. *Pharmacology Review, 51*, 83–133.

French, J., Hannon, P. and Brainard, G. (1990) Effects of bright illuminance on body temperature and human performance. *Annual Review of Chronopharmacology Volume 7, France*, 37–40.

French, J., Whitmore, J., Hannon, P., Brainard, G. and Schiflett, S. (1991) Photic effects on sustained performance. *Fifth Annual Workshop on Space Operations Applications and Research, NASA, 3127*, 482–486.

French, J., Neville, K., Eddy, D.R., Storm, W.F., Cardenas, R., Flynn, C. and Miller, J.C. (2006) *Sensitivity of the Space Flight Cognitive Assessment Tool (S-CAT) to Sleep Deprivation* (Technical Report No. AFRL-HE-BR-TR-2006–0004). Air Force Research Laboratory, Brooks City-Base, TX.

Frey, D.J., Badia, P. and Wright, K.P. Jr. (2004) Inter and intra-individual variability in performance near the circadian nadir during sleep deprivation. *Journal of Sleep Research, 13*, 305–315.

Frey, D.J., Badia, P. and Wright, K.P. Jr. (2005) Analysis of inter and intra-individual variability. *Journal of Sleep Research, 14*, 205–206.

Froberg, J.E. (1977) Twenty-four hour patterns in human performance, subjective and physiological variables and differences between morning and evening active subjects. *Biological Psychology*, *5*, 119–134.

Fuster, J.M. (1989) The prefrontal cortex: anatomy, physiology, and neuropsychology of the frontal lobe. New York: Raven.

Fuster, J.M. (1995) Memory in the cerebral cortex. Cambridge (MA): MIT Press.

Gillberg, M. and Akerstedt, T. (1981) Possible measures of 'sleepiness' for the evaluation of disturbed and displaced sleep. In: A. Reinberg, N. Vieux and P.Andlauer (Eds). *Night and Shift Work: Biological and Social Aspects*. Pergamon, Oxford, England (pp.155–160).

Goldman-Rakic, P. (1987) Circuitry of primate prefrontal cortex and regulation of behavior by representational memory. In: F. Plum (Ed.), Handbook of Physiology, Sect 1: The Nervous System, Vol V: Higher Functions of the Brain (pp. 373–417). Bethesda, MD: American Physiological Society.

Goldman-Rakic, P.S. (1987) Circuitry of the frontal association cortex and its relevance to dementia. Archives of Gerontology and Geriatrics, 6, 299–309.

Grasby, P.M, Frith, C.D., Friston, K., Frackowiak, R.S., Dolan, R.J. (1993) Activation of the human hippocampal formation during auditory-verbal long-term memory function. Neuroscience Letters, 163, 185-188.

Gross, J.B., Gunzelmann, G., Gluck, K.A., Van Dongen, H.P.A. and Dinges, D.F. (2006) Computational modeling of the combined effects of circadian rhythm and sleep deprivation. *Proceedings of the 28th Annual Conference of the Cognitive Science Society*, 297–302.

Habeck, C., Rakiti, Moeller, J., Scarmeas, N., Zarahn, E., Brown, T. and Stern, Y. (2004) An event related fMRI study of the neurobehavioral impact of sleep deprivation on performance of a delayed match-to-sample task. *Cognitive Brain Research*, *18*, 306–321.

Halsband, U., Krause, B.J., Schmidt, D., Herzog, H., Tellmann, L., and Muller-Gartner, H-W. (1998) Encoding and retrieval in declarative learning: a positron emission tomography study. Behavioral Brain Research, 97, 69–78.

Hannon, P., Brainard, G., Childs, R., Gibson, W., French, J., Hanifin, J. and Rollag, M. (1992) Bright light suppresses melatonin and improves cognitive performance during nighttime hours in humans. *Proceedings of the 22nd Annual Meeting of the Society for Neuroscience, USA, 195.*

Harrison, Y. and Espelid, E. (2004) Loss of negative priming following sleep deprivation. *Quarterly Journal of Experimental Psychology*, *57*, 437–446.

Harrison, Y., and Horne, J.A. (1998) Sleep loss impairs short and novel language tasks having a prefrontal focus, *Journal of Sleep Research*, 7, 95–100.

Harrison, Y. and Horne, J.A. (2000a) Sleep loss and temporal memory. *Quarterly Journal of Experimental Psychology*, *53A*, 271–279.

Harrison, Y. and Horne, J.A. (2000b) The impact of sleep deprivation on decision making: A review. *Journal of Experimental Psychology: Applied*, *6*, 236–49.

Heuer, H., Kohlisch, O., and Klein, W. (2005) The effects of total sleep deprivation on the generation of random sequences of key-presses, numbers and nouns. Quarterly Journal of Experimental Psychology A, 58, 275–307.

Hichwa, R.D. (1995) II. PET studies of memory: Novel versus practiced free recall of word lists. *NeuroImage*, *2*, 296–305.

Honey, G.D., Bullmore, E.T. and Sharma, T. (2000) Prolonged reaction time to a verbal memory task predicts increased power of a posterior parietal cortical activation. *NeuroImage*, *12*, 495–503.

Horne, J.A. (1988) *Why We Sleep: The Functions of Sleep in Humans and Other Mammals.* Oxford University Press, Oxford, England (pp. 57–60; 89–90; 130–137).

Hunt, A.E., Al-Ghoul, W.M., Gillette, M.U., and Dubocovich, M.L. (2001). Activation of MT2 melatonin receptors in rat suprachiasmatic nucleus phase advances the circadian clock. American Journal of Physiology, 280, C110–C118.

Jacobson, B.H. and Edgley, B.M. (1987) Effects of Caffeine on simple reaction time and movement time. *Aviation, Space and Environmental Medicine*, *58*, 1153–1156.

Jennings, J.R., Monk, T.H., and van der Molen, M.W. (2003) Sleep deprivation influences some but not all processes of supervisory attention. Psychological Science, 14, 473–479.

Jessen, A.B., Toubro, S., and Astrup A. (2003) Effect of chewing gum containing nicotine and caffeine on energy expenditure and substrate utilization in men. American Journal of Clinical Nutrition, 77, 1442–1447.

Johnson, L.C. (1982) Sleep deprivation and performance. In: W.B. Webb (Ed). *Biological Rhythms, Sleep and Performance*. Wiley, New York (pp. 11–141).

Johnson, M.P., Duffy, J.F., Dijk, D.J., Rhonda, J.M., Dyal, C.M. and Czeisler, C.A. (1992) Short-term memory, alertness and performance: A reappraisal of their relationship to body temperature. *Journal of Sleep Research*, *1*, 24–29.

Jonides, J., Smith, E.E., Koeppe, R.A., Awh, E. Minoshima, S., and Mintun, M.A. (1993) Spatial working memory in humans as revealed by PET. Nature, 363, 623–625.

Just, M., and Carpenter, P. (1992) A capacity theory of comprehension: individual differences in working memory. Psychological Review, 99, 122–149.

Kajimura, N., Uchiyama, M., Takayama, Y., Uchida, S., Uema, T., Kato, M., Sekimoto, M., Watanabe, T., Nakajima, T., Horikoshi, S., Ogawa, K., Nishikawa, M., Hiroki, M., Kudo, Y., Matsuda, H., Okawa, M., and Takahashi K. (1999) Activity of midbrain reticular formation and neocortex during the progression of human non-rapid eye movement sleep. Journal of Neuroscience, 19, 10065–10073.

Kelly, G.S. (2007) Body temperature variability (part 2): Masking influences of body temperature variability and a review of body temperature in disease. *Alternative Medicine Review*, *12*, 49–62.

Kjellberg, A. (1977) Sleep deprivation and some aspects of performance lapses and other attentional effects. Waking and Sleeping, 1, 145–148.

Koslowsky, M. and Babkoff, H. (1992) Meta-analysis of the relationship between sleep deprivation and performance. *Chronobiology International*, *9*, 132–136.

Krause, B. J., Schmidt, D., Mottaghy, F. M., Taylor, J., Halsband, U., Herzog, H., Tellmann, L., and Müller-Gärtner, H-W. (1999) Episodic retrieval activates the precuneus irrespective of the imagery content of word pair associates. A PET study. Brain, 122, 255–263.

Krauchi, K., Knoblauch, V., Wirtz-Justice, A. and Cajochen, C. (2005) Challenging the sleep homeostat does not influence the thermoregulatory system in men: Evidence from a nap vs. sleep-deprivation study. *American Journal of Physiology: Regulatory, Integrative and Comparative Physiology*, *29*, R1052–R1061.

Kubota, T., Uchiyama, M., Suzuki, H., Shibui, K., Kim, K., Tan, X., Tagaya, H., Okaya, M. and Inoue, S. (2002) Effects of nocturnal bright light on saliva melatonin, core body temperature and sleep propensity rhythms in human subjects. *Neuroscience Research*, *42*(8), 115–122.

Landis, C.A., Savage, M.V., Lentz, M.J. and Brengelmann, G.L. (1998) Sleep deprivation alters body temperature dynamics to mild cooling and heating not sweating thresholds in women. *Sleep*, *21*, 101–108.

Lieberman, H.R., Waldhauser, F., Garfield, G., Lynch, H. and Wurtman, R. (1984) Effects of melatonin on human mood and performance. *Brain Research*, *323*, 201–207.

Lieberman, H.R., Wurtman, R.J., Emde, G.G. and Coviella, I.L. (1987) The effects of caffeine and aspirin on mood and performance. *Journal of Clinical Psychopharmacology*, *7*(5), 315–320.

Lieberman, H.R., Tharion, W.J., Shukitt-Hale, B., Speckman, K.L. and Tulley, B. (2002) Effects of caffeine, sleep loss and stress on cognitive performance and mood during U.S. Navy SEAL training. *Psychopharmacology*, *164*, 250–261.

Lim, J., Choo, W.C. and Chee, M.W. (2007) Reproducibility of changes in behavior and fMRI activation associated with sleep deprivation in a working memory task. *Sleep*, *30*, 61–70.

Liu, C., Weaver, D.R., Jin, X., Shearman, L.P., Pieschl, R.L., Gribkoff, V.K. and Reppert, S.M., (1997) Molecular dissection of two distinct actions of melatonin on the suprachiasmatic circadian clock. Neuron, 19, 91–102.

Loke, W.H. (1988) Effects of caffeine on mood and memory. *Physiology and Behavior*, *44*, 367–372.

Long, M.A., Jutras, M.J., Connors, B.W. and Burwell, R.D. (2005). Electrical synapses coordinate activity in the suprachiasmatic nucleus. Nature Neuroscience, 8, 61–66.

Lynch, H.J., Wurtman, R.J., Moskowitz, M.A., Archer, M.C., and Ho, M.H. (1975) Daily rhythm in human urinary melatonin. Science, 187, 169–171

Macchi, M.M and Bruce, J.M. (2004) Human pineal physiology and functional significance of melatonin. *Frontiers of Neuroendocrinology*, *25*, 177–195.

Mason, R., and Brooks, A. (1988) The electrophysiological effects of melatonin and a putative melatonin antagonist (N-acetyltryptamine) on rat suprachiasmatic neurones in vitro. Neuroscience Letters, 95, 296–301.

McCarthy, G., Blamire, A.M., Puce, A., Nobre, A.C., Bloch, G., Hyder, F., Goldman-Rakic, P.S., and Shulman, R.G. (1994) Functional magnetic resonance imaging of human prefrontal cortex activation during a spatial working memory task. Proceedings of the National Academy of Sciences, USA, 91, 8690–8694.

McIntyre, I.M., Norman, T.R., Burrows, G.D. and Armstrong, S.M. (1989) Human melatonin suppression by light is intensity dependent. *Journal of Pineal Research*, *6*, 149–156.

McLellan, T.M., Ducharme, M.B., Canini, F., Moroz, D., Bell, D.G., Baranski, J.V., Gil, V., Buguet, A., and Radomski, M.W. (2002) Effect of modafinil on core temperature during sustained wakefulness and exercise in a warm environment. Aviation, Space and Environmental Medicine, 73, 1079–1088.

Mikulincer, M., Babkoff, H., Caspy, T. and Sing, H.C. (1989) The effects of 72 hours of sleep loss on psychological variables. *British Journal of Psychology*, *80*, 145–162.

Mikulincer, M., Babkoff, H., Caspy, T. and Weiss, H. (1990) The impact of cognitive interference on performance during prolonged sleep loss. *Psychological Research*, *52*, 80–86.

Minors, D.S. and Waterhouse, J.M. (1981) *Circadian Rhythms and the Human*. Wright, Bristol, England.

Mitler, M.M., Carskadon, M.A., Czeisler, C.A., Dement, W.C., Dinges, D.F. and Graeber, R.C. (1988) Catastrophes, sleep and public policy: Consensus report. *Sleep*, *11*, 100–109.

Monk, T.H. (1987) Subjective ratings of sleepiness: The underlying circadian mechanisms. *Sleep*, *10*, 343–353.

Monk, T.H. (1991) Circadian aspects of subjective sleepiness. In: T.H. Monk (Ed). *Sleep, Sleepiness and Performance*. Wiley, New York (pp. 39–63).

Monk, T.H., and Carrier, J. (1997) Speed of mental processing in the middle of the night. Sleep, 20, 399–401.

Montgomery, V.L. (2007) Effect of fatigue, workload, and environment on patient safety in the pediatric intensive care unit. *Pediatric Critical Care Medicine*, *8*(Suppl 2), S11–S16.

Murray, E.J., Williams, H.L. and Lubin, A. (1958) Body temperature and psychological ratings during sleep deprivation. *Journal of Experimental Psychology*, *56*, 271–273.

Muzur, A., Pace-Schott, E.F., and Hobson, J.A. (2002) The prefrontal cortex in sleep. Trends in Cognitive Science, 6, 475–481.

Ninokura, Y., Mushiake, H. and Tanji, J. (2003) Representation of the temporal order of visual objects in the primate lateral prefrontal cortex. Journal of Neurophysiology, 89, 2868–2873.

Owens, J. (2007) Sleep loss and fatigue in healthcare professionals. *Journal of Perinatal and Neonatal Nursing*, *21*, 92–100.

Pandi-Perumal, S.R., Smits, M., Spence, W., Srinivasan, V., Cardinali, D.P., Lowe, A.D. and Kayumov, L. (2007) Dim light melatonin onset (DLMO): A tool for the analysis of circadian phase in human sleep and chronobiological disorders. *Progress in Neurospychopharmacology and Biological Psychiatry*, *31*, 1–11.

Patrick, G.T.W. and Gilbert, J.A. (1896) On the effects of loss of sleep. *Psychological Review*, *3*, 469–483.

Penetar, D., McCann, U. and Thorne, D. (1993) Caffeine reversal of sleep deprivation effects on alertness and mood. *Psychopharmacology*, *112*, 359–365.

Perlstein, W.M., Carter, C.S., Noll, D.C. and Cohen, J.D. (2001) Relation of prefrontal cortex dysfunction to working memory and symptoms of schizophrenia. *American Journal of Psychiatry*, *158*, 1105–1113.

Petrides, M., Alivisatos, B., Meyer, E., and Evans, A.C. (1993) Functional activation of the human frontal cortex during the performance of verbal working memory tasks. Proceedings of the National Academy of Sciences, USA, 90, 878–882.

Philibert, I. (2005) Sleep loss and performance in residents and nonphysicians: A meta-analytic examination. *Sleep*, *28*, 392–402.

Pigeau, R., Naitoh, P., Buguet, A., McCann, C., Baranski, J., Taylor, M., Thompson, M. and Mack, I. (1995) Modafinil, d-amphetamine and placebo during 64 hours of sustained mental work. I. Effects on mood, fatigue, cognitive performance and body temperature. *Journal of Sleep Research*, *4*, 212–228.

Pilcher, J.J., and Huffcutt, A.I. (1996) Effects of sleep deprivation on performance: a meta-analysis. Sleep, 19, 318–326.

Pilcher, J. J., McClelland, L. E., Moore, D. D., Haarmann, H., Baron, J., Wallsten, T. S., and McCubbin, J. A. (2007). Language performance under sustained work and sleep deprivation conditions. Aviation, Space and Environmental Medicine, 78 (5, Suppl.), B25–38.

Porkka-Heiskanen, T., Strecker, R.E. and McCarley, R.W. (2000) Brain site specificity of extracellular adenosine concentration changes during sleep deprivation and spontaneous sleep: An in vivo microdialysis study. *Neuroscience*, *99*, 507–517.

Portas, C.M., Thakkar, M., Rainnie, D.G., Greene, R.W., McCarley, R.W. (1997) Role of adenosine in behavioral state modulation: A microdialysis study in the freely moving cat. *Neuroscience*, *79*, 225–235.

Rainnie, D.G., Grunze, H.C., McCarley, R.W. and Greene, RW. (1994) Adenosine inhibition of mesopontine cholinergic neurons: Implications for EEG arousal. *Science*, *263*, 689–692.

Radulovacki, M., Miletich, R.S., and Green, R.D. (1982) N6 (L-phenylisopropyl) adenosine (L-PHA) increases slow-wave sleep (S2) and decreases wakefulness in rats. Brain Research, 246,178–180.

Rosenthal, L., Roehrs, T., Zwyghuizen-Doorenbos, A., Plath, D. and Roth, T. (1991) Alerting effects of caffeine after normal and restricted sleep. *Neuropsychopharmacology*, *4*(2), 103–108.

Rudge, P., and Warrington, E.K. (1991) Selective impairment of memory and visual perception in splenial tumours. Brain, 114, 349–360.

Savourey, G. and Bittel, J. (1994) Cold thermoregulatory changes induced by sleep deprivation in men. *European Journal of Applied Physiology*, *69*, 216–220.

Sewitch, D.E., Kittrell, E.M.W, Kupfer, D.J. and Reynolds, C.F. (1986) Body temperature and sleep architecture in response to a mild cold stress in women. *Physiology and Behavior*, *36*, 951–957.

Shallice, T. (1988) From Neuropsychology to Mental Structure. Cambridge University Press, Cambridge.

Shimamura, A. P. (1994). Frontal lobes and memory. In M. Gazzaniga (Ed.), The cognitive neurosciences (pp. 803-813). Cambridge, MA: MIT Press.

Sing, H.C., Thorne, D.R., Hegge, F.W. and Babkoff, H. (1985) Trend and rhythm analysis of time-series data using complex demodulation. *Behavior Research Methods, Instruments and Computers*, *17*, 623–629.

Strecker, R.E., Morairty, S., Thaker, M.M., Porkka-Heiskanen, T., Basheer, R., Dauphin, L.J., Rainnie, D.G., Portas, C.M., Greene, R.W. and McCarley, R.W. (2000) Adenosinergic modulation of basal forebrain and preoptic anterior hypothalamic neural activity in the control of behavioral state. *Behavior and Brain Research*, *115*, 183–204.

Stricker, J.L, Brwon, G.G., Wetherell, L.A. and Drummond, S.P.A. (2006) The impact of sleep deprivation and task difficulty on networks of fMRI brain response. *Journal of the International Neuropsychological Society*, *12*, 591–597.

Stuss, D. T., Eskes, G. A., and Foster, J. K. (1994). Experimental neuropsychological studies of frontal lobe functions. In F. Boiler & Gazzaniga (Ed.), The cognitive neurosciences (pp. 1341-1356). Cambridge, MA: MIT Press.

Summala, M. and Mikkala, T. (1994) Fatal accidents among car and truck drivers: Effects of fatigue, age and alcohol consumption. *Human Factors*, *36*, 315–326.

Swartz, B. E., Halgren, E., Fuster, J. M., Simpkins, E., Gee, M., and Mandelkern, M. (1995). Cortical metabolic activation in humans during a visual memory task. Cerebral Cortex, 5, 205–214.

Szatkowska, I., Szymanska, O. and Grabowska, A. (2004) The role of the human ventromedial prefrontal cortex in memory for contextual information. *Neuroscience Letters*, *364*, 71–75.

Thomas, M., Sing, H., Belenky, G., Holcomb, H., et al. (2000) Neural basis of alertness and cognitive performance impairment during sleepiness. I. Effects of 24 hours of sleep deprivation on human regional brain activity. *Journal of Sleep Research*, *9*, 335–352.

Tzischinsky, O., Shlitner, A. and Lavie, P. (1993) The association between the nocturnal sleep gate and nocturnal onset of urinary 6-sulfaoxymelatonin. *Journal of Biological Rhythms*, *8*, 199–209.

Valenstein, E., Bowers, D., Verfaellie, M,, Heilman, K.M., Day, A., and Watson, R.T. (1987) Retrosplenial amnesia. Brain, 110, 1631–1646.

Van Asselen, M., Kessels, R.P.C., Neggers, S.F.W., Kappele, L.J., Frijns, C.J.M. and Postma, A. (2006) Brain areas involved in spatial working memory. *Neuropsychologia*, *44*, 1185–1194.

Van Dongen, H.P., and Dinges, D.F. (2003) Investigating the interaction between the homeostatic and circadian processes of sleep-wake regulation for the prediction of waking neurobehavioural performance. Journal of Sleep Research, 12, 181–187.

Van Dongen, H.P.A., Maislin, G., Mullington, J.M. and Dinges, D.F. (2003) The cumulative cost of additional wakefulness: Dose-response effects n neurobehavioral functions and sleep physiology from chronic sleep restriction and total sleep deprivation. *Sleep*, *26*, 117–126.

Van Dongen, H.P., Baynard, M.D., Maislin, G. and Dinges, D.F. (2004) Systematic inter-individual differences in neurobehavioral impairment from sleep loss: Evidence of trait-like differential vulnerability. *Sleep*, *27*, 423–433.

Van Dongen, H.P., Vittelaro, K.M. and Dinges, D.F. (2005) Individual differences in adult human sleep and wakefulness: Leitmotif for a research agenda. *Sleep*, *28*, 479–496.

Wang, Z., Smith, C.E. and Atchley, W.R. (2007) Application of complex demodulation b Zip and b HLH-PAS protein domains. *Mathematical Biosciences*, *207*, 204–218.

Warrington, E.K., and Weiskrantz, L. (1982) Amnesia: a disconnection syndrome? Neuropsychologia, 20, 233–248.

Wilkinson, R.T. (1965) Sleep deprivation. In Edholm, O.G. and Bacharach, A.L. (Eds.). *Physiology of Human Survival*. Academic, London (pp. 99–430).

Wilkinson, R.T. (1990) Response-stimulus intervals in choice serial reaction time: Interaction with sleep deprivation, choice and practice. *Quarterly Journal of Experimental Psychology*, *42*, 401–423.

Williams, H.L. and Williams, C.L. (1966) Nocturnal EEG profiles and performance. *Psychophysiology*, *3*, 164–175.

Williams, H.L., Lubin, A. and Goodnow, J.J. (1959) Impaired performance with acute sleep loss. *Psychological Monograph*, *73*, 1–26.

Wright, K.P. Jr., Badia, P., Myers, B.L., Plenzler, S.C. and Hakel, M. (1997a) Caffeine and light effects on nighttime melatonin and temperature levels in sleep-deprived humans. *Brain Research*, *747*, 78–84.

Wright, K.P. Jr., Badia, P., Myers, B.L. and Plenzler, S.C. (1997b) Combination of bright light and caffeine as a countermeasure for impaired alertness and performance during extended sleep deprivation. *Journal of Sleep Research*, *6*, 26–35.

Wright, K.P. Jr., Badia, P., Myers, B.L.,Plenzler, S.C. and Drake, C.L. (1997c) Enhancement of nighttime alertness in women with bright light and caffeine during 48 hr of sleep deprivation. *Sleep Research*, *26*, 635.

Wright, K.P. Jr., Badia, P., Myers, B.L., Plenzler, S.C. and Drake, C.L. (1997d) Caffeine and bright light effects on nighttime melatonin and body temperature in women during sleep deprivation. *Sleep Research*, *26*, 636.

Wright, K.P. Jr., Hull, J.T. and Czeisler, C.A. (2002) Relationship between alertness, performance and body temperature in humans. *American Journal of Physiology, Regulation, Integration and Comparative Physiology*, *283*, R1370–R1377.

Wu, J.C., Gillin, J.C., Buchsbaum, M.S., Chen, P., Keator, D.B., Khosla Wu, N., Darnall, L.A., Fallon, J.H., and Bunney, W.E. (2006) Frontal lobe metabolic decreases with sleep deprivation not totally reversed by recovery sleep. Neuropsychopharmacology, 31, 2783–2792.

Wyatt, J.K., Cajochen, C., Ritz-De Cecco, A., Czeisler, C.A., and Dijk, D.J. (2004). Low-dose repeated caffeine administration for circadian-phase-dependent performance degradation during extended wakefulness. Sleep, 27, 374–381.

Yoshihisa, N., Mushiake, H. and Tanji, J. (2003) Representation of the temporal order of visual objects in the primate lateral prefrontal cortex. *Journal of Neurophysiology*, *89*, 2868–2873.

Zhdanova, I.V. (2005) Melatonin as a hypnotic: pro. Sleep Medicine Reviews, 9, 51–65.

Zhdanova, I.V. and Tucci, V. (2003) Melatonin, circadian rhythms and sleep. *Current Treatment Options in Neurology*, *5*, 225–229.

Zhdanova, I.V., Wurtman, R.J., Morabito, C., Piotrovska, V.R., and Lynch, H.J. (1996) Effects of low oral doses of melatonin, given 2-4 hours before habitual bedtime, on sleep in normal young humans. Sleep, 19, 423–431.

Zukerman, G. (2004). Cognitive performance decrement as function of sleep deprivation and circadian (diurnal) rhythm. Unpublished doctoral dissertation, Bar-Ilan University, Ramat-Gan, Israel.

Zukerman, G., Goldstein, A., and Babkoff, H. (2007). The effect of 24-40 hours of sleep deprivation on the P300 response to auditory target stimuli. Aviation, Space, and Environmental Medicine, 78 (S1), B216–B223.

Zully, J., Wever, R. and Aschoff, J. (1981) The dependence of onset and duration of sleep on the circadian rhythm of rectal temperature. *Pflugers Archiven*, *391*, 314–318.

Chapter 16
Open Questions on Mind, Genes, Consciousness, and Behavior: The Circadian and Ultradian Rhythms of Art, Beauty, and Truth in Creativity

E.L. Rossi[1] and K.L. Rossi[2]

An earlier companion volume to this text ended with an epilogue on "The Unification Hypothesis of Chronobiology from Molecule to Mind" wherein we traced the evolution of life and mind as manifest in our circadian and ultradian psychobiology (Lloyd and Rossi, 1992). This chapter extends that broad scenario by reviewing the circadian and ultradian rhythms of gene expression, brain plasticity, and creative experience as we pursue the eternal verities of art, beauty, and truth in science as well as everyday life. This will be a highly speculative and lofty philosophical pursuit, which we will survey as a provocative series of "open questions" about mind, genes, consciousness, and behavior. We will explore our understanding of "the meaning of it all" by balancing the traditional reductionism of chronobiological research with the constructive approaches of the new *in silico* bioinformatic databases of functional genomics, neuroscience, medicine, and psychology. We will balance the traditional "Bottoms-Up Approach" of the physical and biological sciences with our "Top-Down Approach" for a unified understanding of the human condition from a deep chronobiological and psychobiological perspective.

Keywords Art, basic rest activity cycle, brain plasticity, gene expression hypnosis, implicit processing heuristic, psychosocial genomics, psychotherapy, truth, ultradian music

[1] New Neuroscience School of Therepeutic Hypnosis and Psychotherapy, San Lorrenzo Maggiore, Italy and Ernest Rossi Foundation for Psychosocial Genomic Research, 125 Howard Avenue, Los Osos, USA

[2] New Neuroscience School of Therepeutic Hypnosis and Psychotherapy, San Lorrenzo Maggiore, Italy and Ernest Rossi Foundation for Psychosocial Genomic Research, 125 Howard Avenue, Los Osos, USA

16.1 Open Question #1: Do *In Silico* Models of Computer Research Bridge the Cartesian Gaps Between Mind, Gene, Brain, and Body?

In silico is a popular expression in the computer and bioinformatic approaches to simulating life processes on all levels from mind and behavior to molecular genomics. These simulations of complex life processes are performed via information processing models *on silicon chips* in computers as a more economical approach to experimentation than traditional wetware biological. *In silico* research is the key to *data mining: the exploration, assessment, and integration of the meaning and implications of the research literature in many biological and psychological disciplines that cannot be integrated in any other way*. Such *interdisciplinary in silico research* appears to bridge the awkward Cartesian Gap between mind, gene, brain, and body via the concept of information. The genomic revolution made the concept of information the common denominator of all the databases in the life sciences. The Allen Brain Atlas of Gene Expression (ABA), for example, makes it possible to integreate biology, genomics, and neuroscience in a single database. Of greatest interest is how the ABA could make it possible to explore the circadian and ultradian dynamics of activity-dependent gene expression and brain plasticity associated with psychosocial stress, memory, learning, behavior, cognition, creativity, and consciousness itself (Rossi, 2007).

The Allan Brain Atlas (http://www.brain-map.org) is available free as a web based database showing the location and activity level of ~21,000 genes in the mouse brain, which shares about 90% homology with the human brain. Plans are now underway for making a complete human brain atlas of gene expression. This anatomical reference for understanding the role of gene expression for ~50 million Americans suffering from brain dysfunctions such as Alzheimer's, epilepsy, and Parkinson's as well as addiction, depression, and stress is already being described as the foundation for a new neuroscience of mind, behavior, rehabilitation, and psychotherapy. The Allen Brain Atlas consists of the data of 250,000 microscope slides, a million brain sections, and 85 million anatomical photo files, which are assembled for viewing gene expression in three dimensional cross sections of the brain. An unexpected finding revealed by the ABA is that ~80% of genes are expressed in brain cells. The high-resolution digital microscopy images of the ABA show the exact location of the activity-dependent genes that are expressed to generate the proteins carrying out the biological functions of mind and behavior in health and illness.

At the present time research and publications involving the Allen Brain Atlas http://www.brainatlas.org/aba/ are dominated by basic biology with a heavy focus on drug and medical applications. Of particular interest for bridging the Cartesian mind-body gap, however, is the ABA potential for tracking the activity-dependent genes expressed generating brain plasticity (synaptogenesis and neurogenesis) in response to the normal activities of life, memory, learning, behavior as well as creative human experiences of art, truth, and beauty. In principle any active psychological experience of art, truth, and beauty that can be located in the brain by functional magnetic resonance imaging fMRI (Liu et al., 2007; Rossi, 2007) can be tracked down to the molecular genomic level via the ABA. Researchers in psychology, however, have had difficulty in recognizing the implications of this mind-molecular bridge because before the advent of fMRI technology there was no obvious and non-invasive way of assessing the deep molecular-genomic sources of cognition, emotion, and behavior. Gene expression is usually measured by complex and very expensive laboratory procedures such as DNA microarrays that involve taking invasive tissue samples from the brain, blood, saliva, and body (Rossi, 2002, 2004, 2005, 2006 et al., 2007). This stumbling block motivates us to outline in Fig. 16.1 how the ABA in association with other currently available technologies may enable us to bypass invasive biological methods with new *In silico* models of exploring the mind-gene connection in psychotherapy.

Figure 16.1 outlines how the bioinformatic technologies of the Allen Brain Atlas of Gene Expression, functional magnetic resonance imaging (fMRI) (Segel et al., 2006), and the Connectivity Map (Lamb et al., 2006) can be integrated into a new *in silico* model of mind, gene, brain, and body communication with nothing more than a personal computer and an internet connection. This type of *In silico* data mining of the existing scientific literature was originally described in a more limited biological context by Blagosklonny and Pardee (2002).

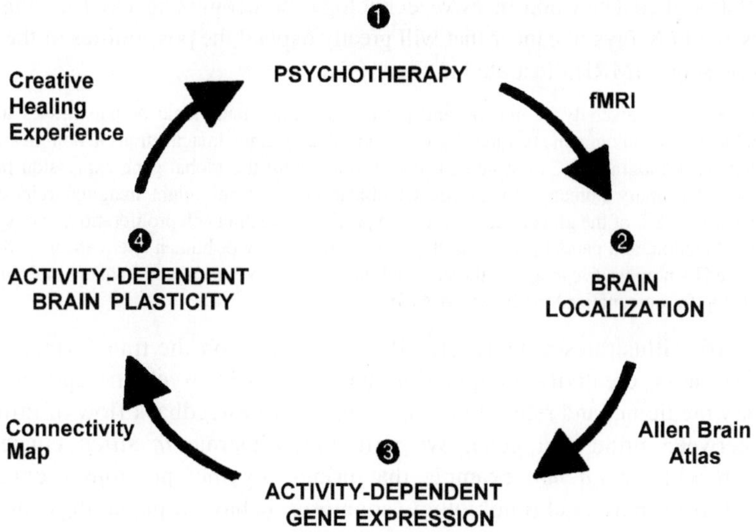

Fig. 16.1 An *In Silico* model for mind-brain-body-gene research that apparently bridges the Cartesian mind-body gap

> Millions of easily retrievable facts are being accumulated in databases, from a variety of sources in seemingly unrelated fields, and from thousands of journals. New knowledge can be generated by 'reviewing' these accumulated results in a concept-driven manner, linking them into testable chains and networks.... Connecting separate facts into new concepts is analogous to combining the 26 letters of the alphabet into languages. One can generate enormous diversity without inventing new letters. These concepts (words), in turn, constitute pieces of more complex concepts (sentences, paragraphs, chapters, books). *We call this process 'conceptual' research, to distinguish it from automated data-mining and from conventional theoretical biology....* Can a review provide new knowledge? A review can constitute a comprehensive summary of the data in the field — this type of writing educates but does not directly generate new knowledge. *But a 'conceptual' review, on the other hand, can generate knowledge by revealing 'cryptic' data and testing hypotheses by published experiments....* Conceptual biology should be recognized and criteria established for its publications — new, testable conclusions, supported by published data. In *[psycho]*biological systems, everything is interconnected, and ostensibly unrelated fields are related — the separation of biology into different disciplines is artificial. Conceptual research *[in psychotherapy]* can encompass many fields without limitation. *In comparison with labour-based research, conceptual research is more cost-effective; indeed, verification of a hypothesis using existing data does not limit research to scientists in well-resourced fields or countries. Hypothesis-driven, experimental research will continue to be a corner stone of biology,* but it should strike up a partnership with the essential components of theoretical and conceptual research. (p. 373, italics added)

As can be seen in Fig. 16.1, *in silico* conceptual research proceeds through four recursive stages to uncover new associations that may never have been considered by the original laboratory researchers who first published their data in fields of biology, bioinformatics, chronobiology, genomics, neuroscience, etc. apparently unrelated to the cognitive-behavioral processes of psychotherapy and the creative creative healing experiences of complementary medicine. Most recently, Segal et al. (2007) described a new non-invasive technology for recognizing and decoding gene expression in X-rays of cancer that will greatly expand the possibilities of the visualization stage (fMRI) illustrated in Fig. 16.1.

> Paralleling the diversity of genetic and protein activities, pathologic human tissues also exhibit diverse radiographic features. Here we show that dynamic imaging traits in non-invasive computed tomography (CT) systematically correlate with the global gene expression programs of primary human liver cancer. Combinations of twenty-eight imaging traits can reconstruct 78% of the global gene expression profiles, revealing cell proliferation, liver synthetic function, and patient prognosis. Thus, genomic activity of human liver cancers can be decoded by noninvasive imaging, thereby enabling noninvasive, serial and frequent molecular profiling for personalized medicine. (p. 675)

Figure 16.1 illustrates a neuroscientific perspective on the transformations of consciousness, creativity, imagination, and healing in psychotherapy, complementary medicine, and rehabilitation as a circular biofeedback flow of information between mind and gene. We will now illustrate *in silico* conceptual research with an *ad hoc* example that brings together previously unrelated research from human chronobiology, cognition, behavior, physiology, psychosocial genomics.

16.2 Open Question #2: Psychosocial Genomics: What are the Research Possibilities of Matching Profiles of Gene Expression with Profiles of Psychobiological Behavior in Circadian and Ultradian Time?

Psychosocial genomics has been described as an interdisciplinary field integrating circadian and ultradian profiles gene expression with their associated profiles of behavioral and psychological experience (Rossi, 2002a, 2004a). The focal issue of psychosocial genomics is to create new models of how profiles of gene expression, brain plasticity, and mind-body relationships are interrelated as a complex adaptive system of human experiencing, behavior, and consciousness. Psychosocial genomics has been applied to a wide range of issues ranging from stress, psychosomatics, psycho-immunology, and psycho-endocrinology to the deep psychobiology of creativity (Rossi, 2002b), optimal performance (Rossi, 2002c, d), slow wave sleep and REM dreaming (Rossi, 2002b), art, ritual, cultural, and spiritual experience (Rossi, 2003, 2005; Rossi et al., 2006; Sanders and Mann, 1955).

Figure 16.2 illustrates an approach to *in silico* research in psychosocial genomics by integrating data from the Allen Brain Atlas of Gene Expression with the psychosocial level. Figure 16.2 was assembled by juxtaposing chronobiological relationships between circadian and ultradian profiles of gene expression with profiles of behavior at the cognitive-behavioral level and the physiological level of core body temperature (Rossi and Lippincott, 1992; Rossi, 2004b, 2007). These profiles were assembled as an *ad hoc* juxtaposing of Aldrich and Bernstein's (1987) circadian profile of hypnotic susceptibility (the cognitive-behavioral level), and a typical circadian profile of core body temperature in humans with the circadian profile of the *Thra* gene (the genomic level) in tissues of the heart and liver of the murine rodent (Storch et al., 2002).

The top cognitive-behavioral profile of Fig. 16.2 illustrates the hourly distribution of hypnotic susceptibility scores for Aldrich and Bernstein's college age subjects. We note that this cognitive-behavioral *distribution is bimodal, with peaks at 12:00 noon and 4:00–6:00 p.m.* and a local minimum at 2:00 p.m. Aldrich and Bernstein hypothesize their results provide preliminary evidence that hypnotizability may be related to the circadian rhythm of body temperature at the physiological level. As may be seen, the circadian profile of core body temperature in Fig. 16.2B is also bimodal and closely approximates the circadian profile of hypnotic susceptibility in Fig. 16.2A.

Figure 16.2C illustrates the circadian expression profile of the *Thra* gene, which is also bimodal and resembles the circadian profiles of hypnotic susceptibility and body temperature. The *Thra* gene, coding for the thyroid hormone receptor α, is itself induced by the thyroid hormones T3 and T4, which are fundamental in regulating the physiological work of metabolism and body temperature (Storch et al., 2002). Figure 16.2D illustrates the circadian expression profile of the clock gene *Period* (*Per 1*), which is associated with many daytime activities in humans, and resembles the circadian profiles of hypnotic susceptibility and body temperature

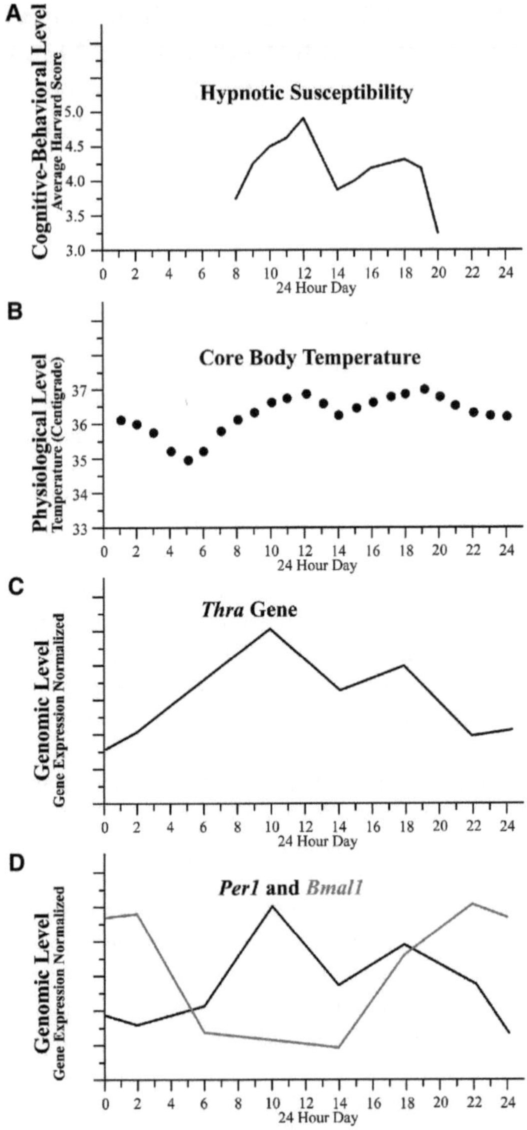

Fig. 16.2 Matching circadian profiles on the cognitive-behavioral, physiological, and genomic levels. This is an *ad hoc* illustration of the psychosocial genomic relationships between the cognitive-behavioral level of hypnosis in (**A**), the physiological level of core body temperature in (**B**), and expression of the *Thra* gene in (**C**) associated with rate of metabolism and body temperature. Figure 16.2**D** illustrates how the circadian profile of the *Per1* gene, typical for the awake state, is similar to the *Thra* gene in (**C**) having a peak of expression about 90–120 min before the peak of hypnotic susceptibility and core body temperature around noon. By contrast notice how the circadian profile of the *Bmal1* gene in (**D**), which is a marker for the sleep state, is in *antiphase* (the opposite of) the awake profiles of *Per1* and *Thra* gene expression associated with peaks of core body temperature and hypnotic susceptibility (Rossi, 2004b)

even more closely than the *Thra* gene. Notice how the circadian profiles of the *Per1* and *Thra* gene are similar in having a peak of expression about 90–120 min *before* the peaks of core body temperature and hypnotic susceptibility around noon. This is consistent with the fact that the 90–120 min ultradian Basic Rest-Activity Cycle (Rossi, 2002a) is typical for many genes to be expressed via gene transcription and translation into the proteins that ultimately generate their physiological and cognitive-behavioral profiles of circadian and ultradian expression (Rossi, 2002b).

Figure 16.1D illustrates the circadian profile of the *Bmal1* gene associated with the sleep state (*the opposite of the Per1* and *Thra* gene profiles associated with being awake). Storch et al. (2002) research on the circadian modulation of gene expression related to body temperature, psychosocial stress (the glucocorticoids), and the immune system (tumor necrosis factor alpha) are of great interest for a psychobiologically oriented approach to therapeutic hypnosis and psychoneuroimmunology at the genomic level. Note the striking similarity of their description of the bimodal circadian profile of gene expression with the bimodal distribution of hypnotic susceptibility scores reported above by Aldrich and Bernstein.

> Many mammalian peripheral tissues have circadian clocks; endogenous oscillators that generate transcriptional rhythms thought to be important for the daily timing of physiological processes. (p. 78) …The *distribution of phases is essentially bimodal, with most genes showing peak expression between circadian time 6 h and 14 h, and a smaller group peaking in phase at about circadian time 20 h.* (p. 81)

Innovative researchers investigating circadian and ultradian profiles of gene expression in the brain, heart, liver, and immune system such as Storch et al. (2002) and others (Panda et al., 2002a,b; Ueda et al., 2002; Rosbash and Takahashi, 2002; Schibler, 2008, this volume), however, do not discuss the implications of their findings for the cognitive-behavioral level of therapeutic hypnosis. Therefore the *ad hoc* assemblage of matched bi-modal circadian profiles of Fig. 16.2 is consistent with *but certainly do not yet prove* that there are causal and reciprocal relationships in the complex interactions between the cognitive-behavioral level of hypnotic susceptibility, physiology, and gene expression. Such proof would require much innovative and integrative psychobiological research to operationalize our next open question (Unterweger et al., 1992).

16.3 Open Question #3: To What Extent Do Top-Down Psychological Processes Modulate Chronobiological Profiles of Gene Expression During Sleeping, Dreaming, and Waking?

Figure 16.3 is a phase histogram, which illustrates the global circadian bi-modal profile of gene expression in the heart and liver, suggests how psychosocial data can be adopted for exploring associations between profiles gene expression, psychophysiology, and complex cognitive-behavioral processes such as therapeutic hypnosis. Panda et al. (2002b) have stated, "The circadian control of transcription

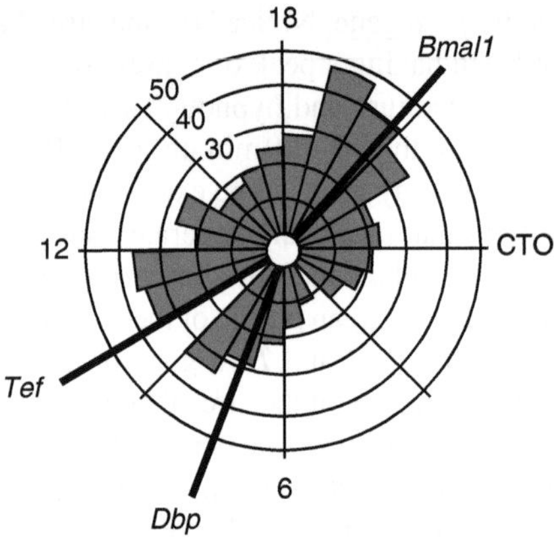

Fig. 16.3 A phase histogram of circadian gene expression in mouse heart and liver. *Bmal1* is a gene with a known robust circadian expression during sleep while the *Dbp* and *Tef* genes serve as phase markers associated with being awake. This circular histogram resembles a circadian 24 h clock face but with CT0 (Circadian Time Zero) on the right representing awakening (or the standardized initial point of data collection such as when movement and physical activity occurs). The nested numbers from the periphery toward the center (50, 40, 30) are the number of genes expressed in each circadian phase of the histogram during this particular experiment (With permission from Storch et al., 2002)

in higher organisms is integrated with the spatial control of *gene expression to target rate-limiting steps in major pathways in their relevant organs, resulting in a systems-level temporal orchestration of behavior and physiology for optimal adaptation of the organism to its environment*" (p. 317, italics added).

Global circadian profiles of gene expression such as Fig. 16.3 may enable us to get a precise answer to a century of controversy regarding the range and limitations of psychological processes and states in modulating the neuro-dynamics of the brain at the molecular-genomic level. Are the circadian and ultradian profiles of gene expression during therapeutic hypnosis, for example, more closely associated with sleep (as proposed by Pavlov, 1927) or the awake state as maintained by Hull (1933/1968)? This question could be answered on the genomic level by investigating the implications of research by Ueda et al. (2002), which they summarize, "Here we demonstrate the role of the Rev-ErbA/ROR response element in gene expression during circadian night, which is in phase with *Bmal1* [during sleep, which Pavlov would predict as associated with hypnosis], and in anti-phase to *Per2* oscillations [while awake, which Hull would predict as associated with hypnosis]" (p. 534).

What is now needed is a more direct assessment of the circadian correlates of therapeutic hypnosis (Rossi, 2002a, 2004a) correlated with core body temperature and profiles of gene expression in high hypnotic susceptible subjects during sleep proposed by Rossi (2004b) somewhat as follows.

> When body temperature is at its lowest point during sleep we could induce hypnosis without awakening the subject (*sleep induced hypnosis*) and keep them occupied with

personally relevant inner work during hypnosis for an ultradian period of 90–120 minutes. If Pavlov is correct (hypnosis is like sleep) their core body temperature should remain low while their "Rev-ErbA/ROR response element in gene expression" and *Bmal1* gene expression remains at a peak illustrated in Fig. 16.2D (indicating that the circadian night profile of gene expression prevails during sleep and hypnosis). If, on the other hand, Hull is correct we would expect the reverse: core body temperature should rise during the hypnosis and hypnotically induced inner work while subjects were sleeping; on the genomic level their Rev-ErbA/ROR response element and *Bmal1* gene expression should go down in anti-phase with their *per1* gene expression should go up toward a peak as is characteristic of the awake state. Such research would be an important step in answering on the genomic level an issue first raised by the author 20 years ago regarding the degree to which therapeutic hypnosis may *modulate*, *entrain* or *shift* the *phase response curves* (Johnson, 1999) of circadian and ultradian rhythms during health, stress, and illness. (Rossi, 2002a)

Extending this type of thought experiment even further, it may be possible to differentiate between waking consciousness and therapeutic hypnosis on the level of gene expression. If a difference is found, for example, between the profiles of gene expression in *sleep induced hypnosis* versus a control group (where subjects are simply awakened from their sleep, for example, and kept awake for a similar 90–120 minute ultradian period), these *circadian difference gene expression profiles* may be useful in distinguishing between gene expression during therapeutic hypnosis, normal waking consciousness, and the spectrum of consciousness in creative states and meditation (Wilber, 1993), and psychotherapy in general. Such distinctions would enable us to determine precisely which profiles of gene expression could be modulated by the many schools of psychotherapy ranging from the cognitive-behavioral and psychoanalysis to therapeutic hypnosis. (pp. 19–20)

Such highly innovative and pioneering research would go a long way to determining the quantitative parameters reifying how top-down (Mehta, 2007) processes modulate genomic processes at the molecular level. This top-down approach would be consistent with neuroscience research over the past decade, which has documented the neuro-anatomical processes and functions of our normally varying profiles of gene expression during circadian and ultradian chronobiological rhythms. We will now review the possible significance of one of the most prominent of these neuro-anatomical processes, which has been described as a dialogue between the neocortex, the hippocampus and other sub-cortical areas during off-line states such as slow wave sleep, REM dream sleep, and the "ultradian healing response" during resting states of normal waking consciousness in daily life (Rossi and Nimmons, 1991; Rossi, 2002a, 2004a; Stickgold et al., 2007).

16.4 Open Question #4: Do the Neocortex-Hippocampus Dialogues Function as Cartesian Bridge Between Mind, Gene, Brain, and Body?

Buzsáki's (1996) early description of a dialogue between the neocortex and the hippocampus set the stage for a new research paradigm investigating the functional dynamics of brain anatomy that has attracted an increasing number of investigators over the past decade (Drosopoulos et al., 2007; Hahn et al., 2007) with these words.

> In gross anatomical terms, the hippocampal archicortex can be conceived as an "appendage" of the large neocortex. In contrast to neocortical areas, the main output targets of the hippocampus are the same as its main inputs (i.e., the entorhinal cortex). *Highly processed information about the external world (the content) reaches the hippocampus via the entorhinal cortex, whereas information about the "internal world" (the context) is conveyed by the sub-cortical inputs. ... From its strategic anatomical position and input-output connections, it may be suggested that the main function of the hippocampal formation is to modify its inputs by feeding back a processed "re-afferent copy" to the neocortex. I hypothesize that neocortico-hippocampal transfer of information and the modification process in neocortical circuitries by the hippocampal output take place in a temporally discontinuous manner and might be delayed by minutes, hours, or days.* Acquisition of information may happen very fast during the activated state of the hippocampus associated with theta/gamma oscillations. *Intra-hippocampal consolidation and the hippocampal-neocortical transfer of the stored representations, on the other hand, is protracted and carried by discrete quanta of cooperative neuronal bursts during slow wave sleep.* (p. 81, Italics added here)

Recently Ji and Wilson (2007) and others (Káli and Dayan, 2004) have confirmed that this bi-directional interaction between the hippocampus and the neocortex is necessary for memory consolidation. These researchers believe this dialogue is initiated by the neocortex because its activity occurs sooner than its synchronized counterparts in the hippocampus. *It appears as if the neocortex is querying the hippocampus to replay its most recent raw sensory-perceptual data and memories. This dialogue is not a simple transfer of memory, however, but represents a more sophisticated processing of data whereby the neocortex selects novel information from the hippocampus.* Researchers suggest that from a top-down perspective, the neocortex is trying to make sense of what is going on in the hippocampus and to build models of the world to understand how and why things happen. These top-down models presumably generate new expectations about the world, which direct cognition, creative planning and adaptive behavior in *the evolutionary memory-prediction framework of intelligence* (Hawkins and Blakeslee, 2004, p. 177; Rossi et al., 2008, in press). We have proposed these top-down neuro-anatomical models as an emerging framework for the deep psychobiological foundation of virtually all schools of psychotherapy and rehabilitation (Rossi, 2007; Rossi et al., 2006, 2007).

16.5 Open Question #5: Is the Offline Replay of Circadian and Ultradian Rhythms the Essence of the Creative Experience in Art, Beauty and Truth?

Figure 16.4 is a profile of the human brain with a focus on the hippocampus, which only makes a temporary recording of new memory, learning or behavior, however. Later, during "offline periods" of sleep, dreaming, and rest when the conscious mind is not actively engaged in dealing with outer realities, the neocortex and hip-

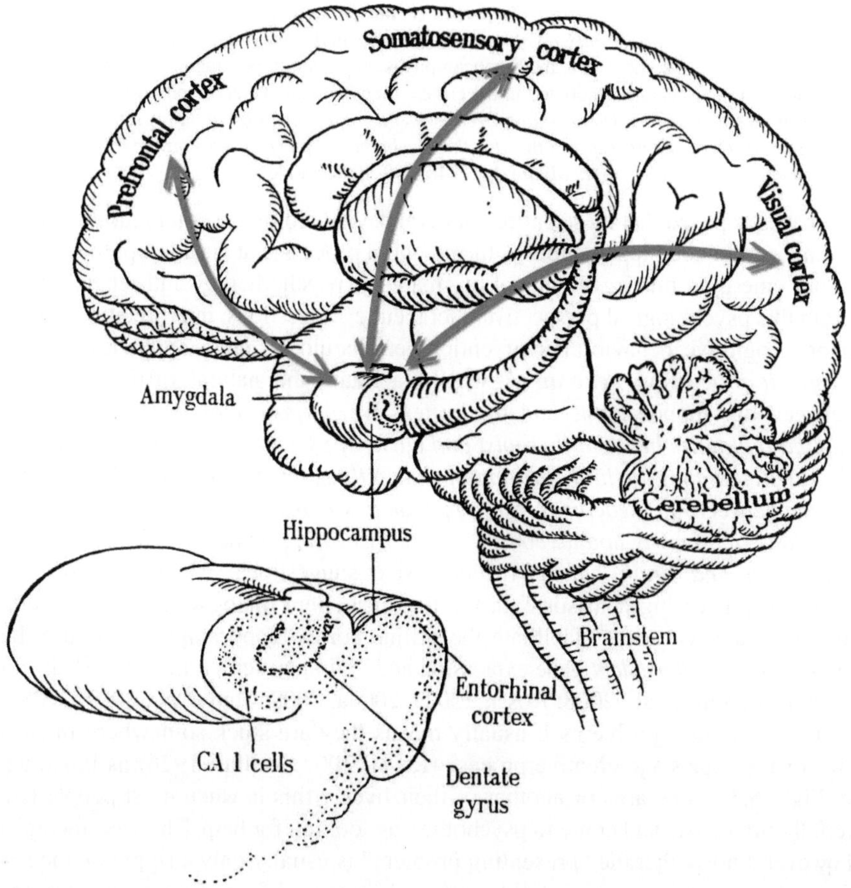

Fig. 16.4 The neocortex and hippocampus dialogues. During "offline" periods of slow wave sleep, dreaming, and rest when the conscious mind is not actively engaged in coping with outer realities, the neocortex and hippocampus (and other sub-cortical brain structures) engage in dialogues to update, replay, and consolidate the new memory and learning in the neocortex where the 'local-global computations' of consciousness are thought to take place (Pereira et al., 2007)

pocampus engage in a neural dialogues to update, replay and consolidate the new memory in more permanent storage locations throughout the neocortex. *These mind-brain-gene dialogues activate and creatively replay the 'local – global computations' of the neocortex (Buzsáki, 2006), which are now believed to be the neural correlates of consciousness long sought by the late Francis Crick.*

Lisman and Morris (2001) describe how this offline dialogue activates and replays novel and significant life experience between the cortex and hippocampus of the brain as follows.

> ... newly acquired sensory information is funneled through the cortex to the hippocampus. Surprisingly, only the hippocampus actually learns at this time — it is said to be online. *Later, when the hippocampus is offline (probably during sleep), it replays stored informa-*

> *tion, transmitting it to the cortex. The cortex is considered to be a slow learner, capable of lasting memory storage only as a result of this repeated replaying of information by the hippocampus.* In some views, the hippocampus is only a temporary memory store — once memory traces become stabilized in the cortex, memories can be accessed even if the hippocampus is removed. *There is now direct evidence that some form of hippocampal replay occurs.... These results support the idea that the hippocampus is the fast online learner that "teaches" the slower cortex offline.* (pp. 248–249, Italics added)

We have proposed that these entirely natural psychobiological "dialogues" between our neocortex and hippocampus is the essential process that we attempt to facilitate in our emerging mind-gene model of creativity, psychotherapy and rehabilitation. From this psychological perspective, therapeutic suggestions, interpretations, metaphors, cognitive behavioral interventions etc. could be more aptly described as *"implicit processing heuristics,"* which facilitate the natural offline dialogues between our hippocampus and the cortex. *We propose that the conscious and explicit dialogues between therapist and client in psychotherapy are efficacious to the extent that they facilitate the appropriate offline, unconscious, and implicit dialogues between the neocortex, the hippocampus and other subcortical structures that daily and hourly update consciousness by turning on activity-dependent gene expression and brain plasticity.* Permissive suggestions, which we now call "Implicit processing heuristics" in the therapist/client dialogue, are creative cues and hints that we use to facilitate the offline cortex/hippocampus dialogues that evokes *activity-dependent* gene expression and brain plasticity for adaptive behavior change (Ribeiro et al., 2008; Rossi, 2002a, 2004a, 2007; Toni et al., 2007, 2008).

If people have problems it usually means they are stuck somewhere in stage two of the four stage creative process (Rossi, 2007; Wallas, 1926) as illustrated in Fig. 16.5 in one area or another of their lives – this is when most people tend to fall into a crisis and come to psychotherapy looking for help. The wise therapist, however, knows that the "presenting problem" is usually only a ripple on the surface of the deeper waters of self-care and creative life management. Our new chronobiological model of the creative process illustrated in Fig. 16.5 implies that every creative individual needs to learn how to breakout of previously learned limitations on all levels from mind to gene expression and brain plasticity (Toni et al., 2007). Facilitating this very general paradigm of the creative process is called, "The Breakout Heuristic" in our growth-oriented model of psychotherapy (Rossi, 2007).

Of all the creative arts, music and appears to be the one most dependent on rhythm and timing. This suggests we might expect to find evidences of ultradian and circadian rhythms in music as illustrated in Fig. 16.6. Figure 16.6 illustrates how the creative experience of the *sonata form*, a four-stage paradigm of musical composition widely in classical music, has essentially the same profile as chronobiological profile of the four-stage creative process in Fig. 16.5 (Kamien, 2006). The term "*sonata form*" refers to the first movement of a symphony. The opening movement of a classical symphony, for example, is usually in sonata form (often called *sonata-allegro*). The sonata form has three main sections (the *exposition*, *development*, and *recapitulation*) that are often followed by the *coda*, a brief concluding section.

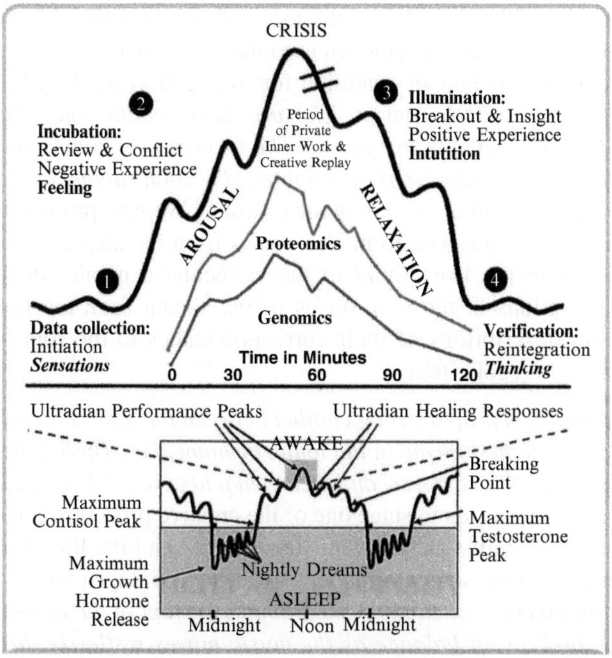

Fig. 16.5 The chronobiology of the four stage creative cycle. The ultradian profile (90–120 min) of the classical four-stage creative process as it is typically experienced on the subjective psychological level is illustrated in the top most portion of the upper curve. The proteomics (protein) profile in middle curve depicts the energy landscape for protein folding within neurons of the brain into the correct structures needed for brain plasticity (Ealch et al., 2008; Cheung et al., 2004). This proteomic profile arises from the functional concordance of co-expressed genes illustrated by the genomics profile below it (Adapted from Levsky et al., 2002). This genomics curve represents the actual gene expression profiles of the immediate-early gene *c-fos* and 10 other genes (alleles) over the typical Basic Rest-Activity (BRAC) period of 90–120 min. The lower diagram illustrates how these ultradian dynamics of the qualia of consciousness are typically experienced as Kleitman's 90–120 min Basic Rest-Activity Cycle within the normal circadian cycle of waking and sleeping (Rossi, 2002a, 2004a; Rossi and Nimmons, 1991)

Fig. 16.6 The Sonata form of classical music. The sonata form of classical music is usually profiled in four movements, which we propose as corresponding to the four stages of the creative process in Fig. 16.5. Recall how the classical octaves of music theory are also made up of two tetrads (four notes in a tetrad). Further research may establish how rudimentary forms of these four stages of the classical creative process are also evident in the quatrains of poetry, lyrics, and song as well as alchemy and psychology (Jung, 1953, 1927)

The sonata form illustrates how composers of the classical period (1750–1820) such as Haydn, Mozart, and Beethoven used the language of music to tell "stories" that resonated with the human condition for over 200 years. In a book on music appreciation, Kamien (2006), states: *"The amazing durability and vitality of sonata form result from its capacity for drama. The form moves from a stable situation toward conflict (in the exposition), to heightened tension (in the development), and then back to stability and resolution of the conflict."* We now propose that the durability and vitality of the sonata form also comes from its integration of multilevels of human experiencing from mind to the molecular-genomic. In the following Kamien's descriptions of the four stages of the sonata form are noted in italics along with our descriptions of their correspondences to the four stages of the creative process in regular font.

1. *"The **Exposition** sets up a strong conflict between the tonic key and the new key. It begins with the first theme in the tonic, or home, key. There follows a bridge, or transition, leading to the second theme, in a new key."* This initial stage of the sonata form corresponds to stage one of the creative process when therapist and patient seek to identify the problem (first theme) and the life changes (second theme) that are at the source of patient's conflicts of stage two.
2. *"The **Development** is often the most dramatic section of the movement. The listener may be kept off balance as the music moves restlessly through several different keys. Through these rapid modulations, the harmonic tension is heightened. In this section, themes are developed, or treated in new ways. They are broken into fragments, or motives, which are short musical ideas developed within the composition. A motive may take on different and unexpected emotional meanings."* This second part of the sonata form corresponds to stage two, the incubation stage of the creative process, with its characteristic conflicts, negative emotions, and symptoms.
3. *"The beginning of the **Recapitulation** brings resolution, as we again hear the first theme in the tonic key.... Earlier in the exposition, there was a strong contrast between the first theme in the home key and the second theme and closing section in a new key; that tension is resolved in the recapitulation.... "* This third section of the sonata form obviously corresponds to stage three of the creative process when there is a resolution of the conflict, problems, and symptoms of stage two.
4. *"An even more powerful feeling is attained by following the recapitulation with yet another section. The coda rounds off a movement by repeating themes or developing them further. It always ends in the tonic [home] key."* Note how this fourth and final part of the sonata form corresponds to stage four of the creative process in illustrated in Fig. 16.5 (Adapted with quotations from Kamien (2006, pp. 163–164) in italics).

Note how well the four stages of the sonata form described in Fig. 16.6 appear to correspond to the four stages of the creative process in Fig. 16.5. This leads us to propose that Fig. 16.6 is a striking illustration of how the creative arts (music, dance, drama, myth, poetry, song, stories, etc.) may be understood as performance modalities for the therapeutic replay, reconstruction, and reframing of negative human experiences into positive perspectives that many cultures have called "healing, therapeutic, rehabilitation and wisdom." Indeed, many of the classical composers such as Beethoven and Schubert were know to use their musical improvisations for healing that "said everything and gave solace to the sufferer in his own language" (Solomon, 2003, p. 229). This leads us to propose that *numinous experiences* (such as *fascination*, *mysteriousness*, and *tremendousness*, Otto, 1923/1950) of art, beauty, and truth could initiate the *activity* of positive healing experiences when they generate the top-down *activity-dependent* gene expression, brain plasticity, and their natural psychophysiological consequences in the psychoneuroimmunology, psycho-endocrinology, etc. (Rossi, 2002a, 2004a, b, 2005, 2008).

16.6 Open Question #6: What Is the Ultimate Molecular-Genomic Source of Kleitman's Ultradian 90–120 min Basic Rest-Activity Cycle?

A review of early conceptions of the relationships between ultradian rhythms, the periodicity of the endocrine system, and psychosomatic problems as fundamental for understanding the implications of Kleitman's 90–120 min Basic-Rest-Activity Cycle (BRAC) for general health and optimal performance was stated as follows (Rossi, 1986/1993).

> One of the most interesting recent theories about the genesis of psychosomatic problems is that they *result from the behavioral disruptions of the ultradian rhythms that modulate both autonomic and endocrine system functioning*. Orr, Hoffman, and Hegge (1974), for example, reported that most humans manifest a stable ultradian rhythm when under quiet conditions. When their subjects were overstressed with extended performance tasks (e.g. monitoring complex panel meters), however, their ultradian rhythms underwent major disruptions in amplitude and patterning. (p. 134)

This conception was generalized into a therapeutic approach to psychosomatic problems by teaching people how to recognize they had a behavioral choice between experiencing an "*Ultradian Healing Response*" by taking an appropriate healing 20 min rest break every 90–120 min or so throughout their busy day, versus their habitual ignoring of their natural mind-body signals for rest that generated a "*Ultradian Stress Response*," which resulted in their psychosomatic symptoms (Rossi and Nimmons, 1991). This therapeutic approach is supported by more recent neuroscience research on new methods of measuring Kleitman's BRAC (Shono et al., 2001) and the recognition of the "protective effect of ultradian rhythms" (Duchniewska and Kokoszka, 2003).

While great progress has been made recently in elucidating the molecular-genomics of the circadian clock (Grimaldi and Sassone-Corsi, 2007; Schibler, 2008, this volume), the mechanisms of Kleitman's ultradian 90–120 min Basic Rest-Activity Cycle remains poorly understood at this level. Lin et al. (2007), for example, have identified the transcriptional coactivator PGC-1 as a key component of the circadian oscillator that integrates the mammalian clock and energy metabolism as follows.

> We have identified PGC-1α, a major metabolic regulator, as a critical component of the mammalian clock. PGC-1α stimulates the expression of *Bmal1* through coactivating the ROR family of orphan nuclear receptors and is essential for normal circadian rhythms. *Because PGC-1α expression is highly responsive to nutritional signals and potentially light, our findings support a mechanism through which energy metabolism and circadian clock can be directly coupled at the transcriptional level. PGC-1*$^{\alpha}$ null mice are resistant to diet-induced obesity and are more insulin-sensitive, whereas *Clock* mutant mice develop obesity. These differences are probably due to the fact that *PGC-1*$^{\alpha}$ null mice are hyperactive and have a higher metabolic rate in the absence of increased food intake. Our findings raise an intriguing possibility that the expression and/or activity of PGC-1α itself may be regulated by components of the clock oscillator. *Because feeding and locomotor activity are regulated by circadian clocks, it is possible that rhythmic PGC-1α expression is controlled by these physiological and behavioural rhythms.* Alternatively, PGC-1α is associated with the SirT1 histone deacetylase complex and may directly sense the metabolic state of the cell, in a similar way to the regulation of the clock homologue NPAS2 by redox status. *Disruption of circadian rhythms has been implicated in the pathogenesis of metabolic disorders. Our studies have uncovered a potential molecular target that could simultaneously modulate circadian clocks and energy metabolism.* (p. 480, italics added here)

The open question now is whether transcriptional coactivator PGC-1 and/or other transcription factors functionally related to it are also key components of Kleitman's ultradian 90–120 min Basic Rest-Activity Cycle. More recently McClung et al. (Talan, 2007) reported that *Clock* mutant mice manifest many of the same symptoms as humans diagnosed with mania and bipolar behavior. Their work implies that the clock gene is a major source of circadian behavior and mood regulation in humans. The *Clock* mutation is now known to disrupt dopamine neurons, which play a important role in mood, activity level, motivation, and pleasure. Continuing research at the molecular-genomic level of transcription factors that activate *Clock* gene expression and related brain plasticity would certainly go a long way toward integrating the deep psychobiology of broad classes of apparently unrelated phenotypic behaviors characteristic of the wave nature of human consciousness and being in creative and motivating experiences of art, beauty and truth (Rossi, 2005, 2007; Rossi & Kleitman, 1992).

16.7 Open Question #7: Will it be Possible to Develop a "Mind-Gene Biofeedback Device" Whereby Voluntary Conscious Mental Activity Could Modulate Activity-Dependent Gene Expression and Brain Plasticity for Mind-Body Healing?

At what levels would a mind-gene biofeedback device or psychological process operate? That is, how would a top down device or psychological process focus consciousness and attention to modulate the dynamics of activity-dependent gene expression? Current concepts and research in epigenetics (Bird, 2007; Feinberg,

2007; Fraser and Blickmore, 2007) suggests that transcription factors are a bridge between environmental stimuli, mental activity, and activity-dependent gene expression, brain plasticity, and a host of other apparently unrelated psychobiological fields of activity such as psychoneuroimmunology, psychoendocrinology, stress, optimal states of creativity and flow in everyday life as well as psychotherapy, rehabilitation, and medicine in general (Rossi, 2008; Rossi and Cheek, 1988).

A brief review of how a mind-gene biofeedback device could be put together has been proposed by Rossi (2004a) and updated here as follows.

> Will it be possible to develop a *mind-gene biofeedback device* in the future that would allow us to modulate activity-dependent gene expression and brain plasticity just as we now use inexpensive biofeedback devices to modulate muscle relaxation? This would be the ultimate kind of mind-body biofeedback that theoretically could facilitate to any type of psychophysiological healing at the molecular-genomic level.
>
> To make a mind-gene biofeedback device we need a *mind-gene transducer*. That is, we need to invent a transducer or "transformer" that converts a highly stimulating and activating subjective psychological experience (novel and numinous *activity* on the cognitive-behavioral level) into a neocortical – sub cortical dialogue (e.g. the neocortical – hippocampus dialogues reviewed above) that would activate series of molecular signals and transcription factors that would turn on *activity-dependent* gene expression and *activity-dependent* brain plasticity. Recent research in nano-technology suggests how this may be possible. (Hahm and Lieber, 2004)
> "Nano" means billionths of a meter, or a hundred-millionths of an inch – much smaller than the usual meaning of small. Individual atoms are just a few nanometers in diameter. This is the scale at which atoms are assembled to form the molecules that create the basic building blocks of life such as DNA, genes, and proteins. With but few exceptions every cell in the body contains the entire genetic code consisting of about 3 billion "base pairs" of nucleotides for making the body's 10 trillion cells. Since these cells replace themselves every few years, the human body makes new DNA at the estimated rate of 10,000 miles an hour! Nature, of course, makes DNA at an implicit or unconscious level. Our proposal for a mind-gene biofeedback machine would be a way of introducing consciousness more directly into the process of modulating DNA production, gene expression, brain plasticity, and mind-body healing. This may be possible via the new "nanolab chips" currently under development.
> Researchers at the California Institute of Technology, UCLA, IBM, and the Institute for Systems Biology in Seattle are creating nanolab chips that greatly improve the speed and efficiency of our current DNA microarray (gene chip) technology. Nanolab chips will be able to analyze what is happening within individual cells in *a few seconds or minutes* what it takes current microarray technology hours to accomplish with much larger tissue samples. These nanolab chips contain nano-mechanical sensors to detect protein and gene interactions (i.e. gene transcription) that could become part of a mind-gene transducer that would be the essence of our proposed mind-gene biofeedback device. The heart of these nano-mechanical mind-gene transducers could be a nano-wire sensor that produces an electric signal when it binds to a gene and/or protein such as a transcription factor and a wide range of other signaling molecules (Hahm and Lieber, 2004). This electric signal can then be amplified to produce an image on a computer screen that would enable *human subjects to use their consciousness to detect when they are in modulating activity-dependent gene expression and brain plasticity*. This is the type of mind-gene transducer that would make the ancient alchemical dream of mind-body communication and healing possible in real time with a practical mind-gene biofeedback device (de Waele et al., 2007; van Hulst, 2007). It already has been proposed that such direct detection and imaging of gene expression could enable researchers use a single drop of blood to screen patients for all the known human genetic disorders (Goho, 2004)." (pp. 304–305)

16.8 Summary and Conclusions

Many fundamental issues must be resolved to facilitate the emerging sciences, technologies, and therapies in response to the open questions of this chapter. We need intensive research in the new sciences of epigenetics and psychosocial genomics to explore the range and limitations of how we can utilize the top-down permissive suggestions (i.e. "implicit processing heuristics") to modulate the cascades of molecular signals between mind and body ranging from the receptors on the surfaces of the cells to the nuclear transcription factors that turn on activity-dependent gene expression and brain plasticity. Conventional image amplifying devices are limited in their resolving power by the diffraction of light with traditional lenses and mirrors. Recent discoveries in nano photonic technology, however, have broken through the diffraction limit of conventional optics to image the molecular-genomic level. This new imaging power down to the molecular-genomic level is what would make it possible to create an in vivo gene sensor and mind-gene biofeedback medical and psychological device for facilitating human performance and healing. We need to energetically monitor websites of neuroscience, chronobiology, epigenetics, functional genomics, nano-technology, etc. to pick up the conceptual breakthroughs, which will implement our emerging sciences, technologies, and therapies to facilitate human consciousness and its potentials in a manner that will be worthy of a Nobel Prize in the future.

References

Aldrich, K., & Bernstein, D. (1987). The effect of time of day on hypnotizability. *International Journal of Clinical & Experimental Hypnosis,* 35, 141–145.

Balch, W., Morimoto, R., Dillin, A., Kelly, J. (2008). Adapting proteostasis for disease intervention. *Science*, 319, 916–919.

Bird, A. (2007). Perceptions of epigenetics. *Nature*, 447, 396–398.

Blagosklonny, M., and Pardee, A. (2002). Conceptual biology: Unearthing the gems. *Nature*, 416, 373.

Buzsáki, G. (1996). The Hippocampo-Neocortical Dialogue. *Cerebral Cortex*, 6, 81–92.

Buzsáki, G. (2006). *Rhythms of the Brain.* New York: Oxford University Press.

Cheung, M., Chavez, L., and Onuchic, J. (2004). The energy landscape for protein folding and possible connections to function. *Polymer*, 45, 547–555.

De Waele, R., Koenderink, A., and Polman, A. (2007). Tunable nanoscale localization of energy on plasmon particale arrays. *Nano Letters*, 7, 2004–2008.

Drosopoulos, S., Windau, E., Wagner U., and Born, J. (2007) Sleep enforces the temporal order in memory. *PLoS ONE* 2(4): e376. doi:10.1371 /journal. pone.0000376.

Duchniewska, K., and Kokoszka, A. (2003). The protective mechanisms of the basic rest-activity cycle as an indirect manifestation of this rhythm in waking: Preliminary report. *International Journal of Neuroscience*, 113, 153–163.

Feinberg, A. (2007). Phenotypic plasticity and the epigenetics of human disease. *Nature*, 447, 433–440.

Fraser, P., and Blickmore, W. (2007). Nuclear organization of the genome and the potential for gene regulation. *Nature*, 447, 413–417.

Goho, A. (2004). Gene screen: Ultrasensitive nanowires catch mutations. *Science News*, 165, 6.
Grimaldi, B., and Sassone-Corsi, P. (2007). Metabolic clockwork. *Nature*, 447, 386–387.
Hahm, J., and Lieber, C. (2004). Direct ultrasensitive electrical detection of DNA and DNA sequence variations using nanowire nanosensors. *Nano Letters*, 4(1), 51–54. doi: 10.1021/nl034853b.
Hahn, T., Sakmann, B., and Mehta, M. (2007). Differential responses of hippocampal subfields to cortical up-down states. *Proceedings of the National Academy of Sciences of the United States of America*, 104, 5169–5174.
Hawkins, J., and Blakeslee, S. (2004). *On Intelligence*. New York: Owl Book/Holt.
Hull, C. (1933/1968). *Hypnosis and suggestibility*. New York: Appleton-Century-Crofts.
Ji, D., and Wilson, M. (2007). Coordinated memory replay in the visual cortex and hippocampus during sleep. *Nature Neuroscience*, 10, 100–107.
Johnson, C. (1999). Fourty years of PRCs – what have we learned? *Chronobiology* International, 16, 711–743.
Jung, C. (1927). *General Index: The Collected Works of C. G. Jung*. Bollingen series, 20. Princeton, NJ: Princeton University Press.
Jung, C. (1953). *Psychology and Alchemy: The Collected Works of C. G. Jung*. Bollingen series, 12. Princeton, NJ: Princeton University Press.
Káli, S., and Dayan, P. (2004). Off-line replay maintains declarative memories in a model of hippocampal-neocortical interactions. *Nature Neuroscience*, 7, 286–294.
Kamien, R. (2006). *Music, an Appreciation*. New York: McGraw-Hill.
Lamb, J., Crawford, E., Peck, D., Modell, J., Blat, I., Wrobel, M., Lerner, J., Brunet, J., Subramanian, A., Ross, K., Reich, M., Hieronymus, H., Wei, G., Armstrong, S., Haggarty, S., Clemons, P., Wei, R., Carr, S., Lander, E., & Golub, T. (2006). The Connectivity Map: Using gene-expression signatures to connect small molecules, genes, and disease. *Science*, 313, 1929–1935.
Levsky, J., Shenoy, S., Pezo, C., and Singer, R. (2002). Single-cell gene expression profiling. *Science*, 297, 836–840.
Liu, C., Li, S., Liu, T., Borjigin, J., and Lin, J. (2007). Transcriptional coactivator PGC-1α integrates the mammalian clock and energy metabolism. *Nature*, 447, 477–481.
Lin, L., Chen, H., Kuang, D., Wang, D., & Tsien, J. (2007). Neural encoding of the concept of nest in the mouse brain. *Proceedings of the National Academy of Sciences USA*, 104, 6066–6071.
Lisman, J. & Morris, G. (2001). Why is the cortex a slow learner? *Nature*, 411, 248–249.
Lloyd, D. & Rossi, E. (Eds.) (1992). *Ultradian Rhythms in Life Processes: An Inquiry into Fundamental Principles of Chronobiology and Psychobiology*. New York: Springer-Verlag.
Mehta, M. (2007). Cortico-hippocampal interaction during up-down states and memory consolidation. *Nature Neuroscience*, 10, 13–15.
Orr, W., Hoffman, H., and Heggey, F. (1974). Ultradian rhythms in extended performance. *Aerospace Medicine*, 45, 995–1000.
Panda, S., Hogenesch, J., and Kay, S. (2002a). Circadian rhythms from flies to humans. *Nature*, 417, 329–335.
Panda, S., Antoch, M., Miller, B., Su, A., Schook, A., Straume, M., Schultz, P., Kay, S., Takahashi, J., and Hogenesch, J. (2002b). Coordinated transcription of key pathways in the mouse by the circadian clock. *Cell*, 109, 307–320.
Pavlov, I. (1927). *Conditioned reflexes: an investigation of the physiological activity of the cerebral cortex.* Anrep, G. (Trans.). Oxford: Oxford University Press.
Pereira, A., Huddleston, D., Brickman, A., Sosunov, A., Hen, R., McKhann, G., Sloan, R., Gage, F., Brown, T., and Small, S. (2007). An *in vivo* correlate of exercise-induced neurogenesis in the adult dentate gyrus. *Proceedings of the National Academy of Sciences*, 104, 5638–5643.
Ribeiro, S., Simões, C., and Nicolelis, M. (2008). Genes, Sleep and Dreams. In Lloyd, D. and Rossi, E. (2008). *Ultradian Rhythms from Molecules to Mind: a New Vision of Life*. New York: Springer.
Rosbash, M., & Takahashi, J. (2002). Circadian rhythms: the cancer connection. *Nature*, 420, 373–374.

Rossi, E. (1986/1993). *The Psychobiology of Mind-Body Healing: New Concepts of Therapeutic Hypnosis*. New York: W. W. Norton.

Rossi, E. (2002a). *The Psychobiology of Gene Expression: Neuroscience and Neurogenesis in Therapeutic Hypnosis and the Healing Arts*. New York: W. W. Norton.

Rossi, E. (2002b). The psychosocial genomics of therapeutic hypnosis and psychotherapy. *Sleep and Hypnosis: An International Journal of Sleep, Dream, and Hypnosis*, 4(1), 26–38.

Rossi, E. (2002c). A conceptual review of the psychosocial genomics of expectancy and surprise: Neuroscience perspectives about the deep psychobiology of therapeutic hypnosis. *American Journal of Clinical Hypnosis*, 45(2), 103–118.

Rossi, E. (2002d). Psychosocial genomics: Gene expression, neurogenesis, and human experience in mind-body medicine. *Advances: In Mind-Body Medicine*, 22–30.

Rossi, E. (2003). Gene expression, neurogenesis, and healing: Psychosocial genomics of therapeutic hypnosis. *American Journal of Clinical Hypnosis*, 45(3), 197–216.

Rossi, E. (2004a). (Salvador Iannotti, Translator and Editor: Saiannot@tin.it,) *Discorso Tra Geni [Italian]. A Discourse with Our Genes: The Psychosocial and Cultural Genomics of Therapeutic Hypnosis and Psychotherapy*. Benevento, Italy: Editris SAS[available in Italian and English editions]. New York: Zeig, Tucker, Theisen.

Rossi, E. (2004b). A bioinformatics approach to the psychosocial Genomics of therapeutic hypnosis. *Hypnos*, 31(1), 15–21.

Rossi, E. (2005). (Laurent Carrer, Translator and Editor). *Cinq essais de psychogénomique – exploration d'une nouvelle démarche scientifique axée sur l'interaction entre l'esprit et la molécule [Five Essays on Psychosocial Genomics: Exploration of a New Scientific Approach to the Interaction Between Mind and Molecule]*. Encinitas, CA: Trance-lations.

Rossi, E. (2007). *The Breakout Heuristic: The New Neuroscience of Mirror Neurons, Consciousness and Creativity in Human Relationships: Selected Papers of Ernest Lawrence Rossi*. Phoenix, AR: The Milton H. Erickson Foundation Press. office@erickson-foundation.org

Rossi, E. (2008). The neuroscience of therapeutic hypnosis, psychotherapy, and rehabilitation. In Rossi, E., Erickson-Klein, R. & Rossi, K., (Eds.), *The Complete Works of Milton H. Erickson, M.D. Volume 1: The Nature of Hypnosis*. Phoenix: The MHE Foundation Press.

Rossi, E., and Cheek, D. (1988). *Mind-Body Therapy: Ideodynamic Healing in Hypnosis*. New York: W. W. Norton.

Rossi, E., and Lippincott, B. (1992). The wave nature of being: Ultradian rhythms and mind-body communication. In Lloyd, D., and Rossi, E. (Eds.), *Ultradian Rhythms in Life Processes: A Fundamental Inquiry into Chronobiology and Psychobiology*. New York: Springer (pp. 371–402).

Rossi, E., and Kleitman, N. (1992). The basic rest-activity cycle – 32 years later: An interview with Nathaniel Kleitman. In Lloyd, D., and Rossi, E. (Eds.), *Ultradian Rhythms in Life Processes: A Fundamental Inquiry into Chronobiology and Psychobiology*. New York: Springer (pp. 303–306).

Rossi, E. and Nimmons, D. (1991). *The Twenty-Minute Break: The Ultradian Healing Response*. Los Angeles, CA: Jeremy Tarcher, New York: Zeig, Tucker, Theisen.

Rossi, E., Rossi, K., Yount, G., Cozzolino, M., and Iannotti, S. (2006). The bioinformatics of integrative medical insights: Proposals for an international psychosocial and cultural bioinformatics project. *Integrated Medicine* Insights. Open access on line: http://www.la-press.com/integmed.htm

Rossi, E., Erickson-Klein, R., and Rossi, K. (2008). The future orientation of constructive memory: An evolutionary perspective on therapeutic hypnosis and brief psychotherapy. *American Journal of Clinical Hypnosis*, 50(4), 341–347.

Rossi, E., Rossi, K., Cozzolino, M., and Iannotti, S. (2007). Expectations of hypnosis future: A new neuroscience school of therapeutic hypnosis, psychotherapy, and rehabilitation. *European Journal of Clinical Hypnosis*, 7(3), 2–9.

Sanders, S., and Mann, B. (1995). The effects of light, temperature, trance length and time of day on hypnotic depth. *American Journal of Clinical Hypnosis*, 37, 43–53.

Segal, E., Sirlin, C., Ooi, C., Adler, A., Gollub, J., Chen, X., Chan B., Matcuk, G., Barry, C., Chang, H., and Kuo, M. (2007). Decoding global gene expression programs in liver cancer by noninvasive imaging. *Nature Biotechnology*, 25, 675–680.

Shono, M., Takasaki, M., Ito, Y., Muro, M., Iwasaka, and T., Sugimori, H. (2001). A new method to analyze basic rest-activity cycle. *Psychiatry Clinical Neuroscience*, 55, 169, 170.

Solomon, M. (2003). *Late Beethoven*. Los Angeles, CA: University of California Press.

Storch, K., Lipan, O., Leykin, I., Viswanathan, N., Davis, F., Wong, W., and Weitz, C. (2002). Extensive and divergent circadian gene expression in liver and heart *Nature*, 417, 78–83.

Stickgold, R., Roar Fosse, R., and Walker, M. (2007). Linking brain and behavior in sleep-dependent learning and memory consolidation. *Proceedings of the National Academy of Sciences*, 99, 16519–16521.

Talan, J. (2007). Calming the manic mind: Mutation in &$$$;'Clock Gene," which helps control body's rhythm, linked to symptoms of mania in mice, humans. *Brain in the News*, Washington, DC: Dana Foundation (p. 5).

Toni, N., Teng, E., Bushong, E., Aimone, J., Zhao, C., Consiglio, A., van Praag, H., Martone, M., Ellisman, M., and Gage, F. (2007). Synapse formation on neurons born in the adult hippocampus. *Nature Neuroscience*, doi:10.1038/nn1908, 1–9.

Ueda, H., Chen, W., Adachi, A., Wakamatsu, H., Hayashi, S., Takasugi, T., Nagano, M., Nakahama, K., Suzuki, Y., Sugano, S., Iino, M., Shigeyoshi, Y., and Hashimoto, S. (2002). A transcription factor response element for gene expression during circadian night. *Nature*, 418, 534–539.

Unterweger, E., Lamas, J., and Bongartz, W. (1992). Heart-rate variability of high and low susceptible subjects during administration of the Stanford scale, Form C. In Bongartz, E. (Ed.), *Hypnosis: 175 Years After Mesmer: Recent Developments in Theory and Application*. Konstanz, Germany: Universitatsvergag.

Van Hulst, N. (2007). Light in chains. *Nature*, 448, 141–142.

Wallas, G. (1926). *The Art of Thought*. New York: Harcourt.

Wilber, K. (1993). The spectrum of consciousness. 2nd Edition. Wheaton, IL: Quest Books.

Chapter 17
Genes, Sleep and Dreams

S. Ribeiro[1,2], C. Simões[1,2], and M. Nicolelis[1,3,4]

Abstract This Chapter considers the continuing debate regarding the mechanisms that underlie the cognitive role of sleep. One theory proposes that the triggering of generalized synaptic downscaling occurs so as to restore homeostatic balance and enable further waking potentiation. The alternative, favored by the present authors, proposes that sleep harbors decreased and increased plasticity in separate circuits. Further work is required to resolve this debate.

Keywords Dream, engram, immediate early gene, learning, memory, differential plasticity, propagation, reverberation, REM sleep, slow wave sleep

17.1 Dreams in Medicine and Psychology

Dream interpretation, religion and medicine were once closely linked. Dream soothsayers prospered during all of the antiquity (Jung et al., 1969; Miller, 1997). In ancient Greece, a ritual known as "Dream Incubation" took place in sanctuaries named *asklepieia* in honor to Asclepius, the god of medicine. Greeks would travel long distances to visit these healing temples in search for cures for their ailments. Each *asklepieion* contained a sacred large hall, the *enkoimeterion*, where Dream Incubation took place. The patient would sleep overnight and the next day a priest

[1] International Institute for Neuroscience of Natal Edmond and Lily Safra (IINN-ELS), Rua Prof. Francisco Luciano de Oliveira, 2460, Natal/RN, 59066-060 Brazil

[2] Universidade Federal do Rio Grande do Norte, Centro de Biociências - Departamento de Fisiologia, Campus Universitário Lagoa Nova | Lagoa Nova, Natal/RN, 59078-970 Brazil, Caixa Postal 1511

[3] Departments of Neurobiology, Biomedical Engineering, Psychology and Neurosciences, Duke University Medical Center, Durham, NC, 27710, USA

[4] Center for Neuroengineering, Duke University Medical Center, Durham, NC 27710, USA

D. Lloyd, E. Rossi (eds.), *Ultradian Rhythms from Molecules to Mind*,

would listen to a description of the patient's dream to diagnose the illness and determine the right treatment (Meier, 2003). Sometimes, Asclepius himself would appear in dreams to prescribe the treatment (Horstmanshoff, 2004). Similar rituals also existed in Ancient Egypt in the temples of the God Serapis and were preserved throughout the Middle Ages in some Byzantine Christian sects (Meier, 2003). Indeed, dreams play a major role in the Christian tradition, as remarkable instruments by which God reveals His wishes to the chosen people. The newborn Jesus was protected a number of times due to advice received in dreams by his father Joseph. The first such episode occurred before Jesus' birth, when Joseph intended to leave Mary because of her pregnancy prior to marriage. An angel of the Lord appeared to Joseph in a dream and told him not to be afraid of taking Mary as his wife, because the baby she carried was the son of God and should be called Jesus. Joseph obeyed, saving the pregnant mother from being stoned (Matthew 1:18–25). When Jesus was born, three magi came to Bethlehem from the East to worship the King of Jews. The local ruler Herod, troubled by the prophecies regarding Jesus, asked the magi to search for the child and bring back news as soon as possible, so that he could honor the new king himself. The magi were warned in their dreams that Herod's request was deceptive, for his real intention was to kill the baby. At the same time, in his dreams, Joseph was advised by an angel to take his family away to Egypt, for Herod had ordered the assassination of all children under the age of two. After Herod's death, the angel visited Joseph's dreams once again to tell him that it was now safe to return home (Matthew 2:1–20).

Given the great importance attributed to the predictive and healing power of dreams in so many ancient cultures, as well as in present-day "primitive" traditions (Kilton, 1951; Cawte, 1984; Shulman et al., 1999; Lincoln, 2003), it is interesting that dreams achieve such poor status in contemporary western society. The arcane relationship between dreams and divinity is likely the explanation for the decay of dream status in a world in which rationalism progressively replaced religiosity. Any effort to rescue the prominent role of dreams in the new rationalist world would require dreaming to be brought inside the scientific arena. Even though dreams are quite accessible as inner reality to most people, they are rather hard to assess empirically. This is related to the high degree of subjectivity involved in dream symbiology. Any introspective dreamer can recall a wide variety of dreams, ranging from fleeting impressions to complex time-evolving narratives. Dreams comprise the reproduction of trivial situations of daily life, such as walking on the street, but they also include impossible feats such as flying like a bird or breathing under water. Familiar and unfamiliar characters can show up in coherent settings or absolutely bizarre contexts and places. Dreams can depict completely unreal situations, as Kafkian as having hair growing on teeth. Anyone who has awakened from a bad dream in the middle of the night knows that dreams can also be diverse in respect to their emotional tone, ranging from the pleasant fulfillment of a secret wish to a gory nightmare. The weird fantasies experienced in dreams are often vivid. Although visual images tend to be prevalent, dreams can also involve combinations of auditory, olfactory, tactile, gustatory, motor, vestibular and linguistic modalities in a mix of senses and emotions.

The first attempt to explore dreams scientifically occurred at the dawn of the 20th century, when Sigmund Freud published "The Interpretation of Dreams" (Freud, 1900). Freud's research program was not yet a quantitative approach of the oneiric phenomenon, but it sought to identify the underlying causes of dreaming in a very objective manner, dissecting the complex oneiric symbology with reference to the patient's own memories and recollections. Freud stated that dreams, far from being nonsense or chaotic, are highly meaningful reflections of the emotions, desires and concerns of the dreamer. He postulated that dream narratives fulfill wishes (or anti-wishes) of the awake subject, simulating the realization (or frustration) of specific desires (Freud, 1920). He also noted that mnemonic images and the wishes that bring them together tend to remain unconscious most of the waking time, surfacing during dreams in a very special manner. Freud's recognition that dreams are usually stamped with impressions from the waking period advanced the important concept of "day residue", i.e. the part of a dream that derives directly from experiences of the previous day(s). The intensity of this waking intrusion varies according to how recent, novel and behaviorally relevant waking events are.

Although implicit in the concept of "day residue", the beneficial role of sleep for learning was not foreseen by Freud himself. The discovery of the cognitive role of sleep occurred more than 20 years later, with pioneering studies of the effect of sleep on the retention of nonsense syllables (Jenkins and Dallenbach, 1924). These first findings remained obscure for several decades, and it was only in the late 1960s that the relationship between sleep and memory consolidation began to be systematically subjected to experimental tests in rodents and humans. The studies concluded that sleep is a powerful learning booster (Fishbein et al., 1966, 1971, 1974; Leconte and Bloch, 1970; Fishbein, 1971; Leconte and Hennevin, 1971, 1973; Leconte et al., 1973, 1974; Smith et al., 1974, 1980; Fishbein and Gutwein, 1977; Hennevin and Leconte, 1977; Smith and Butler, 1982; Hennevin and Hars, 1985, 1992; Smith and Lapp, 1986, 1991; Smith and Kelly, 1988; Hennevin et al., 1989, 1993, 1995a, b; Smith and Wong, 1991; Smith and MacNeill, 1993). Freud's ideas concerning the importance of dreams for mental health remain controversial to this day, mostly due to rifts within the field of psychology. While the Freudian psychoanalytic tradition focused on wish fulfillment and dream meaning (Freud, 1900; Jung, 1953a, b, 1974; Jung et al., 1969; Fosshage and Loew, 1978; Solms, 2004), experimental psychologists pursued the more objective dream aspects related to sleep-dependent memory consolidation (Bryson and Schacher, 1969; Pearlman, 1969; Leconte and Bloch, 1970; Lucero, 1970; Fishbein, 1971; Leconte and Hennevin, 1971; Smith and Butler, 1982; Smith and Lapp, 1986; Karni et al., 1994; Smith, 1995, 1996, 2001; Maquet, 1996, 2001; Maquet et al., 1996, 1997, 2000; Maquet and Phillips, 1998; Stickgold, 1998, 2001, 2003; Stickgold et al., 1999, 2000a, b, 2001, 2002; Mednick et al., 2002, 2003; Walker et al., 2002a, b, 2003; Born and Wagner, 2004a, b; Walker and Stickgold, 2004; Stickgold and Walker, 2005; Born et al., 2006).

If dreams are normally directed by the anxieties and expectations of the dreamer and tend to recapitulate the past and the present (Winson, 1985), dream narratives sometimes contain anticipatory simulations of expected challenges. A good example is provided by the dreams of students facing difficult exams in the near future.

In these cases, the occurrence of dreams related to test questions in the night preceding the exam is common. Such simulations of future events can simply deliver emotional readiness, or sometimes bring forth the solution for some puzzling situation. This unique way of learning is called insight (Kohler, 1947) or abduction (Peirce, 1958) and corresponds to the creation of new memory traces through the restructuration of pre-existing ones. Although insights occur often during waking (Jung-Beeman et al., 2004), they are greatly facilitated by sleep (Wagner et al., 2004). Notorious examples of sleep-dependent insight can be drawn from both science and art (Barrett, 2001). August Kekulé was puzzled by the number of carbon and hydrogen atoms in benzene, which did not fit any linear diagram he could conceive. He then saw in a dream the famous alchemical symbol called *Ouroboros*, the snake that eats its own tail. After giving thought to this dream, he inferred that benzene has a circular structure, the benzene ring that is so important in organic chemistry. Dimitri Mendeleyev, the great discoverer of the periodic table, visualized his breakthrough concept in a dream. Intense dreams have greatly inspired artists like Albrecht Dürer, William Blake, Salvador Dalí, Frida Kahlo and many others (Barrett, 2001). Learning from insight gained through dream evaluation is reminiscent of the old view of dreams as oracles, a strong belief even to this day among lay people with little education. Why would dreams appear to predict the future? Studies comparing dream reports of children exposed to different degrees of environmental threats suggest that dreaming simulates actions that lead to undesirable consequences and therefore should be avoided in the real world (Revonsuo, 2000). This "threat simulation" theory has been generalized to include dreams regarding actions that lead to a desirable outcome and therefore should be performed in the real world (Ribeiro and Nicolelis, 2006). The main function of these simulations would be to test specific novel behaviors against a memory replica of the world, rather than the real world itself, thereby leading to learning without risk. An investigation of REM sleep mentation found that over 70% of the reports included emotions, with a balanced proportion of positive and negative ones (Fosse et al., 2001). The notion that nightmares evolved as a way to negatively modulate particularly dangerous behavior simulations, while blissful dreams correspond to the association of pleasure (reward) with dream simulations of especially adaptive behaviors, is analogous to the concepts of Eros and Thanatos proposed by Freud as life and death drives (Freud, 1920).

17.2 The Search for Mechanisms: Active Versus Passive Sleep

Even if both branches of psychology agree that sleep and dreaming are active mental processes that play a key role in cognition, such certainty is not held by all biologists, nor by all philosophers. The past several decades of investigation have witnessed a fierce clash between two major notions regarding the existence of active, adaptive and cognitive processes during sleep and dreams. While the biological mechanisms underlying the generation and maintenance of the sleep-wake cycle are fairly understood (Aserinsky and Kleitman, 1953; Dement and Kleitman, 1957a, b; Dement, 1958; Jouvet et al., 1959; Grastyan and Karmos, 1961; Roffwarg et al., 1962; Jouvet, 1967; Rechtschaffen and Kales, 1968; Timo-Iaria et al., 1970; Siegel, 1990;

Steriade, 1992; Steriade et al., 1993; Sutcliffe and de Lecea, 2002; Lee and Jones, 2004; Luppi et al., 2004), no consensus has been achieved about the biological function of sleep and dreams (Maquet, 2001; Siegel, 2001; Stickgold et al., 2001). Some authors even stand for the opinion that dreams play no relevant biological role at all, being nothing but an epiphenomenon of sleep. The anti-Freudian philosopher Owen Flanagan, for instance, argues that dreams cannot possibly be the result of a biological adaptation, based on his failure to recognize fitness-enhancing elements on his own dreams. In his opinion, "dreams are the spandrels of sleep" (Flanagan, 2000). The neural corollary of this point of view is that the bizarreness and hyper-associativeness of dreams can be trivially explained as a side effect of random engram activation during sleep, due to a generalized bombardment of the cerebral cortex by neurons located in deep brain nuclei (Crick and Mitchison, 1983, 1995) Dreams would in this case reflect some sort of "house cleaning" process by which randomly activated memories would be washed away. If this was the case, dreams would have no relevant meaning, and would serve no function *per se*. Despite its ingenuity, the "random cortical activation" theory does not survive confrontation with the fact that dreams can sometimes be very repetitive. This phenomenon occurs in healthy subjects but becomes much more frequent in patients with post-traumatic stress disorder (PTSD). Recursive nightmares related to major trauma are indeed an important symptom of this syndrome, and are very common among war veterans (Ross et al., 1994, 1999). Typically, PTSD dream narratives represent battle events that may recur decades after the end of combat (Neylan et al., 1998; Schreuder et al., 1998; Esposito et al., 1999). Given the colossal number of neurons and synapses in the human neocortex, it is clearly impossible to explain the activation of nearly identical neuronal firing patterns over several consecutive dreams by way of random neocortical activation.

To consider sleep as nothing more than an opportunity for the body and mind to rest is very intuitive, since most of sleep coincides with behavioral quiescence and the slumber experience of slow-wave sleep (SWS). The typical mental content of a person woken up during SWS is a dark visual scene plus thoughts related to waking preoccupations (Kales et al., 1967; Fosse et al., 2004). The subjective experience of being in SWS is in line with the fact that SWS is concomitant with slow neural oscillations under 4 Hz, and decreased firing rates in the cerebral cortex (Timo-Iaria et al., 1970; Steriade et al., 1993; Gervasoni et al., 2004; Ribeiro et al., 2004a). On the other hand, the discovery of a sleep state characterized by dreaming and high neuronal activity in most of the cerebral cortex (Aserinsky and Kleitman, 1953; Dement and Kleitman, 1957b), supported the active sleep paradigm. This second sleep state, called rapid-eye-movement (REM) (Aserinsky and Kleitman, 1953) or paradoxical sleep (Jouvet et al., 1959; Jouvet, 1967), is highly correlated in humans with widespread atonia, except for fast eye movements and occasional localized muscle twitches (Dement and Kleitman, 1957a; Dement, 1958; Jouvet et al., 1959; Roffwarg et al., 1962; Kleitman, 1963; Jouvet, 1967; Rechtschaffen and Kales, 1968; Timo-Iaria et al., 1970). Despite the motionless body, increased neural activity during REM sleep gives life to a succession of mental representations that constitutes the vivid subjective experience of dreaming. Conceived by Freud as "a conglomerate of psychic formations" (Freud, 1900), the dream therefore seems to reflect the fragmented activation of the very stuff the unconscious is made of, i.e. latent memories (Freud, 1915). It is amazing that such high levels of brain activity fail to be converted into behavior. The explanation lies in the activation of glycinergic neurons in the pons that inhibit the efferent control of muscles during REM sleep (Jouvet, 1994). Cats with lesions on these brainstem cells sleep quietly during SWS, but become agitated during REM sleep by vigorous species-specific behaviors, such as meowing and pouncing (Jouvet and Delorme, 1965).

Neural activity during REM sleep is grossly similar to that of waking in intensity and spectral content, with a predominance of high oscillatory frequencies above 30 Hz (Cantero et al., 2004; Gervasoni et al., 2004). A selected set of forebrain areas becomes activated during human REM sleep, including portions of the hypothalamus, amygdala, septum and ventral striatum, as well as the orbitofrontal, anterior cingulate, entorhinal and insular cortices (Maquet et al., 1996; Braun et al., 1997; Nofzinger et al., 1997). In addition, dreaming ceases upon lesion of mesolimbic pathways connecting reward centers with the cerebral cortex, striatum and thalamus (Solms, 1997, 2000). This suggests that dreams promote the "integration of neocortical function [...] with motivational and reward mechanisms"(Nofzinger et al., 1997). Deactivation during REM sleep of the dorsolateral prefrontal cortex (Maquet et al., 1996), a brain region essential for the planning, execution and evaluation of goal-directed behaviors (Tanji and Hoshi, 2001; Schultz, 2002), is the likely basis for the limited volitional power of the self-representation during dreams.

The discovery that waking patterns of neuronal activity reverberate throughout both phases of subsequent sleep provided a mechanism to explain memory reactivation during REM sleep dreams, as well as SWS mentation (Pavlides and Winson, 1989; Wilson and McNaughton, 1994; Skaggs and McNaughton, 1996; Qin et al., 1997; Nadasdy et al., 1999; Dave and Margoliash, 2000; Maquet et al., 2000; Hirase et al., 2001; Louie and Wilson, 2001; Hoffman and McNaughton, 2002; Lee and Wilson, 2002; Peigneux et al., 2003; Ribeiro et al., 2004a, 2007). But SWS and REM sleep are different states when it comes to the stability of neuronal reverberation. While SWS exhibits a steady reactivation of memory traces that is likely to strengthen the networks involved, REM sleep exhibits less stationary reverberation (Winson and Abzug, 1977; Pavlides and Winson, 1989; Ribeiro et al., 2004a, b; Pereira et al., 2007). Such 'noisy' reverberation during REM sleep, long postulated by psychology (Hartmann, 1967, 1998), has been proposed to promote the reconfiguration of neuronal networks, assembling pre-existing memory fragments to give origin to novel engrams underlying sleep-dependent insight (Ribeiro et al., 2004a; Ribeiro and Nicolelis, 2006).

17.3 Long-Lasting Memories and Gene Expression During Sleep

Altogether, the electrophysiological studies of neuronal reverberation performed in animal models support the theory that sleep is an active state linked to mnemonic processing and cognitive work (Pavlides and Winson, 1989; Wilson and McNaughton,

1994; Skaggs and McNaughton, 1996; Qin et al., 1997; Nadasdy et al., 1999; Dave and Margoliash, 2000; Hirase et al., 2001; Louie and Wilson, 2001; Hoffman and McNaughton, 2002; Lee and Wilson, 2002; Ribeiro et al., 2004a, 2007). Equivalent results were obtained in human subjects using brain scanning techniques such as positron emission tomography (Maquet et al., 2000; Peigneux et al., 2003). Still, increased blood oxygenation and neuronal depolarization *per se* cannot produce long-lasting memories. The long-term storage of newly acquired memories depends crucially on *de novo* protein synthesis related to neuronal plasticity, resulting from the activation of gene expression programs that promote durable changes in synaptic abundance, localization and efficacy (Bliss and Collingridge, 1993; McGaugh, 2000). Mnemonic changes are triggered by sustained neuronal depolarization and mediated by calcium-dependent kinases (Frankland et al., 2001; Lisman et al., 2002; Bozon et al., 2003). The first wave of gene expression after memory formation corresponds to the transcriptional upregulation of fast response genes that couple membrane depolarization to genomic regulation inside the cell nucleus. The proteins coded by such immediate early genes (IEG) may be direct effectors of plasticity within the cell, or indirect modulators of cellular change, functioning as transcriptional regulators for other genes. An example of memory-related effector gene is the activity-regulated cytoskeleton-associated protein (arc), a calcium-dependent IEG that interacts with glutamatergic AMPA receptors, actin and calcium-calmodulin kinase II to promote synaptic remodeling; arc mRNA is transported to dendrites for local translation, leading to retrograde effects with respect to the cell soma (Lyford et al., 1995; Steward et al., 1998; Guzowski et al., 2000; Waltereit et al., 2001). The calcium-dependent gene zif-268 (a.k.a. egr-1, krox-24, NGFI-A and ZENK) is an example of memory-related gene with indirect regulatory function. Zif-268 encodes a transcription factor potentially involved in the expression of hundreds of different genes (Milbrandt, 1987; Christy and Nathans, 1989; Wisden et al., 1990). There is *in vitro* evidence that the zif-268 protein controls the transcription of synapsins (Thiel et al., 1994; Petersohn et al., 1995), the most abundant protein constituent of synapses, required for synaptic vesicle release (Hilfiker et al., 1999). The control of synapsin levels by the zif-268 protein illustrates how neuronal depolarization can effect anterograde synaptic remodeling with respect to the cell soma (Ribeiro and Nicolelis, 2004).

An early attempt to address the role of *de novo* protein synthesis for the cognitive role of sleep found that the blockade of protein synthesis during sleep impairs memory consolidation (Gutwein et al., 1980). When calcium-dependent IEG were discovered in the late 1980s, many researchers envisioned a link between neuronal reverberation at the electrophysiological level and memory consolidation at the genomic and proteomic level. If sleep benefits memory consolidation and IEG are required for long-lasting plasticity, at least some of those genes should be upregulated during sleep. But the first measurements of the c-fos and zif-268 gene expression in the rat telencephalon across the sleep-wake cycle detected upregulation during waking and downregulation during sleep (Pompeiano et al., 1994). This result was soon corroborated by another other team (O'Hara et al., 1993) and extended to other IEG (Tononi et al., 1995; Cirelli and Tononi, 1998, 1999a, b, 2000a, b; Cirelli et al., 2004, 2006), as well as calcium-dependent kinases and other molecular markers of plasticity (Vyazovskiy et al., 2008). Further studies in flies by the same group detected a sleep-related increase in transcripts encoding membrane trafficking proteins, and a general decay in the expression levels of metabolic enzymes, such as cytochromes necessary for respiration inside mitochondria (Shaw et al., 2000; Huber et al., 2004; Cirelli et al., 2005a, b; Cirelli, 2006; Bushey et al., 2007). The fly studies seemed to confirm the experiments in mammals, i.e. sleep correlated with a transcriptional downregulation of plasticity-related genes such as arc, brain-derived neurotrophic factor (BDNF), homer and zif-268 (Cirelli et al., 2004). Taken together, these results produced an unexpected revival of the passive sleep paradigm, leading to a synaptic downscaling theory of sleep (Tononi and Cirelli, 2003) which proposes that "periods of wakefulness are associated with a net increase in cortical synaptic strength and periods of sleep are associated with a net decrease" (Vyazovskiy et al., 2008). Sleep, in this case, would play a role in "an overall balance of synaptic strength" by favoring "global synaptic depression" (Vyazovskiy et al., 2008).

But other research groups have generated divergent results, based on different assumptions and consequently different experimental strategies. One such assumption is the necessity to compare animals with and without exposure to novel experience in order to investigate the cognitive role of sleep (Giuditta, 1985). Another key point is the need to distinguish the contributions of SWS and REM sleep for the biological variable of interest, since the two main sleep states show very different neural dynamics. In addition, SWS is considerably more abundant than REM sleep (Gervasoni et al., 2004), and therefore experiments that do not sort sleep states tend to be dominated by SWS (Cirelli et al., 2004). Using an experimental strategy that comprised sleep state sorting and pre-sleep novel experience, we found that zif-268 mRNA levels in the cerebral cortex and hippocampus are upregulated during the first REM sleep episode that follows exposure to a novel environment (Ribeiro et al., 1999). In a follow-up study (Ribeiro et al., 2002), we found similar results when exposure to novel environment was replaced by the induction of long-term potentiation (LTP) in the hippocampus, a well-known neurophysiological model of memory (Bliss and Collingridge, 1993). Our experiments revealed a sequence of three spatiotemporally distinct waves of *zif-268* expression, beginning locally at the hippocampus 30 min after stimulation, still during waking, and proceeding to distal extrahippocampal areas during the two subsequent REM sleep episodes. Each *zif-268* upregulation wave was interrupted by the next SWS episode, indicating the existence of recurrent plasticity cycles as the two sleep states alternate.

In 2005, our reports of experience-dependent upregulation of *zif-268* mRNA during REM sleep were extended to other plasticity-related molecules by an independent research team. The study, which employed the active avoidance learning task as behavioral paradigm, linked REM sleep and pontine waves typical of that state to the experience-dependent upregulation of arc and brain-derived nerve growth factor (BDNF) levels, as well as to the increased phosphorylation of the cyclic AMP response element-binding (CREB) protein (Ulloor and Datta, 2005). In

2006, a study of sleep in flies investigated the effects of enriched environment exposure on sleep-related gene expression. The researchers found evidence that sleep is increased in flies exposed to socially enriched environment. Most importantly, during sleep these flies showed increased expression of 17 genes related to long-term memory (Ganguly-Fitzgerald et al., 2006). More recently, our research team showed that the mRNA levels of *zif-268* and *arc* are upregulated in the cerebral cortex during late REM sleep episodes (Ribeiro et al., 2007). Taken together, these findings corroborate the notion that sleep harbors active experience-dependent processes related to neural plasticity.

Karl Popper prescribed that scientists should constantly attempt to falsify their own theories, subjecting them to critical tests in order to best advance knowledge (Popper, 1963). The refutation of a conjecture derived from a well established theory produces anomalous results with respect to that theory. According to Popper, it is the accumulation of such anomalies that allows other scientists to challenge and eventually substitute a faulty theory with a better one. However, Thomas Kuhn showed that when different theories collide, they tend to coexist for a while (Kuhn, 1962), because scientists become partisan about their own theories. Rather than convincing each other, believers of opposite theories tend to build separate edifices, sometimes even reaching the point of completely ignoring the contribution of others. When additional experiments performed by competitors produce results that do not fit the theory, its proponents often neglect to seriously consider these anomalies. Far from trying to kill their own theories, real life scientists tend to protect their own theories from critical tests, and may avoid performing experiments that could directly refute their theories (Kuhn, 1962).

The issue of molecular markers of plasticity during sleep constitutes a good example of paradigm clash in neuroscience. Despite the fact that three independent groups reported an increase in learning-related gene expression during sleep (Ribeiro et al., 1999, 2002, 2007; Ulloor and Datta, 2005; Ganguly-Fitzgerald et al., 2006), these results have been completely ignored by the proponents of the "synaptic downscaling" theory. The neglect was so strong that it even precluded citation of any divergent results in the 40 publications regarding sleep and plasticity produced in the last decade by the proponents of the "synaptic downscaling" theory. In fact, none of these studies attempted to run the critical experiments involving learning and separation of SWS and REM sleep. Only in their most recent article has this issue begun to be contemplated, by way of measurements of evoked electrical responses after novel object exposure and post-exposure sleep. However, post-exposure sleep was simply not investigated at the molecular level, as if the question was not pertinent to contemporary neuroscience (Vyazovskiy et al., 2008).

Remarkably, synaptic upscaling and downscaling are not mutually exclusive. Experience-dependent IEG expression during REM sleep is compatible with the notion of sleep-dependent synaptic downscaling, as long as the latter happens more strongly in neuronal circuits that were not activated by waking experience (Ribeiro and Nicolelis, 2004). Synaptic downscaling during SWS is likely to be essential for learning, but if it occurred with equal strength in activated and non-activated circuits, it should promote generalized forgetting, not learning. We have proposed that the combination of synaptic upscaling in activated circuits and synaptic downscaling in non-activated circuits should markedly increase the signal-to-noise ratio of memory consolidation during sleep. In principle, such differential plasticity should be sufficient to carve high-relief memory traces in a background of disengaged synapses. Recent results comparing stimulated and non-stimulated regions of the cerebral cortex indicate that this is exactly the case, at least with regard to different sensory modalities (Ribeiro et al., 2007).

17.4 Conclusion

Can passive quiescence fully explain the cognitive role of sleep? The ancient Greeks were surely unaware of the neural mechanisms activated during sleep, but in a very intuitive way they knew that a powerful learning tool was effective at night. They believed that dreams could predict the future as a result of divine intervention. The abundance of votive offers at *asklepieion* archaeological sites is intriguing, suggesting that Dream Incubation often resulted in healing (Meier, 2003). During this ritual, the god of medicine Asclepius had the assistance of Hypnos, the god of quiescent sleep who was the twin brother of Thanatos, god of death. Hypnos would first induce the subject into gentle sleep, and then he would request the help of his son Morpheus, god of dreams, to open the mental window that allowed Asclepius to bring prescription of the right treatment for the subject's illness. The dichotomy between Hypnos and Morpheus illustrates the ongoing scientific debate regarding the mechanisms underlying the cognitive role of sleep. One of the competing theories proposes that the function of sleep is to trigger generalized synaptic downscaling, so as to restore homeostatic balance and enable further waking potentiation. This theory focuses on SWS, disregards REM sleep, and does not take into account experience-dependent changes (Tononi and Cirelli, 2003). Our theory proposes that the cognitive role of sleep derives from the cooperative interaction of its two major states: While SWS reverberates waking memories in the absence of distracting stimuli, REM sleep elicits plasticity-related IEG expression in previously activated neurons (Ribeiro and Nicolelis, 2004). According to this view, sleep harbors decreased and increased plasticity in separate circuits, leading to experience-dependent differential plasticity. Further experimentation is required to convincingly refute either theory.

References

Aserinsky E, Kleitman N (1953) Regularly occurring periods of eye motility, and concomitant phenomena, during sleep. Science 118:273–274.

Barrett D (2001) The Committee of Sleep: How Artists, Scientists, and Athletes Use Dreams for Creative Problem-Solving, 1st ed. Edition. New York: Crown Publishers.

Bliss TV, Collingridge GL (1993) A synaptic model of memory: long-term potentiation in the hippocampus. Nature 361:31–39.

Born J, Wagner U (2004a) Memory consolidation during sleep: role of cortisol feedback. Ann NY Acad Sci 1032:198–201.

Born J, Wagner U (2004b) Awareness in memory: being explicit about the role of sleep. Trends Cogn Sci 8:242–244.

Born J, Rasch B, Gais S (2006) Sleep to remember. Neuroscientist 12:410–424.

Bozon B, Kelly A, Josselyn SA, Silva AJ, Davis S, Laroche S (2003) MAPK, CREB and zif268 are all required for the consolidation of recognition memory. Philos Trans R Soc Lond B Biol Sci 358:805–814.

Braun AR, Balkin TJ, Wesenten NJ, Carson RE, Varga M, Baldwin P, Selbie S, Belenky G, Herscovitch P (1997) Regional cerebral blood flow throughout the sleep-wake cycle. An H2(15)O PET study. Brain 120:1173–1197.

Bryson D, Schacher S (1969) Behavioral analysis of mammalian sleep and learning. Perspect Biol Med 13:71–79.

Bushey D, Huber R, Tononi G, Cirelli C (2007) Drosophila Hyperkinetic mutants have reduced sleep and impaired memory. J Neurosci 27:5384–5393.

Cantero JL, Atienza M, Madsen JR, Stickgold R (2004) Gamma EEG dynamics in neocortex and hippocampus during human wakefulness and sleep. Neuroimage 22:1271–1280.

Cawte J (1984) The 'ordinary' dreams of the Yolngu in Arnhem Land. Aust N Z J Psychiatr 18:236–243.

Christy B, Nathans D (1989) DNA binding site of the growth factor-inducible protein Zif268. Proc Natl Acad Sci USA 86:8737–8741.

Cirelli C (2006) Sleep disruption, oxidative stress, and aging: new insights from fruit flies. Proc Natl Acad Sci USA 103:13901–13902.

Cirelli C, Tononi G (1998) Differences in gene expression between sleep and waking as revealed by mRNA differential display. Brain Res Mol Brain Res 56:293–305.

Cirelli C, Tononi G (1999a) Differences in brain gene expression between sleep and waking as revealed by mRNA differential display and cDNA microarray technology. J Sleep Res 8 Suppl 1:44–52.

Cirelli C, Tononi G (1999b) Differences in gene expression during sleep and wakefulness. Ann Med 31:117–124.

Cirelli C, Tononi G (2000a) Gene expression in the brain across the sleep-waking cycle. Brain Res 885:303–321.

Cirelli C, Tononi G (2000b) Differential expression of plasticity-related genes in waking and sleep and their regulation by the noradrenergic system. J Neurosci 20:9187–9194.

Cirelli C, Gutierrez CM, Tononi G (2004) Extensive and divergent effects of sleep and wakefulness on brain gene expression. Neuron 41:35–43.

Cirelli C, LaVaute TM, Tononi G (2005a) Sleep and wakefulness modulate gene expression in Drosophila. J Neurochem 94:1411–1419.

Cirelli C, Faraguna U, Tononi G (2006) Changes in brain gene expression after long-term sleep deprivation. J Neurochem 98:1632–1645.

Cirelli C, Bushey D, Hill S, Huber R, Kreber R, Ganetzky B, Tononi G (2005b) Reduced sleep in Drosophila Shaker mutants. Nature 434:1087–1092.

Crick F, Mitchison G (1983) The function of dream sleep. Nature 304:111–114.

Crick F, Mitchison G (1995) REM sleep and neural nets. Behav Brain Res 69:147–155.

Dave AS, Margoliash D (2000) Song replay during sleep and computational rules for sensorimotor vocal learning. Science 290:812–816.

Dement W, Kleitman N (1957a) Cyclic variations in EEG during sleep and their relation to eye movements, body motility, and dreaming. Electroenceph Clin Neurophysiol Suppl 9:673–690.

Dement WC (1958) The occurrence of low voltage, fast, electroencephalogram patterns during behavioral sleep in the cat. Electroenceph Clin Neurophysiol 10:291–296.

Dement WC, Kleitman N (1957b) The relation of eye movements during sleep to dream activity: an objective method for the study of dreaming. J Exp Psychol 53:339–346.

Esposito K, Benitez A, Barza L, Mellman T (1999) Evaluation of dream content in combat-related PTSD. J Trauma Stress 12:681–687.

Fishbein W (1971) Disruptive effects of rapid eye movement sleep deprivation on long-term memory. Physiol Behav 6:279–282.
Fishbein W, Gutwein BM (1977) Paradoxical sleep and memory storage processes. Behav Biol 19:425–464.
Fishbein W, Schaumburg H, Weitzman ED (1966) Rapid eye movements during sleep in dark-reared kittens. J Nerv Ment Dis 143:281–283.
Fishbein W, McGaugh JL, Swarz JR (1971) Retrograde amnesia: electroconvulsive shock effects after termination of rapid eye movement sleep deprivation. Science 172:80–82.
Fishbein W, Kastaniotis C, Chattman D (1974) Paradoxical sleep: prolonged augmentation following learning. Brain Res 79:61–75.
Flanagan O (2000) Dreaming is not an adaptation. Behavioral & Brain Sciences 23:936–939.
Fosse R, Stickgold R, Hobson JA (2001) The mind in REM sleep: reports of emotional experience. Sleep 24:947–955.
Fosse R, Stickgold R, Hobson JA (2004) Thinking and hallucinating: reciprocal changes in sleep. Psychophysiology 41:298–305.
Fosshage JL, Loew CA (1978) Dream Interpretation: A Comparative Study. New York: SP Medical & Scientific Books.
Frankland PW, O'Brien C, Ohno M, Kirkwood A, Silva AJ (2001) Alpha-CaMKII-dependent plasticity in the cortex is required for permanent memory. Nature 411:309–313.
Freud S (1900) The Interpretation of Dreams, 1952 Edition. London: Encyclopaedia Britannica.
Freud S (1915) The Unconscious, Republished in 1952 Edition. London: Encyclopaedia Britannica.
Freud S (1920) Beyond the Pleasure Principle. New York: W. W. Norton.
Ganguly-Fitzgerald I, Donlea J, Shaw PJ (2006) Waking experience affects sleep need in Drosophila. Science 313:1775–1781.
Gervasoni D, Lin SC, Ribeiro S, Soares ES, Pantoja P, Nicolelis M (2004) Global forebrain dynamics predict rat behavioral states and their transitions. J Neurosci 24:11137–11147.
Giuditta A, ed (1985) A Sequential Hypothesis for the Function of Sleep. Stuttgart: Fisher-Verlag.
Grastyan E, Karmos G (1961) A study of a possible "dreaming" mechanism in the cat. Acta Physiol Acad Sci Hung 20:41–50.
Gutwein BM, Shiromani PJ, Fishbein W (1980) Paradoxical sleep and memory: long-term disruptive effects of Anisomycin. Pharmacol Biochem Behav 12:377–384.
Guzowski JF, Lyford GL, Stevenson GD, Houston FP, McGaugh JL, Worley PF, Barnes CA (2000) Inhibition of activity-dependent arc protein expression in the rat hippocampus impairs the maintenance of long-term potentiation and the consolidation of long-term memory. J Neurosci 20:3993–4001.
Hartmann E (1967) The Biology of Dreaming. Springfield, IL: Charles Thomas.
Hartmann E (1998) Dreams and Nightmares: The New Theory on the origin and Meaning of Dreams. New York: Plenum.
Hennevin E, Leconte P (1977) [Study of the relations between paradoxical sleep and learning processes (author's transl)]. [French]. Physiol Behav 18:307–319.
Hennevin E, Hars B, eds (1985) Post-learning Paradoxical Sleep: a Critical Period When New Memory Is Reactivated. New York: Plenum.
Hennevin E, Hars B (1992) 2nd-order conditioning during sleep. Psychobiology 20:166–176.
Hennevin E, Hars B, Bloch V (1989) Improvement of learning by mesencephalic reticular stimulation during postlearning paradoxical sleep. Behav Neural Biol 51:291–306.
Hennevin E, Hars B, Maho C (1995a) Memory processing in paradoxical sleep. Sleep Res Soc Bull 1:44–50.
Hennevin E, Maho C, Hars B, Dutrieux G (1993) Learning-induced plasticity in the medial geniculate nucleus is expressed during paradoxical sleep. Behav Neurosci 107:1018–1030.
Hennevin E, Hars B, Maho C, Bloch V (1995b) Processing of learned information in paradoxical sleep: relevance for memory. Behav Brain Res 69:125–135.

Hilfiker S, Pieribone VA, Czernik AJ, Kao HT, Augustine GJ, Greengard P (1999) Synapsins as regulators of neurotransmitter release. Philos Trans R Soc Lond B Biol Sci 354:269–279.

Hirase H, Leinekugel X, Czurko A, Csicsvari J, Buzsaki G (2001) Firing rates of hippocampal neurons are preserved during subsequent sleep episodes and modified by novel awake experience. Proc Natl Acad Sci USA 98:9386–9390.

Hoffman KL, McNaughton B (2002) Coordinated reactivation of distributed memory traces in primate neocortex. Science 297:2070–2073.

Horstmanshoff HF (2004) Asclepius and temple medicine in aelius Aristides' Sacred Tales. Stud Anc Med 27:325–341.

Huber R, Hill SL, Holladay C, Biesiadecki M, Tononi G, Cirelli C (2004) Sleep homeostasis in Drosophila melanogaster. Sleep 27:628–639.

Jenkins JB, Dallenbach KM (1924) Oblivescence during sleep and waking. Am J of Psychology 35:605–612.

Jouvet M (1967) The states of sleep. Sci Am 216:62–68.

Jouvet M (1994) Paradoxical sleep mechanisms. Sleep 17:S77–S83.

Jouvet M, Delorme F (1965) Locus coeruleus et sommeil paradoxal. CR Soc Biol (Paris) 159:895–899.

Jouvet M, Michel F, Courjon J (1959) Sur un stade d'activité électrique cérébrale rapide au cours du sommeil physiologique. CR Soc Biol (Paris) 153:1024–1028.

Jung CG (1953a) General Aspects of Dream Psychology. In: Collected Works, pp 237–280. Princeton, NJ: Princeton University Press.

Jung CG (1953b) On the nature of Dreams. In, pp 281–297. Princeton, NJ: Princeton University Press.

Jung CG (1974) Dreams, 1st Edition. Princeton, NJ: Princeton University Press.

Jung CG, Von Franz ML, Henderson J (1969) Man and His Symbols. New York: Doubleday.

Jung-Beeman M, Bowden EM, Haberman J, Frymiare JL, Arambel-Liu S, Greenblatt R, Reber PJ, Kounios J (2004) Neural activity when people solve verbal problems with insight. PLoS Biol 2:500–510.

Kales A, Hoedemaker FS, Jacobson A, Kales JD, Paulson MJ, Wilson TE (1967) Mentation during sleep: REM and NREM recall reports. Percept Mot Skills 24:555–560.

Karni A, Tanne D, Rubenstein BS, Askenasy JJ, Sagi D (1994) Dependence on REM sleep of overnight improvement of a perceptual skill. Science 265:679–682.

Kilton S (1951) Dream theory in Malaya. Complex 6:21–33.

Kleitman N (1963) Sleep and Wakefulness. Chicago, IL: University of Chicago Press.

Kohler W (1947) Gestalt Psychology: An Introduction to New Concepts in Modern Psychology, Reissue Edition. New York: Liveright.

Kuhn TS (1962) The Structure of Scientific Revolutions. Chicago, IL: University of Chicago Press.

Leconte P, Bloch V (1970) Déficit de la rétention d'un conditionnement après privation de sommeil paradoxal chez le rat. Comptes Rendus de l'Académie des Sciences (Paris) 271D:226–229.

Leconte P, Hennevin E (1971) Augmentation de la durée de sommeil paradoxal consécutive à un apprentissage chez le rat. C R Acad Sci (Paris) 273:86–88.

Leconte P, Hennevin E (1973) [Temporal characteristics of the augmentation of paradoxal sleep following learning in the rat]. [French]. Physiol Behav 11:677–686.

Leconte P, Hennevin E, Bloch V (1973) Analyse des effets d'un apprentissage et de son niveau d'acquisition sur le sommeil paradoxal consécutif. Brain Res 49:367–379.

Leconte P, Hennevin E, Bloch V (1974) Duration of paradoxical sleep necessary for the acquisition of conditioned avoidance in the rat. Physiol Behav 13:675–681.

Lee AK, Wilson MA (2002) Memory of sequential experience in the hippocampus during slow wave sleep. Neuron 36:1183–1194.

Lee MG, Jones BE (2004) Role of basalo-cortical system in modulating cortical activity and sleep-wake states. In: Sleep: Circuits & Functions (Luppi P-H, ed). New York: CRC.

Lincoln JS (2003) The Dream in Native American and Other Primitive Cultures. New York: Dover Publications.
Lisman J, Schulman H, Cline H (2002) The molecular basis of CaMKII function in synaptic and behavioural memory. Nat Rev Neurosci 3:175–190.
Louie K, Wilson MA (2001) Temporally structured replay of awake hippocampal ensemble activity during rapid eye movement sleep. Neuron 29:145–156.
Lucero MA (1970) Lengthening of REM sleep duration consecutive to learning in the rat. Brain Res 20:319–322.
Luppi P-H, Boissard R, Gervasoni D, Verret L, Goutagny R, Peyron C, Salvert D, Léger L, Barbagli B, Fort P (2004) The network responsible for paradoxical sleep onset and maintenance: a new theory based on the head-restrained rat model. In: Sleep: Circuits & Functions (Luppi P-H, ed). New York: CRC.
Lyford GL, Yamagata K, Kaufmann WE, Barnes CA, Sanders LK, Copeland NG, Gilbert DJ, Jenkins NA, Lanahan AA, Worley PF (1995) Arc, a growth factor and activity-regulated gene, encodes a novel cytoskeleton-associated protein that is enriched in neuronal dendrites. Neuron 145:433–434.
Maquet P (1996) [Paradoxical human sleep]. Rev Med Liege 51:632.
Maquet P (2001) The role of sleep in learning and memory. Science 294:1048–1052.
Maquet P, Phillips C (1998) Functional brain imaging of human sleep. J Sleep Res 7:42–47.
Maquet P, Peters J, Aerts J, Delfiore G, Degueldre C, Luxen A, Franck G (1996) Functional neuroanatomy of human rapid-eye-movement sleep and dreaming. Nature 383:163–166.
Maquet P, Degueldre C, Delfiore G, Aerts J, Peters JM, Luxen A, Franck G (1997) Functional neuroanatomy of human slow wave sleep. J Neurosci 17:2807–2812.
Maquet P, Laureys S, Peigneux P, Fuchs S, Petiau C, Phillips C, Aerts J, Del Fiore G, Degueldre C, Meulemans T, Luxen A, Franck G, Van Der Linden M, Smith C, Cleeremans A (2000) Experience-dependent changes in cerebral activation during human REM sleep. Nat Neurosci 3:831–836.
McGaugh JL (2000) Memory–a century of consolidation. Science 287:248–251.
Mednick SC, Nakayama K, Stickgold R (2003) Sleep-dependent learning: a nap is as good as a night. Nat Neurosci 6:697–698.
Mednick SC, Nakayama K, Cantero JL, Atienza M, Levin AA, Pathak N, Stickgold R (2002) The restorative effect of naps on perceptual deterioration. Nat Neurosci 5:677–681.
Meier CA (2003) Healing Dream and Ritual: Ancient Incubation and Modern Psychotherapy. Einsiedeln, Switzerland: Daimon Verlag.
Milbrandt J (1987) A nerve growth factor-induced gene encodes a possible transcriptional regulatory factor. Science 238:797–799.
Miller PC (1997) Dreams in Late Antiquity. Princeton, NJ: Princeton University Press.
Muzur A, Pace-Schott EF, Hobson JA (2002) The prefrontal cortex in sleep. Trends Cogn Sci 6:475–481.
Nadasdy Z, Hirase H, Czurko A, Csicsvari J, Buzsaki G (1999) Replay and time compression of recurring spike sequences in the hippocampus. J Neurosci 19:9497–9507.
Neylan TC, Marmar CR, Metzler TJ, Weiss DS, Zatzick DF, Delucchi KL, Wu RM, Schoenfeld FB (1998) Sleep disturbances in the Vietnam generation: findings from a nationally representative sample of male Vietnam veterans. Am J Psychiatr 155:929–933.
Nofzinger EA, Mintun MA, Wiseman M, Kupfer DJ, Moore RY (1997) Forebrain activation in REM sleep: an FDG PET study. Brain Res 770:192–201.
O'Hara BF, Young KA, Watson FL, Heller HC, Kilduff TS (1993) Immediate early gene expression in brain during sleep deprivation: preliminary observations. Sleep 16:1–7.
Pavlides C, Winson J (1989) Influences of hippocampal place cell firing in the awake state on the activity of these cells during subsequent sleep episodes. J Neurosci 9:2907–2918.
Pearlman CA (1969) Effect of rapid eye movement (dreaming) sleep deprivation on retention of avoidance learning in rats. Report US Naval Submarine Medical Center 22:1–4.
Peigneux P, Laureys S, Fuchs S, Destrebecqz A, Collette F, Delbeuck X, Phillips C, Aerts J, Del Fiore G, Degueldre C, Luxen A, Cleeremans A, Maquet P (2003) Learned material content and

acquisition level modulate cerebral reactivation during posttraining rapid-eye-movements sleep. Neuroimage 20:125–134.
Peirce CS (1958) Collected Papers of Charles Sanders Peirce. Cambridge, MA: Harvard University Press.
Pereira A, Ribeiro S, Wiest M, Moore LC, Pantoja J, Lin SC, Nicolelis MA (2007) Processing of tactile information by the hippocampus. Proc Natl Acad Sci USA 104:18286–18291.
Petersohn D, Schoch S, Brinkmann DR, Thiel G (1995) The human synapsin II gene promoter. Possible role for the transcription factor zif268/egr-1, polyoma enhancer activator 3, and AP2. J Biol Chem 270:24361–24369.
Pompeiano M, Cirelli C, Tononi G (1994) Immediate-early genes in spontaneous wakefulness and sleep: expression of c-fos and NGIF-A mRNA protein. J Sleep Res 3:80–96.
Popper K (1963) Conjectures and Refutations: The Growth of Scientific Knowledge. London: Routledge.
Qin YL, McNaughton BL, Skaggs WE, Barnes CA (1997) Memory reprocessing in corticocortical and hippocampocortical neuronal ensembles. Philos Trans R Soc Lond B Biol Sci 352:1525–1533.
Rechtschaffen A, Kales A (1968) A manual of standardized terminology, techniques and scoring system for sleep stages in human subjects. Washington, DC: National Institutes of Health.
Revonsuo A (2000) The reinterpretation of dreams: an evolutionary hypothesis of the function of dreaming. Behav Brain Sci 23:877–901.
Ribeiro S, Nicolelis MAL (2004) Reverberation, storage and postsynaptic propagation of memories during sleep. Learn Mem 11:686–696.
Ribeiro S, Nicolelis M (2006) The evolution of neural systems for sleep and dreaming. In: Evolution of Nervous Systems (Kaas J, ed), pp 451–464. New York: Elsevier.
Ribeiro S, Goyal V, Mello CV, Pavlides C (1999) Brain gene expression during REM sleep depends on prior waking experience. Learn Mem 6:500–508.
Ribeiro S, Mello CV, Velho T, Gardner TJ, Jarvis ED, Pavlides C (2002) Induction of hippocampal long-term potentiation during waking leads to increased extrahippocampal zif-268 expression during ensuing rapid-eye- movement sleep. J Neurosci 22:10914–10923.
Ribeiro S, Gervasoni D, Soares ES, Zhou Y, Lin SC, Pantoja P, Lavine M, Nicolelis M (2004a) Long-lasting novelty-induced neuronal reverberation during slow-wave sleep in multiple forebrain areas. PLoS Biology 2:126–137.
Ribeiro S, Pereira A, Ross J, Gao Z, Lin S-C, Gervasoni D, Marriott P, Lavine M, Nicolelis MAL (2004b) REM sleep is a noisy mode of offline neuronal processing. In: 34th Meeting of the Society for Neuroscience, p 324.310. San Diego, CA: Society for Neuroscience.
Ribeiro S, Xinwu S, Engelhard M, Zhou Y, Zhang H, Gervasoni D, Lin SC, Wada K, Lemos NAM, Nicolelis MAL (2007) Novel experience induces persistent sleep-dependent plasticity in the cortex but not in the hippocampus. Frontiers in Neuroscience 1:43–55.
Roffwarg HP, Dement WC, Muzio JN, Fisher C (1962) Dream imagery: relationship to rapid eye movements of sleep. Arch Gen Psychiatry 7:235–258.
Ross RJ, Ball WA, Dinges DF, Kribbs NB, Morrison AR, Silver SM, Mulvaney FD (1994) Rapid eye movement sleep disturbance in posttraumatic stress disorder. Biol Psychiatry 35:195–202.
Ross RJ, Ball WA, Sanford LD, Morrison AR, Dinges DF, Silver SM, Kribbs NB, Mulvaney FD, Gehrman PR, McGinnis DE (1999) Rapid eye movement sleep changes during the adaptation night in combat veterans with posttraumatic stress disorder. Biol Psychiatry 45:938–941.
Schreuder BJ, van Egmond M, Kleijn WC, Visser AT (1998) Daily reports of posttraumatic nightmares and anxiety dreams in Dutch war victims. J Anxiety Disord 12:511–524.
Schultz W (2002) Getting formal with dopamine and reward. Neuron 36:241–263.
Shaw PJ, Cirelli C, Greenspan RJ, Tononi G (2000) Correlates of sleep and waking in Drosophila melanogaster. Science 287:1834–1837.
Shulman D, Stroumsa GG, Stroumsa GAG (1999) Dream Cultures: Explorations in the Comparative History of Dreaming. Oxford: Oxford University Press.
Siegel JM (1990) Mechanisms of sleep control. J Clin Neurophysiol 7:49–65.

Siegel JM (2001) The REM sleep-memory consolidation hypothesis. Science 294:1058–1063.
Skaggs WE, McNaughton BL (1996) Replay of neuronal firing sequences in rat hippocampus during sleep following spatial experience. Science 271:1870–1873.
Smith C (1995) Sleep states and memory processes. Behav Brain Res 69:137–145.
Smith C (1996) Sleep states, memory processes and synaptic plasticity. Behav Brain Res 78:49–56.
Smith C (2001) Sleep states and memory processes in humans: procedural versus declarative memory systems. Sleep Med Rev 5:491–506.
Smith C, Butler S (1982) Paradoxical sleep at selective times following training is necessary for learning. Physiol Behav 29:469–473.
Smith C, Lapp L (1986) Prolonged increases in both PS and number of REMS following a shuttle avoidance task. Physiol Behav 36:1053–1057.
Smith C, Kelly G (1988) Paradoxical sleep deprivation applied two days after end of training retards learning. Physiol Behav 43:213–216.
Smith C, Wong PTP (1991) Paradoxical sleep increases predict successful learning in a complex operant task. Behav Neurosci 105:282–288.
Smith C, Lapp L (1991) Increases in number of REMS and REM density in humans following an intensive learning period. Sleep 14:325–330.
Smith C, MacNeill C (1993) A paradoxical sleep-dependent window for memory 53–56-H after the end of avoidance training. Psychobiology 21:109–112.
Smith C, Young J, Young W (1980) Prolonged increases in paradoxical sleep during and after avoidance-task acquisition. Sleep 3:67–68.
Smith C, Kitahama K, Valatx JL, Jouvet M (1974) Increased paradoxical sleep in mice during acquisition of a shock avoidance task. Brain Res 77:221–230.
Solms M (1997) The Neuropsychology of Dreams. New York: Lawrence Erlbaum.
Solms M (2000) Dreaming and REM sleep are controlled by different brain mechanisms. Behav Brain Sci 23:843–850.
Solms M (2004) Freud returns. Sci Am 290:82–88.
Steriade M (1992) Basic mechanisms of sleep generation. Neurology 42:9–18.
Steriade M, McCormick DA, Sejnowski TJ (1993) Thalamocortical oscillations in the sleeping and aroused brain. Science 262:679–685.
Steward O, Wallace CS, Lyford GL, Worley PF (1998) Synaptic activation causes the mRNA for the IEG Arc to localize selectively near activated postsynaptic sites on dendrites. Neuron 21:741–751.
Stickgold R (1998) Sleep: off-line memory reprocessing. Trends Cogn Sci 2:484–492.
Stickgold R (2001) Toward a cognitive neuroscience of sleep. Sleep Med Rev 5:417–421.
Stickgold R (2003) Memory, cognition, and dreams. In: Sleep and Brain Plasticity (P. Maquet CS, R. Stickgold, ed), pp 17–39. Oxford: Oxford University Press.
Stickgold R, Walker MP (2005) Memory consolidation and reconsolidation: what is the role of sleep? Trends Neurosci 28:408–415.
Stickgold R, James L, Hobson JA (2000a) Visual discrimination learning requires sleep after training. Nat Neurosci 3:1237–1238.
Stickgold R, Fosse R, Walker MP (2002) Linking brain and behavior in sleep-dependent learning and memory consolidation. Proc Natl Acad Sci USA 99:16519–16521.
Stickgold R, Scott L, Rittenhouse C, Hobson JA (1999) Sleep-induced changes in associative memory. J Cogn Neurosci 11:182–193.
Stickgold R, Hobson JA, Fosse R, Fosse M (2001) Sleep, learning, and dreams: off-line memory reprocessing. Science 294:1052–1057.
Stickgold R, Whidbee D, Schirmer B, Patel V, Hobson JA (2000b) Visual discrimination task improvement: a multi-step process occurring during sleep. J Cogn Neurosci 12:246–254.
Sutcliffe JG, de Lecea L (2002) The hypocretins: setting the arousal threshold. Nat Rev Neurosci 3:339–349.
Tanji J, Hoshi E (2001) Behavioral planning in the prefrontal cortex. Curr Opin Neurobiol 11:164–170.

Thiel G, Schoch S, Petersohn D (1994) Regulation of synapsin I gene expression by the zinc finger transcription factor zif268/egr-1. J Biol Chem 269:15294–15301.

Timo-Iaria C, Negrao N, Schmidek WR, Hoshino K, Lobato de Menezes CE, Leme da Rocha T (1970) Phases and states of sleep in the rat. Physiol Behav 5:1057–1062.

Tononi G, Cirelli C (2003) Sleep and synaptic homeostasis: a hypothesis. Brain Res Bull 62:143–150.

Tononi G, Cirelli C, Pompeiano M (1995) Changes in gene expression during the sleep-waking cycle: a new view of activating systems. Arch Ital Biol 134:21–37.

Ulloor J, Datta S (2005) Spatio-temporal activation of cyclic AMP response element-binding protein, activity-regulated cytoskeletal-associated protein and brain-derived nerve growth factor: a mechanism for pontine-wave generator activation-dependent two-way active-avoidance memory processing in the rat. J Neurochem 95:418–428.

Vyazovskiy V, Cirelli C, Pfister-Genskow M, Faraguna U, Tononi G (2008) Molecular and electrophysiological evidence for net synaptic potentiation in wake and depression in sleep. Nat Neurosci 11:200–208.

Wagner U, Gais S, Haider H, Verleger R, Born J (2004) Sleep inspires insight. Nature 427:352–355.

Walker MP, Stickgold R (2004) Sleep-dependent learning and memory consolidation. Neuron 44:121–133.

Walker MP, Liston C, Hobson JA, Stickgold R (2002a) Cognitive flexibility across the sleep-wake cycle: REM-sleep enhancement of anagram problem solving. Brain Res Cogn Brain Res 14:317–324.

Walker MP, Brakefield T, Morgan A, Hobson JA, Stickgold R (2002b) Practice with sleep makes perfect: sleep-dependent motor skill learning. Neuron 35:205–211.

Walker MP, Brakefield T, Seidman J, Morgan A, Hobson JA, Stickgold R (2003) Sleep and the time course of motor skill learning. Learn Mem 10:275–284.

Waltereit R, Dammermann B, Wulff P, Scafidi J, Staubli U, Kauselmann G, Bundman M, Kuhl D (2001) Arg3.1/Arc mRNA induction by Ca2 + and cAMP requires protein kinase A and mitogen-activated protein kinase/extracellular regulated kinase activation. J Neurosci 21:5484–5493.

Wilson MA, McNaughton BL (1994) Reactivation of hippocampal ensemble memories during sleep. Science 265:676–679.

Winson J (1985) Brain and Psyche. New York: Anchor.

Winson J, Abzug C (1977) Gating of neuronal transmission in the hippocampus: efficacy of transmission varies with behavioral state. Science 196:1223–1225.

Wisden W, Errington ML, Williams S, Dunnett SB, Waters C, Hitchcock D, Evan G, Bliss TV, Hunt SP (1990) Differential expression of immediate early genes in the hippocampus and spinal cord. Neuron 4:603–614.

Chapter 18
Epilogue: A New Vision of Life

D. Lloyd[1] and E.L. Rossi[2]

Abstract This Chapter presents a brief overview of recent progress and indications of future trends in ultradian rhythm research.

Keywords Networks, coherence, homeodynamics

Breathtaking developments in techniques over the past 15 years enable us to do experiments now with a depth and elegance of which we could only have dreamed then. Every chapter in this volume proposes novel concepts acquired via the new technologies, and reinforces the view that auto-dynamic principles provide the most basic of all living characteristics. As emphasised by Yates, ultradian rhythms are the dynamic signature of life. Many other chapters provide evidence for the evolutionary antiquity of these short-term rhythms; they surely predated circadian rhythms ("an evolutionary afterthought"). Most chronobiologists have emphasised the importance of the diurnal phenomena to the neglect of the more fundamental and central internal timekeepers.

In yeast, the self-synchronous continuous culture of Kuriyama has provided a highly accessible inroad into the inner working of a superb model organism, and it is certain that this system of study will continue to provide new insights for the foreseeable future. It is this system that has revealed the insistently repetitive sequential mode of operation of the circahoralian platform of metabolism, biosynthetic activities, assembly of membranes, organelles, and the cellular reproductive machinery. A highly complex interconnected network switches autonomously between oxidative and reductive states. Commonalities and highly conserved features of this network occur, not only with respect to other unicellular eukaryotes, but all through the plant and animal kingdoms to human biology. Thus hepatocytes

[1] Microbiology (BIOSI 1), Main Building, Cardiff School of Biosciences, PO Box 915, Cardiff CF10 3TL, Wales, UK

[2] New Neuroscience School of Therepeutic Hypnosis and Psychotherapy, San Lorrenzo Maggiore, Italy and Ernest Rossi Foundation for Psychosocial Genomic Research, 125 Howard Avenue, Los Osos, USA

(Brodsky, 1966), as well as the cultured mammalian cell lines studies in the laboratories of Klevecz and of Gilbert and Hammond since the late 1960s (Klevecz and Ruddle, 1968; Gilbert, 1966) are now joined by the cardiomyocytes of the Baltimore Laboratory (O'Rourke et al., 2005), as exciting subjects for study.

In the past 15 years, new ways of analysing data have made it possible to view both older and newer experiments through new eyes. The proposal that time structure has a fractal nature, tentatively proposed for the first time by Brodsky (1993) in the earlier version of this book, was being independently formalized by Yates (1992, 1993). Even before that Hildebrandt (1979) had spent a lifetime emphasising that "textbooks of biology fail to mention the important fact that every vital process must take place in certain set time structures and rhythmic functions". His "spectrum of human rhythmic functions according to their period duration." Figure 18.1 represents a prescient forerunner of current developments. Now we can appreciate that this "hidden order" of life can be appreciated right down to the molecular and even sub-molecular levels. Self-similarity of temporal order may well extend downwards to the smallest and upwards to the largest spatio-temporal scales. What occurs on one time scale may cascade downwards or escalate upwards? Thus a very fast event involving a few small molecules can be amplified from microscopic to macroscopic levels. This statistically self-similar (fractal) time-structure implies a heterarchical (rather than strictly hierarchical) system (Yates, 1993) in which high frequency events can produce far-reaching consequences over long time intervals

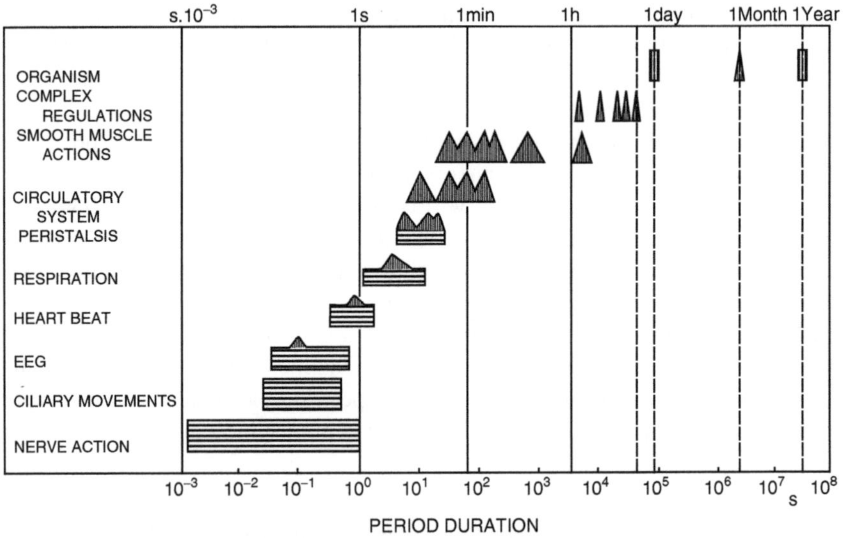

Fig. 18.1 Whole spectrum of human rhythmical functions according to their period duration. Horizontally hatched areas indicate the range of frequency change under functional effort. Vertically hatched triangles indicate the statistical variability of frequencies in repose (Hildebrandt, 1979)

and where molecular changes reverberate through to whole organ function (West, 1999; Noble, 2006). These ideas epitomize new ways of looking at the whole organism, "system biology", nothing more than physiology in a reinvented guise, but with all the power of the new methods of study. This list contains the ultra-fast or high spatial resolution imaging techniques from single molecule fluorescence to *in vivo* electroencephalography of the functioning brain to nuclear magnetic resonance imaging of skeletal muscle (Chance et al., 2006) and of the whole human body.

Another theme repeated through many chapters of this book concerns the likelihood that time measurement in biological systems, as in almost all man-made clocks, involves counting repeated high frequency cycles. Ultradian timekeeping even on millisecond scales is the basis that provides the ability of organisms to measure the days, months and years. This idea is not new; it was one of Bünning's most fundamental contributions to our understanding of photoperiodism (Bünning and Chandrashekaran, 1975). Now experiments with *Drosophila* confirm the deep relationships between millisecond (Kyriacou and Hall, 1980), hourly (Dowse and Ringo, 1987) and circadian timekeeping. Tidal rhythms (Chandrashekaran and Sharma, this volume) too, remind us that 24 h is a long time in the lives of many invertebrates living on the shore. Infradian phenomena, the lunar, annual and longer (e.g. the 17 year life-cycle of cicadas), although outside the scope of the present volume, provide further evidence for the precision of biological clocks as constructed from shorter period rhythms.

Intensive studies of the NOX systems by Morré, Morré and co-workers remind us of earlier work on single enzymes oscillators (Queiroz-Claret et al., 1988). But the startling demonstration that the plasma membrane enzyme shows a temperature-compensated period elevates its status from that of an oscillator to that of a biological clock. This is an even simpler system than the three-protein oscillator (circadian clock) of Kondo's group (Nakajima et al., 2005) extracted from a cyanobacterium. However, that a pure solution of a Cu (11) salt demonstrates clock-like properties is a real revelation (Morré et al., 2007). The redox core of the 40 min yeast clock provides the centrepiece of ultradian rhythmicity (Lloyd and Murray, 2000, 2007). The importance of intracellular redox states and of their non-invasive continuous readout measurement using NAD(P)H and flavin fluorescence has been long appreciated in bacteria (Harrison and Chance, 1970), heart and brain (Chance, 2004) as well as in amoeba and yeast (Bashford et al., 1980). Recently, mitochondrial dysfunction in patients has been monitored using NADH redox state in combination with blood flow and haemoglobin oxygenation states (Mayevsky and Chance, 2007). Smith et al. (2007) have shown that Gts1p-dependent metabolic oscillations have a 40 min period and are involved in glutathione-mediated stress resistance in yeast. Two-photon excitation now enables separation of signals from mitochondria and the extra mitochondrial (cytosolic) spaces in real time and *in vivo* (Aon et al., 2007). That the redox core mechanism in yeast can be tuned to output a 24 h envelope of respiratory quotient in light/dark conditions (12:12 L:D) makes the important point that circadian rhythms are easily constructed from ultradian units (Murray et al., 2007). Six possible ways in which this could be achieved (Fig. 18.2) include "elastic coupling", whereby intimate coupling

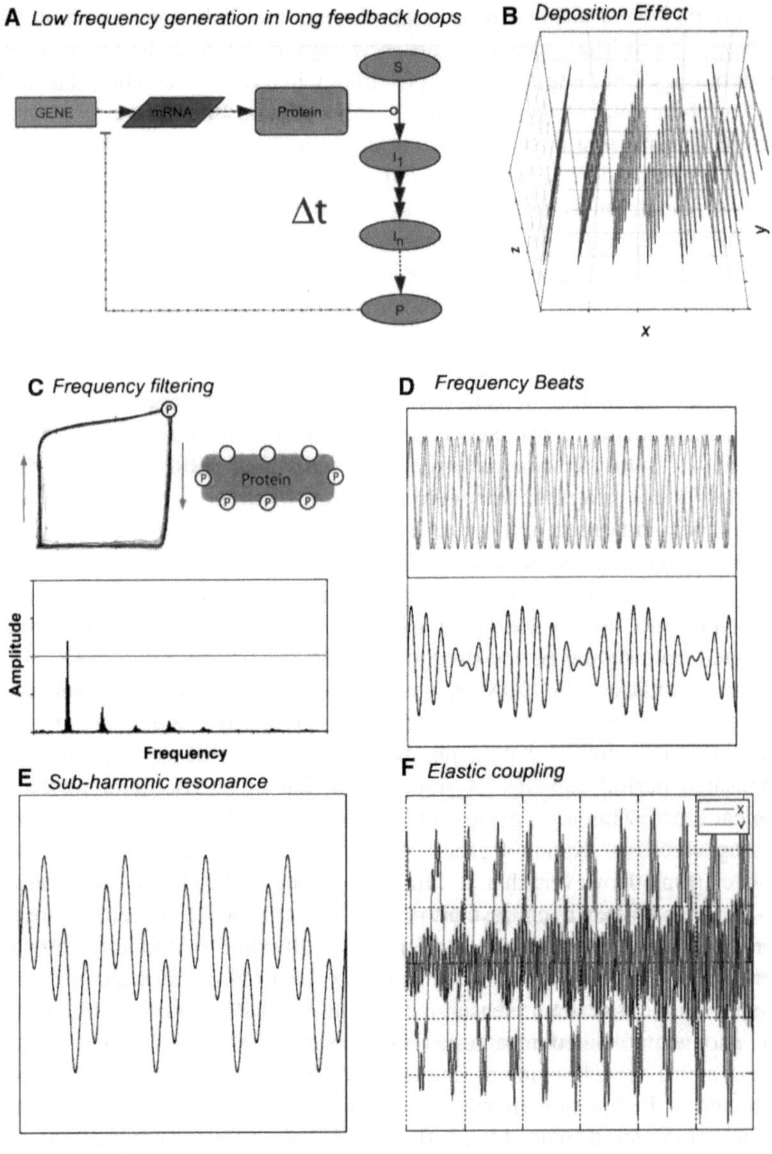

Fig. 18.2 Frequency reduction mechanisms: **A**: Long feedback loop. The period of oscillation in a negative feedback loop is much longer than the half-life of the most stable element of the loop. Δt represents the time delay in the feedback, S represents the substrate, I represents an intermediate and P represents the activated product. Here the feedback loop inhibits transcription, but feedback may also occur in any other step. **B**: Depot effect. Simulated phase plane plot of fructose 6-phosphate concentration (σ_1) versus fructose 1,6 bisphosphate concentration (σ_2) versus glycogen concentration (σ_d) gives highly spiralized stable limit cycles due to nonlinear interaction between two self-oscillatory mechanisms one of short and one of long period. **C**: Frequency filter or counter. **D**: Frequency beats by superposition of two sinusoids of differing periods. **E**: Frequency demultiplication or subharmonic resonance. **F**: Elastic coupling; the output of several coupled oscillators gives limit cycle with long period (x) (Lloyd and Murray, 2007)

through a small set of highly interconnected small molecules (ATP, glutathione, NAD(P)H occurs (Lloyd and Murray, 2007).

In all of these studies, parallel, interactive and iterative mathematical modelling (Fuentes-Pardo et al., 2003) extends our hypotheses and continues to make valuable predictions to be tested in further experiments. The crayfish system presents an unique opportunity for studies of ultradian-circadian relationships, and such a close collaboration between experimental investigators and mathematicians is unusual and highly effective. This is in contrast to the many previous model-building exercises where the biological reality is ignored. Network theory is essential to the ability to express the connectivity of the cellular network in all its complexity (Hütt and Lüttge, 2005).

We should however, be mindful that biology has developed vastly different mechanisms for measuring time on different scales both in animals (Buonomano, 2007) and in plants (Mancuso and Shabala, 2007), It is not realistic to search for a universal clock mechanism. The frenetic race for that prize in the circadian time domain has only indicated diversity and multiplicity of evolutionary origins (Paranjpe and Sharma, 2005). Clearly the obsession with the transcriptional feedback model has obscured the central and fundamental in chronobiology (Lüttge, 2003; Lüttge and Hütt, 2004).

Mechanisms for microsecond and millisecond clocks await investigation Lillo et al. (2001) have shown how multiple oscillatory loops can generate circadian rhythms without a central chronometer. Clearly a clock mechanism with a period of about a minute resides in mitochondria, and the probable mechanisms were (inadvertently) elucidated more than 40 years ago (Gooch and Packer, 1974). New information on the messenger roles of superoxide (O_2^-), a product of mitochondrial electron transport and the opening of IMAC channels enlarges our understanding of cardiac function (O'Rourke et al., 2005), and it is likely that a similar mechanism operates in yeast (Aon et al., 2007, Aon M.A. and Lloyd D., unpublished experiments).

The general public, philosophers, and even most researchers in the life sciences still believe that circadian rhythms are the dominant organizing factor in the timing of biological, sociological, and even psychological processes. Ultradian rhythms are frequently regarded as something of an "afterthought," tossed in by empirical pedants to account for some empirical date that they can deal with in no other way. The grand scope of the theory and research presented in this volume, however, suggests the exact reverse may be true. In the inimitable words of Dowse (this volume), "To explain multiple periods [of ultradian rhythms as] resulting from interacting circadian periods requires 'Ptolemaic reasoning,' multiplying the number of components to ridiculous complexity." Generalizing the research of this volume from such a broad and deep perspective leads naturally to the emerging insights about ultradian rhythms as dynamic signatures of life with their thermodynamic foundation in "open systems" and "homeodynamics" as outlined by Yates and Yates. To many people raised in the Nobel Prize winning generation of Watson and Crick that trace the source of life to genes and information, it can come as a fresh realization to review evidence that the start up of life may have begun with the basics of energy dynamics rather than informational processes.

The data and analyses of this volume are clearly within the realm of "complex systems" in the fullest meaning of that idea in the physical and life sciences. The spirit of the contributors to this volume has been admirably expressed by the mathematician, Richard Foote (2007), in this way:

> Contemporary researchers strive to understand complex physical phenomena that involve many constituents, may be influenced by numerous forces, and may exhibit unexpected or emergent behaviour. Often such "complex systems" are macroscopic manifestations of other systems that exhibit their own complex behaviour and obey more elemental laws. This article proposes that areas of mathematics, even ones based on simple axiomatic foundations, have discernable layers, entirely unexpected "macroscopic" outcomes, and both mathematical and physical ramifications profoundly beyond their historical beginnings.
>
> In these examples, the complex systems are physical phenomena that researchers attempt to model mathematically. But areas of mathematics itself may be viewed as complex systems

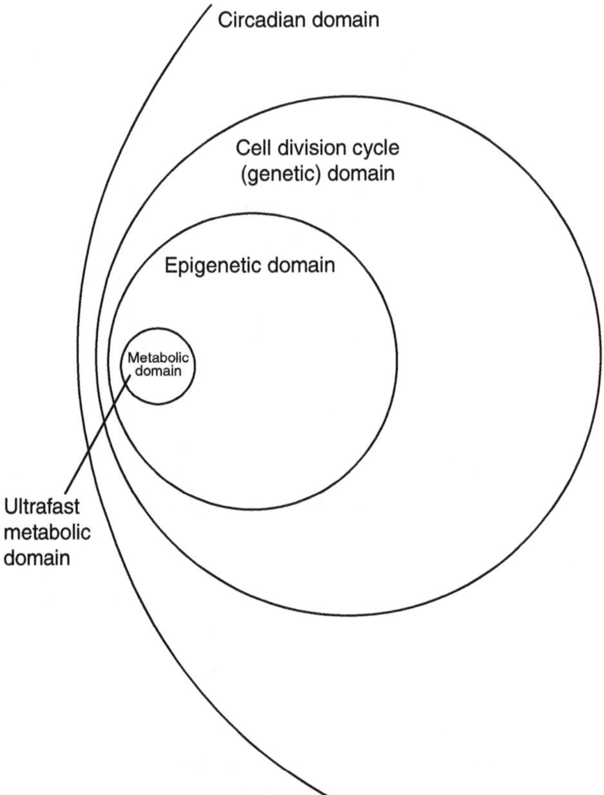

Fig. 18.3 A heterarchy of biological clocks. Although the clocks are mechanistically different, the fractal self-similarity of biological time indicates that interactions within the cellular network and within the organism occurs not only within time domains but also between widely separated time scales

> exhibiting many of the characteristics of the physical structures, including discernable "layers" closely analogous to microscopic or macroscopic strata in physics, biology, and other sciences. As we struggle to understand and systematize precise notions embodying physical complex systems, and ultimately to fathom the potential for mathematics to model these, a closer look at "abstract" complex mathematical systems is in order. We may then hope to glean from the theoretical setting a more rigorous grasp of some mathematical underpinnings that characterize complex systems in the natural world, and thereby make a chink in the Wigner conundrum [regarding the "unreasonable effectiveness of mathematics" in the scientific understanding of nature].
>
> Finally, the study of complex systems should be the exclusive purview of no one but the responsibility of everyone: Each scientist, mathematician, or researcher unfurls the mysteries of nature and humankind in small, deliberate steps. Science embodies the ability to verify, reproduce, and convince others of the veracity of one's discoveries, so the work of scientists is inherently incremental and precise. On the other hand, it is incumbent on us all to work toward enhancing the understanding of the "big picture" issues within our own disciplines and beyond: each of our disciplines must itself exhibit the inherent facets of a complex system, or our research is surely nugatory. (p. 410)

The complex systems and frameworks of this volume conceptualise ultradian and circadian rhythms in a new vision of life ranging from the deep theoretical problems of the Cartesian mind-body gap to the most practical issues of time and rhythm in sleep, mental health, human performance, and the nature of creativity itself. On each and every time scale from the ultrafast to the extended duration of the human life-span there must be a wider ranging necessity for timekeeping and therefore for many types of clocks (Fig. 18.3). We are currently finding internal time keeping is a key factor that we thought we could ignore in overstepping nature's constraints in our ambitions for a continuous 24/7 society around the globe. To the contrary, we now find that a new respect for the normal parameters of ultradian and circadian rhythms is required for our physical and mental health and well being. Internal time keeping is closely tied to cognitive skills, emotions, moods, and performance in everything from the safety issues of personnel in medicine, civic situations, and war to human depression, bipolar disorder, and the neurodegenerative diseases such as Alzheimer's (Bhattacharjee, 2007). Exciting open questions also abound in our pursuit of ultradian and circadian rhythms in a positive vision of the human experience of art, music, beauty, and truth that we hope to learn more about in our next version of this volume in the future.

References

Aon MA, Cortassa S, Lemar KM, Hayes AJ, Lloyd D (2007) Single cell and population oscillations in yeast. A 2-photon scanning laser microscopy study. FEBS Lett 581: 8–14

Bashford CL, Chance B, Lloyd D, Poole RK (1980) Oscillations of redox states in synchronously dividing cultures of *Acanthamoeba castellanii* and *Schizosaccharomyces pombe*. Biophys J 29: 1–12

Bhattacharjee V (2007) Is internal timing the key to mental health? Science 317: 1488–1490

Brodsky VY (1966) Protein Synthesis and Cell Functions (in Russian). Nauka, Moscow

Brodsky VY (1993) Rhythms of Protein Synthesis and Other Circahoralian Oscillations: The Possible Involvement of Fractals. In: D Lloyd and ER Ross (Eds.), Ultradian Rhythms in Life Processes. Springer, London, pp. 23–40

Bünning E, Chandrashekaran MK (1975) Pfeffer's views on rhythms. Chronobiologia 2: 160–167
Buonomano DV (2007) The biology of time across different time scales. Nature Chem Biol 3: 594–597
Chance B (2004) Mitochondrial redox state, monitoring, discovery and deployment in tissue. Meth Enzymol 386: 361–370
Chance B, Im J, Nioka S, Kushmeric M (2006) Skeletal muscle energetics with PNMR: personal views and historic perspectives NMR. Biomed 19: 904–926
Dowse HB, Ringo J (1987) Further evidence that the circadian clock in *Drosophila* is a population of coupled ultradian oscillators. J Biol Rhythm 2: 65076
Foote R (2007) Mathematics and complex systems. Science 318: 410–412
Fuentes-Pardo B, Guzman-Gomez AM, Lara-Aparicio M, Lopez de Medrano S (2003) A qualitative model of a motor circadian rhythm. BioSystems 71: 61–69
Gilbert DA (1966) Isoenzymes and cell regulation. Discovery 27: 23–26
Gooch VD, Packer L (1974) Oscillatory systems in mitochondria. Biochem Biophys Acta 346: 245–260
Harrison DEF, Chance B (1970) Fluorimetric technique for monitoring changes in the level of reduced nicotinamide nucleotides in continuous cultures. Appl Microbiol 19: 446–450
Hildebrandt G (1979) Rhythmical functional order and man's emancipation from the time factor. In KE Schaefer, G Hildebrandt and N Macbeth (Eds.), A New Image of Man in Medicine. 11 Basis of an Individual Physiology. Futura, Mount Kisco, New York, pp. 15–43
Hütt MT, Lüttge U (2005) Network dynamics in plant biology: current progress in historical perspective. Prog Bot 66: 277–310
Klevecz RR, Ruddle FH (1968) Cyclic changes in enzyme activity in synchronized mammalian cell cultures. Science 159: 634–636
Kyriacou CP, Hall JC (1980) Circadian mutations in Drosophilia melanogaster affect short-term fluctuations in the male courtship song. Proc Natl Acad Sci USA 77: 6729–6733
Lillo C, Meyer C, Ruoff P (2001) The nitrate reductase circadian system. The central clock dogma contra multiple oscillatory feedback loops. Plant Physiol 125: 1554–1557
Lloyd D, Murray DB (2000) Redox cycling of intracellular thiols: state variables for ultradian, circadian cell division cycle and circadian rhythms. In T Van den Driessche et al. (Eds.), The Redox State and Circadian Rhythms. Kluwer, Amsterdam
Lloyd D, Murray DB (2007) Redox rhythmicity: clocks at the core of temporal coherence. Bioessays 29: 465–473
Lüttge U (2003) Circardian rhythmicity: is the "biological clock" hardware or software? Prog-Bot 64: 277–319
Lüttge U, Hütt M-T (2004) High frequency or ultradian rhythms in plants. Prog Bot 65: 235–263
Mancuso S, Shabala S (2007) Rhythms in Plants Phenomenology, Mechanisms and Adaptive Significance. Springer, Berlin
Mayevsky A, Chance B (2007) Oxidation-reduction states of NADH *in vivo*: from animals to clinical use. Mitochondrion 7: 330–339
Morré DJ, Orczyk J, Hignite H, Kim C (2007) Regular oscillatory behaviour of aqueous solutions of Cu(11) salts related to effects on equilibrium dynamics of ortho/para hydrogen spin isomers of water. J Inorg Biochem 102: 260–267
Murray DB, Lloyd D, Kitano H (2007) Frequency modulation of the yeast reaction network. FEBS J (Suppl 1): D4–D1
Nakajima M, Imai K, Nishiwaki T, Murayama Y, Iwasaki H, Oyama T, Kondo T (2005) Reconstitution of circadian oscillation of cyanobacterial Kai C phosphorylation *in vitro*. Science 308: 414–415
Noble D (2006) The Music of Life. Oxford University Press, Oxford
O'Rourke B, Aon M, Cortassa S (2005) Mitochondrial ion channels: gatekeepers of life and death. Physiol (Bethesda) 20: 303–315

Paranjpe DA, Sharma VK (2005) Evolution of temporal order in living organisms. J Circadian Rhythms 3: 7–12

Queiroz-Claret C, Valon C, Queiroz O (1988) Are spontaneous conformational interconversions a molecular basis for long-period oscillations in enzyme activity? Chronobiol Int 5: 301–309

Smith MCA, Sumner ER, Avery SV (2007) Glutathione and Gts1p drive beneficial variability in cadmium resistances of individual yeast cells. Mol Microbiol 66: 699–712

West BJ (1999) Physiology, Promiscuity and Prophesy of the Millennium. A Tale of Tails. World Scientific, Singapore

Yates FE (1992) Fractal applications in biology: scaling time in biochemical networks. Meth Enzymol 210: 636–676

Yates FE (1993) Self-organizing systems. In CAR Boyd and D Noble (Eds.), The Logic of Life. Oxford University Press, Oxford, pp. 189–218

[illegible]

Index

K

L

M

N

MIX
Papier aus verantwortungsvollen Quellen
Paper from responsible sources
FSC® C105338
www.fsc.org

If you have any concerns about our products,
you can contact us on
ProductSafety@springernature.com

In case Publisher is established outside the EU,
the EU authorized representative is:
Springer Nature Customer Service Center GmbH
Europaplatz 3, 69115 Heidelberg, Germany

Printed by Libri Plureos GmbH
in Hamburg, Germany